Materials for Advanced Semiconductor Packaging and Heterogeneous Integration

Advanced semiconductor packaging and heterogeneous integration have become a critical area to enable semiconductor devices with high performance, high functional density, high reliability, and low cost. Advancement in semiconductor packaging technology, in turn, depends critically on innovative materials for miniaturization and high *I/O* density, signal integrity, power integrity, thermal integrity, reliability, as well as manufacturing quality and cost. Sustainability, another important consideration for semiconductor and microelectronics devices and for the industry as a whole, also depends directly on innovative materials solutions.

This book offers a comprehensive review of the key topics and the latest advancements in innovative materials for advanced semiconductor packaging and heterogeneous integration. Chapter 1 is an overview of semiconductor packaging technologies, outlining the role of materials in various packaging configurations and heterogeneous integration solutions. Chapter 2 reviews substrate and interposer materials (organic, glass, and ceramic) and printed circuit boards. Chapter 3 discusses innovative solder materials for miniaturized interconnects, and Chapter 4 focuses on materials for flexible electronics. Chapter 5 covers materials for thermal management, which is critical for high-performance semiconductor devices, and Chapter 6 covers materials for encapsulation, including underfill, molding compound, conformal coating, etc. Chapter 7 discusses how materials can affect the reliability of semiconductor packaging – particularly for AI and HPC applications, and Chapter 8 provides an overview of materials analysis and testing methodologies and techniques.

This book is aimed at academic and industry researchers as well as industry practitioners. It covers fundamental aspects as well as practical considerations, and it serves as a comprehensive source of the most up-to-date knowledge and a review of the state of the art of the technology and applications for innovative materials for advanced semiconductor packaging and heterogeneous integration.

Materials for Advanced Semiconductor Packaging and Heterogeneous Integration

Edited by
Dongkai Shangguan

CRC Press
Taylor & Francis Group
Boca Raton London New York

CRC Press is an imprint of the
Taylor & Francis Group, an **informa** business

First edition published 2027
by CRC Press
2385 NW Executive Center Drive, Suite 320, Boca Raton FL 33431

and by CRC Press
4 Park Square, Milton Park, Abingdon, Oxon, OX14 4RN

CRC Press is an imprint of Taylor & Francis Group, LLC

ISBN: 978-1-032-95943-6 (hbk)
ISBN: 978-1-032-96027-2 (pbk)
ISBN: 978-1-003-58763-7 (ebk)

DOI: 10.1201/9781003587637

Typeset in Times
by codeMantra

Contents

Foreword

Advanced semiconductor packaging and heterogeneous integration are now where the future of semiconductors is being written.

As traditional scaling slows, it is the ingenuity in architectures and materials that determines how far we can push performance, functional density, reliability, and sustainability. Materials innovation is no longer a supporting function—it is a primary design axis for the entire semiconductor eco-system.

This volume, *Materials for Advanced Semiconductor Packaging and Heterogeneous Integration*, invites the reader directly into that frontier. It sets out a compelling premise: advances in semiconductor packaging depend critically on innovative materials, not only to enable miniaturization, but perhaps even more importantly, to safeguard signal, power, and thermal integrity in an era of AI and high-performance computing.

Dr. Dongkai Shangguan has assembled a formidable cohort of collaborators, from industry and academia, who contribute deep technical insights from their respective areas of expertise. This collaborative approach ensures that the book is grounded in the practical challenges currently facing the global supply chain, while looking to future semiconductor devices that will enrich our lives.

One of the greatest strengths of this work is its balance between a panoramic overview and a carefully structured deep dive, as it spans the critical spectrum of the industry, addressing foundational inter-connectivity, to critical bottlenecks and emerging frontiers.

Indeed, it is written for the full community that must collaborate to move the field forward. Academic researchers will find a rigorous record of state-of-the-art developments, while industry practitioners will recognize the emphasis on practical considerations: quality, cost, and the rising bar of environmental responsibility.

This is not merely a catalog of current techniques; it is a lens through which to view where the field is heading. For students, it is an accessible entry point; for experienced professionals, it is a structured synthesis to guide strategic technical decisions.

For me, this volume carries a personal resonance. I recognize in these pages the qualities that defined Dongkai's leadership and accomplishments during the many years we worked together at Flex: intellectual curiosity, technical depth, and an unflinching focus on translating innovation into reliable, sustainable solutions.

It is a pleasure and an honor to commend this work to you. I invite you to read it not only as a record of where the field stands today but also as a springboard for the innovations that will define the next generation of technology.

Nicholas E. Brathwaite
Chancellor, McMaster University, Canada
Former CTO, Flex

Foreword

For decades, I have had the privilege of working alongside Dr. Dongkai Shangguan on some of the most significant transitions in modern electronics manufacturing: eliminating CFCs, adopting lead-free solders, relentlessly miniaturizing electronic assemblies, and enabling high-performance semiconductor devices through heterogeneous integration. Each inflection point has reinforced a central truth: materials science is the foundation upon which electronics innovation is built.

In recent years, we have seen the convergence of many previously distinct assembly processes into today's complex integrated systems. Advanced packaging architectures – 2.5D, 3D, chiplets, and system-in-package solutions – have blurred the traditional boundaries between semiconductor fabrication, packaging, and board-level assembly. Performance is no longer defined solely by transistor scaling, but by how effectively materials enable signal integrity, power delivery, thermal management, reliability, manufacturability, and sustainability – all within a manageable cost structure and with consistent quality for high volume deployment throughout the world. In this environment, materials are not merely a supporting consideration; they are a primary driver of system capability.

At Indium Corporation, our sole mission is to enable the most advanced electronics through new materials innovations. We see firsthand how rapidly the demands on semiconductor packaging materials are evolving – particularly in AI and high-performance computing applications, where power density, reliability, and integration complexity continue to escalate.

Dongkai's passion echoes our own – *Materials Science Changes the World.* His deep technical insights and global perspective are essential in a field where progress depends on collaboration among system designers, device manufacturers, materials suppliers, and equipment providers. It is the close partnership and collaboration across the entire eco-system, globally, that makes progress possible in the semiconductor and electronics industry, and materials are a critical pillar of the eco-system. Indeed, innovations in advanced materials are often a catalyst and key enabler for new semiconductor packaging technologies.

Today, more than ever, the industry demands innovative materials solutions to accelerate the pace of advancements in semiconductor packaging and heterogeneous integration. This book captures both the fundamental science and the practical engineering considerations that define this moment in our industry. By bringing together leading experts from academia and industry, this volume serves as a timely and authoritative reference for researchers, practitioners, and students alike. *Materials for Advanced Semiconductor Packaging and Heterogeneous Integration* provides a holistic view of the materials innovations that have shaped the electronics industry and offers insightful guidance on the future directions of materials innovations to shape the next generations of complex electronic and photonic systems.

Ross B. Berntson
President and CEO
Indium Corporation

Preface

My journey to discover materials began when I entered Tsinghua University in Beijing at the age of 15, with a major in foundry technology. While I did not become a foundryman after graduation, that training taught me the fundamental interplay between the materials composition, processes, and microstructures, properties, and performance. Graduate work at the University of Oxford further led me into the fascinating arena of microstructural evolution, governed by fundamental principles.

My professional journey took an unexpected turn when Ford Motor Company offered me a job in 1991 in the Electronics Division, where I got to see electronics "under the hood" – literally and figuratively – for the first time. Since then, I have had the opportunity to witness the evolution of materials technology to enable advanced semiconductor packaging over the past three decades – toward miniaturization and functional densification, with increasing performance and reliability, and reduced cost. Indeed, materials technology must continue to evolve to enable the next generation of technologies for semiconductor packaging and heterogeneous integration.

This book offers a comprehensive review of the key topics and the latest advancements in innovative materials for advanced semiconductor packaging and heterogeneous integration. Chapter 1 is an overview of semiconductor packaging technologies, outlining the role of materials in various packaging configurations and heterogeneous integration solutions. Chapter 2 reviews substrate and interposer materials (organic, glass, and ceramic) and printed circuit boards. Chapter 3 discusses innovative solder materials for miniaturized interconnects, and Chapter 4 focuses on materials for flexible electronics. Chapter 5 covers materials for thermal management, which is critical for high-performance semiconductor devices, and Chapter 6 covers materials for encapsulation, including underfill, molding compound, conformal coating, etc. Chapter 7 discusses how materials can affect the reliability of semiconductor packaging – particularly for AI and HPC applications, and Chapter 8 is an overview of materials analysis and testing methodologies and techniques. Business, economic, and supply chain considerations are discussed, and the current status and future outlook are outlined throughout the chapters.

This book is aimed at academic and industry researchers, as well as industry practitioners. It covers fundamental aspects as well as practical considerations, and it will serve as a comprehensive source of the most up-to-date knowledge and a review of the state of the art of the technology and applications for innovative materials for advanced semiconductor packaging and heterogeneous integration.

It is my sincere hope that this book can play a meaningful role in stimulating further exploration, discovery, and innovation in the field of materials for advanced semiconductor technology and heterogeneous integration, thus enabling future generations of semiconductor devices.

Dongkai Shangguan
San Jose, California
December 2025

Acknowledgments

First, I would like to thank the chapter authors – from many companies and institutions – for their contributions and for their confidence in me.

My industry colleagues worldwide deserve special thanks, as they have been the source of inspiration for my work in this area.

I would also like to thank Thermal Engineering Associates, Inc. (TEA) for supporting me in this endeavor.

Finally, it is the love from my family and friends that constantly motivates me to keep marching forward on this truth-seeking journey that may, in some small way, benefit our society.

Dongkai Shangguan
San Jose, California
December 2025

Editor

Dongkai Shangguan, IEEE Fellow and IMAPS Fellow, is President of Thermal Engineering Associates, Inc., a provider of application-specific silicon chips for semiconductor thermal characterization. He is also a Strategic Advisor to Indium Corporation, a provider of innovative materials for advanced semiconductor packaging and electronics assembly. Previously, he served as Vice President for Technology and Engineering at Flex (formerly Flextronics), a leading global player for electronics design and manufacturing, and as Vice President at STATS ChipPAC (currently JCET), a leading provider of semiconductor assembly and test services. Early in his career, he held various technical and management responsibilities at Ford and Visteon on automotive electronics.

Shangguan has published several books, authored or co-authored over 200 technical papers, and has been issued 34 U.S. patents. He is an IEEE EPS Distinguished Lecturer and has given numerous keynotes, seminars, and lectures globally.

Shangguan has served on the iNEMI Board of Directors, the IEEE EPS Board of Governors, and the IPC Board of Directors. He has received a number of recognitions for his contributions to the industry, including the Electronics Manufacturing Technology Award and the Outstanding Sustained Technical Contribution Award from IEEE EPS, the Daniel C. Hughes Jr. Memorial Award and the William D. Ashman Achievement Award from IMAPS, the President's Award from IPC, and the Total Excellence in Electronics Manufacturing Award from the Society of Manufacturing Engineers, among others.

Shangguan earned a BSc in Mechanical Engineering at Tsinghua University, China, an MBA at San Jose State University, and a PhD in Materials at the University of Oxford, UK. He conducted postdoctoral teaching and research at the University of Cambridge and the University of Alabama, and he has served as a guest professor at several universities.

Shangguan is based in San Jose, California, and travels globally in service of the industry.

Contributors

Gaurang Choksi

Ashay Dani

Timothy Fisher

Subramanian Iyer

Min Jong Kil

Bhaskar Jyoti Krishnatreya

Tae-Kyu Lee

Yan Li

Zihao Lin

Weifeng Liu

Ravi Mahajan

Debendra Mallik

Tarak A. Railkar

Dongkai Shangguan

Naarendharan Meenakshi Sundaram

Ching-Ping Wong

Hongwen Zhang

1 Advanced Packaging Architectures and the Role of Materials in Their Evolution

Ravi Mahajan, Ashay Dani, Debendra Mallik, Gaurang Choksi, and Bhaskar Jyoti Krishnatreya

1.1 INTRODUCTION

For nearly six decades, Moore's law [1–3] (Figure 1.1) has been the engine driving an exponential growth in computing and communications capabilities, fueling the digital revolution and driving fundamental, technological, and economic changes in modern society [4,5]. Moore's law-driven evolution of the transistor to create increasingly powerful devices on a single wafer platform is essentially homogeneous integration. In 2022, the 75th anniversary of the transistor was celebrated by a series of articles that highlighted the foundational societal impacts of this invention [6,7]. In recent years, it has become increasingly clear to broader

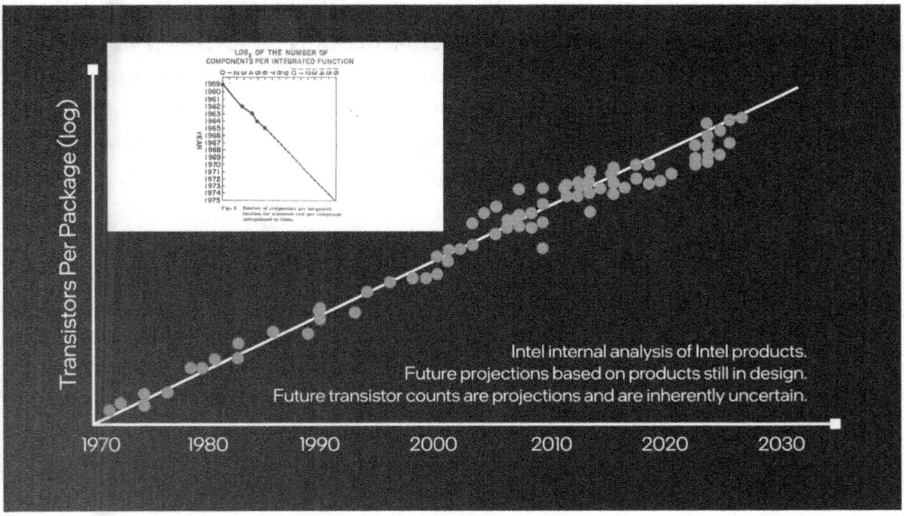

FIGURE 1.1 Moore's law—original (inset) and current projections. (Acknowledgment: Intel Corporation.)

DOI: 10.1201/9781003587637-1

groups of technologists and policymakers that a systems approach, referred to as **S**ystem-**T**echnology-**Co**-**O**ptimization (STCO), is needed to complement homogeneous integration and drive performance [3,8,9]. The most visible driver for this performance demand is the dramatically increased interest in artificial intelligence (AI), with similar trends predicted more broadly in high-performance computing (HPC) [10,11]. The hardware-specific aspect of STCO argues for collaborative co-development and optimization of transistors, interconnects, process technology, hard IP, memory technologies, advanced packaging architectures, and physical system architectures. System hardware optimization needs to be complemented and enhanced by software optimization to deliver energy-efficient, cost-effective performance.

In this chapter, we focus on two key elements of STCO, i.e., interconnects in advanced packaging architectures and the role of materials in driving their evolution. Advanced packaging architectures for AI/HPC represent the technology apex for many aspects of the heterogeneous integration (HI) of microelectronics. Consistent with previous work, for this chapter, advanced packaging architectures are defined only from an interconnect perspective, as architectures with (die-to-die and die-to-package) interconnect pitches of ≤ 100 μm and interconnect lines and spaces $\leq (9/12)$ μm [12]. HI refers to the integration of separately manufactured components into a higher-level assembly that, in the aggregate, provides enhanced functionality and improved operating characteristics [13]. HI has consistently complemented homogeneous integration for system performance and is recognized as being crucially important for the continued evolution of computing and communication [14–19].

The structure of the chapter is as follows. Key attributes of advanced packaging architectures and a projection of how they should scale are first described. Pitch scaling trends for die-to-die and die-to-wafer interconnects are used to establish the need to transition from flip-chip interconnects to **H**ybrid **B**ond **I**nterconnects (HBI) [12,20,21]. The current landscape for HI demand and applications that need advanced packaging with scaled interconnects is described next. High-level overviews of both the flip-chip and HBI process are then discussed, followed by a description of the materials needed for these two architectures and for overall advanced packaging. The chapter concludes with a discussion of opportunities in materials development to enable the evolution of advanced packaging architectures.

1.2 ADVANCED PACKAGING ARCHITECTURES FOR HI

Several roadmaps, technology portfolios, and product announcements attest to the significant importance, influence, and impact of HI technologies on the evolution of computing [22–29]. One such HI architecture is shown in Figure 1.2. Some of the noteworthy aspects of HI architectures are given as follows [22]:

- The different system application environments they influence (high-performance computing/artificial intelligence (HPC/AI)), mobile devices, aerospace and defense devices, systems for medical/health management, autonomous automotives, and Internet of Things (IoT))
- The different environments these systems have to operate in reliably, and the associated reliability standards that need to be satisfied

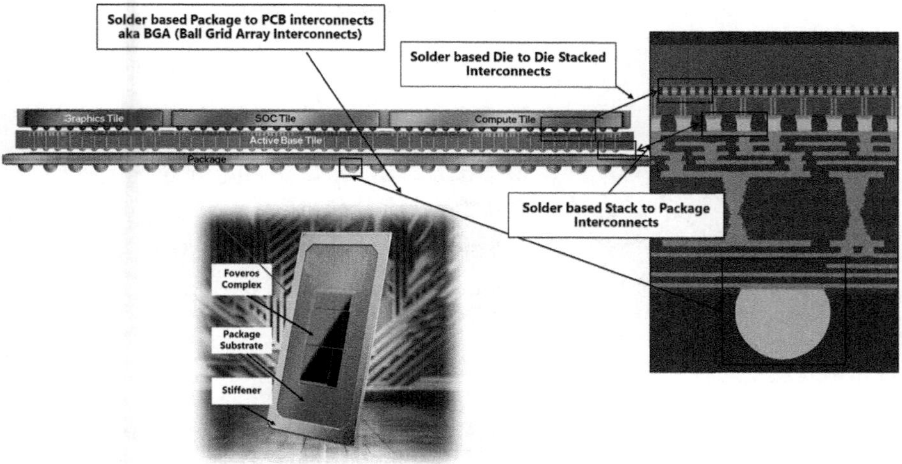

FIGURE 1.2 Interconnect stack in the Intel® Core™ Ultra Processor illustrating the use of advanced packaging-based HI. (Acknowledgment: Intel Corporation.)

- Manufacturing processes for high yield
- Test processes to evaluate and ensure stable performance
- The need for composable, modular architectures, connected by standardized links [30–33]
- The demand for efficient, cost-effective power delivery and cooling
- The need for co-design and physics-aware modeling

All these topics are worthy of significant focus and discussion. In this chapter, we narrow down our scope to projected interconnect evolution and the current and future materials needed for the evolution of advanced packaging.

1.3 INTERCONNECT SCALING

Package interconnects can be classified into six categories [34]:

a. **Die-Die Interconnects**: Interconnects between stacked die for vertical connectivity between multiple dies in a 3-D stack (Figure 1.3a)
b. **On-package Die-Die Interconnects**: Interconnects between die (and/or die stacks/pre-packaged die) within the package for lateral connectivity (Figure 1.3b)
c. **Die-to-Package Interconnects**: Interconnects between the die and the package, typically known as the first-level interconnect (FLI) (Figure 1.3a and b)
d. **Within-Package Interconnects**: Interconnects within the package that enable lateral connectivity between two or more dies and to the system or microsystem
e **Package-to-Board Interconnects**: Interconnects between the package and the next level, which is typically the motherboard, referred to as the

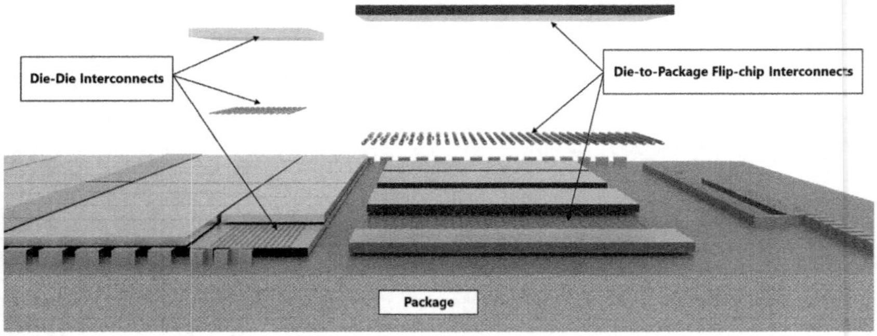

(a)

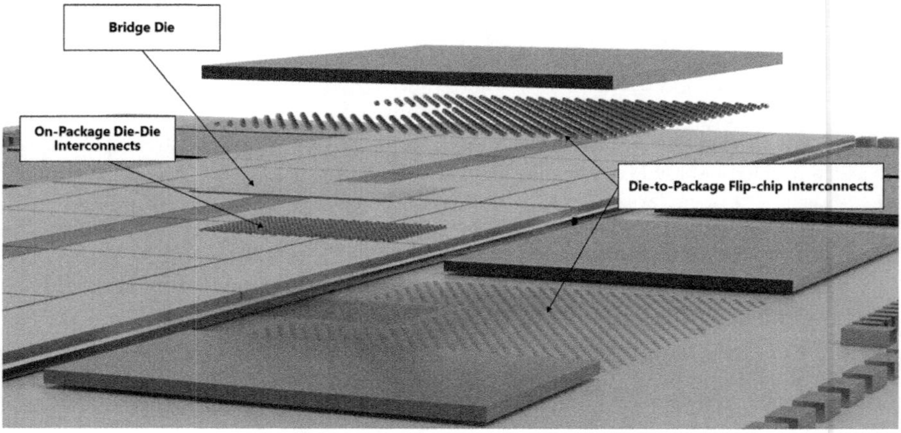

(b)

FIGURE 1.3 Schematic illustrations of (a) Die-Die interconnects and Die-to-Package Interconnects and (b) On-Package Die-Die and Die-to-Package Interconnects.

second-level interconnect (SLI). SLI connections are either socketed or ball grid array (BGA) and may be combined with on-package cabling.

f. **PoP (Package-on-Package) Interconnects**: The PoP construction [35] allows for packages to be placed on top of other packages using peripheral package interconnects, also referred to as VI (**V**ertical **I**nterconnects)

The die-die interconnects (a), the on-package die-die interconnects (b), and die-to-package flip-chip interconnects in most products in the market today are primarily solder based. The HI Roadmap, using bandwidth scaling as a technical driver, shows how pitch scaling can be used to deliver increasing interconnect density (Figure 1.4) and raw bandwidth (Figure 1.5) [34]. It should be noted that the projection shown in Figure 1.4 is a non-unique projection, and since it uses low signaling speeds, it represents a reasonable lower bound on the bandwidth scaling opportunities. It is generally understood (and evidenced in one instantiation [36]) that as pitches continue to scale down, a transition from solder-based interconnects to non-solder-based interconnects is needed. The exact lower scaling bound for

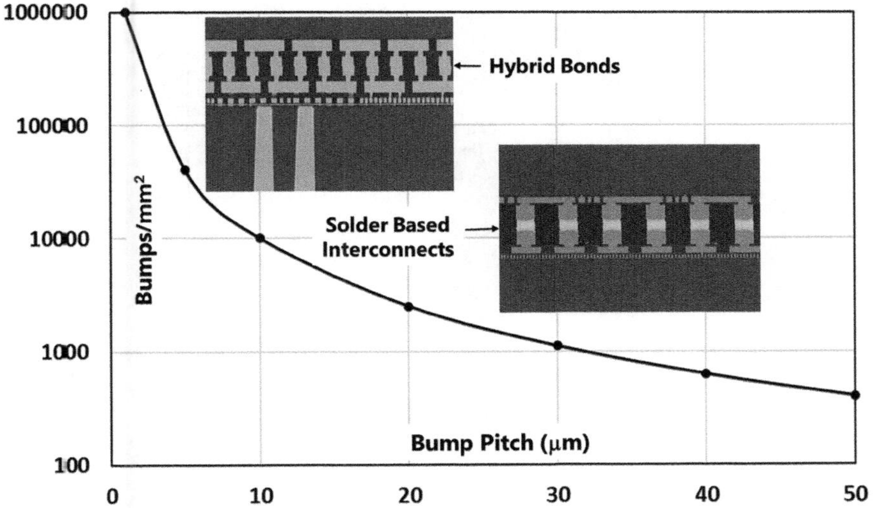

FIGURE 1.4 3D interconnect density scaling as a function of bump pitch. (Acknowledgment: Intel Corporation for images of solder and Cu–Cu interconnects.)

solder-based interconnects is dependent on multiple factors including die size, die warpage, and die thickness; and there have been studies that show pitch shrinks as low as 10 μm [37]. Scaling bump pitches below an approximate range of (25–10 μm) will require a transition to Cu–Cu bonding, and one leading technology for this is the use of hybrid bonding [38].

In the next section, we examine the current landscape for the economic value of HI architectures.

1.4 CURRENT LANDSCAPE—VALUE AND OPPORTUNITIES IN HI

As the High-End Performance Packaging (HEPP) 2025 Market and Technology Report from Yole Group highlights, the high-end packaging (HEP) market was worth \$8B in 2024 and is projected to exceed \$28B by 2030, with a $CAGR_{2024-2030}$ of 23% [39]. Figure 1.6 shows HEPP revenue projections as a function of broad market segments. Key packaging architecture constructions are shown in Figure 1.7. Two specific classes of architecture use silicon backend metallization for achieving high-density interconnects, i.e., silicon bridge-based interconnects [40,41] and silicon interposer-based interconnects [27,42]. Additionally, Figure 1.8 shows how technology leaders are introducing key advanced packaging architectures for applications that integrate compute, memory, and communication functions. While technical, economic, and geopolitical factors will influence the scope and rate of technological progress, it is clear that HI will continue to be a crucial vector, with interconnect scaling described in the previous section highlighting the right directions and technology roadmaps, research and development efforts across the industry, consortia, and academia continue to coalesce and enable advanced interconnect scaling [8].

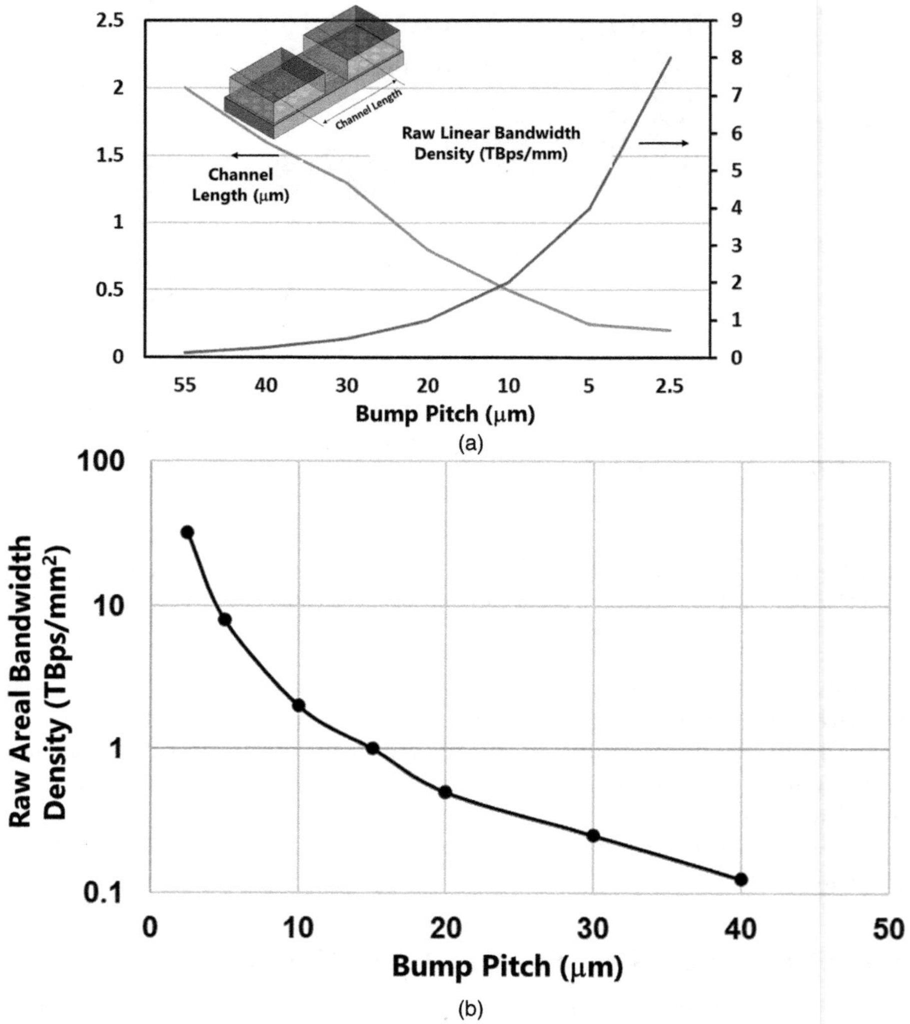

FIGURE 1.5 (a) Raw linear bandwidth density scaling as a function of bump pitch for lateral interconnects. (b) Raw areal bandwidth density scaling as a function of bump pitch for vertical interconnects. ((a) and (b) are reproduced with permission from the HI Roadmap [22].)

1.5 SOLDER AND HYBRID BONDING ASSEMBLY PROCESSES

The two types of interconnects of interest for high-density area-array interconnects are solder-based interconnects and Cu–Cu, i.e., hybrid interconnects (Figure 1.4). As shown in Figure 1.9, solder-based interconnects have evolved over the past three decades in conjunction with major changes in packaging technologies [14,43–46]. Some of the important elements of this evolution are the introduction of Cu pillars, lead-free solders, a reduction in joint pitch, introduction of advanced assembly processes, and a mature understanding of polymeric encapsulation for reliability and manufacturability [47].

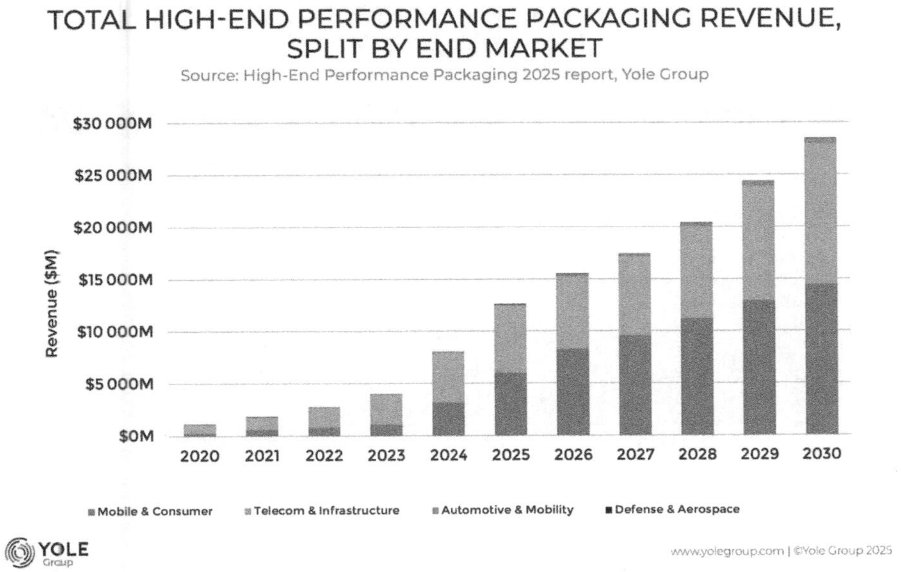

FIGURE 1.6 Current and projected revenues for high-end performance packaging (described in this chapter as *Advanced Packaging*). (*Source*: High-End Performance Packaging 2025 report, Yole Group, www.yolegroup.com [39].)

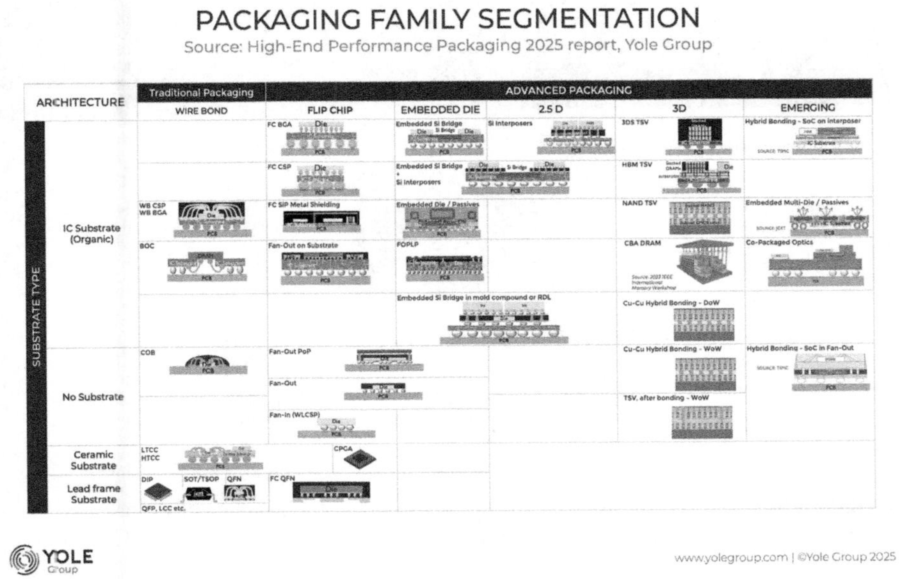

FIGURE 1.7 Schematics showing different traditional and advanced packaging architectures. (*Source*: High-End Performance Packaging 2025 report, Yole Group, www.yolegroup.com [39].)

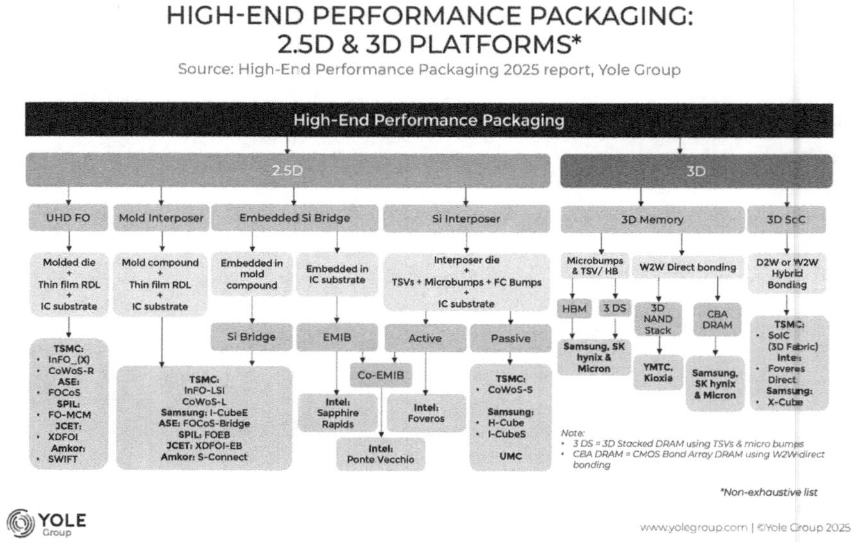

FIGURE 1.8 Packaging architectures provided by key players in the industry. (*Source*: High-End Performance Packaging 2025 report, Yole Group, www.yolegroup.com [39].)

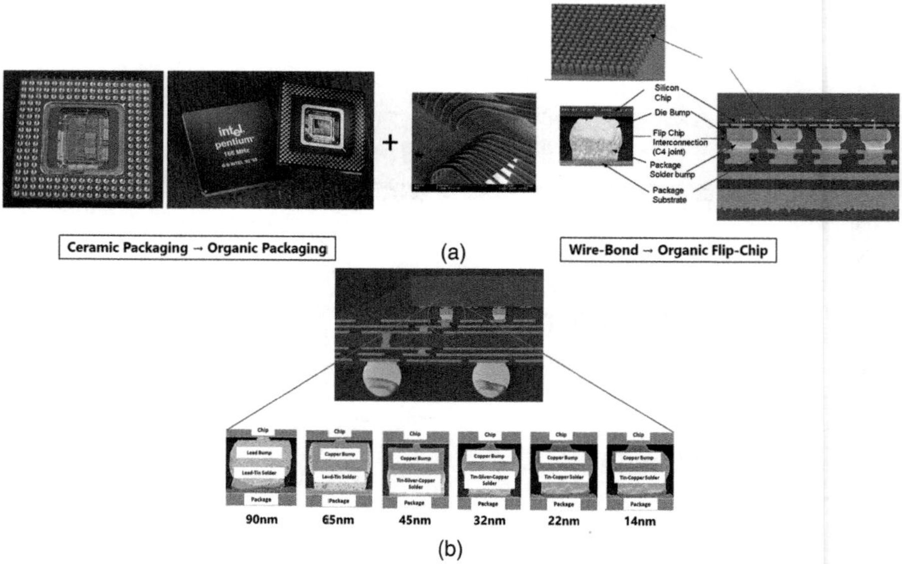

FIGURE 1.9 (a) Major transitions in advanced packaging. (b) Evolution of solder-based interconnects from lead-based solders to lead-free solders. (Acknowledgment: Intel Corporation.)

Figure 1.10 shows the 3D interconnect pitch scaling R&D roadmap from imec and comparison with the industry for leading configurations and attach processes.

Hybrid bonding is the next generation of interconnect assembly process for sub-10 μm die-to-die direct Cu–Cu interconnects with acceptable process thermal

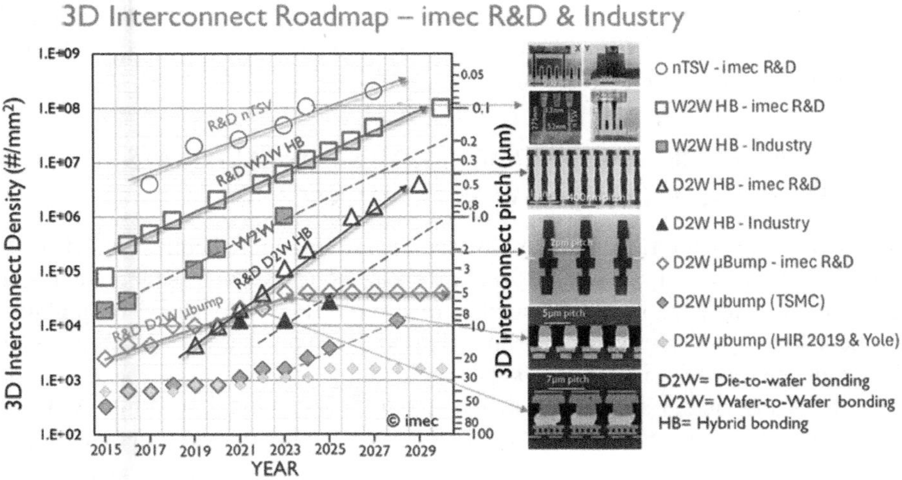

FIGURE 1.10 3D interconnect scaling roadmap of imec and comparison with industry, reproduced with permission from imec.

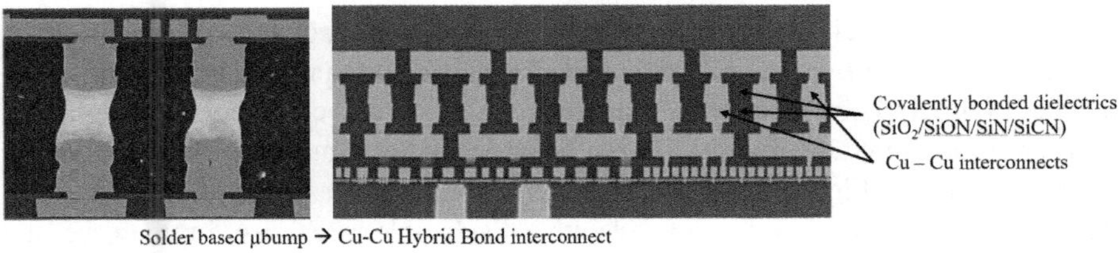

FIGURE 1.11 Transition from solder based on-die interconnects to Cu–Cu interconnects using hybrid bonding. (Acknowledgment: Intel Corporation.)

budgets of 200°C–400°C and low CTE mismatch. These fine interconnect pitches open up many avenues for die disaggregation and improved energy efficiency [8] (Figure 1.11).

Fine-pitch HBI significantly reduces the latency and power compared to the looser pitch of solder interconnects, allowing much higher bandwidth density and improved power efficiency [34]. Wafer-to-wafer hybrid bonding for 3D packaging applications typically has greater homogeneity and less design flexibility compared to chip-to-wafer hybrid bonding, where the ability to select known-good-die (KGD) prior to bonding enables cost-effective HI for 3D IC packaging [48–57]. Hybrid bonding involves the formation of a strong interconnection between two surfaces where each surface has two or more materials.

Typically, each surface contains an array of conducting Cu pads surrounded by a sea of insulating inorganic dielectric such as SiO_2, SiN, SiON, and/or SiCN [12,58]. Recently, there have been demonstrations of successful hybrid bonding using high-temperature curable organic polymers replacing the inorganic dielectrics as the insulating interface [59–62]. When organic polymers are used as the bonding dielectric, an additional thermocompression bonding process step at high temperature

(200°C–250°C) and pressure (1–4 MPa) is required to ensure polymer curing and covalent bond formation before high temperature anneal, which is not required when inorganic dielectrics are used as the bonding dielectric.

1.6 CHALLENGES AND OPPORTUNITIES FOR MATERIALS TECHNOLOGIES IN ADVANCED PACKAGING

We have so far made the case that advanced packaging architectures for HI, with AI/HPC as the primary demand driver, are projected to use increased interconnect densities to enable an increased number of diverse IPs implemented on multiple semiconductor processes and hence an increased die count. These architectures will be challenged to meet increasing power and cooling demands while meeting tight manufacturing yield, reliability, cost, and system fit constraints [8]. While AI/HPC packaging supporting mostly CMOS technologies is today a poster child for defining primary challenges, the overall scope of HI includes the need to include significantly more technologies [63,64]. The advanced package, especially for AI/HPC applications, is a complex composite that needs to deliver cost, manufacturability, and reliability optimized performance within system form-factor constraints (Figure 1.12) using environmentally friendly materials.

Materials are crucial to the current and future evolution of advanced packaging [65]. Direct and indirect organic materials (direct materials are present in the final product and indirect materials are used during processing but they do not show up in the final product) are coupled with different semiconductor materials, glass, and metal in the composite (Figure 1.13). These materials and their associated interfaces are designed, characterized, and validated with considerable care and effort to ensure the success of the HI architectures.

It is very important to recognize that intrinsic materials performance in isolation is not sufficient, and their value must be demonstrated in a relevant and integrated configuration, i.e., at the right dimensions and with industry-relevant interfaces and system configurations. For instance, while it is necessary to develop

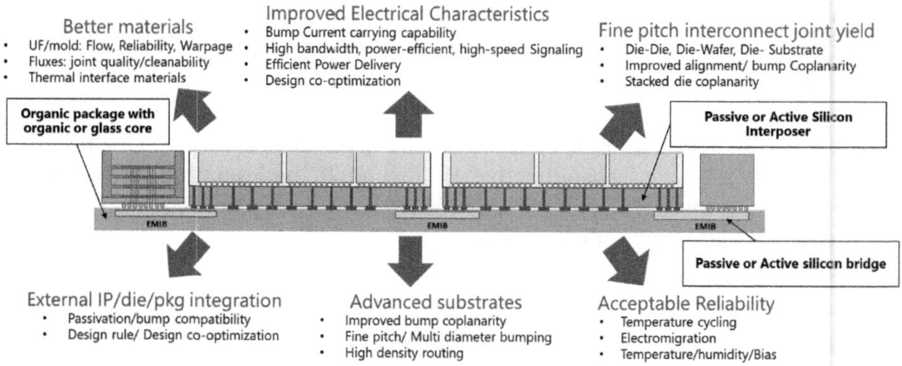

FIGURE 1.12 The package is a complex composite: schematic showing some key optimization attributes (Heat spreader lid not shown).

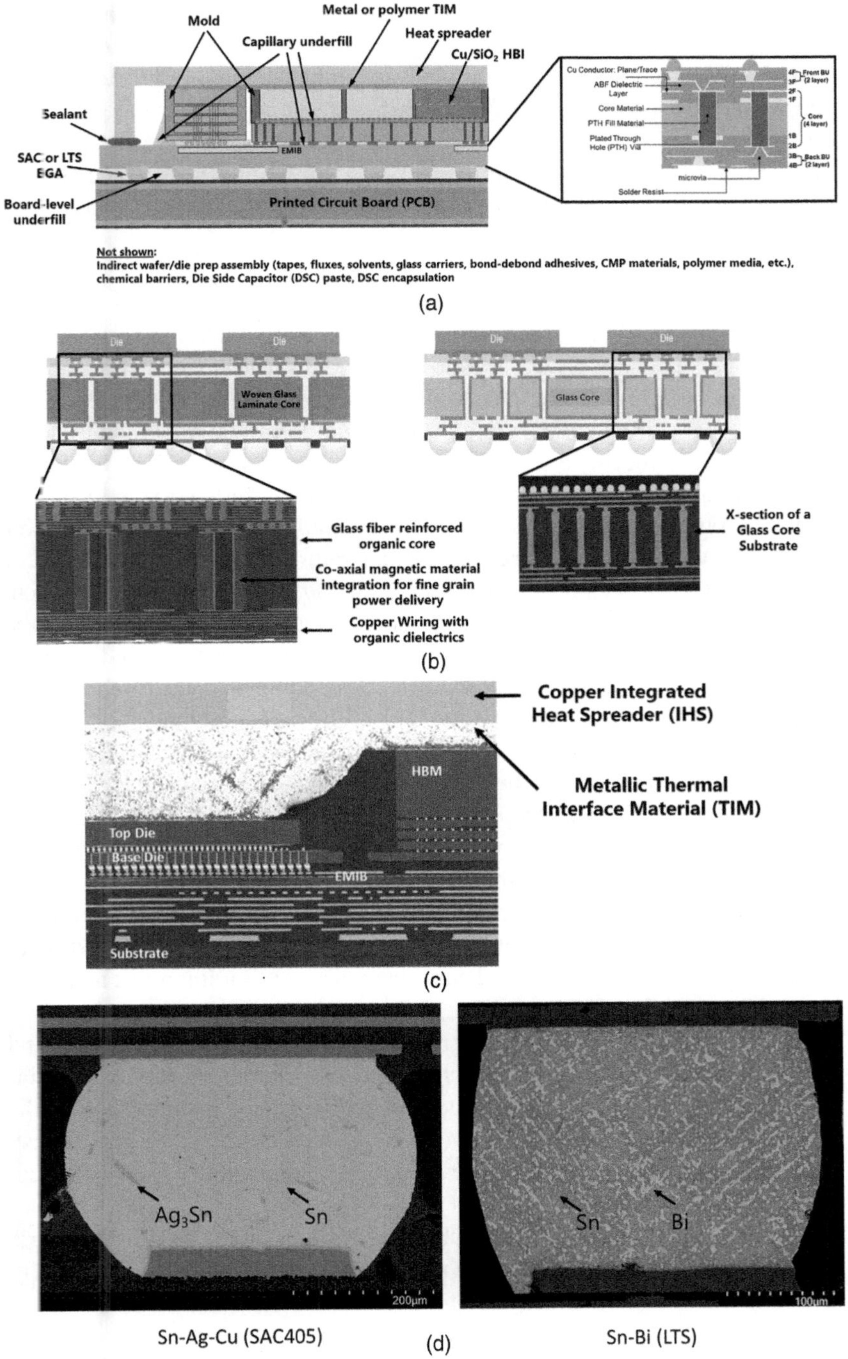

FIGURE 1.13 (a) Typical materials types used in advanced packaging technology. (b) Structural details of organic and glass core package substrates. (c) Thermal management technologies. (d) Materials technologies used in typical ball grid array joints. (Acknowledgments: Mukul Renavikar (Intel Corporation) and Nilesh Badwe (Indian Institute of Technology, Kanpur).)

improved thermal interface materials and understand their intrinsic behavior, it is just as important to demonstrate improved, reliable thermal performance when the material is integrated within a relevant advanced packaging architecture, as interfacial thermal resistance often plays a significant role in the thermal performance of the system.

There are a number of opportunities for new and improved materials to enable significantly improved advanced packaging architectures, some of which are listed below:

- New interconnect materials and interfaces with better electro-migration resistance and tailorable melting points. Electro-migration is a very important concern, especially for composite solder joints where geometry and the evolution of joint composition under current loading will affect the reliability of the joint during the lifetime of a product [66–69]. Additionally, there is significant recent interest in low-temperature solders (LTS) for lower-temperature ball grid array (BGA) joint formation [70,71]. This interest is due to lower energy costs and improved manufacturing yield due to lower warpage variations as a function of temperature change. The reliability of such joints is a function of the solder, pad, and surface finish composition. As the power demand for AI/HPC applications trends upwards, and the BGA pitches in general tend to shrink and interconnect counts increase, there is a greater need for focus on understanding manufacturability and reliability for both LTS and traditional lead-free solder joints.
- New encapsulants that offer flexibility in tailoring properties. Polymer composites (essentially a polymer matrix with non-conductive fillers to modulate their physical properties) are used as protective encapsulants for solder joints and stiffening media for packaging. There is a strong need to be able to tailor their composition to modulate multiple attributes, e.g., thermo-mechanical properties such as modulus, CTE, adhesion, glass transition temperature for improved reliability and warpage control, and other attributes, such as viscosity and cure characteristics, for improved manufacturability [72–75].
- Improved thermal interface materials (TIMs) tailored for component and system integration. Increasing cooling demand will require improved intrinsic TIM thermal conductivity and reduced thermal resistance across the associated interfaces. Immersion cooling is another area of recent interest, and this has sparked interest in developing environmentally friendly dielectric fluids [76–78].
- Improved materials in package substrates, such as novel low-loss dielectric materials [79,80], next-generation resist technologies, plating chemistries to enable redistribution layers, and interconnect metallurgies, etc.
- High-permeability, low-loss magnetic materials for improved inductance and improved capacitors to enable high-efficiency power delivery [81–83].
- Non-PFAS (per- and polyfluoroalkyl substances) materials for sustainability [84,85].
- Materials for reliable, low-loss optical components integration [86].

1.7 SUMMARY

Packaging technologies, traditionally used to provide mechanical protection and structural integrity for fragile semiconductor devices and cost-effective space transformation, have significantly evolved to become crucial platforms for the heterogeneous integration of microelectronics. Advanced packaging technologies enable significant functionality and form-factor optimization to drive the continued evolution of computing and communications. The heterogeneous integration of microelectronics components on advanced packaging substrates constitutes a technological apex in many ways driving advances in interconnect scaling.

The package as an HI platform is a complex composite enabled by a number of direct and indirect materials. The continued development of the materials system is essential to be able to meet anticipated future demand. In this chapter, two important aspects of the current state of the art in advanced packaging architectures, i.e., interconnects and materials, are explored. Interconnect scaling projections and the role and opportunities that materials advancements offer are discussed.

ACKNOWLEDGMENTS

The authors would like to gratefully acknowledge help from Mukul Renavikar, Nitesh Nimkar, Daniel Bills, Zhiguo Qian, Kemal Aygun, Zhichao Zhang, Je-Young Chang, Bob Brennan, Tanay Karnik, Vivek Rajan, Anand Iyer, Dipankar Das, Edi Roytman, Josh Fryman, Debendra Das Sharma, Robert Munoz, Adel Elsherbini, and Rajiv Mongia from Intel Corporation.

REFERENCES

1. G. E. Moore, Cramming more Components onto Integrated Circuits, *Proceedings of the IEEE*, vol. 86, no. 1, pp. 82–85, Jan. 1998, doi: 10.1109/JPROC.1998.658762.
2. G. E. Moore, No Exponential is Forever: But Forever Can be Delayed! [Semiconductor Industry], *2003 IEEE International Solid-State Circuits Conference, 2003*, San Francisco, CA, USA, 2003, vol. 1, pp. 20–23, doi: 10.1109/ISSCC.2003.1234194.
3. A. B. Kelleher, Celebrating 75 Years of the Transistor a Look at the Evolution of Moore's Law Innovation, *2022 International Electron Devices Meeting (IEDM)*, San Francisco, CA, USA, 2022, pp. 1.1.1–1.1.5, doi: 10.1109/IEDM45625.2022.10019538.
4. P. D. Azar, Moore's Law and Economic Growth, Federal Reserve Bank of New York, Staff Reports, no. 970, October 2022, https://www.newyorkfed.org/medialibrary/media/research/staff_reports/sr970.pdf?sc_lang=en
5. State of the Semiconductor Industry, 2025 Semiconductor Industry Association Report, https://www.semiconductors.org/wp-content/uploads/2025/07/SIA-State-of-the-Industry-Report-2025.pdf
6. IEEE Spectrum, 31 Dec 2022, https://spectrum.ieee.org/tag/75th-anniversary-of-the-transistor
7. J. Goodrich, IEEE Spectrum, 01 Nov 2022, https://spectrum.ieee.org/celebrate-transistor-anniversary-with-ieee
8. AI Era Innovation Matrix - Navid Shahriari (Intel), 19 Feb 2025, Plenary talk at ISSCC 2025, https://www.youtube.com/watch?v=jloGZltzgjI&t=439s
9. Semiconductor Industry; Present and Future, K, Zhang (TSMC), 19 Feb 2024, Plenary talk at ISSCC 2024, https://www.youtube.com/watch?v=D9cBvoBFGuk

10. Investigating the trajectory of AI for the benefit of society, https://epoch.ai/
11. M. B. Giles and I. Reguly, Trends in High-Performance Computing for Engineering Calculations, *Philosophical Transactions of the Royal Society A*, vol. 372, p. 20130319, 2014, doi: 10.1098/rsta.2013.0319.
12. R. Mahajan, S. Agraharam, D. Mallik and V. Boddu, Advanced Packaging Landscape for Heterogeneous Integration. Shangguan, D. (Ed.). (2024). *Direct Copper Interconnection for Advanced Semiconductor Technology* (1st ed., p. 462). CRC Press, Boca Raton, FL, doi: 10.1201/9781003408673.
13. Chapter 1: HIR Overview and Executive Summary, 2019 Edition, https://eps.ieee.org/images/files/HIR_2021/ch01_overview.pdf
14. R. Mahajan, Quiet Revolutions: How Advanced Microelectronics Packaging Continues to Drive Heterogeneous Integration, *2020 19th IEEE Intersociety Conference on Thermal and Thermomechanical Phenomena in Electronic Systems (ITherm)*, Orlando, FL, USA, 2020, pp. 1408–1412, doi: 10.1109/ITherm45881.2020.9190247.
15. J. Shalf, The Future of Computing Beyond Moore's Law, *Philosophical Transactions of the Royal Society A*, vol. 378, p. 20190061, 2020, doi: 10.1098/rsta.2019.0061.
16. R. Patti, Advanced Packaging, Heterogeneous Integration, and Foundry 2.0, *2024 IEEE 67th International Midwest Symposium on Circuits and Systems (MWSCAS)*, Springfield, MA, USA, 2024, pp. 873–874, doi: 10.1109/MWSCAS60917.2024.10658948.
17. D. Das Sharma and R.V. Mahajan, Advanced Packaging of Chiplets for Future Computing Needs, *Nature Electronics*, vol. 7, pp. 425–427, 2024, doi: 10.1038/s41928-024-01175-3.
18. J. H. Lau, Recent Advances and Trends in Multiple System and Heterogeneous Integration with TSV Interposers, *IEEE Transactions on Components, Packaging and Manufacturing Technology*, vol. 13, no. 1, pp. 3–25, Jan. 2023, doi: 10.1109/TCPMT.2023.3234007.
19. J. H. Lau, Recent Advances and Trends in Advanced Packaging, *IEEE Transactions on Components, Packaging and Manufacturing Technology*, vol. 12, no. 2, pp. 228–252, Feb. 2022, doi: 10.1109/TCPMT.2022.3144461.
20. C. -K. Hsiung and K. -N. Chen, A Review on Hybrid Bonding Interconnection and Its Characterization, *IEEE Nanotechnology Magazine*, vol. 18, no. 2, pp. 41–50, April 2024, doi: 10.1109/MNANO.2024.3358714.
21. S. Kim, P. Kang, T. Kim, K. Lee, J. Jang, K. Moon, H. Na, S. Hyun and K. Hwang, Cu Microstructure of High Density Cu Hybrid Bonding Interconnection, *2019 IEEE 69th Electronic Components and Technology Conference (ECTC)*, Las Vegas, NV, USA, 2019, pp. 636–641, doi: 10.1109/ECTC.2019.00101.
22. Heterogeneous Integration Roadmap, 2024 Edition, https://eps.ieee.org/technology/heterogeneous-integration-roadmap.html (2024).
23. International Roadmap for Devices and Systems, https://irds.ieee.org/ (2023).
24. Microelectronics and Advanced Packaging Technologies Roadmap, Semiconductor Research Corporation, Version 2, https://www.src.org/about/mapt-roadmap/ (2024).
25. Manufacturing Roadmap for Heterogeneous Integration and Electronics Packaging (MRHIEP), https://www.chips.ucla.edu/page/MRHIEP/MRHIEP%20Final%20Report (2023).
26. TSMC 3D Fabric, https://3dfabric.tsmc.com/english/dedicatedFoundry/technology/3DFabric.htm
27. Intel Foundry Semiconductor Chip Packaging, https://www.intel.com/content/www/us/en/foundry/packaging.html
28. Intel Leads the Way with Advanced Packaging, Intel Newsroom Sep 6, 2023, https://newsroom.intel.com/manufacturing/intel-leads-the-way-with-advanced-packaging
29. A. Shilov, Nvidia shifts to CoWoS-L packaging for Blackwell GPU production ramp-up, Tom's Hardware, Jan 16, 2025, https://www.tomshardware.com/tech-industry/nvidia-shifts-to-cowos-l-packaging-for-blackwell-gpu-production-ramp-up
30. Universal Chiplet Interconnect Express 3.0 Specification, https://www.uciexpress.org/

31. Open Compute Project, Bunch of Wires (BoW) PHY Specification, Jan 2, 2023, https://opencomputeproject.github.io/ODSA-BoW/bow_specification.html
32. D. Das Sharma, G. Pasdast, S. Tiagaraj and K. Aygün, High-Performance, Power-Efficient Three-Dimensional System-in-Package Designs with Universal Chiplet Interconnect Express, *Nature Electronics*, vol. 7, pp. 244–254, 2024, doi: 10.1038/s41928-024-01126-y.
33. D. D. Sharma, G. Pasdast, Z. Qian and K. Aygun, Universal Chiplet Interconnect Express (UCIe): An Open Industry Standard for Innovations with Chiplets at Package Level, *IEEE Transactions on Components, Packaging and Manufacturing Technology*, vol. 12, no. 9, pp. 1423–1431, Sept. 2022, doi: 10.1109/TCPMT.2022.3207195.
34. Heterogeneous Integration Roadmap Chapter 22, 2024 Edition, https://eps.ieee.org/wp-content/uploads/2025/11/HIR_2024_ch22_2D-3D.pdf
35. R. Co et al., "High-Volume-Manufacturing (HMV) of BVA Enabled Advanced Package-on-Package (PoP)," ICEP, April 14-17, 2015
36. A. Smith, G. H. Loh, S. Naffziger, J. Wuu, N. Kalyanasundharam, E. Chapman, R. Swaminathan, T. Huang, W. Jung, A. Kaganov, H. McIntyre and R. Mangaser, Interconnect Design for Heterogeneous Integration of Chiplets in the AMD Instinct MI300X Accelerator, *IEEE Micro*, vol. 45, no. 1, pp. 57–66, Jan.-Feb. 2025, doi: 10.1109/MM.2024.3462351.
37. Z. Li, Y. Tomita, A. A. Elsherbini, P. Liu, H. A. Sawyer, J. M. Swan and S. M. Liff, Scaling Solder Micro-Bump Interconnect Down to 10 μm Pitch for Advanced 3D IC Packages, *2021 IEEE 71st Electronic Components and Technology Conference (ECTC)*, San Diego, CA, USA, 2021, pp. 451–456, doi: 10.1109/ECTC32696.2021.00082.
38. K. Sahoo, V. Harish, H. Ren and S. S. Iyer, A Review of Die-to-Die, Die-to-Substrate and Die-to-Wafer Heterogeneous Integration, *IEEE Electron Devices Reviews*, vol. 2, pp. 6–31, 2025, doi: 10.1109/EDR.2024.3501214.
39. High-End Performance Packaging 2025 report, Yole Group, www.yolegroup.com
40. R. Mahajan, Z. Qian, R. S. Viswanath, S. Srinivasan, K. Aygün, W. -L. Jen, S. Sharan and A. Dhall, Embedded Multidie Interconnect Bridge—A Localized, High-Density Multichip Packaging Interconnect, *IEEE Transactions on Components, Packaging and Manufacturing Technology*, vol. 9, no. 10, pp. 1952–1962, Oct. 2019, doi: 10.1109/TCPMT.2019.2942708.
41. R. Mahajan, R. Manepalli, K. Aygün, K. Radhakrishnan and S. Sharan, Scaling EMIB—A Comprehensive Architecture Portfolio for Localized Interconnects, *IEEE Transactions on Electron Devices*, vol. 72, no. 7, pp. 3911–3914, July 2025, doi: 10.1109/TED.2025.3574287.
42. TSMC's Chip on Wafer on Substrate with Silicon Interposer (CoWoS®-S), https://3dfabric.tsmc.com/english/dedicatedFoundry/technology/cowos.htm
43. R. Tummala, *Fundamentals of Device and Systems Packaging: Technologies and Applications*, 2nd Edition, McGraw-Hill, New York, August 2019.
44. V. K. Nagesh, R. Peddada, S. Ramalingam, B. Sur and A. Tai, Challenges of Flip Chip on Organic Substrate Assembly Technology, *1999 Proceedings. 49th Electronic Components and Technology Conference*, San Diego, CA, USA, 1999, pp. 975–978, doi: 10.1109/ECTC.1999.776304.
45. T. Wang, F. Tung, L. Foo and V. Dutta, Studies on a Novel Flip-Chip Interconnect Structure. *Pillar bump, 2001 Proceedings. 51st Electronic Components and Technology Conference (Cat. No.01CH37220)*, Orlando, FL, USA, 2001, pp. 945–949, doi: 10.1109/ECTC.2001.927911.
46 H. Balkan D. Patterson, G. Burgess, C. Carlson, P. Elenius, M. Johnson, B. Rooney, J. Sanchez, D. Stepniak and J. Wood, Flip Chip Reliability: Comparative Characterization of Lead Free (Sn/Ag/Cu) and 63Sn/Pb Eutectic Solder, *52nd Electronic Components and Technology Conference 2002 (Cat. No. 02CH37345)*, San Diego, CA, USA, 2002, pp. 1263–1269, doi: 10.1109/ECTC.2002.1008268.

47. A. Eitan and K. -Y. Hung, Thermo-Compression Bonding for Fine-Pitch Copper-Pillar Flip-Chip Interconnect: Tool Features as Enablers of Unique Technology, *2015 IEEE 65th Electronic Components and Technology Conference (ECTC)*, San Diego, CA, USA, 2015, pp. 460–464, doi: 10.1109/ECTC.2015.7159631.
48. S. Lhostis, A. Farcy, E. Deloffre, et al., Reliable 300mm Wafer Level Hybrid Bonding for 3D Stacked CMOS Image Sensors, *2016 IEEE 66th Electronic Components and Technology Conference (ECTC)*, Las Vegas, NV, USA, 2016, pp. 869–876, doi: 10.1109/ECTC.2016.202.
49. D. B. L. Yolanda, Wafer to Wafer Bonding to Increase Memory Density, *2022 China Semiconductor Technology International Conference (CSTIC)*, Shanghai, China, 2022, pp. 1–4, doi: 10.1109/CSTIC55103.2022.9856846.
50. A. Elsherbini, K. Jun, R. Vreeland, et al., Enabling Hybrid Bonding on Intel Process, *2021 IEEE International Electron Devices Meeting (IEDM)*, San Francisco, CA, USA, 2021, pp. 34.3.1–34.3.4, doi: 10.1109/IEDM19574.2021.9720586.
51. D. W. Fischer, S. Knickerbocker, D. Smith, R. Katz, J. Garant, J. Lubguban, V. Soler and N. Robson, Face to Face Hybrid Wafer Bonding for Fine pitch applications, *IEEE 70th Electronic Components and Technology Conference*, Orlando, FL, USA, 2020, pp. 595–600.
52. H. Ren, Y. -T. Yang, G. Ouyang and S. S. Iyer, Mechanism and Process Window Study of Die-to Wafer (D2W) Hybrid Bonding, *ECS Journal of Solid State Science and Technology*, vol. 10, no. 6, p. 064008, June 2021, DOI:10.1149/2162-8777/ac0a52.
53. J. A. Theil, L. Mirkarimi, G. Fountain, G. Gao and R. Katkar, Recent Developments in Fine Pitch Wafer-to-Wafer Hybrid Bonding with Copper Interconnect, *2019 International Wafer Level Packaging Conference (IWLPC)*, San Jose, CA, USA, 2019, pp. 1–6, doi: 10.23919/IWLPC.2019.8913862.
54. B. Brandstätter D. Aschenwald, B. Auer, N. Bilewicz, R. Boomsma, C. Kröll, A. Mayr, R. Neumayr, H. Rieser, M. Schernthaner, H. Selhofer, F. Speer, P. Unterwaditzer, A. Vidéki, M. Widauer and T. Widmann, High-Speed Ultra-Accurate Direct C2W Bonding, *2020 IEEE 70th Electronic Components and Technology Conference (ECTC)*, Orlando, FL, USA, 2020, pp. 1943–1949, doi: 10.1109/ECTC32862.2020.00303.
55. S. Moreau, J. Jourdon, S. Lhostis, D. Bouchu, S. Ayoub, L. Arnaud and H. Fremont, Review- Hybrid Bonding-Based Interconnects – A Status on Last Robustness and Reliability Achievements, *ECS Journal of Solid State Science and Technology*, vol. 11, p. 024001, 2022.
56. S. C. Chong, I. C. Daniel, V. N. Sekhar, S. Lim and V. Srinivas, Chip-to-Wafer Hybrid Bonding for High Performance 2.5D Applications, *2022 IEEE 24th Electronics Packaging Technology Conference (EPTC)*, Singapore, Singapore, 2022, pp. 19–23, doi: 10.1109/EPTC56328.2022.10013161.
57. R. Hung, G. See, Y. Wang, C. B. Yong, K. Zheng, Y. Chang, A. Shantaram, R. Wang, A. Sundarrajan, J. Abdilla, N. Hegde, S. Schmid, D. Bikaljevic and M. Glantschnig., Enabling Die-to-Wafer Hybrid Bonding for the Next Generation Advanced 3D Packaging, *2024 IEEE 74th Electronic Components and Technology Conference (ECTC)*, Denver, CO, USA, 2024, pp. 778–783, doi: 10.1109/ECTC51529.2024.00127.
58. K. Ma, N. Bekiaris, S. Ramaswami, T. Ding, G. Probst, J. Burggraf and T. Uhrmann, 0.5 ¼m Pitch Wafer-to-wafer Hybrid Bonding with SiCN Bonding Interface for Advanced Memory, *2023 IEEE 73rd Electronic Components and Technology Conference (ECTC)*, Orlando, FL, USA, 2023, pp. 1110–1114, doi: 10.1109/ECTC51909.2023.00190.
59. Y. Nakamura, K. Tamura, Y. Kayaba, W. Okada, T. Shikama and S. Inada, A Study on Multi-chip Stacking Process by Novel Dielectric Polymer Adhesive for Cu-Cu Hybrid Bonding, *2023 IEEE 73rd Electronic Components and Technology Conference (ECTC)*, Orlando, FL, USA, 2023, pp. 1129–1134, doi: 10.1109/ECTC51909.2023.00194.

60. Y Kayaba, Y. Nakamura, W. Okada, T. Shikama, K. Tamura and S. Inada, A New Adhesive for CoW Cu-Cu Hybrid Bonding with High Throughput and Room Temperature pre-Bonding, *2023 IEEE 73rd Electronic Components and Technology Conference (ECTC)*, Orlando, FL, USA, 2023, pp. 571–578, doi: 10.1109/ECTC51909.2023.00101.
61. S Nagayama, R. Sugano, T. Ogawa, H. Nakagawa, H. Deguchi and E. Ishikawa, Novel Polymer for Hybrid Bonding with Precise Tunable Crosslink Density, *2025 IEEE 75th Electronic Components and Technology Conference (ECTC)*, Dallas, TX, USA, 2025, pp. 87–91, doi: 10.1109/ECTC51687.2025.00021.
62. T. Shirasaka, T. Okuda, T. Shibata, S. Yoneda, D. Matsukawa, M. Mariappan, M. Koyanagi and T. Fukushima, Comprehensive Study on Advanced Chip on Wafer Hybrid Bonding with Copper/Polyimide Systems, *2022 IEEE 72nd Electronic Components and Technology Conference (ECTC)*, San Diego, CA, USA, 2022, pp. 317–323, doi: 10.1109/ECTC51906.2022.00059.
63. NGMM, Next-Generation Microelectronics Manufacturing, DARPA, https://www.darpa.mil/research/programs/next-generation-microelectronics
64. DARPA Minitherms3D Program, https://www.darpa.mil/research/programs/Minitherms3D
65. V. Zhirnov, M. E. Chen, M. Malakoutian, H. R. M. Margavio, E. Pawliczak, K. Reidy, W. Yanez and T. Younkin, SRC-Led Materials Research: 40 Years Ago, and Now, *MRS Advances*, vol. 8, pp. 751–762, 2023, doi: 10.1557/s43580-023-00665-4.
66. C. Jois, Sharp Interface Simulations and Experimental Characterization of Surface Diffusion Driven Phase Evolution in Bonded Interconnects, Ph.D. Thesis, Purdue University, 2025.
67. S. P. Prasad, C. Jois, Y. Singh, G. Subbarayan, B. Penmecha and P. Raghavan, Void Growth and Intermetallic Bridging in Microscale Solder Interconnects Under Thermal Annealing, *IEEE Transactions on Components, Packaging and Manufacturing Technology*, vol. 14, no. 7, pp. 1308–1318, July 2024, doi: 10.1109/TCPMT.2024.3416430.
68. S. Sadasiva, Simulations of Diffusion Driven Phase Evoluation in Heterogeneous Solids, Ph.D. Thesis, Purdue University, 2014.
69. S. Sadasiva, G. Subbarayan, L. Jiang and D. Pantuso, Numerical Simulations of Electromigration and Stressmigration Driven Void Evolution in Solder Interconnects, *ASME Transactions Journal of Electronic Packaging*, vol. 134, no. 2, pp. 020907, 2012, doi:10.1115/1.4006707.
70. P. Singh, L. Palmer, M. Hamid and T. Wassick, In Situ Study of Electromigration in Eutectic Tin-Bismuth Planar Solder Joints, *2024 Pan Pacific Strategic Electronics Symposium (Pan Pacific)*, Kona, Big Island, HI, USA, 2024, pp. 1–6, doi: 10.23919/PanPacific60013.2024.10436519.
71. R. Lasky, I. Tevis and R. Lasky, The Status of Low Temperature Solders, *2024 Pan Pacific Strategic Electronics Symposium (Pan Pacific)*, Kona, Big Island, HI, USA, 2024, pp. 1–7, doi: 10.23919/PanPacific60013.2024.10436444.
72. Encapsulation Technologies for Electronic Applications, 2nd Edition - October 11, 2018, Haleh Ardebili, Jiawei Zhang, Michael G. Pecht, Paperback ISBN: 9780128119785, eBook ISBN: 9780128119792, https://shop.elsevier.com/books/encapsulation-technologies-for-electronic-applications/ardebili/978-0-12-811978-5
73. The application of organic materials used in IC advanced packaging: A review; Liu Jikang, Department of Physics and Mechanical and Electrical Engineering, Hubei University of Education, Wuhan, 430205, PR China, https://doi.org/10.1016/j.memori.2025.100124
74. Y. Wen, C. Chen, Y. Ye, Z. Xue, H. Liu, X. Zhou, Y. Zhang, D. Li, X. Xie and Y. Mai, Advances on Thermally Conductive Epoxy-based Composites as Electronic Packaging Underfill Materials – A Review, *Advanced Materials*, vol. 34, no. 52, p. 2201023, 2022.

75. S. P. Phansalkar, C. Kim and B. Han, Effect of Critical Properties of Epoxy Molding Compound on Warpage Prediction: A Critical Review, *Microelectronics Reliability*, vol. 130, p. 114480, 2022, doi: 10.1016/j.microrel.2022.114480.
76. I. Fofana, U. M. Rao, P. Rozga, A. Beroual, M. Lashbrook and J. Acosta, Dielectric Liquids for Insulation and Cooling: An Electric Vehicle Perspective by the IEEE DEIS Technical Committee on Liquid Dielectrics, *2025 IEEE International Conference on Dielectric Liquids (ICDL)*, Lodz, Poland, 2025, pp. 1–4, doi: 10.1109/ICDL63868.2025.11068293.
77. A. F. Farzaneh, M. Olofsson and E. Gustavsson, Aging and Air Infusion Effects on the Performance and Longevity of Polyol Ester-Based Immersion Cooling Fluids for Data Centers, *2025 IEEE International Conference on Dielectric Liquids (ICDL)*, Lodz, Poland, 2025, pp. 1–4, doi: 10.1109/ICDL63868.2025.11068238.
78. Intel and Shell Advance Immersion Cooling in Xeon-Based Data Centers Intel Newsroom, May 13, 2025, https://newsroom.intel.com/data-center/intel-shell-advance-immersion-cooling-xeon-based-data-centers
79. M. -Y. Hsu, C. -H. Lin and K. -C. Chen, Low Dielectric Materials for High Frequency Semiconductor Packaging Application, *2024 International Conference on Electronics Packaging (ICEP)*, Toyama, Japan, 2024, pp. 279–280, doi: 10.23919/ICEP61562.2024.10535560.
80. S. -M. Kang, H. Kweon, S. Park and B. -S. Bae, Low Dk/Df Siloxane Hybrid Laminates for Advanced Packaging Substrate, *2025 IEEE 75th Electronic Components and Technology Conference (ECTC)*, Dallas, TX, USA, 2025, pp. 1113–1118, doi: 10.1109/ECTC51687.2025.00193.
81. J. Baek, B. Choi, S. Kulasekaran, H. Do, B. Marin, J. Chavarria, L. Wojewoda and K. Radhakrishnan, Fully Integrated Voltage Regulators (FIVRs) with Package In-situ Coupled CoaxMIL Inductor for High Power Density Microprocessor Applications, *2025 IEEE Applied Power Electronics Conference and Exposition (APEC)*, Atlanta, GA, USA, 2025, pp. 491–497, doi: 10.1109/APEC48143.2025.10977257.
82. K. Radhakrishnan, M. Swaminathan and B. K. Bhattacharyya, Power Delivery for High-Performance Microprocessors—Challenges, Solutions, and Future Trends, *IEEE Transactions on Components, Packaging and Manufacturing Technology*, vol. 11, no. 4, pp. 655–671, April 2021, doi: 10.1109/TCPMT.2021.3065690.
83. H. Ren, Z. Chen, J.-H. Liu, C. T. Yang, B. Yang, B. Vaisband and I. Subramanian, Integrated Passive Devices in the Silicon-Interconnect Fabric, *2025 IEEE 75th Electronic Components and Technology Conference (ECTC)*, Dallas, TX, USA, 2025, pp. 1994–1998, doi: 10.1109/ECTC51687.2025.00341.
84. R. -S. Yu, H. -C. Yu, Y. -F. Yang and S. Singh, A Global Overview of Per- and Polyfluoroalkyl Substance Regulatory Strategies and Their Environmental Impact, *Toxics*, vol. 13, no. 4, p. 251, 2025, doi: 10.3390/toxics13040251
85. W. -C. Chen, H. -G. Chen, F. Liu, L. -Y. Hung, A. Kang and Y. -P. Wang, PFAS-Free Assembly Material Development, *2025 International Conference on Electronics Packaging and iMAPS All Asia Conference (ICEP-IAAC)*, Nagano, Japan, 2025, pp. 329–330, doi: 10.23919/ICEP-IAAC64884.2025.11003020.
86. M. Kajiwara, Y. Yamada, T. Yamaguchi and N. Sakai, New PFAS-Free Optical Adhesive with High Heat Resistance and Low Refractive Index, *2024 IEEE CPMT Symposium Japan (ICSJ)*, Kyoto, Japan, 2024, pp. 102–105, doi: 10.1109/ICSJ62869.2024.10804700.

2 Substrate and Interposer Materials for Semiconductor Packaging and Heterogeneous Integration

Tarak A. Railkar

2.1 INTRODUCTION

It was December 16, 1947, when John Bardeen and Walter Brattain successfully demonstrated the world's first functioning transistor. Progress was somewhat slow until 1958, when Jack Kilby demonstrated the first working integrated circuit (Figure 2.1).

Robert Noyce then invented the first monolithic integrated circuit in 1959 on a silicon chip, followed by IBM's invention of the world's first flip-chip in the early 1960s, called the Controlled Collapse Chip Connection (or "C4"). Since then, miniaturization of feature sizes on the chip has supported as well as required higher integration at the package and sub-system levels. This led to the birth and rapid development of the substrate and interposer technologies.

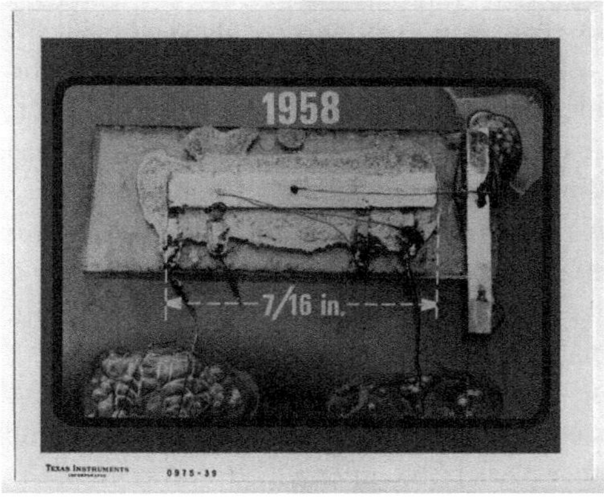

FIGURE 2.1 First integrated circuit invented by Jack Kilby [1].

DOI: 10.1201/9781003587637-2

Substrates and interposers perform a critical role in the functionality and reliability of practically all advanced semiconductor devices. These materials serve as the mechanical support as well as interconnection within a package or in a system application. They enable efficient signal transmission and support an effective heat removal path for various thermal management strategies.

Technological advancements in other areas of microelectronics have placed expanding and escalating demands on substrates and interposers as well. These are in the form of higher routing densities, better thermal and mechanical performance as well as enhanced reliability. In response, the industry has developed newer and advanced solutions to address demands from advanced applications such as high-performance computing (HPC), artificial intelligence (AI) and others. RF applications present challenges that are often in contrast to those presented by digital and mixed-signal considerations, thus requiring careful assessment for identifying the appropriate solution for the requirements of any given advanced application.

This chapter addresses several of those considerations, related materials and their properties, as well as fabrication techniques, and offers a glimpse into the future possibilities as the industry continues in its effort of driving innovation and the advancement of modern electronics.

2.1.1 Terminology and Nomenclature

The same word often means different things for individuals with different backgrounds. For example, to a semiconductor wafer-fab professional, the word "substrate" typically means a semiconductor wafer, such as single-crystal silicon, that is used as the building block for the fabrication of electronic circuits and devices. However, to a microelectronic packaging and assembly professional, the same term "substrate" usually refers to a laminate, ceramic or a similar carrier onto which semiconductor die, surface-mounted devices (SMD), etc. are assembled and interconnected with other parts of the microelectronic ecosystem. As such, to minimize any potential confusion, the following outlines some of the terms that are commonly used in the microelectronics industry and will be used in this chapter.

- **Wafer**: A thin slice of semiconductor material such as silicon, GaAs or GaN, that is used for the fabrication of microelectronic devices. From a crystallographic perspective, wafers are usually single-crystal materials.
- **Chip** or **Die**: A singulated piece of the wafer that contains integrated circuits (IC) or a building block of the circuitry.
- **Interposer**: An electrical interface with advanced routing capability that routes connections between different chips. Interposers can also provide a thermal path for heat removal and are usually made of silicon, glass, ceramics or occasionally of laminate materials.
- **Substrate**: Substrate refers to a laminate, ceramic or a similar carrier onto which components such as semiconductor die are assembled and interconnected with other parts of the microelectronic system.
- **Package**: In simple terms, a microelectronic "package" is a microelectronic component. A package also protects the die and elements inside of it

from the external environment and vice versa, while providing a means of mechanically, electrically and/or thermally connecting it to the rest of the electronic system. A package can have a diversity of forms such as ball grid array (BGA), line-grid array (LGA) and others.

- **Printed Circuit Board (PCB)**: This is a part of the electronic sub-system that mechanically supports and electrically connects various components to each other and to other parts of the system. PCBs are usually much larger and thicker than substrates.
- **Flip-Chip**: This is a semiconductor die in which the die is mechanically and electrically connected to the substrate with the active side facing "down" towards the substrate surface, by means of a solder or a metal interconnect.
- **Wire Bonding**: This is another configuration of interconnecting the die to the substrate by means of a wire, typically made of gold, copper or aluminum. The active side of the die faces "up" in this configuration away from the substrate surface.

2.1.2 Packaging Hierarchy and Levels of Interconnection

Packaging hierarchy in microelectronics assembly is a reference to the sequence in which parts of an integrated circuit are assembled, packaged and interconnected.

- **First-Level Assembly** involves the direct attachment of the die to the substrate, either with wires for a wire-bondable die or with a metal interconnect for a flip-chip die.
- **Second-Level Assembly** refers to the attachment of the packaged die or component onto a PCB. Depending on the type of package, this typically includes additional process steps such as printing flux on the PCB for BGA packages or printing solder paste on the PCB for LGA packages, followed by component placement and reflow.
- **Next-Level Assembly (NLA)** refers to the integration of several assembled PCBs and in many cases, other components into a final product or system.

This is shown schematically in Figure 2.2.

There are two key reasons why the packaging hierarchy is significant. Firstly, it is important to assemble the components – which are the building blocks of any microelectronic system – then progressively assemble the PCB and finally the system itself.

Secondly, given that the component, the PCB and the NLA are typically assembled sequentially, it becomes important to address the so-called solder hierarchy

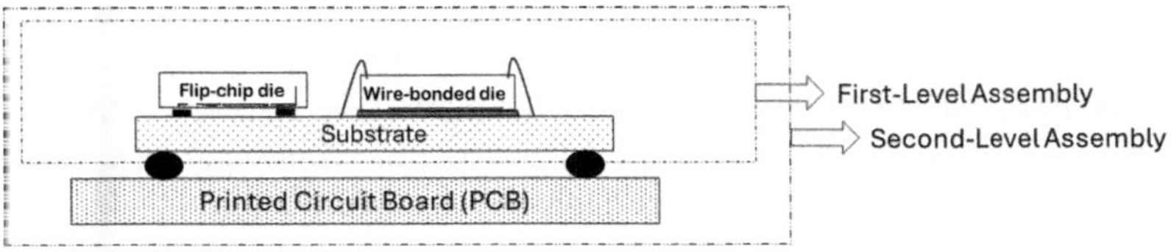

FIGURE 2.2 Example of packaging hierarchy and levels of interconnect.

considerations. This refers to the temperatures that the respective components and solder interconnects attain during solder reflow as part of the assembly process. When multiple reflows are required for the assembly of the complete sub-system or system, it is important that the earliest reflow is performed at the highest temperature, and the last reflow should be performed at the lowest temperature. This ensures that the solder used to attach components during the previous assembly steps does not reflow again during a subsequent reflow process step. These considerations are called "solder hierarchy", and it requires the design and assembly teams to collectively choose an appropriate solder metallurgy at the various steps of the assembly process or a clever use of land-pad metallurgies that alter the reflow temperatures during a second or subsequent reflow due to intermetallic formation at the first reflow. Use of mold compounds or similar means of containing any re-reflow solder is also commonly leveraged in the assembly industry.

2.2 KEY CRITERIA FOR PACKAGING TECHNOLOGY SELECTIONS

2.2.1 General Considerations

Selecting an appropriate packaging technology is important for ensuring optimum electrical, thermal and reliability performance while being mindful of its implications to manufacturability, throughput and cost. Accordingly, some of the important considerations include the following:

- **Performance Requirements**: Understanding the electrical, thermal and reliability performance requirements is one of the most important considerations in selecting the appropriate packaging technology.
- **Type of Application**: Digital, mixed-signal and RF systems can have quite different system requirements, and it is important to know this in the early phases when selecting the appropriate packaging technology. For example, RF chipsets – especially those operating at frequencies around or beyond ~25 GHz – often are sensitive to the dielectric properties of the assembly materials, specifically the dielectric constant and the loss tangent.
- **Use-Life Environment**: Selecting the packaging technology and assembly materials can strongly depend on the environmental conditions where the microelectronic component is intended for use. Components intended for use in cellphones have different expectations compared to components that are intended for automotive under-the-hood applications. Microelectronics intended for harsh environments such as those with extreme heat, ionizing or nuclear radiation, corrosive environments, vibration and shock, or extreme cold, require specialized design, materials and assembly to ensure the reliability and longevity of microelectronic systems.
- **Manufacturability Considerations**: Assembly and test yield, throughput, cost and automation are some of the important considerations. Special applications will typically require additional considerations.
- **Interconnect Count and Pitch**: Interconnect pitch and interconnect count play an important role in the selection process. One such example is outlined in a report by Yole Developpement [2] and shown in Figure 2.3.

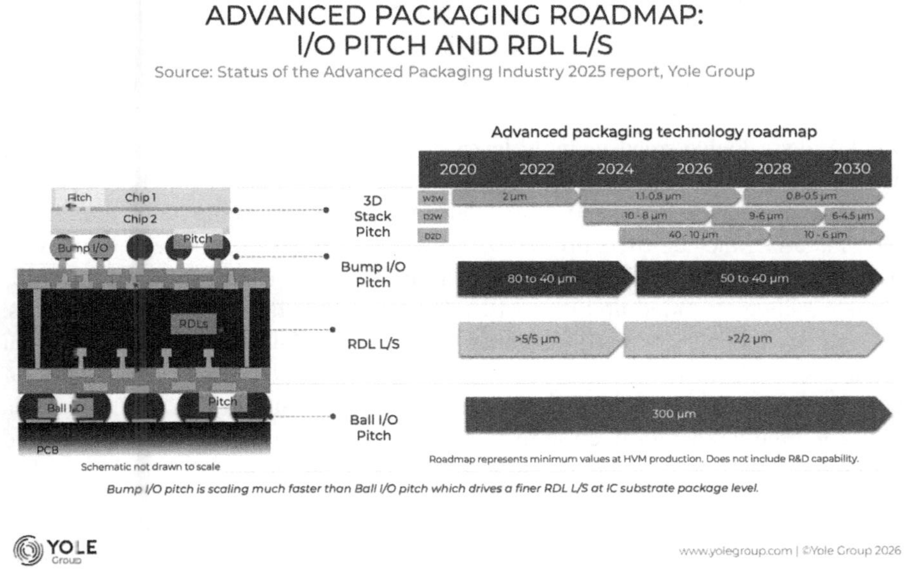

FIGURE 2.3 Advanced substrates and package selection considerations.

These are some of the most important considerations that need careful attention in achieving the required functionality and performance of the microelectronic systems. In the context of selecting appropriate materials for microelectronics, the following sections outline additional details.

2.2.2 PCB, Substrates and Interposers

Printed Circuit Boards (PCBs), substrates and interposers have some similarities and common functions, but they generally serve different functions.

2.2.2.1 Printed Circuit Board (PCB)

A PCB is typically found in most electronic devices. It serves multiple functions, including providing mechanical support to the core components within the device, such as integrated circuits (ICs), resistors and capacitors, etc. PCBs enable electrical interconnectivity between such components that are assembled on it. This interconnectivity is typically established by means of copper traces fabricated on the surface of the PCB, as well as copper traces that are embedded within several routing layers in the PCB. PCBs can be of several different types. A single-layer or single-sided PCB has one layer of conductive material. A double-layer or double-sided PCB supports two layers of conductive material, while a multi-layered PCB consists of three or more layers of conductive material. PCBs that support complex applications can have as many as 50 routing layers or more. The thickness of PCBs can vary across a few hundred micrometers to a few millimeters.

2.2.2.2 Substrates

In the context of microelectronics packaging, "substrate" refers to the base material on which semiconductor die and/or surface mount components such as resistors, capacitors or similar components are assembled. Like PCBs, substrates too can be single-sided, double-sided or multi-layered. However, substrates can support finer trace-widths and better alignment capabilities than those offered by traditional PCBs.

Substrates can be fabricated from a variety of materials. Some of the more common materials include FR4 (fire retardant) which is a glass-reinforced epoxy laminate material that is used for laminates and PCBs. More recent standard materials include Ajinomoto Build-up Film (ABF). Others include BT-resin based dielectrics, as well as Teflon® for high-frequency applications. Advanced applications that require specialized performance such as those for high-frequency or high-power, use more advanced materials like ceramics – Aluminum Nitride (AlN), Alumina (Al_2O_3), high-temperature co-fired ceramic (HTCC), low temperature co-fired ceramic (LTCC) or other high-performance materials. These will be addressed in more detail elsewhere within this chapter.

2.2.2.3 Interposers

The term interposer comes from the Latin word "interpōnere", which approximately means "to put between". In the context of microelectronics, the role of an interposer is that of an advanced substrate that serves as an intermediate layer or layers that primarily facilitate electrical connections between multiple chips and with the next level of assembly within a package. Interposers thus support "2.5D" and "3D" packaging, details of which are described elsewhere within this book. Although their primary function is to provide electrical connectivity, interposers can also provide help with mechanical robustness as well as support heat removal for enhancing thermal management for complex architectures. Some of the commonly used interposer materials include silicon, glass, and advanced laminates such as high-density build-up (HDBU). Materials with a higher thermal conductivity such as silicon carbide (SiC) and chemical vapor deposited (CVD) diamond (i.e., synthetic diamond), are also used in some specialty applications, based on the electrical, mechanical and thermal demands of advanced chips.

2.2.3 Materials Properties and Tradeoffs

Identifying the correct interposer in microelectronics involves important considerations and tradeoffs to ensure that any specific requirements of the targeted application are met or exceeded, including the following:

- **Electrical Performance**: Silicon makes a compelling case for use as an interposer material – especially for digital applications, due to its tailorable electrical conductivity, ability to support fine pitch interconnects, robust and mature infrastructure in the industry, as well as abundant availability. Silicon has good applicability to RF applications as well; however, materials such as SiC or glass might be preferred due to their low loss characteristics especially for high-frequency applications such as radar, phased antenna, and others.

- **Thermal Performance**: Thermal conductivity and the coefficient of thermal expansion (CTE) become an important consideration for interposers, especially for high-power applications. Silicon continues to be a preferred choice; however, SiC and glass are important considerations for high-performance applications. Organic materials bring value for consumer or other cost-sensitive applications; however, such materials typically bring higher CTE, which can lead to reliability tradeoffs, especially their performance under thermal cycling. Need for fluidic cooling, if any, drives additional considerations.
- **Mechanical and Reliability Performance**: Young's modulus, CTE and T_g (glass transition temperature) are some of the key mechanical and reliability considerations. CTE compatibility is essential in ensuring that the thermo-mechanical stress due to the CTE mismatch at the interface of two or more dissimilar materials is manageable and does not cause reliability concerns. Silicon and glass – and to a lesser extent, SiC as well – can also crack under mechanical shock. In contrast, interposers fabricated from laminates or organic materials have a lower modulus and can withstand mechanical stresses (such as mechanical shock) better than the so-called rigid materials.
- **Routing Density**: Laminates or organic materials usually come second to rigid interposers when it comes to thickness, line/space and number of routing layers within the interposer.
- **Moisture Absorption**: Applications requiring hermeticity or fine-leak performance typically prefer the rigid interposers such as Si, SiC or glass. Organic materials are much more likely to absorb moisture, which can affect their electrical properties and long-term reliability. Organic materials are also semi-permeable to air and other gases, which makes it difficult for them to support applications that demand fine-leak compliance.
- **Cost**: Interposers made from organic materials are generally the cheapest, while rigid interposers such as Si, SiC or glass are more expensive.
- **Other Considerations**: Impact of lead times and geo-political considerations on the materials selection is duly acknowledged but not addressed further here.

2.2.4 Routing Layer Characteristics

Routing layers provide a means of electrical connectivity between various functional blocks within the chip and between chips and components within the microelectronics package as well as enable a means of communicating with the external world. Digital applications such as memory and processor call for a high number of interconnects at a high density compared to RF or mixed-signal applications, and hence require the routing layers to support fine line/space capabilities that can be as tight as 2 µm/2 µm [3,4]. Many digital applications require the availability of multiple routing layers, to support the required number of interconnections.

In contrast, RF and mixed-signal applications involve fewer connections than those for digital applications. However, RF applications often require ground planes within the routing layers of the substrate/interposer, for performance reasons. This requirement typically can work with reasonably coarse line/space requirements; 25 µm/25 µm is often adequate. However, the need for a dedicated ground plane can place specific demands on the via type and dimensions as well as on the dimensional and registrational tolerance available at the fabricator with the materials selected for the substrate or interposer.

The following summarizes some of the key considerations when selecting the appropriate packaging materials and technology, based on routing layer considerations:

- **Signal Integrity (SI)**: While this is especially true for RF and mixed-signal applications, it is also important for digital applications to ensure that the signal loss and cross-talk is managed adequately.
- **Power Integrity (PI) and Distribution**: SI/PI is an important consideration of all power components, RF, mixed-signal and digital alike. Routing layer strategy and capabilities must ensure that adequate and reliable power delivery is available for all components.
- **Thermal Management**: Routing layers typically are not a part of the thermal management scheme of components or sub-system; however, it is important to ensure that their configuration does not significantly impede the heat removal pathways.
- **Density and Miniaturization**: With shrinking die and interconnect sizes, the ability to route denser signals without compromising performance is critical. Impact of miniaturization on any increase in thermal density and/or current density should be assessed early in the design stages.
- **Mechanical Robustness**: Selected package materials and technologies are required to withstand physical handling and various types of stress during manufacturing as well as through the useful life of the product in operation.

2.3 SUBSTRATE MATERIALS

In this chapter, the term "substrate" refers to the base material of IC packages, wherein semiconductor die and other components are assembled – or sometimes embedded. By far, laminate substrates constitute a vast majority of the volume of microelectronic packages fabricated. Some estimates [5,6] suggest that the global semiconductor substrate market size was valued at over USD 12 billion in 2024. During this time, the global ceramic substrate market was valued at about USD 8.8 billion; however, this market translates into a much lower volume of products, given the significantly higher cost of ceramic materials, compared to the laminates.

2.3.1 PCBs and SLPs

A Printed Circuit Board (PCB) is an important building block of most microelectronic systems. PCBs provide the mechanical and electrical foundation for

TABLE 2.1
Typical Characteristics of PCBs

Type of Substrate	Typical Number of Routing Layers	Typical Substrate Thickness (mm)	Line/Space Capability (μm/μm)	Fabrication Processes
"Standard" PCB	20+	1–3	50–75/50–75	Various
High-density interconnect (HDI) PCB	16–20	1–2	40/40	UV-LDI
Substrate-like PCB (SLP)	10–12	1–1.5	25–30/25–30	mSAP, UV-LDI

interconnecting electronic components to realize a functional and operating system. Over the years, demand has continually increased on the number of components that need to be integrated on or into PCBs. This has led to an increase in the number of routing layers within the PCB as well as advancements in the form of finer line/space capabilities. The resultant increases in the component density and in the power density have required due considerations towards integrating thermal management solutions into the PCB designs, fabrication and implementation. With the wider proliferation of components from diverse packaging platforms – such as ball grid array (BGA), chip-scale packaging (CSP), lead-frame packages such as QFNs, and many others – that need to be integrated onto a PCB, the PCB technology has had to evolve as well. This has led to the growth of Substrate-Like PCB (SLP) solutions. Table 2.1 outlines some of the typical characteristics of PCBs [7].

2.3.2 Laminates

Laminate substrates – or simply laminates – are one of the most commonly used materials in high-volume consumer electronics applications. This is due to their versatility in materials property offerings, routability for highly integrated microelectronic components and modules, acceptable reliability performance and low cost. Historically, some of the earliest consumer electronics used substrates made by impregnating layers of paper with phenolic resin. This resulted in a reasonably strong and stiff material that had good electrically insulating properties. Based on the grade of phenolic resins used in their fabrication, the resultant PCBs were labeled as FR-1 and FR-2 and rated for their temperature tolerances and mechanical properties.

Current state-of-the-art laminate substrates use a combination of more advanced materials for achieving the desired properties for specific applications. One of these specialized materials is epoxy resin, which is a thermosetting polymer that has excellent adhesion properties, chemical resistance, and ability to withstand high temperatures in the range of about 270°C–290°C. Fiberglass is combined with epoxy resin for reinforcing the laminate substrates. Applications requiring a higher heat resistance typically use polyimides, which are known for their high thermal stability. Further, although this is not common for laminate substrates, ceramic fillers are sometimes integrated to improve thermal conductivity, electrical insulation or dielectric performance of specialized laminates.

A laminate can also be distinguished based on the number of routing layers that it supports. The simplest of laminates, historically, was a single-layered board with traces that enabled interconnection between components that were mounted on its surface. On the other side of the spectrum, modern laminates can support in excess of 20 routing layers. Also, laminate substrates can have one or two “cores” or can be coreless.

2.3.2.1 Thermally Enhanced Laminates

Ever-increasing functionality integration in almost all applications has increased the packaging density in microelectronic devices, leading to an increase in power handling and heat dissipation. Correspondingly, the demand for thermally enhanced laminates is also on the rise, especially for high-power applications such as those in automotive and industrial electronics where efficient thermal management is crucial together with cost-effectiveness of the solution.

There are two main technologies that enhance the thermal handling capabilities of laminate substrates, namely the use of metals and that of thermally conductive non-metals within the laminate.

Incorporation of metals into laminate substrates is usually accomplished by integrating one or more of the following:

- **Thermal Vias**: These are vias or through connections between routing layers of a laminate, primarily to provide a path for heat removal. Thermal vias are typically plated and filled with copper or other suitable thermally conductive materials. “Via-farm” or close packing of thermal vias within a laminate, can significantly improve heat management within a component or the sub-system that uses them. Figure 2.4 shows a cross-sectional view of thermal vias within a laminate as an example.
- **Metal Inlays**: Metal inlay technology is based on integrating large metal elements into a laminate or PCB. Typically, the metal is copper; however, the use of aluminum can also be supported. Metal inlays are relatively large and can range from about 5 mm to as much as a few centimeters in dimensions. Inlays are typically round or rectangular in shape with thickness ranging from sub-1mm to well in excess of 2.5 mm [8,9]. These inlays are integrated into laminates or PCBs either through lamination or by drilling/punching holes into the PCB. Components that dissipate significant heat can be directly mounted on the topside of such inlays with the bottom-side enabling heat dissipation through suitable means, such as a heat sink. Metal inlays are commonly used in applications such as car headlights or charging electronics for electric vehicles. Figure 2.5 shows a schematic of how metal inlays are used in high-power applications [9].

2.3.3 Ceramics

As seen from the preceding sections, laminate materials have found widespread use in a variety of microelectronic applications. However, ceramic materials are preferred for applications that require hermeticity, or high temperature/high power,

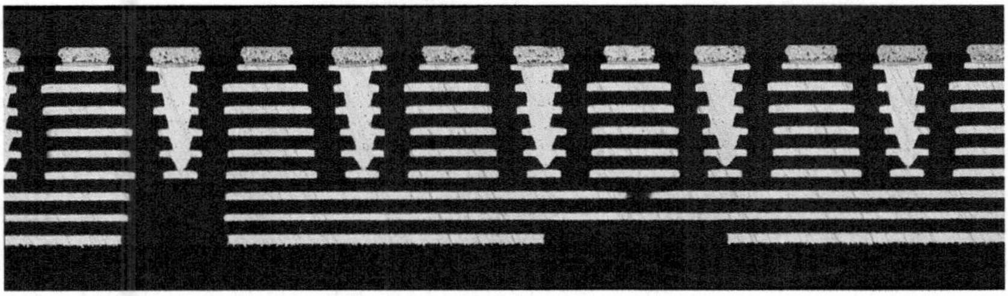

FIGURE 2.4 Example of a thermal via farm within a laminate substrate.

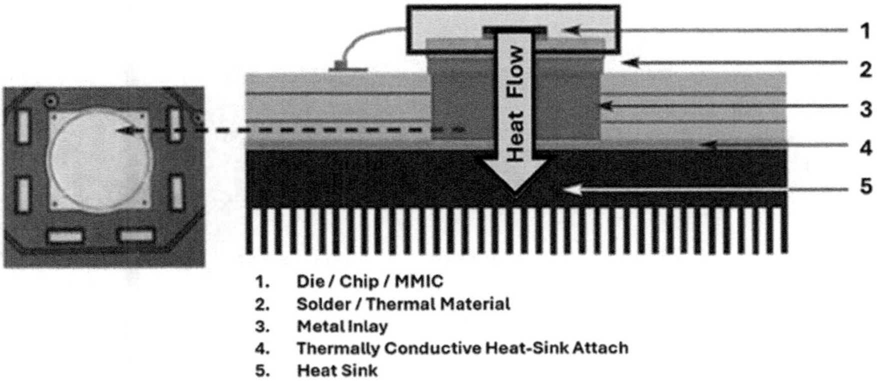

FIGURE 2.5 Example of the use of metal inlay for thermal management.

or applications intended for harsh environments. Excellent electrical insulation, thermal conductivity and mechanical strength are some of the key properties that make ceramic substrates attractive for several demanding applications.

"Ceramic" is a class of substrates that include the following materials with some of their key characteristics:

- **Aluminum Oxide (Al_2O_3)**: High electrical insulation, excellent mechanical strength
- **Aluminum Nitride (AlN)**: Excellent thermal conductivity, thin substrates possible due to high strength
- **Beryllium Oxide (BeO)**: Excellent thermal conductivity and electrical insulation; however, there are toxicity concerns associated with its manufacturing process
- **Silicon Nitride (Si_3N_4)**: High fracture toughness and thermal stability, suitable for harsh environments
- **Zirconium Oxide (ZrO_2)**: High strength and fracture toughness, another material for harsh environments
- **LTCC (Low Temperature Co-fired Ceramic)**: Multi-layered technology, processed around 850°C–900°C, with tight *L/S*, high thermal and mechanical characteristics, and superior high-frequency performance

TABLE 2.2
Key Properties of Ceramic Substrate Materials

Material	Alumina (99.5%)	Alumina (96%)	Beryllium Oxide (99.5%)	Aluminum Nitride	Fused Silica	Silicon	Silicon Carbide
Chemical formula	Al_2O_3	Al_2O_3	BeO	AlN	SiO_2	Si	SiC
Melting point (°C)	2,072	2,072	2,514–2,626	2,397–2,507	1,713	1,414	1,478–1,722
Maximum use temperature (°C)	1,750	1,700	800–2,045	1,027–1,727	1,100	477–527	297–697
Coefficient of thermal expansion (um/°C)	7–8.4	8.2	7.4–9	4.3–5.6	0.55	2.56–3	5 (7.9–11)
Specific heat (J/kg.K)	880	880	750–1,020	740–820	740	668–715	510–750
Thermal conductivity (W/m.K)	25.5–35	20–25	209–330	60–177	1.38	84–156	70–120
Dielectric constant @ 1 MHz	9.8	9	6.1–7.5	8.3–9.3	3.82	11–12	–
Bulk resistivity (ohm/cm)	> 1e14	>1e14	>1e16	>1e14	>1e10	1e6–1e10	1e2–1e6

- **HTCC (High Temperature Co-fired Ceramic)**: Somewhat LTCC-like applications but processed around 1,600°C and with different compositions

Table 2.2 outlines some of the key properties of ceramic materials [10].

Some of the challenges with the use of ceramic substrates include higher materials cost and longer fabrication lead times, compared with laminate substrates. Ceramic substrates are typically heavier than their laminate counterparts – an important consideration for applications such as space and aerospace where the total payload is a key parameter, or even hybrid/electric cars, where lighter vehicles are preferred for higher fuel efficiency or longer driving range capabilities.

2.3.4 Flexible Materials

Flexible electronics involves the assembly of electronic components on mechanically flexible substrates, thereby allowing electronic circuits to bend and conform to various shapes, for applications where flexibility and space savings are crucial. This includes consumer applications such as print heads on inkjet printers, foldable displays in smartphones or in wearable devices such as fitness trackers, etc. Flexible substrates are an important enabler for implantable and healthcare monitoring devices in medicine, where electronics are required to conform to the shape of the human body in addition to being highly reliable and biocompatible.

The following are some of the leading materials used in flexible electronics together with their key advantages:

- **Polyimide**: This is a thermally stable and mechanically robust material [11]
- **PET (Polyethylene Terephthalate)**: Flexibility and cost-effectiveness of PET are attractive for a range of applications [12]
- **Flexible Glass**: Suitable for applications in healthcare, fitness tracking, and smart textiles [13]
- **Liquid Crystal Polymer (LCP)**: Ideal for miniaturized and hermetic sensor module applications due to its high thermal and mechanical stability [14]

Earliest references to the use of flexible materials in microelectronics go back to the early 20th century when in 1903 Albert Hanson filed a British patent for a device described as a flat, foil conductor on an insulating board with multiple layers, and the next year Thomas Edison experimented with various chemical methods to plate conductors onto linen paper [15]. NASA has used flexible substrates in many of its space missions – including the mission to the planet Mars [16] – and leveraged its characteristics of low weight, high flexibility, and ability to withstand hot and cold environments without mechanical degradation. Flexible substrates are expected to continue to be an important material choice for the foreseeable future.

Further discussions on flexible electronics can be found elsewhere in this book.

2.3.5 Fabrication Technologies

One of the following three processes are typically used for the manufacture of PCB and laminate substrates:

- Subtractive Process
- Additive or Fully Additive Process (FAP), and;
- Semi-Additive Process (SAP) or modified Semi-Additive Process (mSAP).

2.3.5.1 Subtractive Laminate Process

As the name suggests, the subtractive process of PCB fabrication is based on selective removal of portions of the copper foil from each layer of a copper-clad laminate to realize the desired pattern of traces and pads for electrical connectivity. This process typically starts with laminating copper foils on both sides of the laminate, although the principles are also applicable in circumstances where only one side carries the copper foil. Unwanted parts of the copper cladding are commonly removed by first applying a resist material over the copper cladding, then patterning, exposing and removing the resist material from areas where the copper needs to be removed, and finally chemically etching off the copper from the exposed areas. Subsequent processes then remove the protective and developed resist and clean the work-in-progress laminate. For multi-layered PCBs and laminates, additional layers of dielectric and copper cladding is then applied, and the process is repeated until the required number of routing layers are realized. Figure 2.6 shows a schematic process flow for the subtractive process of laminate manufacturing. A more in-depth description of the subtractive laminate manufacturing process may be found in the literature [17].

The subtractive process is the oldest of the PCB manufacturing processes and is also one of the most mature processes. As such, it is also among the cheapest

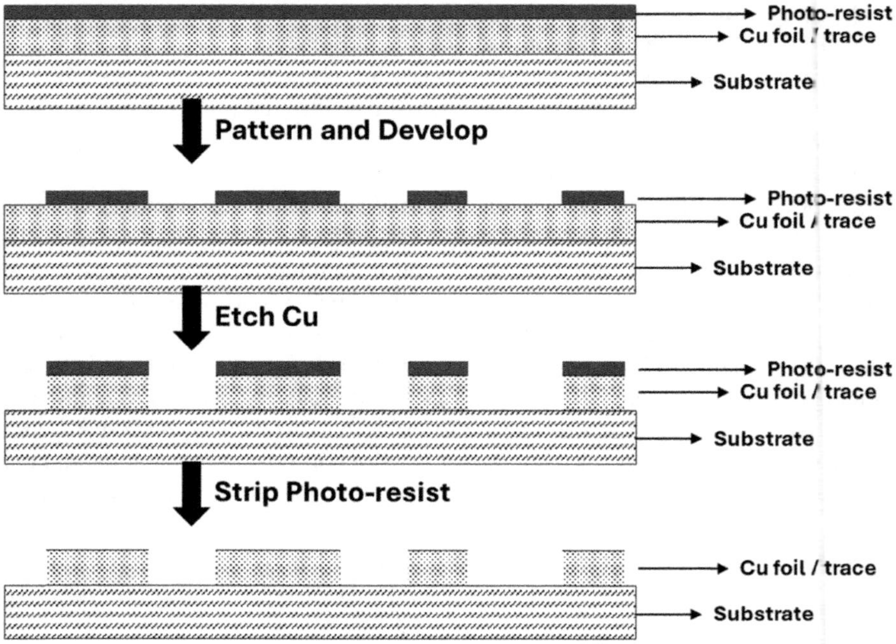

FIGURE 2.6 Schematic process flow for the subtractive process of laminate manufacturing.

process technologies for laminate fabrication. Subtractive technologies generally use photo-chemical processes for realizing patterns for electrical connectivity and routing within the copper foil. As shown schematically in Figure 2.7, two panels are attached back-to-back in the manufacturing process for a laminate that has a central core. It should be noted that this process is slightly different for the fabrication of laminates that are coreless. In both scenarios though, the symmetry of material removal from the etchant used and the resulting undercut of the copper trace under the photoresist material typically results in a taper along the edges of the etched copper features due to the isotropic nature of the chemical etch. These etch characteristics lend their unique shape as well as reliability characteristics to the etched vias that help establish electrical connectivity across routing layers within laminates.

2.3.5.2 Semi-Additive and Modified-Semi-Additive Laminate Process

Although the subtractive laminate technology is mature and is among the cheapest, it also brings some limitations with it. These include challenges in achieving fine line/space capabilities, especially at less than about 25 μm/25 μm, where the side-etching of the traces reduces consistency and precision of the resultant traces. This adversely impacts the laminate yield as well as the signal integrity required for high-performance components.

The semi-additive process (SAP) and modified semi-additive process (mSAP) address and resolve many of these challenges [18]. Both SAP and mSAP technologies deliver fine lines and spaces that are as thin as 10 μm or lower [19]. Applications that drive these requirements include advanced consumer electronics such as

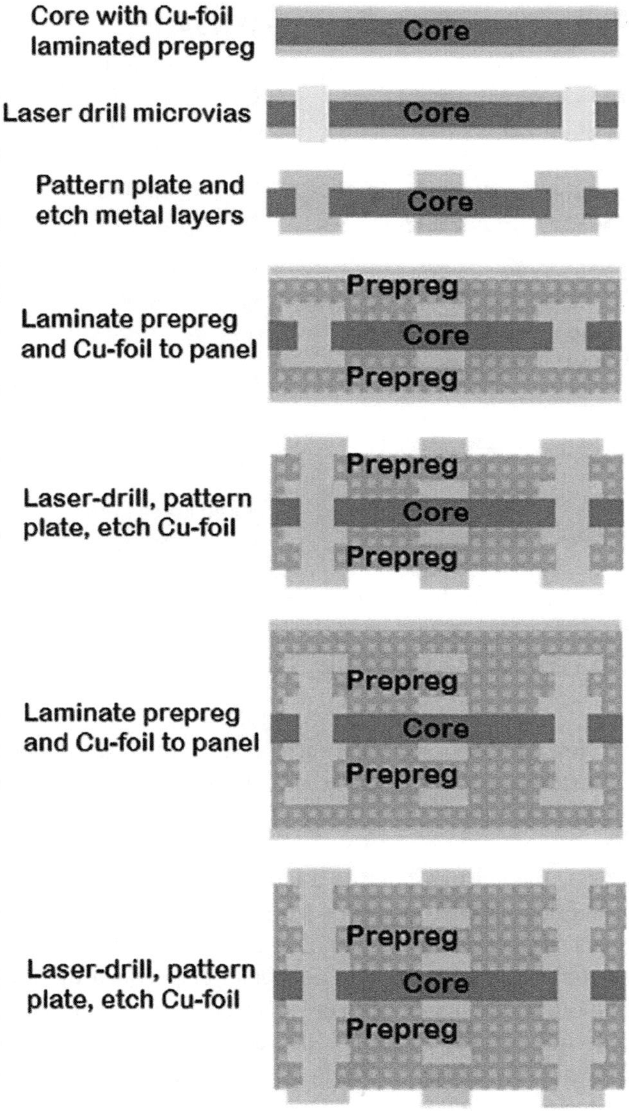

FIGURE 2.7 Another example of process flow for subtractive laminate manufacturing.

smartphones, medical devices such as ultrasound and magnetic resonance imaging (MRI), telecommunications for 5G infrastructure, as well as high-density digital applications such as memory and logic. The standard SAP process leverages roughening or texturing of the dielectric substrate for adhesion promotion. A thin layer of copper is then applied onto the dielectric substrate. Thicker copper is then selectively plated in select areas of the laminate that are defined by using photoresist patterns. After developing and stripping the cured photoresist, the laminated copper layer remaining between the plated areas is removed, leaving behind the desired

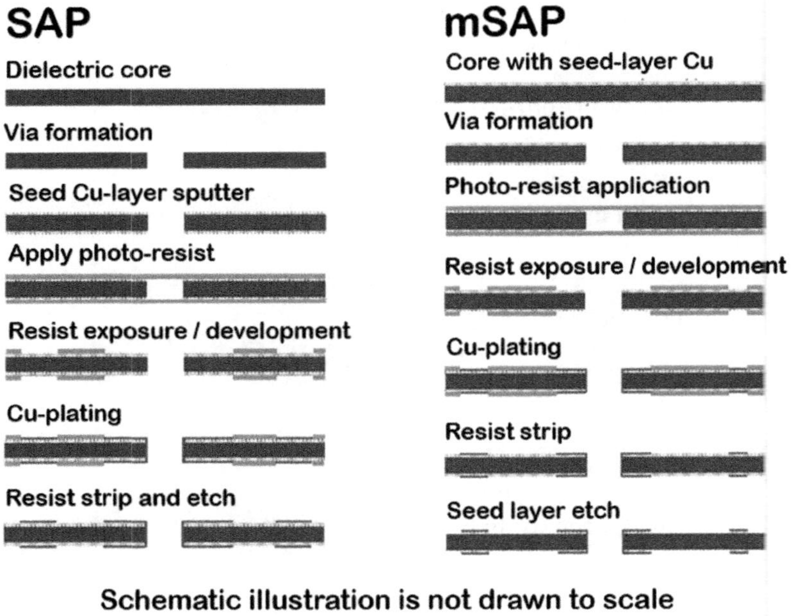

FIGURE 2.8 SAP and mSAP – A schematic comparison.

copper pattern. However, the roughened surface at the interface of resin and copper increases the transmission loss, especially at high signal frequencies or for high-performance RF applications.

This is where the mSAP processes offer advantages over the SAP process. In the mSAP process, the starting point is a laminate that already has a thin copper layer. Circuit pattern is defined onto this starting material using photolithography. Additional copper is plated onto the exposed areas to build up the desired traces. After stripping the photoresist, the thin copper layer between the tracks is etched away, leaving the desired copper structure that has a much smoother characteristics at the copper-dielectric interface. This characteristic makes the mSAP substrates more suitable for high frequency and high electrical performance applications [18]. Figure 2.8 shows a schematic process flow for the SAP and the mSAP processes respectively.

2.3.5.3 Additive Substrate Process

Additive manufacturing was invented in the 1980s and matured as a viable manufacturing option through the 1990s and early-2000s [20]. This technology prints conductive traces, pads and other features directly onto the substrate. Techniques such as inkjet printing and aerosol jet printing are often employed to deposit conductive materials selectively on each routing layer. Usually, the conductive materials are deposited in the form of paste which also includes a suitable bonding ingredient. The bonding ingredient is activated typically using heat (e.g. conventional curing) or laser/uv radiation and is cured. If the application requires multiple routing layers, then layers of dielectric materials are laminated onto the additively fabricated layer, and the process is repeated until the substrate fabrication is complete. Since conductive and other materials are added only where needed, additive processes can be cost-effective relative to subtractive processes – especially for complex designs. Additive processes also can be more environmentally friendly due to reduced waste. Production cycles can be shorter. Designs can be customized for specific applications to include multiple variants, which is especially advantageous in early stages of product development. These characteristics are particularly important in applications such as medical devices, aerospace, consumer electronics and others where precision and miniaturization are critical. Accordingly, appropriate substrate fabrication technology should be selected, based on the requirements of the application and its working environment. Further discussions can be found elsewhere in this book.

2.3.6 Via Types and Considerations

Vias primarily provide interconnectivity across routing layers within microelectronic substrates and interposers. Vias can also be a part of the thermal management strategy and can be strategically used to remove heat away from the die or other hot-spot areas. Performance of the component and sub-system, manufacturability and of course cost, depends strongly on the type of vias selected for the application. A brief outline is followed below, while a more detailed assessment is available in the literature [21,22].

Plated Through-Hole (PTH) Vias: PTH are probably the simplest of interconnects that are realized by drilling a hole through the substrate or PCB, then plating the interior walls with a conductive material such as copper. PTHs are relatively simple and cheap when fabricated in volume and they are quite suitable for most standard applications. However, PTHs are quite large in their *x*–*y* footprint relative to some other via types. Also, when PTH vias are not filled with conductive materials, they can suffer from lower conductivity and limited thermal performance compared to filled vias.

Filled Vias: As the name suggests, they are filled with conductive materials. Filled vias bring higher electrical and thermal conductivities and are suitable for high-power, high-frequency applications. Their disadvantage is that filled vias are more complex and expensive than PTH to manufacture.

Blind Vias: Blind vias connect an outer routing layer to one or more inner routing layers within a substrate or an interposer. However, blind vias do not go through the entire substrate or interposer. Blind vias are typically fabricated with a laser process and can have fine features, which support high-density designs as well as reduce signal loss to improve electrical performance. As a limitation, blind vias are more complex to manufacture and can incur higher costs to fabricate.

Buried Vias: Buried vias are similar to blind vias in their construction but connect inner layers without reaching the outer layers. Like blind vias, buried vias are usually fabricated with laser tools and have fine features that support compact designs. They are complex to manufacture and can incur a higher cost to fabricate.

Microvias: As their name indicates, these vias are small and are used in high-density interconnect (HDI) substrates. Microvias enable miniaturization and are used in applications that require dense interconnections or for high-frequency RF, digital or mixed-signal applications. They do require advanced manufacturing capabilities and hence they can be more expensive to manufacture.

2.3.7 Surface Finishes

The copper traces that provide connectivity, power and electrical bias, need to be protected from oxidation and corrosion. This function is performed by solder-mask on the top and bottom layers of the laminate substrate and by the dielectric layers for the interior routing layers. However, this protection on the exterior traces needs to be broken, so that the traces can contact the electrical connections on the die or SMD to provide interconnection. A variety of pad finishes are available to protect such exposed copper portions of the traces or land pads, depending on the specifics of the application. Some of the commonly used pad finishes include the following:

Organic Solderability Preservative (OSP): OSP is a thin, organic layer applied to copper pads to prevent oxidation and maintain solderability. It is a cost-effective pad finish that works well for flip-chip interconnects and for pads where SMT components are soldered. OSP is a generic name for a finish that is typically composed of azole compounds like benzimidazole or benzotriazole [23]. The coating is usually clear and very thin, hence the impact to the overall laminate thickness is negligible. However, OSP is not suitable for wire-bond applications. OSP pad finishes have some inherent challenges, including poor wetting during soldering. High-temperature degradation of OSP coatings can be accelerated by exposures to lead-free reflow temperatures. Interaction with flux residue too can result in degradation of the OSP finish [24]. Poor wetting can lead to electrical opens, which can be detected relatively easily. However, poor wetting can also result in mechanically weak connections, which typically degrade over time, thus negatively impacting long-term reliability. It is for these reasons that the OSP finish is typically not preferred for applications with long (>~5-years) product lifetimes or for components intended for harsh environments. OSP coated laminates also have a relatively short shelf-life of about six months; coating needs to be re-applied thereafter using appropriate processes. Survivability through multiple heat exposures and reworkability of OSP is also limited, due to degradation of the OSP coating over time and temperature.

Electroless Nickel Immersion Gold (ENIG): This pad finish has been used in the industry since the late-1980s. This is a two-layered metallurgy. In contact with copper is a 3–6 μm thick layer of electroless nickel, which acts as a diffusion barrier for copper migration. Over the layer of Ni is a thin layer of gold, usually in the range of about 0.05–0.1 μm that is applied by the immersion process [25]. ENIG offers excellent solderability to several flip-chip technologies such as Cu-pillar bumps and solder bumps, and for various solder metallurgies such as SAC305, eutectic PbSn, pure-Sn and others. ENIG works well with Al-wire bonding; however, it is not suited for wirebonding gold wires. Cost of ENIG pad finish is strongly dependent on the price of gold, hence it can be expensive. Further, if the Ni-layer is not properly protected with gold, oxidation of the Ni-layer can cause reliability concerns (the so-called "Black Pad" phenomenon). On the other hand, excessive gold can cause gold embrittlement, which can be a concern as the solder volume in the joint reduces with tightening flip-chip pitch and even more so, for Cu-pillar bumps. However, the industry has matured the technology for realizing good ENIG finish, which consistently has provided good reliability performance.

ENEPIG: Electroless Nickel-Electroless Palladium-Immersion Gold (or ENEPIG) is a versatile pad finish metallurgy. It is known for its excellent reliability performance, which makes it suitable for high-reliability applications. ENEPIG can be used with wire-bondable die as well as flip-chip die, thus lending significant design flexibility to designers in their choice of die configurations.

The standard ENEPIG metallurgy is made up of the following three metal layers. In immediate contact with the underlying Cu pad is nickel (Ni), which is applied using the electroless process and is about 3–6 μm thick. The purpose of the nickel layer in the ENEPIG metallurgy, as with ENIG, is to provide a barrier to copper diffusion. This is followed by a 0.05–0.15 μm thick layer of palladium (Pd), which functions as a protective layer to prevent nickel corrosion and enhances wire bonding capabilities. The final layer is of immersion gold (Au) that is 0.03–0.05 μm thick and it lends surface planarity as well as protection from oxidation [26].

The key challenges associated with the ENEPIG finish are the relative complexity of the fabrication process, as it is a three-layered metallurgy – and the relative higher cost from palladium and gold.

Solder-on-Pad (SoP): This pad finish involves printing, plating or otherwise depositing solder on the land pads of the substrate. A variety of solder metallurgies are available with this technology, including SAC305, SAC105, eutectic PbSn and others. SoP technology enables several advantages for flip-chip bonding, including improved solder joint reliability because of increased solder volume in the solder joint, improved wettability and near-elimination of electrically open contacts [27,28]. To realize SoP, techniques such as stencil printing, solder jetting, electroplating or micro ball placement are commonly used [29].

Key advantages of the SoP pad finish include its relatively low cost, good shelf life and usually excellent solderability. The main challenge associated with SoP is the occasional non-uniformity of the solder volume as well as a lower thermal shock resistance due to the formation of excessive intermetallic compounds at the interface, which can be brittle, thus adversely impacting the reliability performance of the solder joints.

2.3.8 Key Material Properties

Selecting appropriate materials for microelectronic packages is critical for performance, reliability and volume manufacturability. Coefficient of thermal expansion (CTE), dielectric properties and moisture absorption are among the most critical parameters; however, there are other considerations too, as are outlined in this section.

Table 2.3 lists some of the important properties of several ceramic materials, including alumina, aluminum nitride, silicon nitride, beryllium oxide, HTCC, LTCC and others, as well as for some metals and semiconductors.

Historically, ceramic substrates have been in commercial use in microelectronics since the early 1970s. Leveraging their mechanical ruggedness, multilayer thin-film packages were manufactured using ceramic substrates [30]. Their ability to form hermetic joints with metal and glass has been leveraged extensively for sensors and

TABLE 2.3
Properties of Some Ceramic, Metal and Semiconductor Materials

	Material Properties					
Material/Form	**Density (kg/m^3)**	**CTE (μm/m/K)**	**Thermal Conductivity (W/m/K)**	**Dielectric Constant**	**Young's Modulus (GPa)**	**Poisson's Ratio**
Aluminum (Al)	2,700	23.1	235	–	70	0.35
Aluminum nitride (AlN)	3,260	4.5	140–180	9	273–346	0.24
Aluminum oxide/ Alumina (Al_2O_3)	3,890	8.4	35	9.8	220–370	0.22
Copper (Cu)	8,960	16.5	401 (@300K)	–	100–128	0.34
Gold (Au)	19,300	14.2	318	–	79	0.44
Silver (Ag)	10,490	18.9	429	–	83	0.37
Silicon carbide (SiC) – 3C(β)	3,210	4.0–4.5	360	9.7	430–450	0.14
Silicon carbide (SiC) – 4H	3,210	4.0–4.5	370	9.7	430–450	0.14
Silicon carbide (SiC) – 6H(α)	3,210	4.0–4.5	490	9.7	430–450	0.14
Gallium arsenide (GaAs)	5,320	5.8	46	12.5 (DC)	85.5	0.31
Gallium nitride (GaN)	6,150	5.59 (along a_0) 7.75 (along c_0)	130	8.9–9.7 (DC) 5.35–5.8 (HF)	150	0.183
Silicon (Si)	2,330	2.6	141.2	11.9 (DC)	155.8	0.17
CVD Diamond	3,515	2 (@300C)	700–1,800	5.68	900–1,100	0.069
PTFE (Polytetrafluoro-ethylene – teflon)	2,150	180–250	5.36	2.1	0.5	0.46
Polyimide kapton	1,420	20–40	0.12	3.5	2.5	0.34

actuators, ultra-high vacuum compatible feedthroughs, microfluidics, MEMS and similar high-performance applications [31]. Ceramic substrates can support integrated fluidics – something that is a distinct advantage over laminate substrate-based solutions. However, the typical lead times for fabricating ceramic substrates can be in excess of 10-week, which is significantly longer than the ~5-week or less lead time that is supported by several high-volume laminate substrate manufacturers. Non-recurring engineering (NRE) costs for ceramic substrates are usually higher than those for similar laminate substrates.

However, technological advancements in ceramics continue to not only keep ceramic substrates relevant but have also opened more applications – especially in areas of high-frequency and high-power applications. Electric vehicles, EV chargers, high-power industrial motor drives, LED lighting and many such consumer as well as commercial applications require ceramic substrate-based electronics [32].

These and similar applications continue to grow the ceramic substrate market. Sustained development of new ceramic materials and improvements to the manufacturing techniques are expected to continue keeping ceramic substrates relevant and an integral part of materials choices for microelectronic substrates.

Although ceramic materials offer several distinct advantages as substrate materials, laminate materials constitute bulk of the high-volume consumer applications, especially in the low-power low-frequency applications such as cellular handsets, electronic wearables, digital cameras and others. This is primarily because laminate substrates are typically less expensive to mass produce and hence make them a cost-effective choice compared to ceramic substrates. Laminates weigh less than their ceramic counterparts and offer fine line/space and tight via-density for high-density routing for demanding applications. Advances in laminate materials now enable a wide range of material properties, which has further enabled their broader use.

Table 2.4 outlines key properties of some common dielectric materials used in laminate substrates [33,34]. Additionally, laminate materials can be customized to meet specific performance requirements such as high electrical isolation, moisture resistance, thermal stability at a higher temperature, or chemical inertness.

TABLE 2.4
Properties of Common Dielectric Materials Used in Laminate Substrates

Property	Teflon (PTFE)	ABF (Ajinomoto Build-up Film) [33]	BT (Bismaleimide-Triazine) [34]
Glass transition temp T_g (°C)	–	150–200	180–220
Dielectric constant (D_k)	2.1	3.0–4.0	3.5–4.5
Dissipation factor (D_f)	0.0002	0.002–0.005	0.005–0.010
Coefficient of thermal expansion CTE (ppm/°C)	100	50–70	40–60
Thermal conductivity (W/mK)	0.25	0.5–0.8	0.7–1.0
Modulus (GPa)	0.5–1.0	2.0–3.0	3.0–4.0

In summary, material properties play a critical role in selecting the appropriate substrate technology and materials that are appropriate for performance requirements of the application, together with pricing and manufacturability considerations.

2.3.9 Fan-Out and Fan-In

"Fan-In" and "Fan-Out" refer to technologies wherein the original interconnect pads on the die are re-routed to other areas within the die or the package.

Fan-out is used when the pad density is large and needs to be spread across to a wider, more relaxed pitch. This requires additional space beyond the extent of the die and hence requires reconstitution. The reconstitution process can be accomplished at a wafer-level but is also implemented at a panel level across the industry [35,36]. One of two contrasting processes are used in manufacturing, namely "die-first" and "die-last", wherein the sequence of placing the die is before or after respectively, relative to creating the interconnectivity and die placement. Figure 2.9 shows a schematic flow for both these approaches [37].

Fan-in is usually used when the number of pads on the die, originally designed for wire bonding, are relatively fewer but at a tight pitch (e.g. 40 μm) and need to be routed to a more relaxed pitch (e.g. 250 μm) for flip-chip attachment. Space on the die is used for fan-in routing. Fan-in is a wafer-level process technology, where final packaged devices are, in most cases, the same size as the die itself. However, this exposes the sides of the die – and the WLP device – to the environments. To address this, the so-called five-side-molded chip-scale package (CSP) was developed and patented in 2011 and described in detail elsewhere [38]. Packages protected with mold compound on all six sides were also developed subsequently [39]. A later section on RDL/BEOL for substrates provides some manufacturing process related perspectives for the fan-in, fan-out and WLP/PLP technologies.

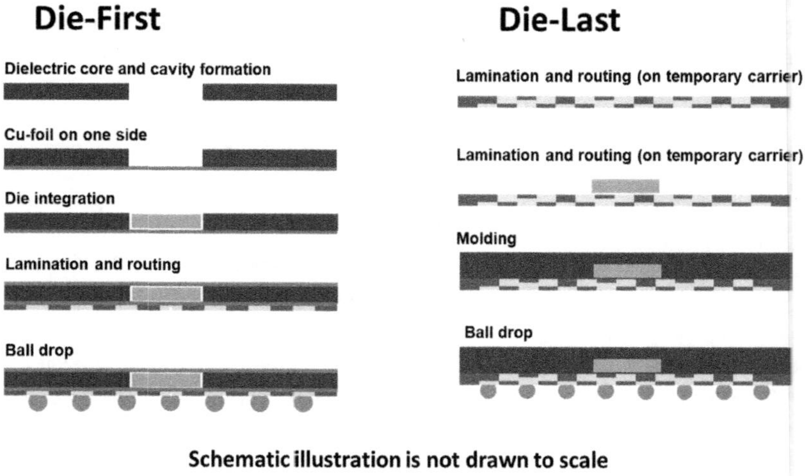

FIGURE 2.9 Schematic Flow for "Die-First" and "Die-Last" approaches for panel-level packaging (PLP).

2.4 INTERPOSER MATERIALS

2.4.1 Glass

In theoretical terms, glass can be described as a supercooled liquid of near-infinite viscosity. Practically, glass is an amorphous solid that is often transparent. It is chemically inert for most chemicals – except for materials such as hydrofluoric acid (HF). For microelectronics applications, glass brings some of the most attractive properties, including the following:

- Chemical inertness
- Low dielectric constant
- Low loss tangent
- Mechanical stiffness
- Zero or near-zero shrinkage, allowing for excellent dimensional tolerances
- Ability to be fabricated in a wide range of thicknesses, sizes and shapes
- Coefficient of Thermal Expansion (CTE) that can be tailored for specific applications
- Thermal stability
- Abundant and relatively low cost

Additionally, the refractive, reflective and transmission properties of glass facilitate manufacturing of lenses, prisms, waveguides and fibers for photonics and optoelectronic applications.

The commonly used glass types for microelectronics applications are borosilicate glass and fused silica. Both types of glasses have excellent electrical insulating properties, as shown in Table 2.5 [40], and customizable coefficients of thermal expansion (CTE). These allow for the fabrication of through-glass vias (TGV), where the CTE differences between conductive metals such as copper and that of glass can be efficiently managed, thus enabling high-density interconnects [41].

TABLE 2.5
Properties of Common Types of Glass Materials Used in Microelectronics

Description	Fused Silica	Borosilicate Glass
Composition	SiO_2 (100%)	SiO_2 (~80%), B_2O_3
Density (g/cm^3)	2.2	2.2
Mohs hardness	5.3–6.5	6.5
Maximum working temperature (°C)	1,400	400–500
Refractive index (600 nm)	1.46	1.47
Spectral transmission range (µm)	0.17–3	0.35–2
Thermal expansion coefficient (10^{-6}/K)	0.54	3
Heat conductivity (W/mK)	1.38	1.2
Dielectric strength (kV/mm)	40	30

Due to their excellent electrical properties and capabilities of supporting high-density interconnections, glass-based substrates are suitable for specialty applications in high-performance computing, high-bandwidth memory, micro-fluidics and opto-electronics. The low dielectric constant and high-frequency performance of glass interposers also enable high-performance RF applications such as specialized antenna [42]. Stability of its mechanical and electrical properties over temperature make glass-based substrates suitable for high-precision test, measurement and calibration equipment [43]. However, the use of glass itself as an interposer material has not yet found widespread use. This is primarily because the current state-of-the-art is not able to fully support the fine via features and tight line/space requirements of advanced applications for routing. Generating multiple layers of routing requires post-processing in the form of not only RDL (redistribution layers), but also fabrication of micro vias through the glass. The brittle nature of glass presents manufacturability challenges that continue to be addressed.

Glass-core substrates offer more favorable alternatives, where the glass core lends rigidity and routing layers separated by appropriate dielectric materials leverage the mature panel processing infrastructures and know-how of laminate substrate manufacturing. These aspects are elaborated in a later section.

2.4.2 Silicon

2.5D and 3D integration enables higher levels of integration. However, vertical interconnection with standard laminate-based modules continues to present significant limitations. Via size and spacing design rules with the organic core substrate are a key bottleneck. Using silicon as an interposer material provides several advantages in advanced semiconductor packaging, including the following [44,45]:

- **Higher Interconnect Density**: Si-interposers can enable shorter signal paths thereby reducing the electrical resistance and RF delays. This helps improve the overall performance of devices assembled with this technology.
- **Superior Signal Integrity**: Si-interposers can allow for consistent signal integrity, especially at high frequencies (20 GHz and beyond), making them particularly suitable for high-speed applications. RF applications can particularly benefit from these characteristics.
- **Efficient Power Distribution**: A structured platform can be enabled by Si-interposers for power routing, which can enhance power efficiency of the distribution network.
- **Heterogeneous Integration**: Si-interposers present lower CTE mismatch to a variety of semiconductors compared to the laminate alternatives. This can allow larger die to be integrated into devices, thus enabling high-density heterogeneous integration.
- **Miniaturization**: Package designs can be made significantly more compact that is enabled by high-density wiring on the Si-interposers.

However, silicon interposers have thus far mostly been limited to high-performance applications and have not yet found widespread adoption in the consumer segments of the industry, as the silicon interposer technology has some important limitations [46]. A silicon interposer can be expensive as it requires a dedicated wafer mask for its fabrication. Even the smallest design changes require a re-spin and re-fabrication, which is not only expensive but also time-consuming. This can present a significant barrier for widespread adoption. Scalability of silicon interposers is somewhat limited by the maximum size of the wafers that ca be used, which is smaller than the panel-level scalability afforded by glass-based interposers or even laminates. Si-interposers can also suffer from higher insertion loss than specialty laminates, especially at high frequencies; this can be highly undesirable for certain high-performance applications. Thermal management can be challenging too.

2.4.3 Glass-Core Substrates

As stated in an earlier section, the use of glass as an interposer material presents some significant challenges. However, using glass as a core material of an otherwise organic substrate opens up more possibilities from a routing capability perspective as well as for manufacturability. Some of the main advantages and characteristics offered by glass-core substrates include the following:

- Glass-core substrates are flatter than standard core or coreless laminate substrates. As such, they can support finer lines/spaces and higher routing density per routing layer. This can be especially beneficial for high *I/O* count applications.
- The higher Young's modulus and stiffness of glass lends mechanical stability to the fabricated substrates. Adding dielectric materials for routing mechanically helps reduce the risk of breaking glass by absorbing mechanical stresses to a certain extent.
- Glass-core substrates can support embedding of active and/or passive components – as can standard laminates. However, the volume of space that remains in the glass-core cavity needs to be filled with an epoxy or a suitable material, whereas in case of standard laminates, the available resin naturally fills such gaps during the lamination processes.
- Since multiple types of glass materials are available commercially [47,48], it is possible to adjust its CTE for the desired application, which is a distinct advantage.
- Glass-core substrates have a higher Young's modulus, making it more resistant to deformation under stress. Stiffer glass is less likely to bend or warp, which makes it an attractive option for demanding applications like aerospace, automotive and those intended for harsh environments.

Some of the latest physical and thermal characteristics of glass-core substrates are outlined in a recent paper published in the ICEP 2024 Proceedings [49] and some of the references therein.

Glass-core laminates also present some significant challenges. These include the following:

- Presently, the glass-core substrates can support up to six or eight routing layers [50]. Laminate substrates can have more than 20 routing layers, which is a distinct advantage over glass-core substrates.
- The infrastructure to support glass-core laminates is not yet robust. Volumes of applications are relatively low, which means that the economies of scale have not yet delivered favorable pricing.
- Forming vias is still complex, especially for the fabrication of high-density vias, which can adversely impact component reliability and possibly, thermal performance.
- Availability of board-level reliability (BLR) data is also very limited.

Nevertheless, much progress is being made in maturing the manufacturability of glass-core substrates as well as in improving our collective understanding of their reliability performance through focused characterization and qualification efforts.

2.4.4 Silicon Core Substrates

There are key challenges associated with conventional 3D stacking of die on standard laminate substrates, including the following:

- Signals and power to the top die have to navigate through the entire stack of die, beginning from the bottom die, as there is no practical alternative path.
- For applications with high *I/O* density and where signal/power from one die to another can use only vias (not traces), the die beneath the top die (e.g. the "middle" and/or the "bottom" die) cannot be smaller than the top die.
- Path for heat removal is typically possible only from the top and/or the bottom, hence thermal management is typically challenging.
- Assembly can be very challenging, especially if restrictions on the overall thickness of the device require handling of thin, large die.

Many of these challenges can be satisfactorily resolved with silicon core substrates. Thermal conductivity of silicon is significantly higher than that of laminate substrates. As such, silicon core substrates are more efficient at removing the heat from components assembled onto them. Silicon is also mechanically stiffer than the traditional core materials used in laminates. Thus, the use of silicon core enables the fabrication of substrates that have similar stiffness but are thinner than traditional laminates. This feature can help with meeting tight thickness budgets for critical applications. Silicon core also supports tighter line/space than conventional laminates, thus easing signal/power routing constraints [51]. Some of the main advantages of the Si-core substrates include the following:

- Core, and hence the substrate itself, is typically stiffer than the standard laminate substrates.
- Core is more homogenous, which can contribute to enhancements to the reliability performance by minimizing issues arising from CTE mismatch

within the core (e.g. between glass fibers and resin, as in case of standard laminates).

- Substrates and especially the core can be virtually free of creep and plastic deformation.
- Si-core substrates can lend dimensional stability and mechanical robustness.
- Substrates with Si-core are usually more heat conductive than their standard laminate counterparts.
- Si-core substrates have a better CTE match with the silicon die.

Like the glass-core substrate technology, silicon core substrate technology is continuing to mature. Substrates are being developed by multiple suppliers across the supply chain [52–54].

Ultimately, the choice of materials and processes used for microelectronic substrates will be determined by the specific technological, business and manufacturability demands of the application.

2.4.5 Emerging Interconnects

Newer forms of interconnects have emerged as an integral part of interposers and substrates for advanced packaging.

2.4.5.1 Intel EMIB

Intel's embedded multi-die interconnect bridge (EMIB) is an example of 2.5D multi-chip package (MCP) bridge interconnect technology. The bridge interconnect is enabled by a relatively small piece of silicon that provides interconnectivity between adjacent edges of two dies within an MCP module. This bridge is embedded inside a cavity within the package, as is detailed elsewhere [55].

In the EMIB architecture, the top metal layer provides an electrical reference plane. Routing vias enable connectivity between this reference plane and the die. An attractive characteristic of this technology is that the EMIB process is built upon the standard package construction flow where some additional steps are incorporated within the laminate substrate to create the EMIB cavities. The silicon bridges are placed into the cavities and secured in place with an appropriate adhesive. Subsequent metal routing build-up layers and the dielectric layers are added. Published materials outline detailed electrical analysis for the EMIB interconnect [56–58].

2.4.5.2 Interconnects for PCIe/UCIe

PCIe (Peripheral Component Interconnect Express) is a high-speed serial computer bus that allows for standardization for connecting peripherals such as graphic cards, sound cards, and hard-disk drive. PCIe is the standard motherboard interface for personal computers since it was created in 2003 [59]. Substrates and interposers that are used for PCIe applications require specialized capabilities for tight line/space, tight, repeatable dimensional and registrational tolerance, and high-performance materials, since they need to support requirements that include the following:

- Each of the PCIe lanes consists of two unidirectional differential pairs operating at 2.5, 5, 8, 16 or 32 Gbit/s, depending on the performance ratings.
- The width of a PCIe connector is about 8.8mm, and its height is about 11.25 mm [60]. The length of the connector is variable, depending on the application or the peripheral.
- The thickness of the laminate/interposer card, as it enters the connector, is 1.6 mm. The thickness and tolerance of the card, together with warpage considerations, are some of the related challenges. Figure 2.10 shows an example of a PCIe card [61].

Universal Chiplet Interconnect Express (UCIe) is an open industry specification or standard for die-to-die interconnects within multi-die systems. It was developed by several corporations and was released on March 2nd, 2022 [62]. The specification defines details of the physical layer, the stack protocol as well as the software model and procedures for compliance testing. Some of the key advantages of the UCIe specification include enablement of large System-on-Chip (SoC) packages that can – and often do – exceed the current maximum reticle size in the wafer fabrication process. The standardization also allows for heterogeneous integration, i.e. the intermixing of components from different silicon vendors within the same package. This allows for selecting appropriate semiconductor technology nodes and manufacturers, as well as helping improve manufacturing yields by supporting the use of smaller die [63].

From an advanced packaging perspective, the demands on the physical layer (including supporting up to 32 GT/s with 16–64 lane connectivity) present stringent

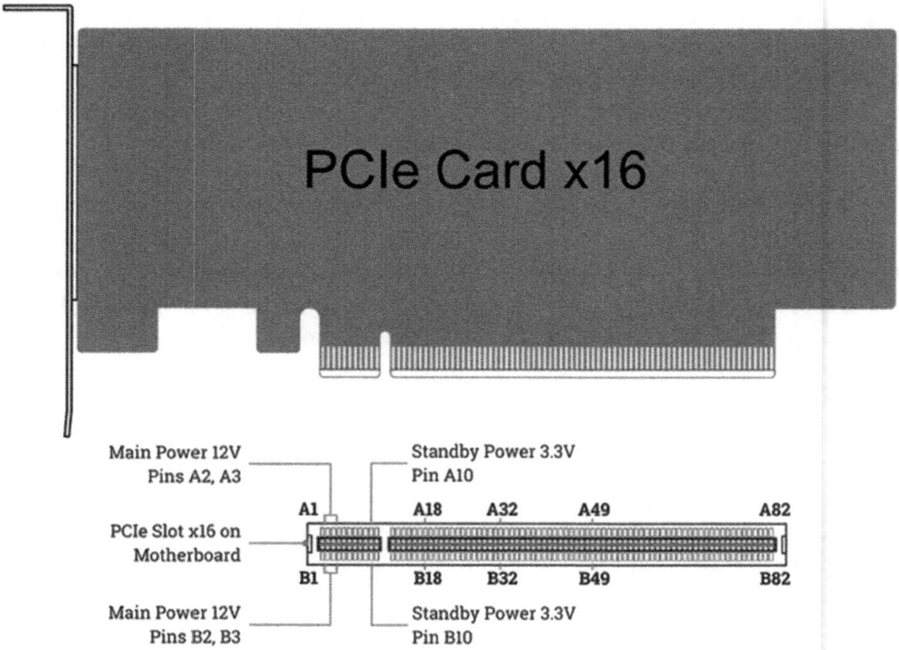

FIGURE 2.10 Example of a PCIe card.

demands on the substrate/interposer materials. Additional details may be found at [64] and the resources therein.

2.4.5.3 Co-Packaged Optics (CPO)

Co-packaged optics (CPO) is a term that represents the integration of optical or photonic functions with electronics within a single module. This technology integrates optical components such as glass fibers, photonic integrated circuit (PIC), electronic integrated circuit (EIC), and additional electronic and/or photonic functionality, passive components and connectors. Integration of laser sources is also supported within a CPO module. Figure 2.11 shows a schematic outline of potential components in a CPO module.

CPO applications demand some key characteristics from the laminate or interposer that are used for their manufacture, including the following [65]:

- **IO Count/Line and Space/Pitch**: Typically, EIC and PIC in a CPO module have hundreds – or even thousands – of interconnects in the form of solder bumps (or copper-pillars) for flip-chip die. These are typically at a center-to-center pitch of around 150 μm; however, some advanced applications can require the center-to-center pitch down to 50 μm.
- **High Bandwidth**: Since CPO modules handle large amounts of photonic and electronic data, substrates need to support a large bandwidth.
- **Latency**: Ability to handle large amounts of data requires efficient transfer with minimal latency.
- **Mechanical Stability and Reliability**: Substrates for CPO modules are required to provide high levels of mechanical and dimensional stability and support high levels of alignment accuracy requirements. Further, the reliability performance of the CPO laminates needs to be very high.
- **Scalability**: CPO substrate materials and designs should be able to support scalability to accommodate potential expansions and future upgrades.
- **Thermal Management**: Systems that use CPO modules and the CPO modules themselves typically generate a lot of heat, especially when the module includes, for example, an on-module laser source, or power amplifiers. An effective thermal management strategy includes a substrate that is able to handle the generated and ambient heat efficiently, so that the performance and reliability of the optical as well as the electrical components are properly supported.

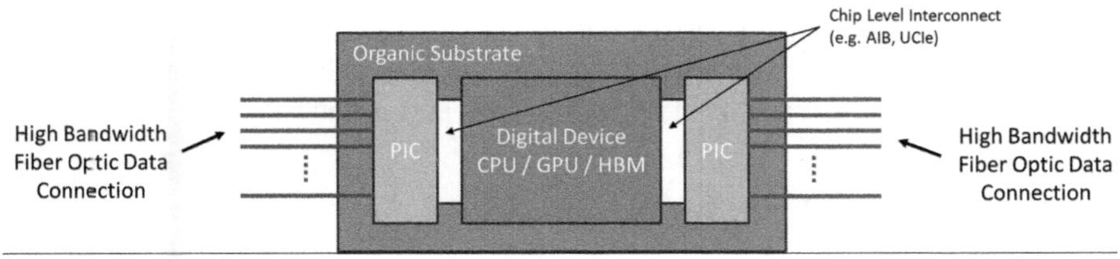

FIGURE 2.11 Schematic of a CPO module.

- **Power Management**: CPO substrates are expected to facilitate power-efficient operation to meet the shrinking power budgets for applications such as data centers, networking, generative artificial intelligence (AI) and others.

Over the recent past, the global demand for chipsets fueling the AI infrastructure has exploded. Amounts of data handled, processed and transmitted are regularly establishing new records. Photonic communications and CPO are critical elements of this technology. Continued innovation and development will be required to support this growth.

2.5 SELECTION CRITERIA AND TRADEOFFS – SUBSTRATES AND INTERPOSERS

In this section, we will qualitatively address factors that influence selection criteria for substrates and interposers. Factors that will be addressed are technical considerations (such as performance and space/footprint requirements) as well as business aspects (such as supply chain and cost). Many of these factors have interdependence and typically cannot be considered in isolation.

2.5.1 Performance Requirements

Understanding the performance requirements of the intended application is critical in selecting the appropriate laminate or interposer. Performance can be electrical (digital, RF or mixed signal), thermal or mechanical. Components used in harsh environments such as high vibration or chemically aggressive, or in critical applications such as nuclear, defense or space applications, require customized considerations. Typical expected lifetime of components is also an important factor.

For applications requiring high thermal conductivity, ceramic materials such as alumina or LTCC/HTCC are preferred for effective heat dissipation and thermal management. Ceramic materials have a better coefficient of thermal expansion (CTE) match with the CTE of the die material than that offered by most laminate substrates. This better CTE match allows for assembly of larger die than those that can be supported on laminate substrates under similar reliability requirements.

Ceramic substrates can support better routing capabilities than their laminate equivalent, when factoring in the line/space, routing layer count, layer-to-layer alignment and coplanarity for large sized substrates. Further, ceramic substrates can withstand higher sustained device operating temperatures than laminates, hence are preferred for applications in harsh environments. Hermeticity or fine-leak compliance is usually extremely difficult to attain with laminate substrates. Ceramic materials and metal are more capable of delivering such performance.

2.5.2 Space and 3D Footprint

With the constant pressure to pack more functionality into high-performance consumer electronics such as cellphones, space within the electronic device is always a premium. For applications such as electronics for aerospace, there is tremendous pressure to keep the payload as small as possible. Tight line/space capabilities enable high-density routing and hence support miniaturization. Technologies capable of supporting more routing layers allow for integration in the *z*-dimension, thus can help shrink the *x*–*y* footprint of the component. Additional factors, including the higher heat density resulting from miniaturization and the corresponding thermal management as well as the mechanical stress that these choices could lead to, also need to be carefully considered.

2.5.3 Supply Chain

Supply chain plays a pivotal role in selecting substrates or interposers for any given application in microelectronics. Availability and lead times play a big role in the selection considerations. Availability of raw materials is not only influenced by their abundance and lead times but can be influenced at times by geo-political factors or natural disasters. Experience, maturity of technology as well as that of the workforce have a huge and sometimes unforeseen impact on yield and viability of a technological solution. Suppliers that have innovative solutions can provide a competitive edge, but this may come with tradeoffs in terms of cost and availability. Existing intellectual property (IP) can also impact who can offer certain materials or solutions. This can be a significant consideration for OEMs who typically do not prefer to be single-sourced, especially for their critical requirements.

In addition, environmental and regulatory compliance can affect the selection of materials, and eco-friendly solutions could be preferred by manufacturers for a variety of reasons.

2.5.4 Cost Considerations

The cost of materials and components is always a very important consideration when selecting for any application. Pricing is strongly dependent on continuity of supply and on the robustness of the sourcing infrastructure. Reliability of the materials as well as that of the suppliers who deliver them can reduce the risk of non-conformance, hence help reduce costs [66].

Additionally, environmental and regulatory factors can also influence the choice of materials.

2.6 A GLIMPSE INTO THE FUTURE

Since the invention of the transistor in December 1947 by John Bardeen, Walter Brattain, and William Shockley, virtually all aspects of modern technology have been revolutionized by electronics. Soon after, Jack Kilby and Robert Noyce developed the first integrated circuit in 1958 [67]. Since then, the density of transistors per unit area on a digital chip has approximately doubled every year – an empirical observation and a prediction made by Gordon Moore [68]. With the passage of time, demands of increasing the functionality within a semiconductor device have continued to sky-rocket. Semiconductor chips are approaching fundamental limits of miniaturization of its building blocks such as the transistor. Diversity of the applications where microelectronics components are used has forced the integration of analog chips and functionality with the digital functions. Photonics and optical aspects are increasingly being integrated into modern components and systems. Advanced Packaging, which was often an after-thought through much of the 20th century, has not only fueled, but enabled much of the continued growth and functionality enhancements of virtually all aspects of our lives today. With the advent of AI/ML and emerging technological advancements such as quantum computing, huge amounts of data need to be processed at ever-increasing speeds. Advanced substrates and interposer materials, together with advanced packaging and continually evolving wafer-fab technologies, are certain to drive innovations into the future.

2.6.1 Substrate-Less Packaging

Substrate-less microelectronic packaging, as the name suggests, is a technology that eliminates the need for a traditional substrate. Accordingly, the mechanical support and interconnects between the various elements of the integrated circuit are supplied by alternative means. These elements, such as semiconductor die, surface-mountable components and others, are typically mounted directly onto an interposer or onto a wafer, which then become a part of the finished component, when the assembly is complete. Alternatively, such a wafer or a suitable alternative material can be used simply as a temporary carrier that supports the building blocks through component assembly.

There are two main drivers for substrate-less packaging. In consumer applications such as cellphones, substrate-less components result in a thinner package; this is an attractive characteristic for applications where space and volume are premium. Another is the advantage of a potentially reduced signal path or reduced impedance for high-frequency applications.

Advantages of substrate-less packaging also include possibilities of improved thermal performance; absence of the traditional substrate can reduce thermal resistance for heat removal, which can be a significant advantage for high-performance applications.

Several wafer-level or panel-level fan-out technologies can be thought of as substrate-less technologies. When this is so, the substrate-less technologies can be especially beneficial for 3D and other applications that require high-density interconnections and multi-die integration.

High-frequency applications can also benefit from substrate-less packaging technologies. Driven by the growing demand for higher data rates, the allowed operating frequency range for wireless communications is increasingly moving towards bands in the higher-frequency range. For example, currently the E-band (71–76/81–86 GHz) is in volume production for point-to-point wireless links that support multi-Gbps data rates [69,70]. Advances in semiconductor technologies and substrate-less packaging have helped realize semiconductor devices at millimeter-wave (mmW) and Terahertz (THz) frequency bands, for a wide range of applications, including communication, sensing and imaging [71]. Substrate-less components have also been reported to be used in some applications to address environmental considerations and recyclability [72].

2.6.2 RDL/BEOL for Substrates

The redistribution layers (RDL) technology realizes routing layers on the die to reroute the electrical connections of a chip to a different location. The routing layers are usually made with thin copper traces and are separated mechanically and electrically using a thin layer of dielectric. RDL technology is presently used extensively in wafer-level packaging (WLP) and System-in-Package (SiP) platforms and is an important tool for substrate-less technologies. Historically, the redistribution layer (RDL) technology was used in the early stages of flip-chip development to convert die designed for wire-bond applications for use as flip-chip die [73]. Figure 2.12 shows a schematic representation of the use of BEOL processes and RDL for transforming an IC designed as a wire-bondable die into flip-chip configuration.

Back-End-of-Line (BEOL) is a generic term that refers to the processes associated with the latter stages of semiconductor device fabrication. These include the formation of routing or interconnect layers on wafers that have devices fabricated previously during the so-called Front-End-of-Line (FEOL) processes. As such, BEOL was often used in the early stages of the technology to transform already fabricated wafers with wire-bondable devices into flip-chip configurations with the use of RDL. BEOL processes also help create the wiring and interconnections between individual devices (transistors, capacitors, resistors, etc.) using multiple layers of dielectric materials and metals. Copper and, to a certain extent, aluminum, are commonly used together with appropriate dielectrics such as BCB, polyimide or others to generate RDL interconnectivity. Further, in conjunction with substrate-less packaging, BEOL processes can leverage RDL for interconnecting multiple dies within a single

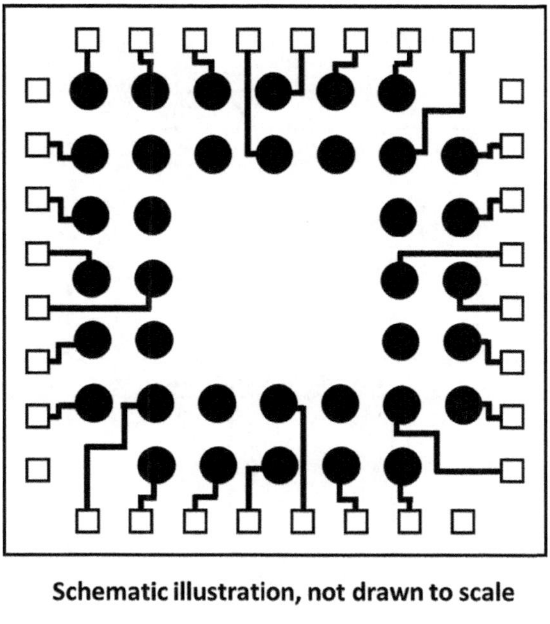

FIGURE 2.12 BEOL/RDL transformation of a wire-bondable die into flip-chip.

substrate-less component. BEOL processes include the formation of vias for vertical connections as well as contacts to allow access for electrical connectivity to different metal layers, where required. A protective passivation layer is often a part of the final BEOL processes, intended to protect the IC and its functionality [74].

2.6.3 2.xD Configurations

2D packaging technology is where the die is assembled on a substrate, while 3D packaging technologies refer to die-on-die (or package-on-package) configurations. An intermediate stage is one where there is a high-density routing layer between the die and the substrate. Mechanically, such a routing layer could be an interposer or a redistribution layer-based technology.

However, with advancements in materials, in processing as well as in system architecture, additional variants have emerged. Some of these are described below:

- As seen schematically in Figure 2.13, 2.1D integration involves fabricating fine metal linewidth and spacing (*L*/*S*) thin-film layers directly on top of a built-up package substrate or high-density fan-out (HDFO) substrate. Fundamentally, 2.1D packaging is an enhancement of the traditional 2D packaging wherein advanced organic substrates are used to improve the performance and miniaturization of System-in-Package (SiP) designs [75]. This approach enables a higher interconnect density and improved electrical performance compared to 2D IC integration. It can incur relatively high non-recurring engineering (NRE) costs but low unit production costs.

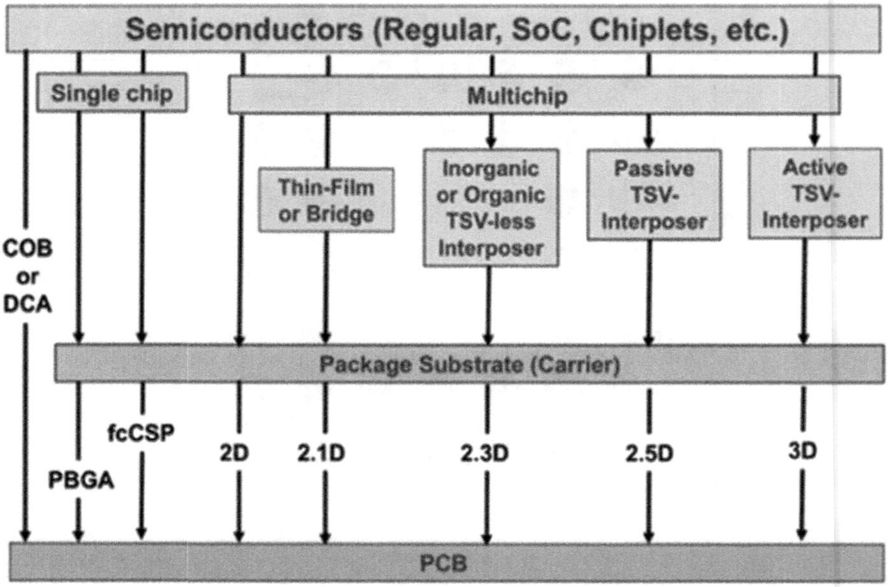

FIGURE 2.13 Levels of component-level packaging integration.

- 2.2D configurations support more advanced interconnects and substrates compared to 2.1D – see Figure 2.13. This technology targets enhancing the electrical performance and thermal management of the package, making it more suitable for high-performance computing and similarly demanding applications.
- 2.3D integration represents an incrementally higher level of integration by integrating multiple chips or components horizontally and vertically, connected through advanced interconnects. This approach is shown schematically in Figure 2.13. It allows for a higher density and better performance, making it ideal for applications requiring significant computational power and efficiency. Multi-chips are supported by an inorganic or organic TSV-less interposer, which is then attached to a package substrate and offers a higher interconnect density and performance compared to 2D and 2.1D integration.

These advanced packaging technologies offer key advantages for applications such as high-performance computing (HPC), autonomous vehicles, and Internet of Things (IoT) devices [76].

2.7 SUMMARY

Electronic devices have enabled practically every aspect of human experience, and microelectronic packaging is finally beginning to command the respect it has always deserved. Innovations and advancements in substrate and interposer technologies are an important foundation that continues to drive microelectronics to higher levels of integration and to capabilities that were not practical until recent times. Revolutionary and evolutionary demands for microelectronics continue to drive the need for higher circuit density and faster circuits. Consequently, substrates and interposer materials continue to simultaneously enable smaller and larger form factors to deliver to the requirements of demanding applications.

This chapter has attempted to provide a broad and qualitative insight into some of the main drivers and related considerations on substrates and interposers. Key capabilities, properties, advantages, and trade-off considerations associated with these materials are discussed. The critical role played by these materials in the electronics industry and their impact on future technological advancements are reviewed. The current applications and future possibilities enabled by these and emerging materials are outlined.

REFERENCES

1. First Integrated Circuit; Jack Kilby, photo courtesy: Courtesy: Texas Instruments, Inc.
2. Emilie Jolivet, "IC Substrate: Key Element of the Packaging Industry", *EPIC Winter Conference*, Villeurbanne, France (2023).
3. Jae-gwon Jang, Kyoung-lim Suk, Seok-hyun Lee, Jin-woo Park, Gwang-jae Jeon, and Jung-ho Park, "Advanced RDL Interposer PKG Technology for Heterogeneous Integration", *2020 International Wafer Level Packaging Conference*, New York, pp. 1–5 (2020). https://doi.org/10.23919/IWLPC52010.2020.9375895

4. Dyi-Chung Hu, Wen-Liang Yeh, Yu-Hua Chen, and Ray Tain, "2/2 μm Embedded Fine Line Technology for Organic Interposer Applications", *2016 IEEE 66th Electronic Components and Technology Conference*, New York, pp 147–152 (2016). https://doi.org/10.1109/ECTC.2016.126
5. "Ceramic Substrate Market – Manufacturers, Share, Analysis and Size"; https://www.mordorintelligence.com/industry-reports/ceramic-substrate-market
6. "Semiconductor Substrate Market Size, Share and Trends Report by 2033"; https://straitsresearch.com/report/semiconductor-substrate-market
7. PCB Material Guide: Types, Properties, and Selection Tips; https://unitypcb.com/pcb-material-guide/
8. Thermally Enhanced Printed Circuit Boards from AT&S Austria Technologie and Systemtechnik Aktiengesellschaft; https://ats.net/en/products/thermally-enhanced-printed-circuit-boards/. Courtesy: AT&S
9. Metal Inlay Technology from Unimicron, Germany; https://unimicron-germany.com/metal-inlay-technology.html
10. https://www.accumet.com/resources/3387-tb-selecting-substrate-materials.pdf
11. https://www.dupont.com/content/dam/electronics/amer/us/en/electronics/public/documents/en/EI-10142_Kapton-Summary-of-Properties.pdf
12. Mikail Olam, "PET Production, Properties and Applications", Chapter 4 in "Advances in Materials Science Research", Volume 48 in Advances in Materials Science Research Series ed. Maryann C. Wythers; BISAC: TEC021000; https://doi.org/10.52305/EXUA6721
13. https://pcbmake.com/semiconductor-substrates/ (March 14, 2024).
14. https://www.mst.com/sites/default/files/2020-12/DYCONEX_Whitepaper_Flexible_Substrates_for_Smart_Sensor_Applications_2020-12.pdf
15. https://chemigraphic.co.uk/2019/01/16/a-short-history-of-the-pcb/ (January 16, 2019).
16. Kendra Short and David Van Buren, "Early Stage Innovation: NASA Innovative Advanced Concepts (NIAC) - Phase-2"; Final Report; Jet Propulsion Laboratory (JPL) September 10, 2014
17. Dominique K. Numakura, "Chapter 63 - Manufacturing of Flexible Circuits", Editor: Clyde F. Coombs, Jr. "*Printed Circuits Handbook*" 6th ed. McGraw-Hill Education, New York (2007).
18. Fei Peng, Naomi Ando, Roger F. Bernards, and Bill Decesare, "Semi-Additive Process for Low Loss Build-Up Material in High Frequency Signal Transmission Substrates". https://api.semanticscholar.org/CorpusID:201858837 (2016).
19. https://www.gs-swiss.com/en/technologies-and-products/msap-and-sap-technology
20. https://prototaluk.com/blog/history-of-additive-manufacturing/
21. Goran Matijasevic, Catherine Gallagher, and Pradeep Gandhi, "High Density Multilayer Substrates Using Patterned Microvia Interposers", *IEEE Aerospace Conference Proceedings*, Taipei, Taiwan, pp. 383–388 (1998).
22. Mark Lefebvre, Elie Najjar, Luis Gomez, and Leon Barstad, "Next Generation Electroplating Process for HDI Microvia Filling and Through Hole Plating", *3rd International Microsystems Packaging Assembly and Circuits Technology Conference*, Taipei, Taiwan, pp. 102–105 (2008).
23. https://flexpcb.org/what-is-osp-finish-on-pcb-pros-and-cons-of-osp/ (2023).
24. Wei Yu, Fa Xing Che, Yeow Chon Ong, Ling Pan, and Wee Gee Cheong, "Study on Substrate Copper Pad Crack through Experiment and Simulation", *26th IEEE Electronics Packaging Technology Conference (EPTC)*, Singapore, pp. 956–961 (2024).
25. https://www.protoexpress.com/kb/enig/
26. https://www.wevolver.com/article/enepig-the-gold-standard-in-surface-finishing-technology; Muhammad Sufyan; 22 September 2024.

27. Kwang-Seong Choi, Teck-Gyu Kang, Ik-Seong Park, Jong-Hyun Lee, and Ki-Bon Cha, "Copper Lead Frame: An Ultimate Solution to the Reliability of BLP Package", *IEEE Transactions on Electronics Packaging Manufacturing*, vol. 23, no. 1, pp. 32–38 (2000).
28. https://fctsolder.com/wp-content/uploads/2018/10/2018-SMTAI-How-Does-Surface-Finish-Affect-Solder-Paste-Performance.pdf; Tony Lentz (2018).
29. Kwang-Seong Choi, Sun-Woo Choo, Jong-Jin Lee, Ki-Jun Sung, Hyun-Chel Bae, Byeong-Ok Lim, Jong-Tae Moon, and Yong-Sung Eom, "Novel Solder-on-Pad (SoP) Technology for Fine-Pitch Flip Chip Bonding", *Proceedings of 11th International Conference on Electronic Packaging Technology and High Density Packaging*, Xi'an, China, pp. 805–809 (2010).
30. R.R. Tummala, "Microelectronics Packaging: Past, Present and Future", *MRS Online Proceedings Library*, vol. 154, pp. 379–386 (1989).
31. Franz Bechtold, "A Comprehensive Overview on Today's Ceramic Substrate Technologies", *European Microelectronics and Packaging Conference at Rimini*, Italy, pp. 1–12 (2009).
32. https://www.ceramtec-group.com/en/ceramtec-us/substrates
33. https://polymerinnovationblog.com/polymers-in-electronic-packaging-build-up-films-for-flip-chip-semiconductor-substrates-part-two/; Jeff Gotro (2018).
34. https://pcbmake.com/bt-substrate/
35. Curry Lin, John Lau, N. Liu, H. Yang, T. Xia, C. Ko, B. Lin, M. Ma, and T. Tseng, "Fan-Out Chip-Last Panel-Level Packaging for Heterogeneous Integration", *IMAPS-2024*, Boston MA (2024).
36. Amkor's Hybrid PLP Technology Based 650 mm×650 mm Platform; Eoin O'Toole (2024), https://www.semiconductorpackagingnews.com/uploads/1/A_Hybrid_PLP_Technology_Based_On_A_650_mm_X_650_mm_Platform-AUG.pdf
37. Yonghoon Kim, Doohwan Lee, Youngsik Hur, Jungsoo Byun, Taeje Cho, and Sayoon Kang, "Samsung Fan-out Panel Level Package Solutions", *2019 Electron Devices Technology and Manufacturing Conference (EDTM)*, Singapore, pp. 77–79 (2019).
38. Y. Lin, K. Chen, K. Heng, L. Chua, and S. Yoon, "Encapsulated Wafer Level Chip Scale Package (eWLCSP) for Cost Effective and Robust Solutions in FlexLine", *Proceedings of IEEE IMPACT*, Taipei, Taiwan, pp. 316–319 (2014).
39. J.H. Lau, C. Ko, T. Tseng, T. Peng, K. Yang, T. Xia, P. Lin, L. Chang, N. Liu, C. Lin, D. Cheng, and W. Lu, "Six Side Molded Panel Level Chip-Scale Package with Multiple Diced Wafers", *IMAPS Transactions, Journal of Microelectronics and Electronic Packaging*, vol. 141, pp. 111–120 (2020).
40. https://www.microchemicals.com/dokumente/application_notes/glass_wafers_borosilicate.pdf
41. Shuhei Nomura, Shigeki Sawamura, Kazutaka Hayashi Naoya Suzuki, and Masaaki Takekoshi, "Development and Evaluation of Carrier Glass Substrate for Fan-out WLP/PLP Process", *IEEE 67th ECTC*, Orlando, FL (2017).
42. Lakshmi Narasimha, Vijay Kumar, and Madhavan Swaminathan, "Ultra-Wide Bandwidth Substrate Integrated Waveguide Fed Vivaldi Antenna in D-Band Using Glass Interposer", *2023 IEEE Radio and Wireless Symposium (RWS)*, Las Vegas, NV, pp. 150–152 (2023).
43. A.B. Shorey and R. Lu, "Progress and Application of Through Glass Via (TGV) Technology", *2016 Pan Pacific Microelectronics Symposium*, Big Island, HI, pp. 1–6 (2016).
44. G. Pares, J.P. Michel, E. Deschaseaux, P. Ferris, A. Serhan, and A. Giry, "Highly Compact RF Transceiver Module using High Resistive Silicon Interposer with Embedded Inductors and Heterogeneous Dies Integration", *Proceedings of the IEEE 69th Electronic Components and Technology Conference*, Las Vegas, NV, pp. 1279–1286 (2019).

45. Seann Ayers, Steven Verhaverbeke, Han-Wen Chen, Liu Jiang, and El Mehdi Bazizi, "Signal & Power Integrity Optimization Utilizing Silicon Core Substrate", *Proceedings of the IEEE 74th Electronic Components and Technology Conference*, Denver, CO, pp. 279–284 (2024).
46. Qiao Chen, Yuya Suzuki, Gokul Kumar, Venky Sundaram, and Rao R. Tummala, "Modeling, Fabrication, and Characterization of Low-Cost and High-Performance Polycrystalline Panel-Based Silicon Interposer with Through Vias and Redistribution Layers", *IEEE Transactions on Components, Packaging and Manufacturing Technology*, vol.4, no.12, pp. 2035–2041 (2014).
47. https://www.agc.com/en/products/electoric/index.html
48. https://www.schott.com/en-gb/product-selector?filter=market%3Asemiconductor-data com%3Aapplication%3Aelectronic-integration%3AsubApplication%3Aic-packaging
49. S. Mitarai, K. Adachi, T. Igarashi, K. Seki, N. Kakoiyama, Y. Tanaka, S. Oka, M. Nakazawa, and H. Iwamoto "Physical and Thermal Characteristics of the Advanced Package with Glass Core Substrate", *Proceedings of the 2024 International Conference on Electronics Packaging (ICEP)*, Toyama, Japan, pp. 173–174 (2024).
50. Koji Fujimoto, Yashuhiro Okawa, Takahiro Tai, and Satoru Kuramochi, "Development of Glass Core Substrate with the Stress Analysis, Transmission Characteristics and Reliability", *74th IEEE Electronic Components and Technology Conference (ECTC)*, Las Vegas, NV, pp. 546–550 (2024).
51. https://www.appliedmaterials.com/us/en/semiconductor/markets-and-inflections/heterogeneous-integration/advanced-substrates.html
52. https://gf.com/blog/differentiated-silicon-starts-differentiated-substrates/
53. https://www.skorpiosinc.com/
54. https://gobridg.com/solutions/
55. Tom Dillinger, "Intel's EMIB Packaging Technology – A Deep Dive" (2021); https://semiwiki.com/semiconductor-manufacturers/intel/298674-intels-emib-packaging-technology-a-deep-dive/
56. Ravi Mahajan, Robert Sankman, Neha Patel, Dae-Woo Kim, Kemal Aygun, Zhiguo Qian, Yidnekachew Mekonnen, Islam Salama, Sujit Sharan, Deepti Iyengar, and Debendra Mallik, "Embedded Multi-Die Interconnect Bridge (EMIB) – A High Density High Bandwidth Packaging Interconnect", *2016 IEEE 66th ECTC conference*, Las Vegas, NV, pp. 557–565 (2016).
57. Ahmet C. Durgun, Zhiguo Qian, Kemal Aygun, Ravi Mahajan, Tim Tri Hoang, and Sergey Yuryevich Shumarayev, "Electrical Performance Limits of Fine Pitch Interconnects for Heterogeneous Integration", *2019 IEEE 69th ECTC conference*, Las Vegas, NV, pp. 667–673 (2019).
58. Ravi Mahajan, Ram S. Viswanath, Sriram Srinivasan, Kemal Aygün, Wei-Lun Jen, Sujit Sharan, and Ashish Dhall, "Embedded Multidie Interconnect Bridge – A Localized, High-Density Multichip Packaging Interconnect", *IEEE Transactions on Components, Packaging, and Manufacturing Technology*, vol. 9, no. 10, pp. 1952–1962 (2019).
59. David Mayhew and Venata Krishnan, "PCI Express and Advanced Switching: Evolutionary Path to Building Next Generation Interconnects", *Proceedings of the 11th Symposium on High Performance Interconnects*, Canada, pp. 21–29 (2003).
60. PCI Sizes and Dimensions; https://www.vme.com/wp/engineers-guide/pci/pci-sizes/
61. Lam Tung Lee; https://commons.wikimedia.org/w/index.php?curid=119129502 (2022).
62. https://www.uciexpress.org/why-choose-us
63. "Universal Chiplet Interconnect Express (UCIe)®: Building an open chiplet ecosystem"; Debendra Das Sharma; https://www.uciexpress.org/_files/ugd/0c1418_c5970a68ab214ffc97fab16d11581449.pdf
64. https://www.uciexpress.org/specifications

65. Antroy Roy Chowdhury, Wahid Rahman, and Vladimir Stojanovic, "Electronic-Photonic Co-Optimization of Linear Drive Laser-Forwarded Coherent Silicon Photonic Transmitters for Co-Packaged Optical (CPO) Links", *Journal of Lightwave Technology*, vol. 43, no. 9, pp. 4338–4351 (2025).
66. Yole Developpment, "Advanced Substrates Overview Report" (February 2017).
67. Charles A. Harper (editor), "*Electronic Materials and Processes Handbook*", 3rd ed. McGraw-Hill Handbooks, New York (2004).
68. Gordon E. Moore, "Cramming more Components onto Integrated Circuits", *Electronics*, vol. 38, no. 8, pp. 114–117 (1965).
69. "FCC Online Table of Frequency Allocations, Federal Communications Commission", Federal Communications Commission; Washington DC, USA (2018). https://transition.fcc.gov/oet/spectrum/table/fcctable.pdf
70. Yinggang Li and Jonas Hansryd, "Fixed Wireless Communication Links Beyond 100 GHz", *2018 Asia-Pacific Microwave Conference (APMC)*, Kyoto Japan pp. 31–33 (2018).
71. Weihua Yu, Abbas Vosoogh, Bowu Wang, and Zhongxia Simon He, "Substrateless Packaging for a D-Band MMIC Based on a Waveguide with a Glide-Symmetric EBG Hole Configuration", Editor: Yuwen Li, "*Sensors*"; 2nd ed. Hyderabad, India (2023).
72. Jinxiao Wei, Jinpeng Cheng, Yuxiang Chen, Li Ran, Hao Feng, Homer Alan Mantooth, Frede Blaabjerg, Xu Zhang, Sudharsan Chinnaiyan, and Philip A. Mawby, "Substrate-Less Power Semiconductor Packaging for the Potential of Recyclability", *IEEE Journal of Emerging and Selected Topics in Power Electronics*, vol. 13, no. 2, pp. 2158–2172 (2025).
73. Jia-Wei Fang, I-Jye Lin, Yao-Wen Chang, and Jyh-Herng Wang, "A Network-Flow-Based RDL Routing Algorithmz for Flip-Chip Design", *IEEE Transactions on Computer-Aided Design of Integrated Circuits and Systems*, vol. 26, no.8, pp. 1417–1429 (2007).
74. Jens Lienig and Juergen Scheible, "*Fundamentals of Layout Design for Electronic Circuits*", 1st ed. Springer, New York (2020).
75. https://www.latitudeds.com/post/chiplet-design-and-heterogeneous-integration-packaging (2024).
76. Eric J. Suh and Joseph Riendeau, "2D, 2.1D, 2.5D, and 3D Package Studies in NEPP", NASA Electronic Parts and Packaging Program (NEPP), 2019 Annual NEPP Electronic Technology Workshop, Washington, DC (2019).

3 Materials for Solder Interconnects

Hongwen Zhang

3.1 INTRODUCTION

Interconnection in electronics refers to the physical linkage between different components in an electronic device, functionalized with electrical connection for signal or power flow, mechanical bonding and/or thermal reliability. Interconnections are essential for device functionality, packaging integration, and performance optimization.

Interconnections in electronics [1–7] can be classified based on scale by (i) nanometers (~10 nm to 1 μm) within a chip connecting transistors and logic blocks, (ii) micrometers (~5 to 300 μm) of first level bonding for connection of IC die to the package substrate; (iii) millimeters (~0.3 to 2 mm) of second level bonding between the package and the board; and (iv) system-level bonding (~1 mm to 10 cm) for assembly between PCBs and system modules. Table 3.1 has summarized the classification of interconnections for electronics.

In power semiconductor applications, interconnections play a vital role in enabling the transfer of large currents and voltages with high reliability, thermal efficiency, and mechanical integrity. These interconnections span from die-level to system-level and must meet stringent demands due to the extreme operating environments typical in power electronics (e.g., automotive, industrial drives, solar inverters) [8–15]. Table 3.2 has summarized the interconnects in power semiconductor application.

For both interconnections of electronics and power devices, various materials are used, including metals, dielectric materials, underfill/encapsulants, solder alloys,

TABLE 3.1
Classification of Electronics Interconnection [1–7]

Level	Scale/Pitch	Function	Examples	Interconnection Materials (Examples)
On-chip	10 nm to 1 μm	Transistor-to-transistor	Metal layers in ICs	Cu, Al, Ni, Au
First-level	5–300 μm	Die-to-substrate	Wire bonding, Microbumps	Au wire, Cu pillar, solders
Second-level	0.3–2 mm	Package-to-PCB	BGA, LGA, solder joints	solders
System-level	1 mm to 10 cm	Board/module interconnect	Edge connectors, high-speed cables	Cu, Au plating, solder

DOI: 10.1201/9781003587637-3

TABLE 3.2
Classification of Power Semiconductor Interconnection [8–15]

Interconnect Level	Technology	Materials (Examples)
Chip-to-substrate	Wire Bonding	Al, Cu wires
	Copper Clip Bonding	Cu clips + Solder/Sintered metal
	Die Soldering	SnAg, PbSn, SnSb, and high-temp lead-free solders (AuSn, BiAgX, ZnAl, Durfuse® HT, etc.)
	Die Sintering	Ag/Cu paste or Sintering Film
Substrate-to-baseplate	Soldering	Sn-based, or PbSn solders
	Sintering	Ag/Cu paste
	Adhesive Bonding	Thermally conductive epoxy
Substrate layer	DBC/AMB	Ceramic (Al_2O_3, AlN, Si_3N_4), Cu layers
Module-to-system	Press-fit	Cu, Sn plating, thermal interface materials
	Busbar Terminals	Cu bars, screws, or clips
	Soldering	Pb-free solders

conductive adhesives, sintering materials and transient liquid bonding materials [16–31]. Table 3.3 has summarized the functions of each material category.

This chapter will focus on the solder interconnections, in which metallurgical bonding joints are formed through (i) melting the solder between two solderable metallic surfaces, (ii) the molten solder wetting on both bonding surfaces, (iii) the intermetallic compound (IMC) layer formation along the interface, and (iv) the joint solidification. Solder joints provide electrical connectivity, mechanical support, and sometimes thermal pathways in electronic assemblies. Table 3.4 has summarized the detailed aspects of solder joints [2,3,6,13,15–17,23–26,31–37].

3.2 EVOLUTION OF LEAD-FREE SOLDER ALLOYS

SnPb alloys have been used as joining materials for thousands of years, primarily in jewelry, pipes, plumbing and decorative objectives, and reached their golden era when SnPb was widely adopted following the rapid growth if electrical and electronics industry from 1960s. After environmental concerns of the toxicity of Pb emerged from 1990s, it took more than a decade to convert from SnPb to lead-free solders, although exemption still exists now for certain applications. Lead-free soldering has been enforced since 2006 by the European Union RoHS (Restriction of hazardous substances in electrical and electronics equipment).

The first generation of lead-free solders end up with high Ag content Sn-rich options, including SAC387 (Sn/3.8Ag/0.7Cu), SAC405 (Sn/4.0Ag/0.5Cu), SAC396 (Sn/3.9Ag/0.6Cu), SAC305 (Sn/3.0Ag/0.5Cu) and Sn3.5Ag, led by multiple organizations and solder suppliers. After multiple-years practice, the IPC Solder Value Product Council as the preferred option endorsed Sn96.5Ag3.0Cu0.5 (SAC305) for SMT assembly and most assemblers have transitioned to this alloy for their solder

TABLE 3.3
Materials and Major Functions Used in Interconnection [16–31]

Category	Function	Common Materials (Examples)
Metals	Electrical conduction in interconnects	• Copper (Cu): PCBs, traces, TSVs • Aluminum (Al): On-chip wiring • Gold (Au): Wire bonding, reliable contacts • Others: Nickel (Ni), Silver (Ag), Tin (Sn), Palladium (Pd)
Dielectric materials	Electrical insulation; substrate formation	• Polyimide, Epoxy, Glass Fiber (PCB base materials) • Low-*k* Dielectrics: Used in advanced IC nodes
Underfills and encapsulants	Mechanical support, stress relief	• Epoxies or resins used beneath flip chips or BGAs to reduce thermal/mechanical stresses
Solder alloys	Electrical and mechanical bonding	• Tin-lead: SnPb, SnPbAg (Legacy, non-RoHS) • Lead-Free: Sn-Ag-Cu (SAC), Sn-Bi, Sn-In • AuSn (Eutectic): High-temperature lead-free • Low-temp: Sn-Bi-based for sensitive devices
Conductive adhesives	Bonding at low temperature	• Isotropic Conductive Adhesives (ICA) • Anisotropic Conductive Films (ACF)
Sintering materials	Pressure-assisted solid-state bonding, usually for die-attach	• Silver (Ag) sintering paste and film • Copper paste (emerging)
Transient liquid phase (TLP) bonding materials	TLP bonding for robust interconnects	• Bi-layer / multi-layer alloys: Sn + Cu, Sn + Ni, Sn + Ag, In-Ag etc. • Paste: Low-melting and high-melting powders combo

paste requirements [17,38,39]. Table 3.5 has summarized the melting behaviors, the mechanical properties and the physical properties of the first generation of lead-free solders and the SnPb counterpart.

The relatively poor mechanical shock performance, the plated-through-hole (PTH) damage during wave-soldering, repair and rework challenges, and the cost of high Ag content, accompanied by the low-cost considerations for the booming smart phone market, drove the development of the second generation of lead-free solders [41–43]. Lower Ag contents and dopant alloy additions dominated the second generation of lead-free solder alloy design, majorly improving the mechanical shock performance and interfacial IMC formation. Table 3.6 has summarized the performance of some commercial low-Ag SAC solders [41–48]. However, most of the low-Ag Sn-rich solders may compromise the long-term thermal creep-fatigue performance. Underfill and corner/edge bonds are widely used to enhance the mechanical shock performance of the joints without changing the solder alloy selection [49–52].

TABLE 3.4
Detailed Aspects of Solder Joints

Aspect	Details (Examples)
Primary functions	• Electrical connection • Mechanical support • Thermal conduction
Common joint types	• Fillet joint (through hole, gull-wing) • Leadless joint (chip components) • Ball joints (ball grid array - BGA) • Microbumps (flip-chip) • Solder columns (high-reliability)
Typical alloys	Sn63Pb37 (eutectic), PbSnAg (Pb > 85wt%) SnAgCu (first Gen lead-free) Low-Ag SnAgCu + dopants (second Gen low-Ag lead-free) SnAgCu + strengthening elements (third Gen high-reliability lead-free) BiSn + strengthening elements (low-temperature lead-free) SnAgCu + In (mid-temperature lead-free) Au80Sn20, SnSbAgCu, BiAgX (high-temperature lead-free)
Defects to avoid	• Cold joints • Voids • Bridging • Head-in-pillow/Head-on-pillow • Non-wetting open • Tombstoning • Cracks
Reliability concerns	• Thermal cycling creep fatigue • Aging coarsening • Mechanical shock/vibration • Electromigration
Inspection methods	• AOI (visual) • X-ray (QFN/BGA/Microbumps) • Cross-section analysis • Dye and Pry
Relevant standards	• IPC-A-610 (Assembly Acceptability) • IPC J-STD-001 (Requirements for soldered assemblies)

The development of the third-generation lead-free solders (high reliability) was shaped by semiconductor package complexity and harsh operation environments with the dramatic increase of electronic contents in automobiles. Seeking higher reliability solders than traditional SnAgCu drove the innovation and the R&D efforts. Additional alloying elements (namely Bi, In and Sb) were used to further strengthen the solder through solid solution, precipitates, and grain refinement [53–58]. Compared to traditional SnAgCu, multiple alloying elements would make the melting behavior more complicated, influence the volatile bubble escape and thus compromise the voiding performance, which may damage the thermal performance and the associated reliability [59]. With the rapid growth of 5G, AI chips, and high-performance

TABLE 3.5
Melting Temperatures, Physical and Mechanical Properties of First Gen Lead-Free Solders [40]

	Sn-3.9Ag-0.6Cu	Sn-3.5Ag-0.7Cu	Sn-3.0Ag-0.7Cu	Sn-3.0Ag-0.5Cu	63Sn-37Pb
Liquidus temperature (°C)	218	218	220	220	183
Solidus temperature (°C)	217	217	217	217	183
Tensile strength (Mpa)	43.7	44	44.1	44	34.7
Yield strength (Mpa)	31.3	35	34.2	33	28.4
Elongation (%)	40.6	43.4	41	41	91.2
Work hardening coefficient	0.083	0.056	0.04	0.04	0.032
Young's modulus (Gpa)	52	51	50	50	40
Poisson's ratio	0.35	0.36	0.36	0.36	0.37
Density (g/mm^3)	7.4	7.4	7.4	7.4	8.4
Specific heat (J/g*K)	0.22	0.22	0.23	0.22	0.18
Thermal conductivity (W/m*K)	61.1	62.1	63.2	63.5	52.8
CTE (ppm/K) 20°C–100°C	21.8	21.7	21.6	21.6	21.6

computing (HPC), the demand for high-reliability solder materials remains strong, driven by solder joint miniaturization, long lifetime and increasing power consumption. Low voiding and high reliability solder paste become a major requirement to satisfy the demand.

The development of low-temperature solders (LTS, with reflow peak temperatures below 200°C) has been primarily driven by goals of energy savings, environmental sustainability, reduced carbon footprint, and improved manufacturing yield [60–65]. Most LTS alloys are based on BiSn eutectic and off-eutectic compositions, typically containing 37–58wt% Bi and exhibiting melting ranges from 139°C to 180°C. However, the Bi-Sn system faces significant challenges: the inherent brittleness of bismuth, the relatively poor electrical resistivity and the enrichment of Bi phases driven by phase coarsening and/or electrical current, compromise both the mechanical shock resistance and the electromigration reliability of solder joints [66–70]. InSn eutectic, with the melting temperature around 118°C, provided a ductile option of LTS. However, InSn has a much lower strength than both SnPb and SAC, which degrades quickly when the temperature rises over 80°C, with dramatically reduced creep resistance [71–73]. Also, the high cost of In further limited its affordability for wide use in the industry.

As an alternative, a Bi-free, In-containing lead-free solder (Durafuse® LT) has been developed using mixed-powder paste technology [74–81]. The reflowed solder joint exhibits a solidus temperature of 189°C and a liquidus temperature of 205°C, eliminating the use of bismuth in design to avoid the early mechanical shock failure and the rapid increase in solder joint electrical resistance driven by the electrical current and temperature. Durafuse® Technology employs a mixed-powder paste approach, where two different solder powders are blended with flux; during

TABLE 3.6
Summary of Second Gen Low-Ag Lead-Free Solders [41–48]

Alloy	Alloy Content	Tensile Strength	Ductility	Creep/ Fatigue	Interfacial IMC	Key Features
SAC105	1.0% Ag, 0.5% Cu	40–50 MPa	25%–35%	Lower thermal fatigue resistance than SAC305	Similar to SAC305	Good drop/shock resistance compared to SAC305; Lower cost; Weaker fatigue resistance
SAC105 + Bi/Sb additions	1.0% Ag, 0.5% Cu, Dopants (Bi, Sb)	40–55 MPa	15%–25%	Enhanced creep-fatigue resistance	Controlled the IMC growth	Bi and Sb improve strength but may reduce ductility
SAC0307	0.3% Ag, 0.7% Cu	35–45 MPa	25%–35%	Lower fatigue and creep resistance	Similar to SAC305	Good drop-shock resistance, high ductility. Very cost-effective; Weaker fatigue resistance
SACX	0.3% Ag, 0.7% Cu, Dopants (Bi, Ni)	40–55 MPa (with dopants)	20%–30%	Improved fatigue vs. plain SAC0307	Controlled the IMC growth	Designed to balance low-Ag cost advantage with reliability through microalloying for strengthening and interfacial IMC modification
SACM	≤1% Ag, ≤1% Cu, Dopant (Mn)	40–55 MPa (with dopants)	20%–30%	Improved fatigue vs. plain SAC105, SAC305	Controlled the IMC growth	Designed to balance low-Ag cost advantage with reliability through microalloying for strengthening and interfacial IMC modification

reflow, one or both powder types melt and coalesce, forming either a homogeneous or composite joint depending on the reflow conditions. The commercial Durafuse® LT product can be reflowed at temperatures as low as 200°C while delivering drop-shock performance comparable to, or even superior to, SAC305. Unlike Bi-Sn low-temperature solders (LTS), which reflow below 200°C, Durafuse® LT are classified as middle-temperature solders (MTS).

High-lead solders (with Pb content above 85 wt%) are still commonly used in power discrete devices and power modules under the current RoHS exemption, which remains valid through the end of 2027 [82]. After that, any renewal of the exemption will be reviewed every other year, rather than on the previous five-year cycle.

Their persistence is largely due to the limited maturity of alternative materials and technologies. The Die-Attach 5 (DA5) consortium—established in 2009 and, since 2017, comprising Infineon, Bosch, STMicroelectronics, NXP, and Nexperia—has been at the forefront of qualifying potential replacement solutions, including conductive adhesives, metal sintering, solders, and transient liquid phase (TLP) materials [9,13,15,83–85].

This chapter will highlight recent advancements in solder alloy development, with a focus on third-generation high-reliability solders, low-temperature solders, mid-temperature solders, and high-temperature lead-free solders.

3.3 HIGH RELIABILITY LEAD-FREE SOLDERS

3.3.1 Overview

High reliability lead-free solders are in demand when the conventional SAC solders cannot meet the requirements majorly for harsh operation environments, solder joint miniaturization, long lifetime and increased power consumption. iNEMI was leading, and HDPUG joined later, the industry efforts to evaluate the thermal fatigue behavior of the high reliability lead-free solder alloys for a decade, which included most of the commercially available alloys from the major solder suppliers [15,53–59,86–89]. Table 3.7 has summarized the high reliability solders with their melting temperature ranges.

3.3.2 Role of Ag_3Sn in Sn-rich SnAgCu Solder

Under long-term service conditions, SnAgCu solder joints typically fail through creep-fatigue mechanisms. The formation of an Ag_3Sn precipitate network surrounding Sn dendrites is widely recognized as one of the most effective methods for enhancing the creep resistance in Sn-rich, lead-free solders [26,91–93]. For this reason, many high-reliability solder alloys contain higher Ag levels (>3 wt%) than standard SAC305. As shown in Figure 3.1, the typical microstructure of SAC305 consists of isolated Sn dendrites embedded within interconnected eutectic regions [94]. Within these eutectic zones, fine sub-micron Ag_3Sn particles are uniformly dispersed, with small inter-particle spacing. These precipitates help inhibit dislocation movement along the boundaries of the eutectic regions, thereby improving creep resistance. However, after aging at 125°C for four weeks, the Ag_3Sn particles undergo coarsening as shown in the right image of Figure 3.1. As they grow, the spacing between the particles increases, and the previously distinct Sn dendrites begin to lose their isolation. This microstructural evolution reduces the alloy's ability to resist creep under long-term operation.

Temperature cycling (TC) continues to drive the microstructural evolution of a BGA solder joint as the solder joint is repeatedly exposed to combined thermal and mechanical stresses (Figure 3.2). These stresses primarily arise from the coefficient of thermal expansion (CTE) mismatch between the package and the PCB as temperatures fluctuate. At elevated temperatures—where the homologous temperature exceeds 0.6—atomic diffusion promotes dislocation motion, accelerating

TABLE 3.7
Compositions and Melting Temperatures of Selected Lead-Free Solders [15,53–59,88–90]

Alloy	Developer	Sn	Ag	Cu	Bi	Sb	In	Other	Solidus (°C)	Liquidus (°C)
405Y	1. Inventec	95.5	4	0.5				0.05 Ni; Zn	218	220
Cyclomax (SAC-Q)	2. Accurus	92.8	3.4	0.5	3.3				208	218
Ecalloy	3. Accurus	97.3		0.7	2			0.05 Ni	–	–
HT1	4. Heraeus	95	2.5	0.5			2	Nd	212	219
Indalloy 272	5. Indium	90	3.8	1.2	1.5	3.5			218	226
Indalloy 277	6. Indium	89	3.8	0.7	0.5	3.5	2.5		214	223
Indalloy 279	7. Indium	89.3	3.8	0.9		5.5	0.5		221	228
Indalloy 292	8. Indium	86.7	3.2	0.7	3.2	5.5	0.5	0.2 Ni	214	229
Innolot	9. Heraeus	91.3	3.8	0.7	3	1.5		0.12 Ni	213	222
LF-C2	10. Nihon	92.5	3.5	1	3				208	213
M794	11. Senju	89.7	3.4	0.7	3.2	3		Ni	211	225
M758	12. Senju	93.2	3	0.8	3			Ni	209	217
M58	13. Senju		3.4	0.7	3	3.2			210	221
MaxRel plus	14. Alpha	91.9	4	0.6	3.5				210	219
PS48BR	15. Harima	Bal.	3.2	0.5	4	3.5		Ni, Co	209	223
REL22	16. AIM	Bal.	3	0.7	3	0.6		0.05 Ni; other	210	212
REL61	17. AIM	Bal.	0.6	0.7	2				208	215
SB6NX	18. Koki	89.2	3.5	0.8	0.5		6		201	210
SN100CV	19. Nihon	97.8		0.7	1.5			0.05 Ni	221	225
SN100CW1	20. Nihon	95.8		0.7	1.5	2			–	–
Violet	21. Indium	91.25	2.25	0.5	6				199	215
Viromet 347	22. Asahi	88.4	4.1	0.5			7		202	207
Viromet 349	23. Asahi	91.4	4.1	0.5			4		205	210
Durafuse® HR	24. Indium	88.8	2.6	0.6	1.6	5.2	0.2	0.05	205	221

creep-plastic deformation. Homologous temperature, *Th*, is defined as the ratio of the operating temperature, in degree Kelvin, relative to the melting temperature, also in degree Kelvin of the alloy. Atomic diffusion assisted creep and viscoelasticity dominate the mechanical deformation behavior of the alloy when *Th* exceeds 0.5. In the eutectic regions, the fine, well-dispersed Ag_3Sn particles initially act as strong pinning sites for dislocations. However, the accumulation of pinned dislocations, combined with high service temperatures, eventually triggers recrystallization in regions experiencing greater mechanical strains. Figure 3.2a shows recrystallization occurring beneath the package pad in zone 1. The polarized optical image in Figure 3.2b reveals a band of newly formed, fine grains with varied orientations,

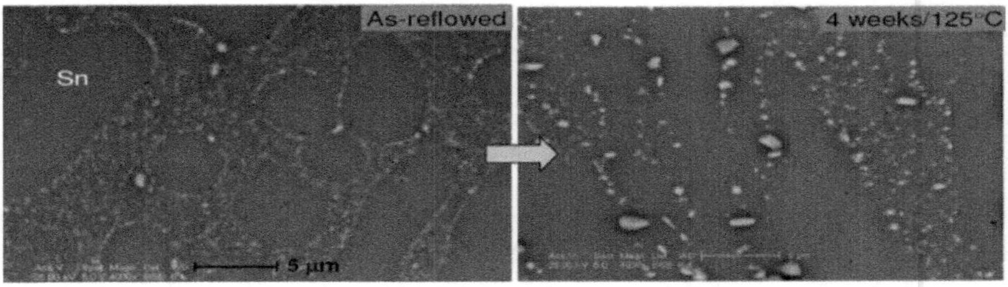

FIGURE 3.1 Typical Microstructure of SAC305, with eutectic zones to isolate Sn dendrite. Inside the eutectic zone, submicron Ag_3Sn particles with fine interparticle spacing are well dispersed. After aging for 4 weeks under 125°C, Ag_3Sn particles grow larger driven by coarsening, and the interparticle spacing becomes wider and the isolation surrounding Sn dendrite becomes less [94]. (Reused with permission from Area Consortium.)

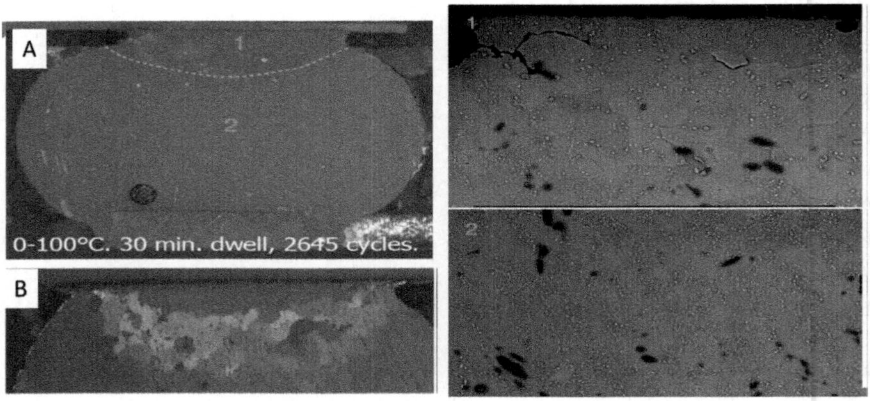

FIGURE 3.2 Recrystallization of a BGA solder joint after 2645 cycles of 0°C–100°C TCT with 30min dwell. Image A shows the overall solder joint morphology. Within the solder joint, zone 1 close to the package side underwent different microstructural evolution from zone 2 (beneath zone 1). Image B (polarized image) shows a band containing numerous fine grains after the recrystallization, and crack initiated from the solder mask corner. Image 1 is the enlarged SEM image of zone 1 showing the loss of the network eutectic zone, existence of grain boundaries and coarse and widely-separated Ag_3Sn particles after recrystallization driven by TCT. Image 2 is the enlarged SEM image showing the well-kept network eutectic zone surrounding the isolated Sn dendrite [91]. (Reused with permission from Springer Nature.)

resulting in differences in contrast and brightness. Cracks were observed to initiate along the solder mask edge and propagate through the boundaries of these fine recrystallized grains. The SEM imaging of zone 1 further indicates that the eutectic network becomes disrupted after recrystallization, with Ag_3Sn particles becoming more widely spaced within the Sn matrix. In contrast, zone 2 retains its distinct Sn dendrites surrounded by intact eutectic regions, indicating minimal microstructural degradation [91].

3.3.3 Effects of Bi, Sb, In Additions

To further improve the creep resistance, additional alloying elements are often introduced to stabilize Ag_3Sn and Cu_6Sn_5 precipitates around the Sn dendrites, dissolve as solute atoms within the Sn matrix, and/or form new in-situ precipitates inside the dendrites. Elements such as Bi, In, and Sb are commonly added to SnAgCu solders because they enhance solder joint strength, ductility, and creep-fatigue performance. Figure 3.3 illustrates the variety of precipitates (Ag_3Sn, Cu_6Sn_5, SnSb,

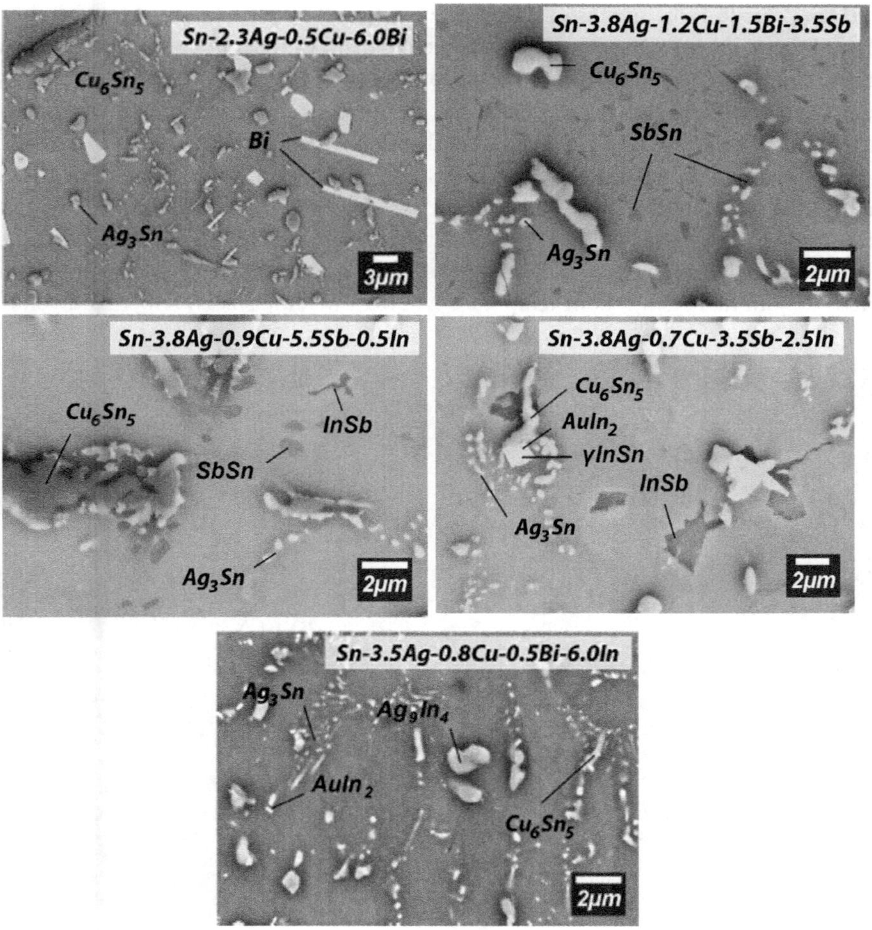

FIGURE 3.3 Selected solder joints demonstrated the strengthening by various precipitates. Sn/2.3Ag/0.5Cu/6Bi presented Ag_3Sn particles, Cu_6Sn_5 rods and Bi platelets. Sn/3.8Ag/1.2Cu/1.5Bi/3.5Sb presented Ag_3Sn particles, Cu_6Sn_5 rods, and fine SnSb particles. Sn/3.8Ag/0.9Cu/5.5Sb/0.5In presented Ag_3Sn particles, Cu_6Sn_5 rods, fine SnSb particles and fine InSb particles. Sn/3.8Ag/0.7Cu/3.5Sb/2.5In presented Ag_3Sn particles, Cu_6Sn_5 rods, InSb particles, InSn particles and possibly SnSb particles. Sn/3.5Ag/0.8Cu/0.5Bi/6.0In presented Ag_3Sn particles, Cu_6Sn_5 rods, and Ag_9In_4 particles [101]. (Reused with permission from SMTA. All rights reserved.)

Bi, SbIn, and Ag_3In_4) formed in five selected solder alloys strengthened with Bi, In, and/or Sb [95–106].

Bi does not form intermetallic compounds (IMCs) with either Ag or Cu, but it has a relatively high solubility in Sn. This solubility decreases significantly—from about 21 wt% at 139°C to roughly 3 wt% at room temperature. As the solubility drops, excess Bi precipitates out of the Sn matrix, forming in-situ Bi phases in addition to the primary Bi-phase (if the Bi content is higher than the solubility at the liquidus temperature of its corresponding composition). As shown in Figure 3.4 for a SAC–4Bi solder joint on a Cu-OSP pad, bright Bi particles nucleate during solidification and continue to grow to tens of nanometers within the first 2 hours. After 24 hours, these Bi precipitates develop into micrometer-scale, well-separated plate-like structures. Coarsening proceeds over the following days (168 and 312 hours). It is also observed that Bi phases located near Cu_6Sn_5 tend to grow larger and exhibit less alignment compared to Bi precipitates embedded within the Sn matrix [99].

The addition of Sb leads to the formation of SbSn precipitates within the Sn matrix following reflow and solidification. As shown in Figure 3.5, these SbSn particles are not distinguishable after 2 hours but become visible after 68 hours of aging, appearing as darker features both adjacent to Cu_6Sn_5 intermetallics and within the Sn matrix. With increasing room-temperature aging time, the size, number, and volume

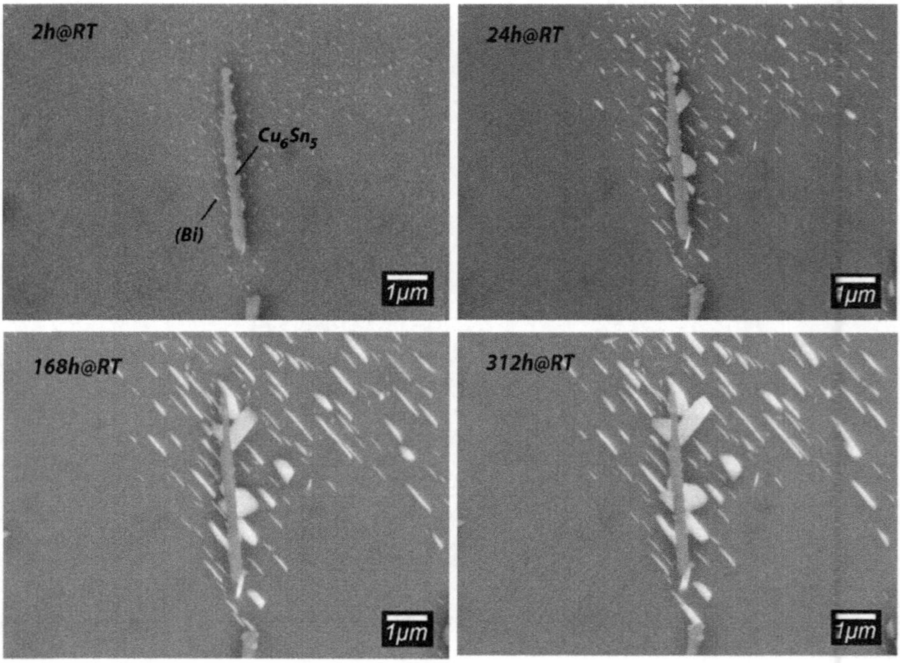

FIGURE 3.4 The Evolution of white Bi phases in SAC305-4Bi/Cu solder joint under room temperature aging. Bi precipitated as submicron particles after reflow and 2 hours room temperature aging. The Bi precipitates grow into platelets with one dimension around microns after 24 hours. Bi platelets grew continuously with additional aging up to 168 and 312 hours [99]. (Reused with permission from SMTA. All rights reserved.)

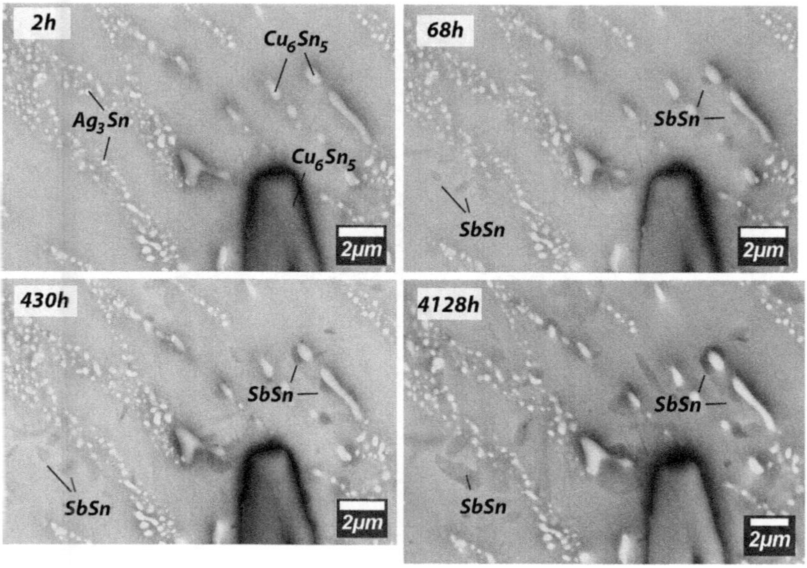

FIGURE 3.5 The evolution of SnSb precipitates in SAC305-5Sb under room temperature aging. After two hours of reflow, there is no clear evidence of SnSb precipitates. SnSb particles become visible inside the Sn dendrite or along the side surface of Cu_6Sn_5 rods after 68 hours. SnSb particles grow slowly upon room temperature aging for 430 hours and even above 4,000 hours [99]. (Reused with permission from SMTA. All rights reserved.)

fraction of SbSn precipitates grow substantially. Although SbSn particles are initially smaller than Ag_3Sn precipitates after 2 hours of aging, they coarsen at a much faster rate. By 430 hours (~18 days), SbSn particles become noticeably larger than Ag_3Sn, while the Ag_3Sn precipitates remain nearly unchanged in size even after 4,128 hours of aging [99].

The formation of In-containing precipitates in SAC solders depends strongly on the amount of In added [105,106]. Indium can react with Ag and Cu to form new phases such as Ag_9In_4 and $Cu_{11}In_9$, or it may incorporate together with Sn to produce mixed IMCs such as $Ag_3(SnIn)$ and $Cu_6(SnIn)_5$. In can also interact with Sb to generate InSb precipitates. Within the Sn matrix, In and Sn may form Sn_4In when the In content is between 15 and 25 wt%, Sn(In) solid solution when the In content is below 10 wt%, or a combination of these phases. However, a comprehensive, systematic investigation of these phase relationships and their evolution is still lacking.

Table 3.8 had summarized the possible impacts of Bi, Sb and In on solder joint morphology. Ni and other trace-level additions ($\leq$0.2 wt%) are mainly intended to modify interfacial IMCs and refine the joint microstructure when possible.

3.3.4 Design of Low Voiding and High-Reliability Solder

These more complex alloy systems discussed above exhibit significantly more complicated melting behaviors and produce a broader range of intermetallic compounds (IMCs) than eutectic or near-eutectic SnAgCu alloys. Consequently, they

TABLE 3.8
Summary of Impacts of Bi, In and Sb in Lead-Free SnAgCu Solders [98–105]

Element	Melting Temperature	Strengthening Mechanism	Ductility and Toughness	Strength and Creep Resistance	IMCs	Additional Notes
Bi (Bismuth)	Lowers melting point	Solid Solution (Bi in Sn matrix) Precipitates (Bi-rich Phase) if excessive	Improves ductility in low content. Reduces ductility if excessive	Improves strength by solid-solution strengthening and by precipitates if excessive. Good creep resistance.	Slightly slows (Ag_3Sn and Cu_6Sn_5) IMC coarsening by pinning atomic diffusion. Do not form IMCs with Cu and Ag.	Improves wettability; Risks brittleness if excessive; Risks electromigration performance if excessive.
In (Indium)	Lowers melting point	Solid Solution (In in Sn matrix) IMC stabilizer ($Ag_3(SnIn)$, $Cu_6(SnIn)_5$) Precipitates ($InSn_4$, $AgIn_2$) if excessive	Greatly improves ductility and toughness	Moderate strengthening through solid solution and precipitates. Good creep resistance.	Stabilizes IMCs (Ag_3Sn and Cu_6Sn_5) by participating into IMC formation and pinning atomic diffusion; Forms $InSn_4$ and Ag_9In_4 IMCs if excessive.	Enhances thermal cycling reliability; Expensive addition.
Sb (Antimony)	Slightly increases melting point	Solid Solution (Sb in Sn matrix) Precipitates (SnSb) if excessive	Neutral on ductility. Reduces ductility if excessive	Improves strength by solid-solution strengthening and precipitates; Excellent creep resistance.	Form in-situ SnSb precipitates; Refines Ag_3Sn and Cu_6Sn_5 IMCs.	Good for high-temperature reliability; Less effect on brittleness.

require tighter process control and more rigorous inspection, and they introduce additional challenges in reducing void formation and optimizing the microstructure. Voids in solder joints can impair heat dissipation, create localized hotspots, reduce mechanical strength, and even lead to cracking. Vacuum reflow is commonly used to mitigate voiding, but it carries drawbacks such as higher equipment cost, extended process cycle time, and the need for carefully tuned thermal profiles to prevent solder beading or balling.

To address these issues, a mixed-powder paste (Durafuse® HR) has recently been developed to achieve low voiding while maintaining high reliability under standard SAC air or nitrogen reflow conditions [59,107–109]. This system employs two solder powders of different compositions—both SnAgCu-based but with distinct strengthening elements—resulting in different melting points and wetting behaviors. During reflow, the lower-melting alloy promotes early wetting and interfacial reactions, while the higher-melting alloy helps preserve joint stand-off until it reaches its melting temperature. This separation in melting behavior increases the time available for flux to escape and for volatile bubbles to dissipate. Ultimately, both powders coalesce during reflow to form a single, homogeneous solder joint—functionally equivalent to solder joints produced from traditional single-powder pastes—while still delivering the targeted improvements in fatigue resistance. Durafuse® HR, formulated within this design framework and containing Bi, Sb, In, and Ni in a SnAgCu matrix, has demonstrated voiding performance comparable to or better than SAC305.As shown in Figure 3.6, DFHR achieves voiding levels of 10% or less—significantly lower than those observed with traditional SAC pastes (>20%)—for a QFN68 package on both ImSn and Cu-OSP surface finishes, validating the effectiveness of the mixed-powder design [110].

Its reliability has been assessed using the same test vehicles that iNEMI has employed for many years in high-reliability solder evaluations, including BGA192, 1206 ceramic chip resistors, and QFN68 components. Table 3.9 compares Durafuse® HR with SAC305 and another high-reliability alloy, further confirming its enhanced voiding performance and long-term reliability across multiple assembly types.

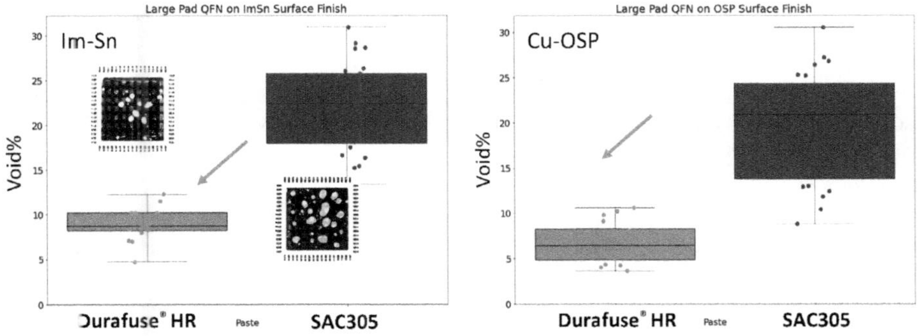

FIGURE 3.6 Voiding performance of Durafsue® HR vs SAC305 in QFN68. On the ImSn surface finish, Durafuse® HR had average voiding below 10% with a range from 5% to 13% while SAC305 presented average voiding of 22% with a range from 10% to 30%. On the Cu-OSP surface, Durafuse® HR had average voiding of 7% with a range from 3% to 11% while SAC305 presented average voiding of 22% with a range from 8% to 30%.

TABLE 3.9
Summary of the Selected High Reliability Lead-Free Solders Compared to SAC [110]

Composition		**Sn/3.0Ag/0.5Cu**	**Sn/2.5Ag/0.5Cu/ 1.6Sb/5.2Bi/0.2In/ 0.06Ni**	**Sn/3.8Ag/0.7Cu/ 1.45Sb/3Bi/ 0.15Ni**
Solidus		217	205	213
Liquidus		220	221	222
Process profiles		230°C–260°C	230°C–260°C	230°C–260°C
		Voiding		
QFN68/Cu-OSP	Average total voids	~15%	~10%	~30%
QFN68/ImSn	Average total voids	~22%	~10%	~22%
	Thermal Cycling –40°C/125°C @ 10 minutes			
Chip-resistor 1206/ Cu-OSP PCB	Average Crack Length after 3,000 cycles	Full Crack	<20%	<75%
Chip-resistor 1206/ ENIG PCB	Average Crack Length after 3,000 cycles	<80%	<5%	<75%
BGA192-SAC305 Ball/Cu-OSP PCB	Characteristic life (N63.2)	2,661	5,998	4,682
BGA192-SAC305 Ball/ImSn PCB	Characteristic life (N63.2)	2,571	5,934	4,512
QFN68/Cu-OSP PCB	Characteristic life (N63.2)	3,232	7,397	5,577
QFN68/ImSn PCB	Characteristic life (N63.2)	3,009	7,142	5,164
	Thermal Cycling –40°C/150°C @ 10 minutes			
BGA192-SAC305 Ball/Cu-OSP PCB	Characteristic life (N63.2)	1,181	2,527	1,676
BGA192-SAC305 Ball/ImSn PCB	Characteristic life (N63.2)	9,97	2,604	2,090
QFN68/Cu-OSP PCB	Characteristic life (N63.2)	1,680	3,435	2,908
QFN68/ImSn PCB	Characteristic life (N63.2)	1,686	3,385	2,640

3.3.5 Summary of High Reliability Solder Alloys

High-reliability lead-free solders are needed when conventional SAC alloys (like SAC305) cannot deliver enough lifetime under harsher conditions—higher temperatures, higher power, finer joints, and longer service. Over the last decade, iNEMI and others have evaluated a broad range of commercial "Hi-Rel" SnAgCu-based alloys that introduce elements such as Bi, Sb, In, and Ni to improve creep-fatigue resistance and thermal cycling performance. A key design principle is controlling the microstructure: in standard SAC, a fine Ag_3Sn/Cu_6Sn_5 eutectic network surrounds Sn dendrites and pins dislocations, giving good creep resistance. But under long-term aging and thermal cycling, these precipitates coarsen, their spacing increases, recrystallization occurs in high-strain regions, and cracks start along newly formed grain boundaries—especially at package corners and solder mask edges.

To counter this, high-reliability alloys use Bi, Sb, and In in different roles: Bi and Sb strengthen by solid solution and fine precipitates (Bi plates, SnSb), In improves ductility and helps stabilize IMCs, and trace Ni refines interfacial layers. The result is a richer mix of precipitates (Ag_3Sn, Cu_6Sn_5, SnSb, InSb, Ag–In phases) that better pin dislocations and slow coarsening. However, the more complex chemistries also create more complicated melting behaviors and IMC populations, demanding tighter process control and making voiding more difficult to manage. Durafuse® HR addresses this by using a mixed-powder design: two SnAgCu-based powders with different melting points and wetting behaviors. The lower-melting powder wets and reacts early, while the higher-melting powder maintains standoff and gives flux and volatiles more time to escape before everything coalesces into a single joint. This architecture has shown significantly lower voiding (≤10% vs. >20% for SAC305 in QFN68) and long-term reliability at least comparable to, and often better than, SAC305 and other Hi-Rel alloys on test vehicles such as BGA192, chip resistors, and QFN68.

3.4 LOW-TEMPERATURE AND MID-TEMPERATURE LEAD-FREE SOLDERS

3.4.1 Low-Temperature Solders (BiSn and InSn)

Low-temperature lead-free solders are generally defined as alloys that can be reflowed below 200°C. BiSn-based LTS materials offer several advantages—particularly for low-heat, cost-sensitive, and highly miniaturized devices—when compared with traditional SAC305. Table 3.10 summarizes the key differences between BiSn-based low-temperature solders and SAC305, highlighting the drivers for adopting lower-temperature soldering [61–64,111–115].

Mainstream BiSn-based low-temperature solders (LTS) are typically eutectic or near-eutectic alloys with melting points in the range of 137°C–139°C, depending on the alloying additions. Off-eutectic BiSn alloys with Bi contents below 40 wt% are also available. Bi has a rhombohedral crystal structure with lattice spacings of 1.19 nm along the c-axis and 0.47 nm along the *a*/*b*-axes, resulting in highly anisotropic mechanical behavior that depends on the loading orientation relative to the (111) plane. This anisotropy leads to overall brittleness especially under a high strain rate and high electrical resistivity when large amounts of Bi phase are present in BiSn solder joints. Lowering the Bi content can modestly improve the mechanical shock resistance and partially mitigate the electrical resistance increases under electromigration [66–68,116–118]; however, it does not eliminate the inherent weaknesses associated with substantial Bi-phase presence.

TABLE 3.10
Comparison of BiSn LTS vs SAC [61–64,111–114]

Driving Force	Benefit of LTS (BiSn-based)	SAC305
Lower reflow temperature	Reflow peak >165°C–190°C Less stress on PCBs and components	Reflow peak ~240°C–250°C; Higher thermal/mechanical stress
Energy and cost savings	Lower reflow temperature and reduced energy and cost	Higher energy use; longer heating cycles
Device compatibility	Suitable for MEMS, LEDs, optical modules, sensors, plastic connectors, etc.	Risk of component warpage/damage in heat-sensitive devices
Reliability (Targeted)	Adequate when alloyed (Bi + Ag/Sb/In) but generally lower fatigue life than SAC305	Well-established high fatigue and creep resistance
Miniaturization support	Reduces warpage, head-in-pillow, and bridging in fine-pitch, high *I/O* packaging	More prone to warpage and head-in-pillow defects in fine pitch devices
Sustainability	Lower energy footprint; BiSn-based solders reduce Ag use (cost and supply benefit)	Contains 3% Ag → higher materials cost; more energy-intensive

As an alternative, eutectic and near-eutectic InSn alloys—containing roughly 50 wt% In and melting at around 118°C—offer much greater ductility along with good thermal and electrical conductivities. However, these alloys exhibit significantly lower strength than SAC or SnPb, and their mechanical properties deteriorate rapidly at elevated temperatures (~70°C–80°C), resulting in poor creep resistance [71–73]. In addition, the high cost of indium limits widespread adoption of eutectic InSn systems.

iNEMI has led extensive research on BiSn LTS, focusing on the mechanical shock behavior, thermal cycling reliability, and electromigration performance [62,66–68,112–114,116–118]. Numerous BiSn and modified BiSn compositions have been evaluated, with alloying additions such as Ag, Cu, and Sb providing measurable benefits. Despite these advances, fundamental challenges remain: the low melting point restricts long-term service temperatures to below 125°C based on homologous temperature considerations; board-level drop shock performance is one to two orders of magnitude worse than SAC; and electrical resistance is higher due to the presence of Bi phases. Under electromigration stress—where current and temperature act simultaneously—excess Bi tends to accumulate at the anode, further increasing solder joint electrical resistance. Moreover, the electromigration rate of BiSn is roughly one to two orders of magnitude higher than that of SAC, underscoring an additional reliability concern.

3.4.2 Mid-Temperature Solder

3.4.2.1 Design of Bi-Free Mid-Temperature Solders

Mid-temperature solder systems were developed to overcome the inherent limitations of BiSn alloys, leading to the creation of a Bi-free, In-containing, Sn-rich paste known as Durafuse® LT (DFLT). Indium offers high ductility and low electrical resistivity (8.4 μΩ·cm), which improves mechanical shock performance and reduces copper (Cu) diffusion, thereby enhancing electromigration resistance. DFLT employs a mixed-powder approach—combining low-melting SnInAg with Sn-rich alloys—enabling reflow at 190°C–210°C while raising the final joint remelting temperature to above 180°C, making it suitable for demanding service environments [11,12]. The Sn–In phase diagram (Figure 3.7) indicates that when the Sn content exceeds 84 wt%, the solidus temperature increases to around 175°C and the liquidus surpasses 205°C. Consequently, ensuring the Sn content remains above 84 wt% guarantees a final joint remelting temperature above 175°C, satisfying the reliability requirements of portable and consumer electronics applications.

To achieve reflow near 200°C while keeping the final solder joint melting temperature above 175°C, the mixed solder paste DFLT was formulated using optimized ratios of low-melting SnInAg and SnAgCu powders. Differential scanning calorimetry (DSC) confirms that the paste exhibits an initial melting range of 154.5°C–205°C. After reflow at a 200°C peak, the solidified joints show a remelting range of

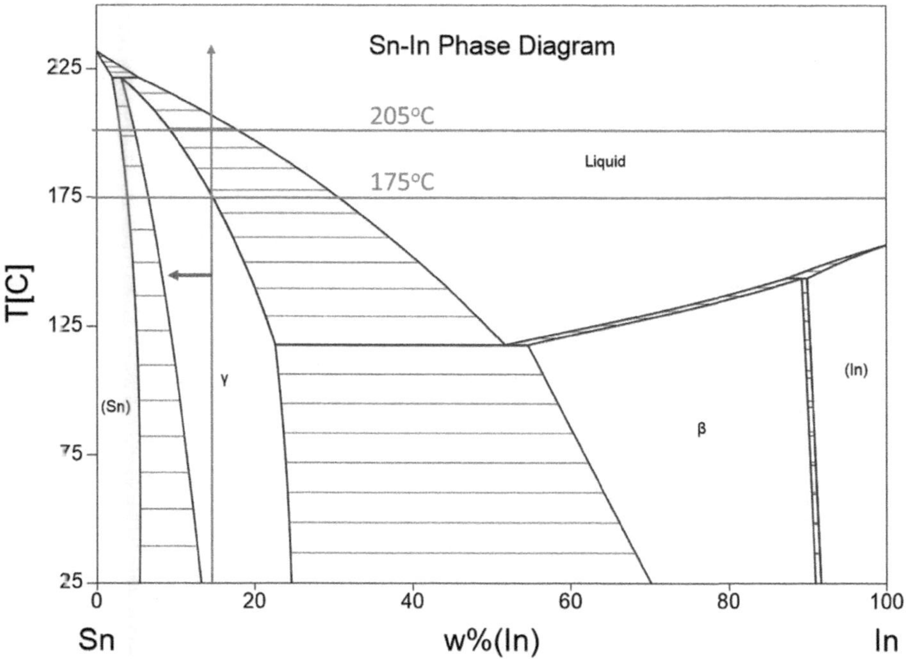

FIGURE 3.7 SnIn phase diagram [123]. (Reused with permission from ASM International. All rights reserved.)

189°C–205°C, demonstrating that DFLT supports low-temperature processing while delivering a high remelting temperature in the finished joint [123] (Figure 3.8).

A plateau-style reflow profile with a 203°C peak, shown in Figure 3.9, was developed, featuring an extended plateau of roughly 140 seconds above 198°C. This prolonged high-temperature dwell promotes uniform solder joint formation from the solder paste and was used to produce the solder spheres evaluated in Figure 3.8 (right). As the peak reflow temperature increases, the required plateau duration can be proportionally reduced. For BGA assemblies using SnAgCu balls, the reflow profile should be further refined according to the package size and warpage behavior to ensure robust joint formation, maximize yield, and enhance overall reliability.

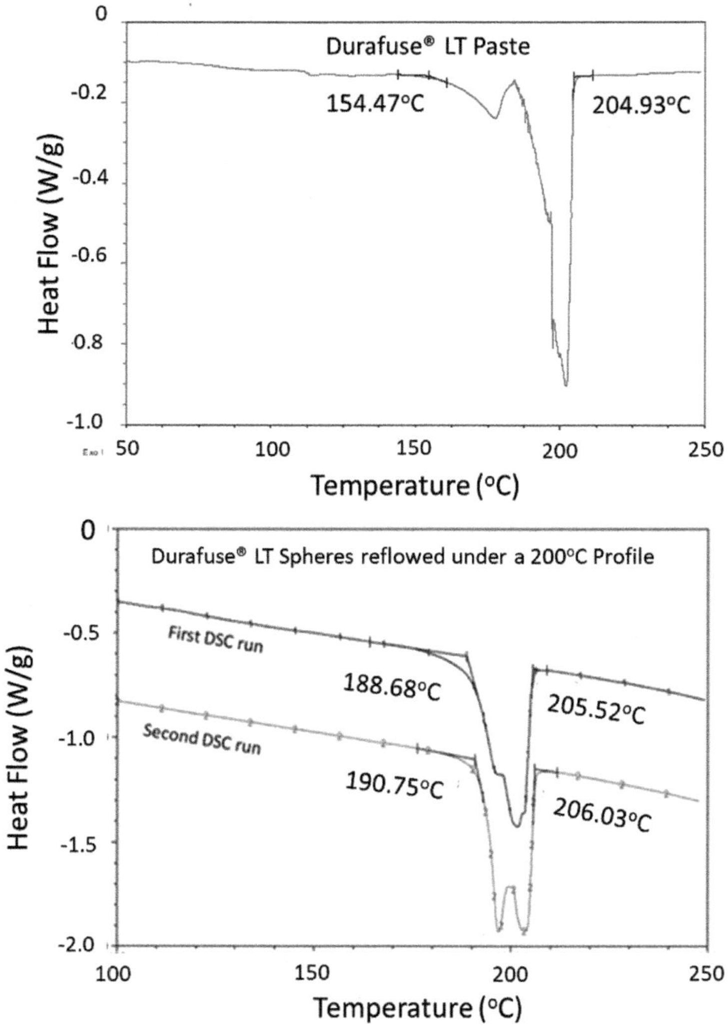

FIGURE 3.8 DSC curves of DFLT paste (a) and reflowed DFLT spheres (b).

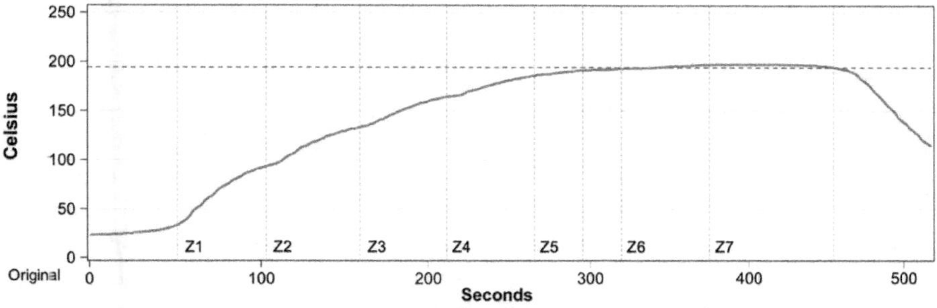

	Speed (in/min)	Zone 1	Zone 2	Zone 3	Zone 4	Zone 5	Zone 6	Zone 7
Original Top	14	95	135	150	200	203	203	200
Orginal Bottom	14	95	135	150	200	203	203	200
Predicted Top	14	95	135	150	200	203	203	200
Predicted Bottom	14	95	135	150	200	203	203	200

FIGURE 3.9 Optimized plateau reflow profile with peak temperature 203°C and 140 seconds above 198°C for DFLT.

3.4.2.2 Mechanical Shear Performance

Room-temperature and elevated-temperature shear tests were conducted to assess (i) whether any solder phases melt below 165°C, and (ii) how solder joint mechanical strength degrades with increasing temperatures, as shown in Figure 3.10. The SnInAg alloy—used as the low-melting powder in the DFLT paste—showed a rapid decline in shear strength, dropping to 10 MPa at 110°C and reaching zero at 125°C due to its low eutectic melting point of 118°C. In contrast, SAC305 retained a shear strength of approximately 15 MPa even at 165°C. DFLT displayed a shear strength of about 23 MPa at 125°C, indicating no melting near 118°C, and maintained roughly 14 MPa at 165°C, demonstrating controlled and predictable mechanical degradation with temperature. For comparison, eutectic BiSnAg exhibited a shear strength of

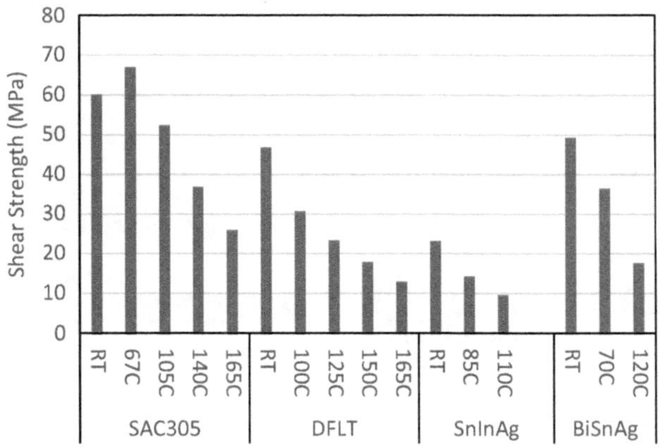

FIGURE 3.10 Shear strength with elevated temperature for SAC305, DFLT, SnInAg and BiSnAg.

18 MPa at 120°C but dropped to nearly zero above 140°C, consistent with its lower melting temperature of 138°C.

3.4.2.3 Drop-Shock Performance

Ball-drop tests were performed to assess the shock resistance of solder joints, as shown in Figure 3.11. A 10×10 OSP BGA package (chip size: 2×2 in^2; board size: 4×4 in^2) was mounted face-down, with a metal block and four steel pins positioned at the package corners to transfer impact. A steel ball was dropped from a fixed height, and drops were repeated until the BGA completely separated from the board. Accelerometer measurements recorded average impact accelerations of approximately 2,907 g-force for a 44-gm ball and 6,398 g-force for a 110-gm ball, both dropped from 500 mm. Each value represents the mean of ten repeated tests [124,125].

Initial ball-drop testing compared two BiSn solders—eutectic BiSnAg and a more ductile BiSnAg variant (both reflowed at a 170°C peak)—against DFLT solder, which was reflowed at 203°C (per Figure 3.10). Using a 44gm steel ball dropped from 500mm (producing ~2,900 g-force acceleration), the hybrid SAC/eutectic BiSnAg solder joints failed after an average of 19 drops, while the hybrid SAC/ductile BiSnAg solder joints failed after 28 drops (average of five samples). In sharp contrast, the hybrid SAC/DFLT solder joints exhibited no failures even after 2,000 drops—representing roughly a two-order-of-magnitude improvement over both BiSn-based materials (Figure 3.12).

In the second test series, a 110-gm steel ball was dropped from 500 mm, producing an impact acceleration of approximately 6,400 g-force. Both eutectic BiSnAg and ductile BiSnAg hybrid joints (using SAC305 balls on the package side) failed after a single drop, showing no measurable difference in performance, as illustrated in Figure 3.13. In comparison, SAC/SnInAg hybrid joints—where SnInAg has a low melting point of 118°C—survived about 125 drops, representing nearly a two-order-of-magnitude improvement over the BiSnAg-based systems shown in Figure 3.12. SAC/DFLT hybrid joints, reflowed using an optimized 200°C profile,

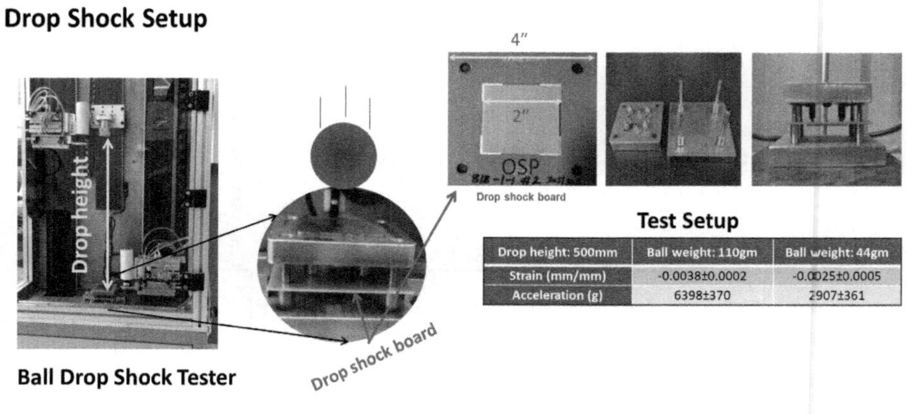

FIGURE 3.11 Ball drop shock tester.

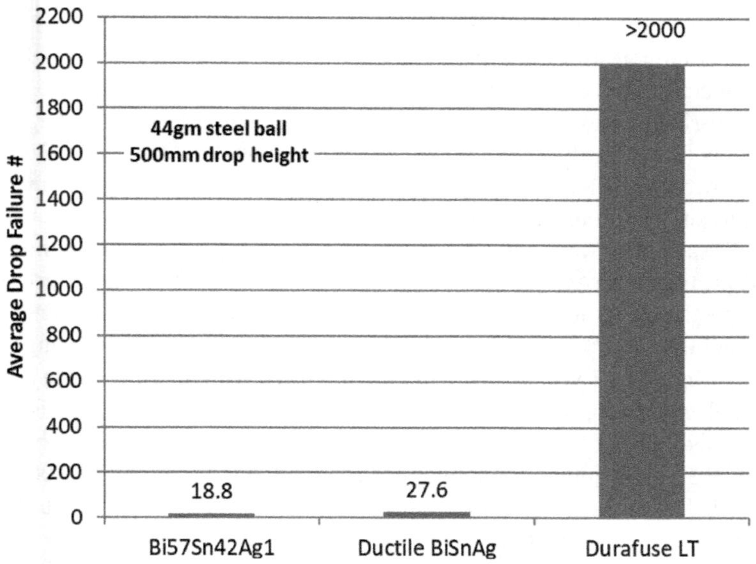

FIGURE 3.12 Drop shock test for $Bi_{57}Sn_{42}Ag_1$, a ductile BiSnAg, and DFLT (SAC305 BGA ball). Ball weight: 44 gm, and drop height: 500 mm.

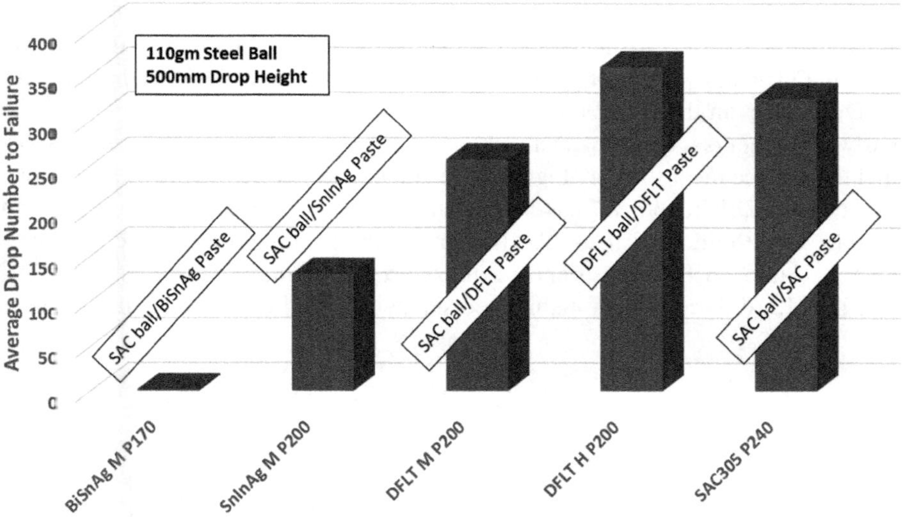

FIGURE 3.13 Drop shock performance of the hybrid joints of SAC/BiSnAg eutectic, SAC/SnInAg, and SAC/DFLT hybrid joints (DF M P200), and the homogeneous joints of DFLT/DFLT, and SAC/SAC305. Drop shock parameters: ball weight 110 g, and drop height 500 mm.

endured approximately 260 drops, twice the performance of the SAC/SnInAg hybrids. Fully DFLT-based joints (DFLT balls on the BGA paired with DFLT paste on the PCB) exhibited the highest robustness, surviving roughly 360 drops—about 12% more than SAC305 (~320 drops) and 38% more than SAC/DFLT hybrid joints (~260 drops).

3.4.2.4 Electromigration Behavior

Electromigration testing was performed on LGA assemblies using DFLT, B37 (an off-eutectic BiSnAg alloy containing 37 wt% Bi, selected to limit the formation of large Bi-rich phases), and SAC305 as the control. The LGA556 package utilized ENIG pads (0.38 mm diameter, 1.00 mm pitch) mounted onto a Cu-OSP PCB with matching pad dimensions. A daisy-chain configuration was incorporated into the test vehicle, as shown in Figure 3.14 [69,126].

DFLT assemblies were reflowed using a 208°C peak profile, B37 assemblies with a 190°C profile, and SAC305 with a standard 240°C profile. During electromigration testing, the daisy-chain test vehicles were stressed with a DC current density of 2.5 kA/cm^2 at 90°C for up to 1,500 hours. The direction of current flow through the solder joints was alternatively reversed to maintain consistent anode/cathode conditions and to enable clear observation of microstructural changes. Electrical resistance was continuously monitored using a data logger. As shown in Figure 3.15, both DFLT and SAC305 solder joints maintained stable resistance levels below 400 μΩ with minimal drift over the test duration. In contrast, B37 solder joints exhibited a higher initial resistance (400–800 μΩ) due to the inherently higher resistivity of Bi-rich phases and showed pronounced resistance increases in several samples.

Figure 3.16a and b illustrate the electrical resistance evolution of DFLT and B37 solder joints during electromigration testing. All DFLT samples maintained electrical resistance values between 250 μΩ and 380 μΩ throughout the 1,500-hour current and thermal stress period, indicating minimal change. In contrast, several B37 solder joints experienced electrical resistance increases of 200–280 μΩ—exceeding 25% of their initial values—while others exhibited more moderate rises of less than 100 μΩ.

Differences in the electromigration behavior of BiSn joints became apparent as early as 500 hours into testing, with outcomes influenced by both the current polarity and the surface metallization. Figure 3.17a compares the electrical resistance evolution of SAC, DFLT, and B37 under opposite electron flow directions: from the LGA chip (ENIG) to the PCB (Cu-OSP) (E-Down), and from the PCB to the LGA chip (E-Up). Although B37 consistently exhibited a higher initial resistance than SAC and DFLT, all three solders maintained stable electrical resistance under E-Down

FIGURE 3.14 LGA556 Layout and PCB design of the electromigration test vehicle [69,126]. (Reused with permission from Area Consortium and SMTA. All rights reserved.)

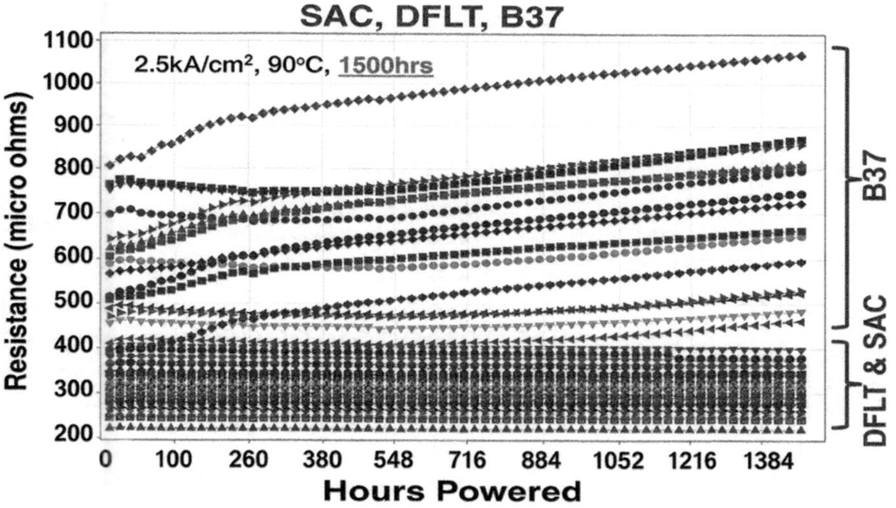

FIGURE 3.15 Resistance evolution under DC current (2.5 kA/cm^2, 90°C, 1,500 hours) for both current polarities of three solder joints (SAC305, DFLT, and B37) [69,126]. (Reused with permission from Area Consortium and SMTA. All rights reserved.)

stressing, as shown in Figure 3.17a. Under E-Up stressing, however, B37 solder joints displayed a significant increase in electrical resistance, whereas SAC and DFLT remained essentially unchanged, as reflected in the right-hand chart of Figure 3.17a. Figure 3.17b summarizes the electrical resistance changes for all three solder pastes, clearly demonstrating the strong influence of current polarity on B37.

Microstructural analysis revealed strong polarity-dependent behavior in B37 solder joints. Under E-Down stressing (electron flow from the chip to the PCB), shown in Figure 3.18a (left), coarse Bi phases aligned along the current direction, with a thin and discontinuous Bi-layer forming near the Cu-OSP pad (anode) and a Bi-depleted region appearing at the ENIG pad (cathode). In contrast, under E-Up stressing (electron flow from the PCB to the chip), Figure 3.18a (right) shows the development of a continuous Bi-rich layer adjacent to the ENIG pad (anode) and a pronounced Bi-depleted zone on the Cu-OSP side (cathode). This accumulation of high-resistivity Bi near the anode likely contributed to the substantial resistance increase observed in B37 solder joints under E-Up conditions. The underlying cause of this polarity effect in BiSn alloys remains under investigation, but it may be related to Bi's strong affinity for nickel (Ni), which could impede Bi transport from ENIG to Cu-OSP under E-Down stressing while promoting Bi enrichment at the ENIG interface under E-Up stressing.

In contrast, DFLT solder joints showed only minor intermetallic compound (IMC) thickening in Figure 3.18b. Being Bi-free and containing In, DFLT behaves similarly to SAC alloys: In is chemically compatible with Sn, participates in IMC formation with Cu, Ni, and Ag, and forms solid solutions with Sn. Consequently, both DFLT and SAC joints exhibited modest IMC growth on the anode side and negligible changes on the cathode. It is also likely that the applied test conditions were not severe enough to induce strong electromigration-driven differences between DFLT and SAC.

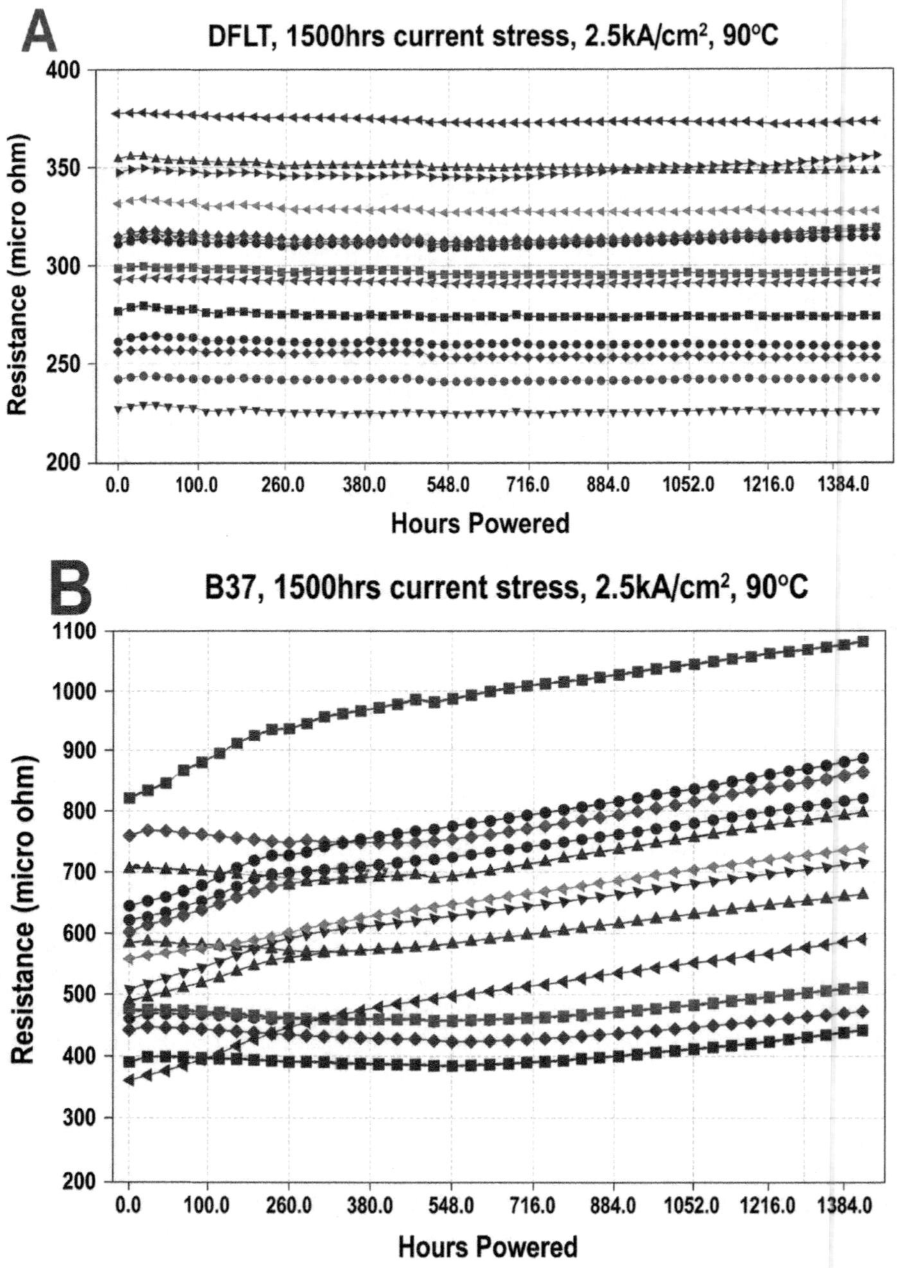

FIGURE 3.16 Joint resistance evolution of (a) DFLT and (b) B37, under the combination of DC current (2.5 kA/cm^2) and 90°C up to 1,500 hours [69,126]. (Reused with permission from Area Consortium and SMTA. All rights reserved.)

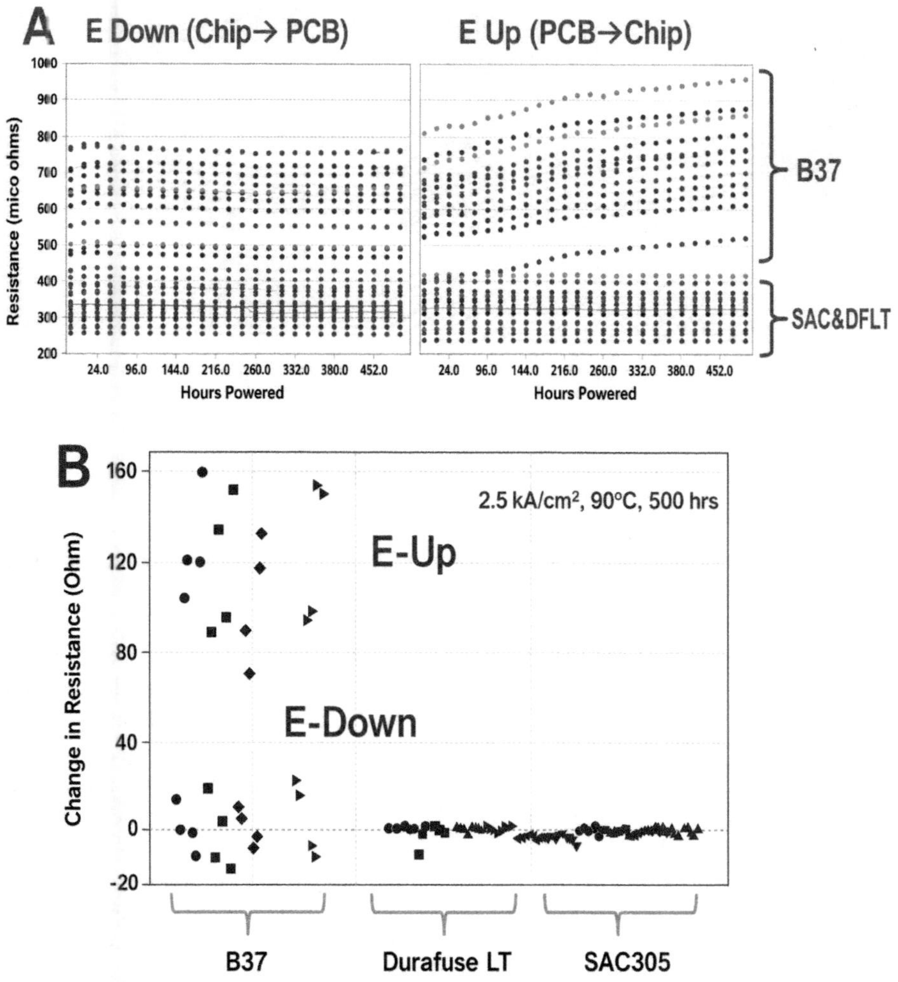

FIGURE 3.17 Change in electrical resistance at a DC of 2.5 kA/cm^2, temperature of 90°C for 500 hours [69,126]. (Reused with permission from Area Consortium and SMTA. All rights reserved.)

These results highlight two major reliability concerns associated with Bi-containing solders: their intrinsically high electrical resistance due to Bi-rich phases, and their tendency for rapid increases of electrical resistance under electromigration. In contrast, DFLT avoids both issues, exhibiting low initial electrical resistance and long-term stability comparable to SAC305, while markedly outperforming B37.

3.4.2.5 Mitigating the Warpage-Caused Defects in Large BGA Assembly

Long-term TCT reliability testing has also been performed using DFLT paste on chip resistors, QFN68, BGA192, and wafer-level packages (WLP256) [75–79,124,125]. These studies demonstrated that reflow profiles strongly influence joint formation

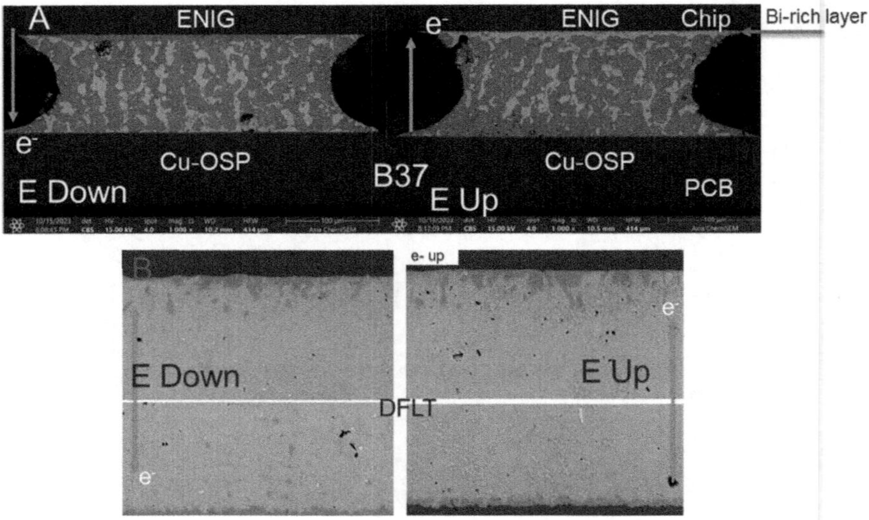

FIGURE 3.18 Morphology of the cross-section of B37 and DFLT joints [69,126]. (Reused with permission from Area Consortium and SMTA. All rights reserved.)

and, consequently, TCT reliability. DFLT achieved thermal cycling performance comparable to—or better than—traditional SAC305 when "healthy" joints were formed. In this context, "healthy" joints are those that exhibit no fatal defects, such as head-in-pillow (HIP), non-wet opens (NWO), cold-welds, non-contact opens (NCO), insufficient fillets, hot tearing, or severe necking. These failure modes are largely process- or design-related and can typically be mitigated through profile optimization and package-level adjustments.

Large BGA assemblies incorporating multiple stacked chips connected through an interposer enable high computing performance for data-intensive applications. However, these advanced architectures introduce significant challenges in package design, qualification, and high-volume manufacturing due to pronounced package warpage [127–129]. Dynamic warpage—particularly that induced by temperature fluctuations during reflow—creates assembly difficulties and increases susceptibility to defects such as HIP, NCO, NWO, necking, and hot tearing [4]. To mitigate these risks, low-temperature soldering strategies that reduce package warpage have been extensively explored [79–81,130,131].

Most BGAs continue to use SAC or SAC-based solder balls, which have melting points above 210°C. When reflow temperatures fall far below the melting point of SAC, the limited dissolution of the SAC ball into the molten low-temperature solder paste may be insufficient to compensate for warpage-induced displacement. Increasing the low-temperature paste volume and/or raising the reflow peak temperature can improve solder joint formation by providing more molten solder to accommodate the displacement, provided that the warpage remains within manageable limits. Hot tearing has been associated with situations where a component's inversion temperature, referring to the temperature at which the warpage of a structure changes sign, transitioning from a convex to a concave shape or the reverse, coincides with solder solidification [62,121], underscoring the importance of selecting solder alloys that align well with the package's warpage behavior. Despite this need, comprehensive studies on solder joint formation in large BGAs experiencing significant dynamic warpage are still limited.

In the present work, both DFLT and eutectic BiSnAg low-temperature solders were evaluated. Reflow profiles spanning low (185°C, 200°C), mid (210°C, 220°C), and high (240°C, standard SAC305) peak temperatures were included in the test matrix. Paste-to-ball volume ratios (PBVRs) of 0.13, 0.25, and 0.5—achieved through various stencil designs—were examined as an additional variable.

The PBGA928 device (40 mm×40 mm) used SAC305 balls (0.63 mm diameter, 1 mm pitch) and exhibited dynamic warpage ranging from –11 mils (concave at 100°C) to +34 mils (convex at 250°C), with an inversion temperature near 140°C. The PCB employed a T_g 130°C laminate and non-solder-mask-defined Cu-OSP pads (0.45 mm diameter). Reflow profiles P185, P200, P210, P220, and P240 incorporated peak temperatures of 185.2°C, 203.8°C, 213.1°C, 222.7°C, and 236.7°C, respectively.

Assemblies produced under each condition were selected for cross-sectioning and metallographic evaluation. Optical microscopy and scanning electron microscopy (SEM) were used to identify and characterize joint-level defects. Remaining assemblies were reserved for thermal cycling testing (TCT). Solder joints were scored based on the presence of defects and overall morphology, as illustrated in Figure 3.19.

Under the P185 reflow profile, BiSnAg solder joints showed severe defects (Score 1), including non-contact opens, open necking, hot tearing, and necking, across all PBVRs, as illustrated in Figures 3.20–3.22. SAC balls retained their spherical shape, indicating minimal melting and limited contribution to joint formation at the 185°C peak temperature. At a PBVR of 0.13, noticeable package tilt caused non-contact

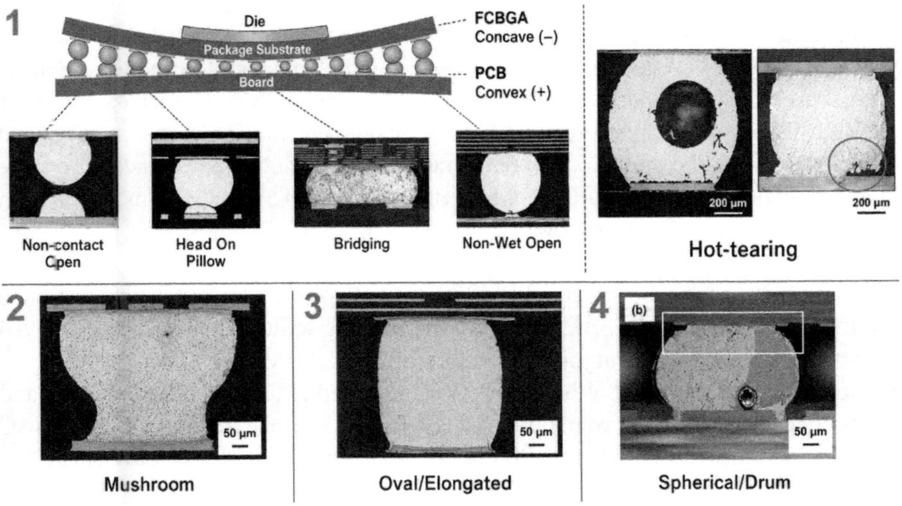

FIGURE 3.19 Demonstration of solder joint scores: (a) "fatal" defects with open, HIP, NWO, hot-tearing, and bridging; (b) necking; (c) oval/elongated/stretched joint; and (d) spherical/drum.

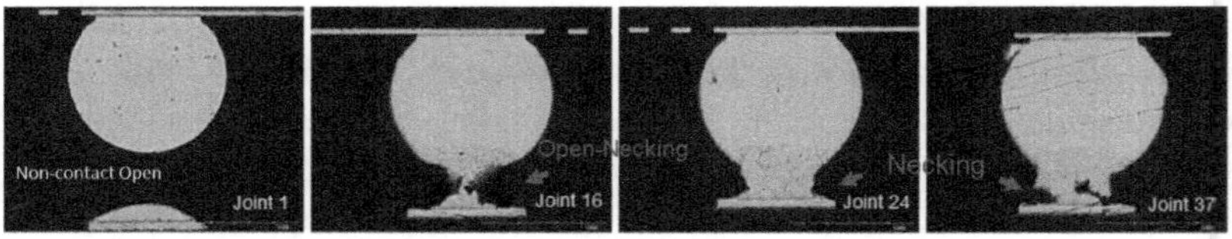

FIGURE 3.20 SAC ball/BiSnAg joints, P185, PBVR of 0.13; package tilt; joint score rank 1.

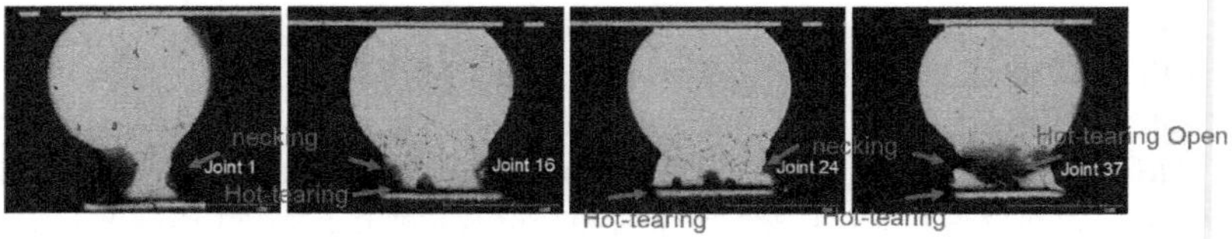

FIGURE 3.21 SAC ball/BiSnAg joints, P185, PBVR of 0.25; package concave; joint score rank 1.

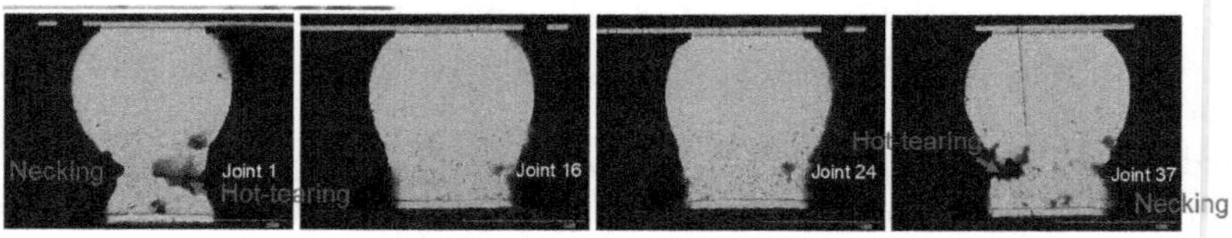

FIGURE 3.22 SAC ball/BiSnAg joints, P185, PBVR of 0.5; package concave; joint score rank 1.

open joints at one corner. Increasing the PBVR to 0.25 and 0.5 reduced tilt but did not eliminate hot tearing and necking defects. The assembled package maintained a concave shape, consistent with the PBGA928 warpage behavior below 140°C.

Under the P200 reflow profile, shrinkage, bubble voids, and hot-tearing defects were observed particularly in corner joints (Figures 3.23–3.25) across all three PBVRs. These issues likely resulted from incomplete mixing between SAC and BiSnAg alloys at the applied peak temperature. The package maintained a concave shape at PBVRs of 0.13 and 0.25, while the PBVR of 0.5 produced more uniform joint heights across the entire row.

Under the P220 reflow profile, spherical or drum-shaped joints with only minor voiding were observed even at a PBVR of 0.13 (Figure 3.26), indicating complete melting and mixing of the SAC balls with the BiSnAg solder. The package displayed a convex shape, with central joints (e.g., Joints 16 and 24) slightly taller than the corner joints (Joints 1 and 37). When the PBVR increased to 0.5, pear-shaped joints and voids beneath the package were observed (Figure 3.27), further confirming full SAC melting and alloy mixing. At this PBVR, the package exhibited a concave shape.

Under the P200 profile with a PBVR of 0.13, DFLT solder joints exhibited pronounced necking and hot tearing, particularly in the corner joints, along with a HIP defect adjacent to the corner (Figure 3.28). When the PBVR increased to 0.25, hot tearing was eliminated and severe necking was greatly reduced (Figure 3.29). At a

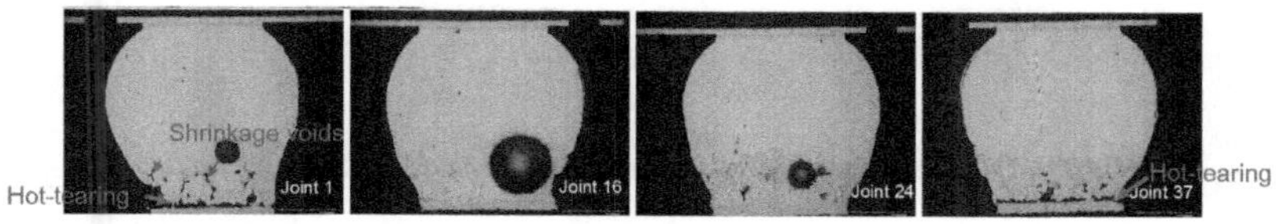

FIGURE 3.23 SAC ball/BiSnAg joints, P200, PBVR of 0.13; package concave; joint score rank 1.

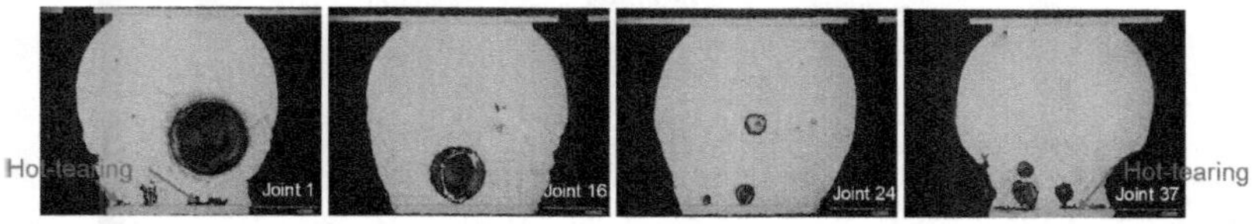

FIGURE 3.24 SAC ball/BiSnAg joints, P200, PBVR of 0.25; package concave; joint score rank 1.

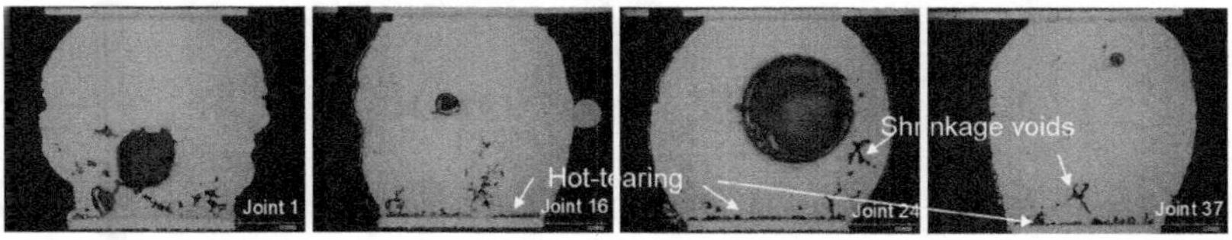

FIGURE 3.25 SAC ball/BiSnAg joints, P200, PBVR of 0.5; package flat; joint score rank 1.

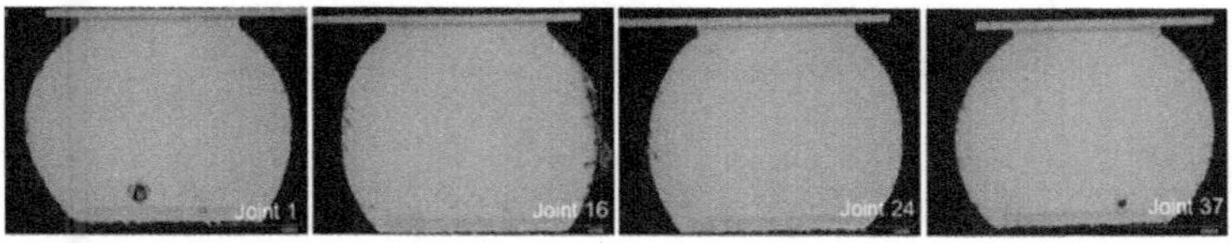

FIGURE 3.26 SAC ball/BiSnAg joints, P220, PBVR of 0.13; package slight convex; joint score rank 4.

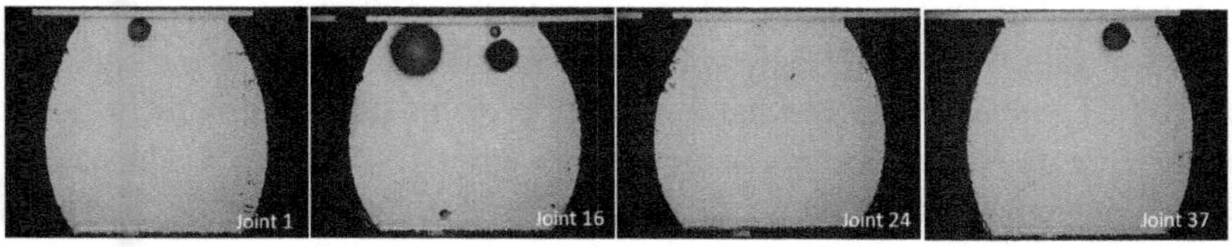

FIGURE 3.27 SAC ball/BiSnAg joints, P220, PBVR of 0.5; package slight concave; joint score rank 4.

PBVR of 0.5, DFLT solder joints showed elongated shapes without significant necking (Figure 3.30). Packages with a PBVR of 0.13 displayed slight convex warpage, while flat warpage was observed at PBVRs of 0.25 and 0.5.

Under the P210 profile, DFLT solder joints with a PBVR of 0.13 showed oval to spherical shapes in the corner joints, while the middle joints still exhibited necking

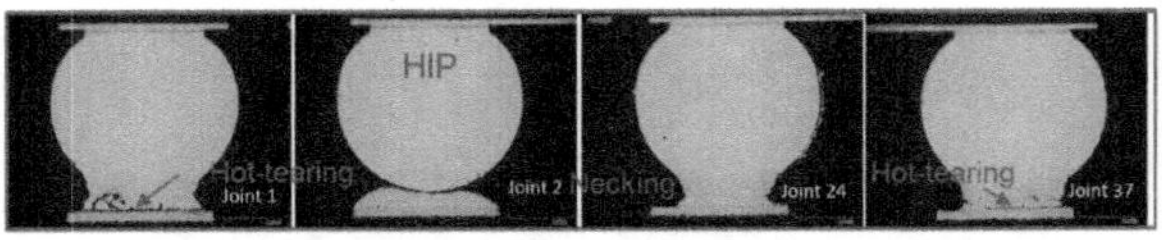

FIGURE 3.28 SAC ball/DFLT joints, P200 joints, PBVR of 0.13; package slight convex; joint score rank 1.

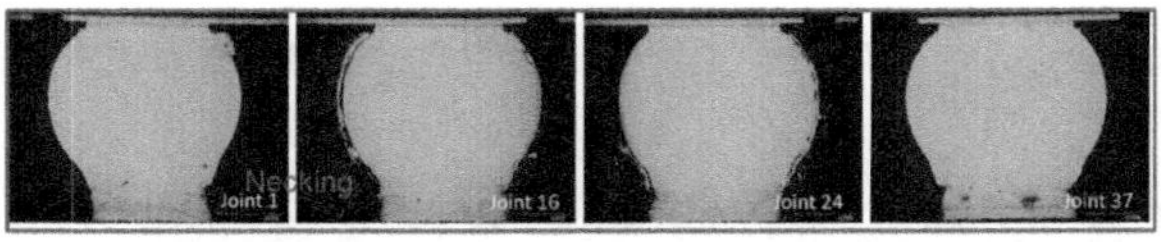

FIGURE 3.29 SAC ball/DFLT joints, P200 joints, PBVR of 0.25; package flat; joint score rank 2.

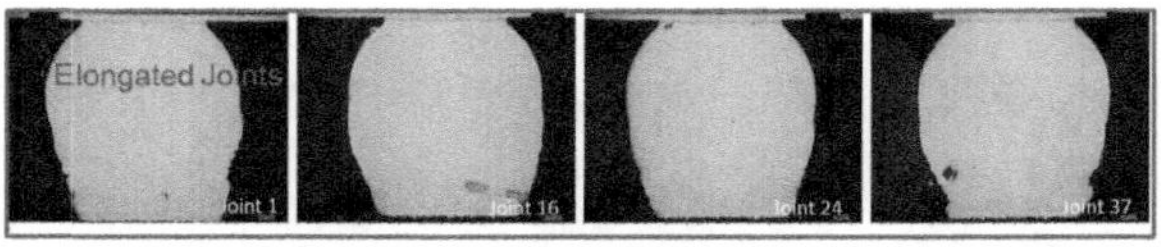

FIGURE 3.30 SAC ball/DFLT, P200 joints, PBVR of 0.5; package flat; joint score rank 3.

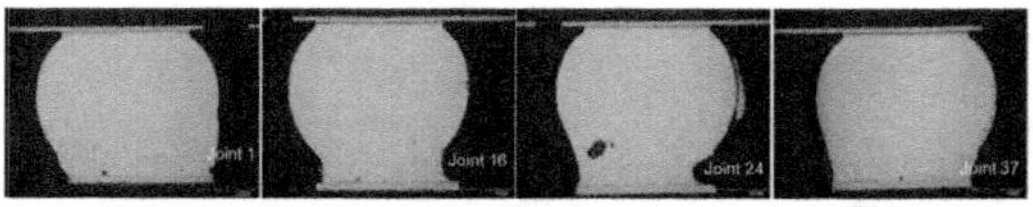

FIGURE 3.31 SAC ball/DFLT joints, P210, PBVR of 0.13; package convex; joint score rank 2.

(Figure 3.31). Increasing the PBVR to 0.25 improved solder joint sphericity and overall integrity across the entire row for both pastes (Figure 3.32). At a PBVR of 0.5, all solder joints appeared fully spherical with no visible defects (Figure 3.33). All packages maintained a convex warpage shape, regardless of PBVR.

Reflow profiles P220 and P240 with a PBVR of 0.13 produced the most favorable results. DFLT solder joints formed short, wide drum-shaped structures with uniform morphology and no observable defects (Figures 3.34 and 3.35). The assemblies exhibited convex warpage, and the difference between the two profiles had minimal effect on the overall joint quality.

Table 3.11 summarizes the package warpage behavior and corresponding solder joint quality scores for each test condition. BiSnAg assemblies predominantly exhibited concave warpage, whereas DFLT assemblies mainly showed convex warpage. Package warpage generally "freezes in" at the temperature where the solder joints solidify. Because solidification occurs below the melting point due to undercooling, BiSnAg—which fully solidifies below 138°C—produces concave warpage in PBGA928. In contrast, DFLT, with a melting point well above the 140°C inversion temperature, tends to solidify in the convex state.

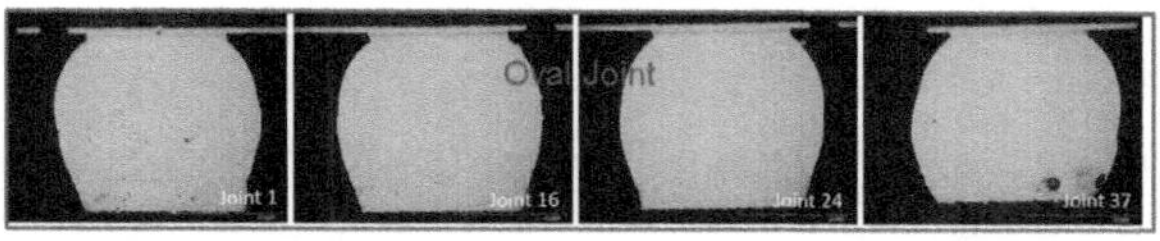

FIGURE 3.32 SAC ball/DFLT joints, P210, PBVR of 0.25; package convex; joint score rank 3.

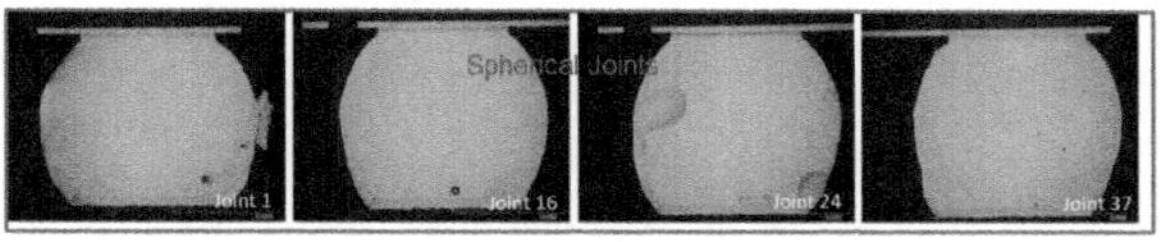

FIGURE 3.33 SAC ball/DFLT joints, P210, PBVR of 0.5; package convex; joint score rank 4.

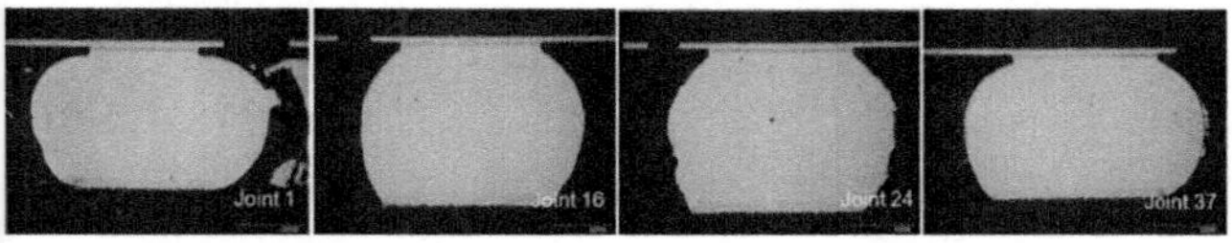

FIGURE 3.34 SAC ball/DFLT joints, P220, PBVR of 0.13; package convex; joint score rank 4.

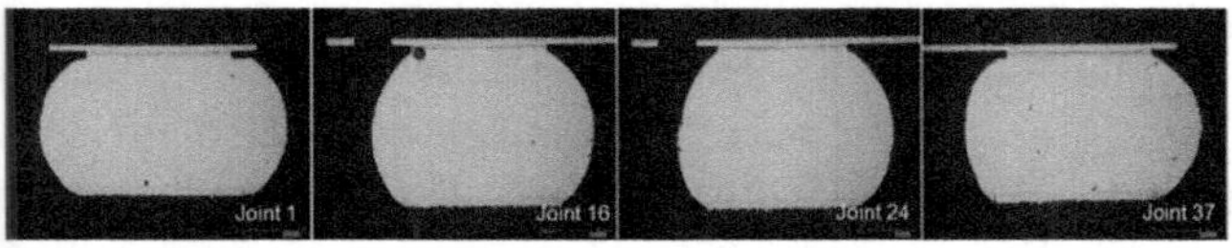

FIGURE 3.35 SAC ball/DFLT joints, P240, PBVR of 0.13; package convex; joint score rank 4

A few exceptions were observed. For example, BiSnAg reflowed at 220°C with a PBVR of 0.13 exhibited convex warpage, while DFLT reflowed at 200°C with PBVRs of 0.25 and 0.5 showed a nearly flat profile. At a 220°C peak, both the SAC balls and solder paste were fully molten; in the BiSnAg case with a PBVR of 0.13, dilution reduced the Bi concentration to below ~4 wt%, raising the solidification temperature well above 140°C and resulting in convex warpage. For DFLT, undercooling could delay final solidification to around the inversion temperature, producing flat warpage.

However, solder joint quality rankings showed no consistent correlation with the warpage type. Poor solder joint scores consistently occurred at lower reflow temperatures (≤200°C), where insufficient SAC melting limited the system's ability to accommodate warpage. Once reflow temperatures reached 220°C or higher, complete SAC melting allowed more effective warpage compensation and improved solder joint robustness. At intermediate conditions around 210°C—where SAC balls only partially melt—higher paste volumes (larger PBVRs) helped mitigate warpage-induced displacement and enhance solder joint integrity, as reflected in Table 3.11.

TABLE 3.11
Summary of Package Warpage Type and Solder Joint Score Rank

Joint Score		Reflow				
Warpage	PBVR	P185	P200	P210	P220	P240
BiSnAg	0.13	1, D	1, B		4, A	
	0.25	1, B	1, B			
	0.5	1, B	1, C		4, B	
DFLT	0.13		1, A	2, A	4, A	4, A
	0.25		2, C	3, A		
	0.5		3, C	4, A		

Code	Wargafe	Joint Score	Description
A	Convex	1	Open/HIP/NWO/Hot-tearing
B	Concave	2	Necking
C	Flat	3	Oval/Elongated
D	Tilt	4	Spherical/Drum

Using the solder joint score rankings from Table 3.11, a data analysis was conducted to determine the factors most strongly influencing solder joint morphology, as shown in Figure 3.36. The results indicate that all three variables—reflow temperature, PBVR, and the solder alloy—have statistically significant effects on the solder joint morphology. Among them, the reflow temperature is the dominant factor, while PBVR and the alloy type exhibit similar levels of influence, both slightly exceeding the significance threshold of 2.228. The interaction between alloy and PBVR falls just below this threshold, indicating a comparatively weaker effect. Overall, the analysis reinforces that higher reflow temperatures—where SAC balls dissolve

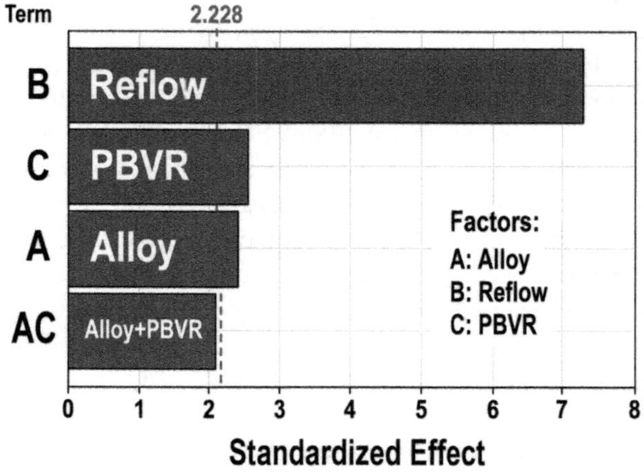

FIGURE 3.36 Response of factors on joint morphology.

extensively or fully melt and thereby better accommodate warpage displacement—are critical for reducing solder joint defects.

BiSnAg solder joints exhibited a higher incidence of hot tearing compared with DFLT. This defect arises when the solder's solidification temperature overlaps with the package's warpage inversion temperature [9]. Undercooling during solidification further reinforces the importance of selecting solder alloys whose melting points lie either well above or well below the inversion temperature, thereby avoiding this critical overlap.

During reflow, partial dissolution or melting of the SAC balls into the lower-melting solder alters the overall solder joint composition, producing a melting behavior that differs from either alloy alone. As shown in Figure 3.37, SEM images of solder joints processed with the same P200 profile and a PBVR of 0.5 reveal pronounced hot tearing within the hybrid region and near the PCB pad in the SAC ball/BiSnAg joint (top). In contrast, the SAC ball/DFLT solder joint (bottom) shows no evidence of hot tearing. These results underscore the importance of selecting solder alloys appropriately to prevent hot tearing and to ensure robust solder joint reliability.

The BiSn phase diagram (Figure 3.38) indicates that as SAC305 continuously dissolves into molten BiSn eutectic solder, the alloy's pasty range initially widens. Only when the Bi concentration is diluted below approximately 21 wt% does the pasty range begin to narrow again. Due to undercooling, solidification initiates at a temperature below the liquidus and completes below the solidus. For SAC ball/

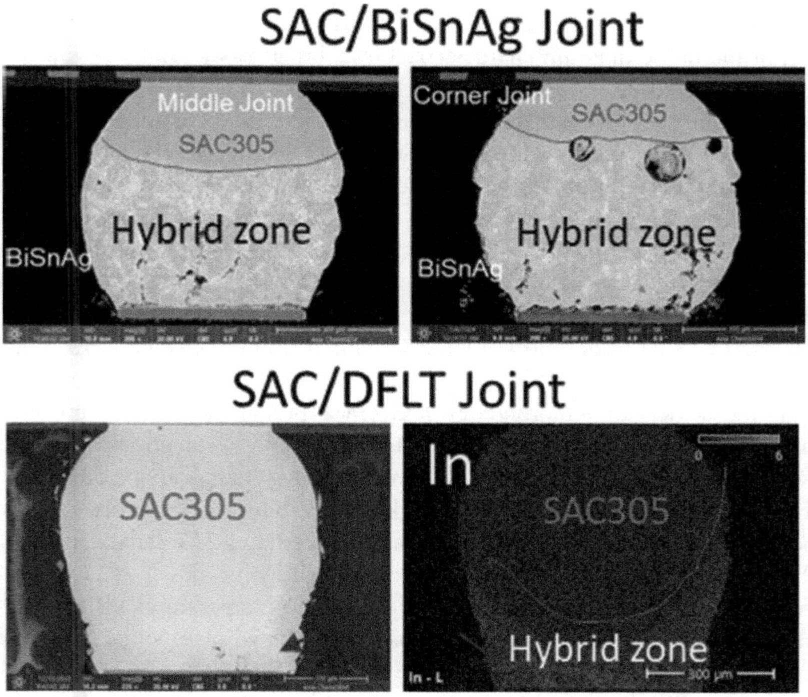

FIGURE 3.37 SEM images of SAC ball/BiSnAg solder joints (a) and SEM and EDX mapping images of SAC ball/DFLT solder joints (b) formed under P200 and a PBVR of 0.5.

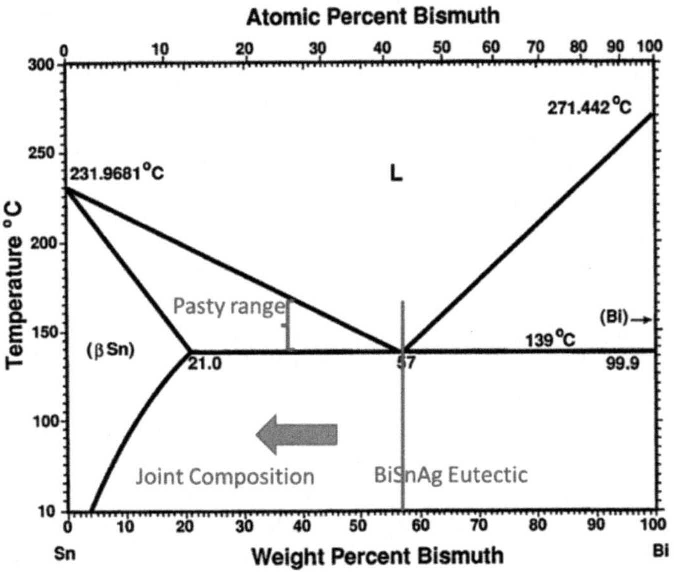

FIGURE 3.38 Binary BiSn phase diagram and the pasty range with composition [123]. (Reused with permission from ASM International. All rights reserved.)

BiSnAg solder joints, the final joint—or hybrid zone—composition lies within the region of the phase diagram characterized by a broad pasty range. This places the solidification window at high risk of overlapping with the package's warpage inversion temperature (~140°C), thereby increasing the likelihood of hot tearing.

In contrast, DFLT starts with a high Sn content (Figure 3.39), and dissolution of the SAC ball during reflow further increases the Sn concentration within the solder joint or hybrid zone. This compositional shift moves the alloy into a region of the phase diagram with a much narrower pasty range. As a result, solidification begins and ends at temperatures safely above the 140°C inversion point, greatly reducing the likelihood of hot tearing.

Thermal cycling tests (–40°C to 125°C with 20-minute dwells) were conducted on selected packages. After 500 cycles, cross-sectional evaluations were performed to examine solder joint morphology and crack propagation. As summarized in Figure 3.40, solder joints were grouped into six crack-ratio categories. Severe cracking was observed only in assemblies reflowed at the lower peak temperatures for both BiSnAg and DFLT. When the reflow peak temperature was increased to 210°C or higher, DFLT solder joints exhibited no cracks exceeding a 25% ratio, demonstrating a substantial improvement in thermal cycling reliability.

Forming defect-free solder joints at low reflow temperatures (P185–P200) is challenging because the limited solder paste volume and insufficient dissolution of SAC305 balls cannot fully accommodate the warpage of the PBGA928 package. Increasing the PBVR can help compensate for this displacement, but doing so also increases the likelihood of solder bridging.

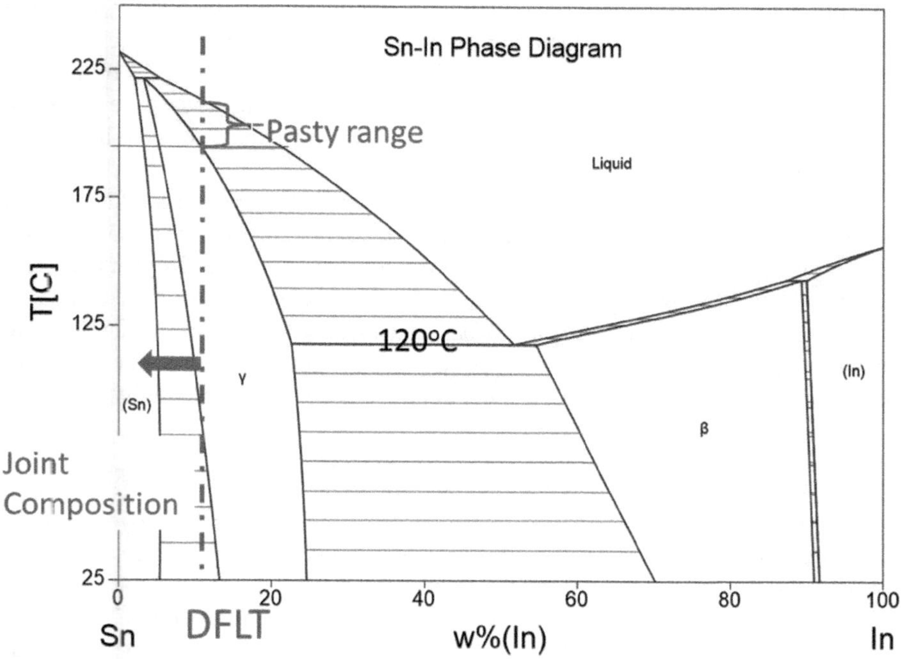

FIGURE 3.39 Binary InSn phase diagram and the pasty range with composition [123]. (Reused with permission from ASM International. All rights reserved.)

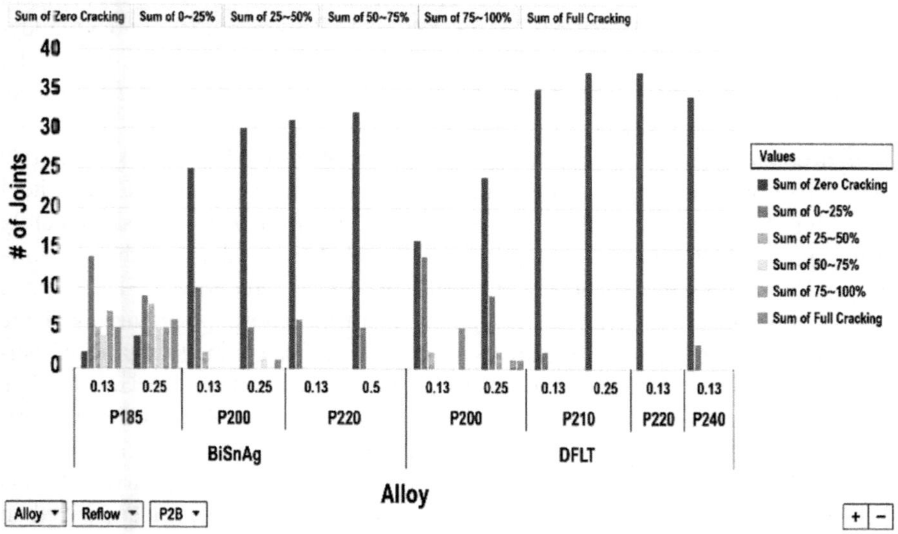

FIGURE 3.40 Number of solder joints with the crack ratio in six categories after a 500-cycle TCT.

Mid- to high-temperature profiles (P210–P240) allow both the solder paste and SAC balls to partially or fully melt, improving solder joint integrity despite warpage. Ensuring adequate solder volume is critical. To prevent hot tearing, solidification must occur either well before or well after the 140°C component warpage inversion point. Overall, higher reflow temperatures promote more robust solder joint formation and improved long-term reliability.

DFLT solder paste has already been validated for low-temperature reflow during the assembly of a GPU module for a data-center server that incorporates a 65×65 mm^2 BGA processor with SAC-Bi solder balls [132]. The process delivered higher manufacturing yield and throughput, and achieved a twofold increase in service life compared with assemblies using SAC balls and other solder pastes. These results demonstrate the advantages of combining low-temperature solder balls with mid-temperature solder (MTS) paste to achieve both excellent manufacturability and reliability.

3.4.3 Summary of Low-Temperature and Mid-Temperature Solders

Low-temperature solders (LTS), primarily based on BiSn and InSn alloys, offer the key advantage of reflowing below 200°C, which reduces thermal stresses, energy consumption, and warpage—especially important for highly miniaturized or heat-sensitive devices. BiSn eutectic alloys, with melting points near 138°C, provide significant process benefits but suffer inherent material limitations: Bi's brittle crystal structure, its high electrical resistivity, and its tendency to coarsen or segregate under current and temperature. These factors degrade drop-shock reliability, increase electromigration risk, and restrict long-term operating temperatures to below ~125°C. InSn eutectic alloys are more ductile and conductive, but their low strength, rapid mechanical degradation above 70°C–80°C, and high indium cost limit their practical use. Extensive iNEMI evaluations [61,66–68,111–122] confirm that although alloying additions such as Ag, Sb, or Cu can improve certain properties, BiSn-based LTS systems fundamentally fall short for applications requiring high shock performance, long lifetime, or high service temperatures.

To overcome these limitations, mid-temperature solders (MTS) such as Durafuse® LT (DFLT) were developed using a mixed-powder Sn-In-Ag + Sn-Ag-Cu architecture. This design enables reflow at 190°C–210°C while delivering a significantly higher post-reflow remelting temperature (189°C–205°C), thus combining low-temperature processing benefits with robust in-service reliability. DFLT exhibits superior mechanical behaviors as compared with BiSn alloys—maintaining stable shear strength at elevated temperatures—and delivers exceptional drop-shock performance, surviving hundreds to thousands of impact cycles where BiSn solder joints fail in only a few drops. Electromigration testing shows that DFLT behaves similarly to SAC305 and avoids the polarity-dependent Bi enrichment that destabilizes BiSn solder joints. In large BGA assemblies, where dynamic warpage drives defects such as HIP, NWO, and hot tearing, DFLT forms far more reliable solder joints than BiSnAg over a broad range of reflow profiles and paste-to-ball ratios. When properly processed, DFLT achieves thermal-cycling performance comparable to or better than SAC305, while providing the manufacturability, energy savings, and warpage mitigation needed for next-generation electronic packaging. Overall, DFLT represents a next-generation lower-temperature solder solution that combines process efficiency, energy savings, and high reliability for advanced electronic assemblies.

3.5 HIGH TEMPERATURE LEAD-FREE BONDING MATERIALS

3.5.1 Overview

Lead-free alternatives to eutectic tin–lead solder have been extensively studied, with Sn–Ag, Sn–Cu, and Sn–Ag–Cu systems emerging as the primary solutions for the electronics industry [17,23,24,26–28,133–136]. In contrast, high-temperature lead-free (HTLF) bonding materials—intended to replace traditional high-lead (Pb>85wt%) solders in die-attach applications for power semiconductor components, modules, and devices—remain in the early stages of development and adoption, despite significant research efforts [13,14,83,136–140].

Die-attach materials for discrete power components must withstand elevated operating conditions and therefore require high-temperature solders or bonding materials capable of preserving joint integrity between the semiconductor die and the lead frame or substrate. Key requirements for die-attach materials in discrete power devices include [13,14,83,136]:

i. a joint softening or remelting temperature of at least 260°C to allow compatibility with subsequent board-level reflows;
ii. reliability at operating temperatures up to 150°C–165°C;
iii. drop-in compatibility with existing high-lead soldering processes;
iv. strong thermal fatigue resistance;
v. electrical and thermal performance comparable to or better than high-lead solders; and
vi. low overall cost of ownership.

Growing demand for power semiconductor modules and devices—particularly high-power, miniaturized designs incorporating SiC or GaN dies—places even greater performance requirements on die-attach materials. These applications demand a higher thermal conductivity, the ability to withstand elevated current and power densities, and reliable operation at junction temperatures approaching or exceeding 200°C. As a result, a broad range of die-attach materials is under active investigation [13,14,83,136–140], including Sn-rich solders, novel HTLF solder systems, transient liquid-phase bonding (TLPB) materials, and sintered materials, all evaluated as potential replacements for high-lead solders.

Figure 3.41 summarizes the available solder systems with melting temperatures above 200°C, as well as transient liquid-phase bonding (TLPB) and sintering materials. High-lead solders (>85 wt%), highlighted in the red box of Figure 3.41, have melting temperatures ranging from 250°C for Pb-5Sb to 310°C for Pb-2Sn. These alloys remain widely used—under RoHS exemption—for die-attachment in discrete power components such as diodes, DrMOS (MOSFET/drivers), PQFN, D^2PAK, and TO packages.

In the 200°C–250°C melting range, tin-rich lead-free solders alloyed with elements such as Ag, Cu, Bi, and Sb dominate. However, these tin-rich alloys are not suitable

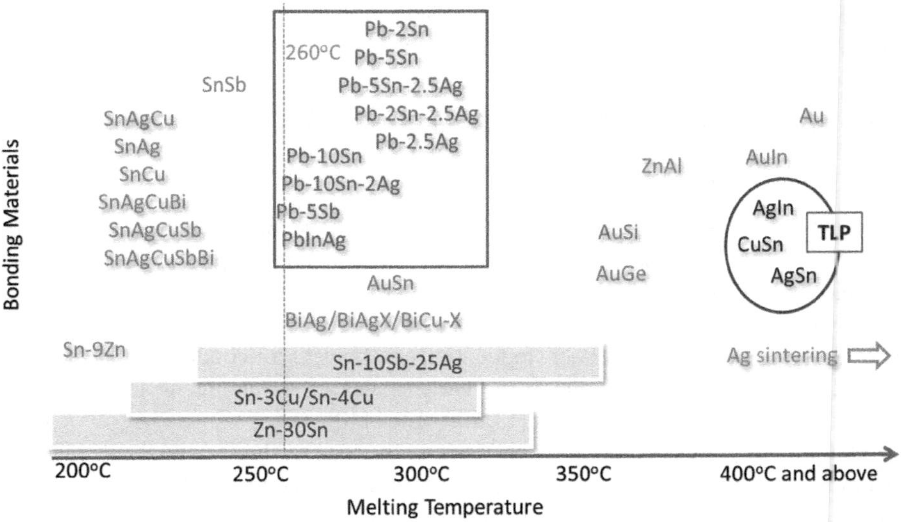

FIGURE 3.41 Summary of die-attach bonding materials [83,136].

for discrete power components that undergo subsequent board-level reflow, as the solder joints would re-melt during typical peak reflow temperatures of 240°C–260°C or higher. They may be acceptable for die-attachment in power devices or modules that do not undergo secondary SMT reflow. Even then, tin-rich solders may struggle to meet long-term reliability requirements at junction temperatures exceeding 175°C, because this range approaches their intrinsic melting temperatures and accelerates joint degradation—both at the interfaces and within the solder bulk.

3.5.2 Candidates of High-Temperature Lead-Free Interconnection Materials

Lead-free alternatives with melting temperatures above 250°C encompass a broad set of materials. BiAg/BiAgX, BiCu/BiCuX, and eutectic AuSn, all melt between 260°C and 280°C. Specifically, BiAg/BiAgX has a solidus near 262°C [123], BiCu/BiCuX near 270°C [123], and Au80/Sn20 near 278°C [123], placing them at the lower end of the high-lead solder temperature range. Tin–antimony alloys with Sb-to-Sn ratios close to 1:1 exhibit solidus temperatures above 310°C—higher even than Pb-2Sn. Other candidates, including Au88/Ge12 [123], Au96.8/Si3.2 [123], and Zn–Al eutectic alloys with 4–6 wt% Al [123], possess melting temperatures above 360°C, well beyond those of high-lead solders.

Beyond solder alloys, both TLPB and sintering materials can be processed between approximately 250°C and 400°C to create die-attach joints in microelectronic packages [14,83,136,141–164]. TLPB systems typically consist of two or more alloy constituents with different melting temperatures. During processing, the low-melting phase liquefies first and reacts with the higher-melting constituents and the surface metallization to form intermetallic compounds, thereby minimizing or eliminating residual low-melting phases.

Sintering materials—traditionally micron-scale silver particles or flakes—have been used for die-attach bonding since the 1980s, generally with the application of pressure [64–66]. Unlike solders or TLPB materials, sintering materials do not melt; instead, joint formation relies on solid-state atomic interdiffusion, which typically requires longer processing times and/or higher temperatures, along with external pressure. The introduction of silver nanoparticles has enabled significant reductions in processing temperature—to around 250°C or even lower—and shortened bonding times to just a few minutes under pressure. Some reports suggest that nano-Ag sintering materials can even be processed under low pressure or completely pressure-free (pressureless sintering) [14,158,160–162].

All potential HTLF bonding materials demonstrate bond shear strengths that are comparable to—or significantly higher than—those of high-lead solders, as shown in Figure 3.42 [161,165–171]. A range of testing vehicles has been used in these evaluations, including (i) SiC die on AMB (active metal brazed) Si_3N_4 substrate, (ii) Si die on Cu substrate, (iii) Si die on DBC (direct-bonded copper) substrate, (iv) Cu die on Cu substrate, and (v) chip capacitors on PCBs.

High-lead solders, such as Pb-10Sn-2Ag, Pb-5Sn-1.5Ag, and Pb-5Sn-2.5Ag, consistently exhibit bond shear strengths of ~30 MPa at room temperature, largely independent of the testing vehicle. When test temperatures increase from 25°C to 250°C, their bond shear strengths drop to ~8 MPa.

In contrast, BiAgX alloys show strong dependence on the test vehicle [167,168, 172–181]. For example, the SiC/AMB package (vehicle a) exhibits a room-temperature shear strength of 54 MPa [167], nearly double the 28.2 MPa measured for the Si/Cu package (vehicle b) [168]. The higher CTE mismatch between Si and Cu likely

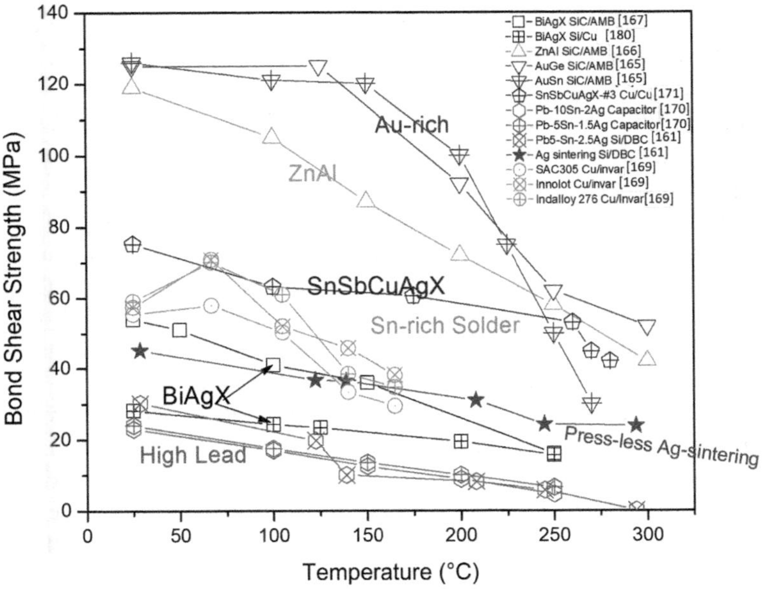

FIGURE 3.42 The bond shear strength of HTLF candidates [171].

generates a greater internal stress in BiAgX joints, reducing their shear strength relative to SiC/AMB. For both vehicles, shear strength decreases linearly with temperature and converges to ~16 MPa at 250°C—corresponding to a homologous temperature of ~0.98 for BiAgX.

Tin-antimony solders are strong and stiff enough that, during shear testing, the silicon die fractures before the bonding material fails [171]. Thus, Cu-die/Cu-substrate assemblies are used when evaluating SnSbCuAgX alloys. These alloys exhibit room-temperature bond shear strengths of ~50–80 MPa, decreasing to ~40 MPa at 260°C–280°C (homologous temperature = 0.92–0.95). Figure 3.42 illustrates the temperature-dependent behavior for a SnSbCuAgX alloy.

Zn-Al eutectic solder [166] provides bond shear strengths near 120 MPa at room temperature (vehicle a: SiC/AMB), comparable to gold-based alloys. Its strength decreases linearly with temperature, reaching ~42 MPa at 300°C (homologous temperature ~0.87).

For comparison, eutectic AuSn and AuGe solders (vehicle a: SiC/AMB) exhibit room-temperature shear strengths above 120 MPa [165]. Their strengths remain high up to 150°C, then decrease nearly linearly to ~50 MPa at 250°C—ranking them among the strongest lead-free candidates tested.

Pressure-less sintered Ag joints between Si dies and DBC substrates exhibit ~45 MPa bond strength at 25°C [161]. With increasing temperature, the strength gradually decreases but remains ~25 MPa even at 300°C.

Conventional solder materials—paste, wire, and preform—are well established for mass production. High-lead solders are available in multiple forms, but not all HTLF candidates share this flexibility. TLPB materials are typically supplied as pastes, as they cannot be fabricated into wires. Composite metal foils (e.g., Sn, Sn alloys, In, or In alloys, laminated onto Ag or Cu cores) have also been evaluated for high-temperature bonding [182]. Silver-sintering materials are commonly paste-based, though film-type formats analogous to conductive adhesive films have been reported [163,164].

Because TLPB uses semi-solid reactions and sintering relies on solid-state diffusion, both diverge significantly from standard reflow processes for high-lead solders. These approaches often require longer processing times, pressure assistance, and vacuum or low-pressure conditions [13,14,141,142,158,161].

Novel HTLF solder candidates are not fully compatible with traditional high-lead reflow profiles. Zn-Al eutectic, for example, is highly reactive and unsuitable for conventional solder pastes due to poor shelf and open-time stability. Its reflow typically requires vacuum, forming gas, or formic acid atmospheres, and even pressure-assisted bonding may be needed to expel oxides and ensure wetting.

This section reviews high-temperature lead-free solder candidates with emphasis on performance, microstructure, and microstructural evolution, with TLPB and sintering materials included for comparison. No single bonding material satisfies all die-attach requirements across processing, reliability, performance, and cost. However, each material may meet the needs of specific niches within the broader landscape of power semiconductor packaging.

3.5.2.1 Gold-Based Solders

Three well-known gold-based eutectics—Au-20Sn (278°C), Au-3.2Si (363°C), and Au-12Ge (361°C)—are widely used for high-temperature joining. Au-20Sn contains the δ-AuSn and ζ-Au_5Sn IMC phases at the eutectic point [123]. Au-Ge and Au-Si consist of (Au) + (Ge) and (Au) + (Si) solid-solution phases, respectively [123].

During soldering, Au-20Sn reacts with the substrate primarily through Sn diffusion, forming Sn-rich IMCs at the interface [123,183–185]. Au-Ge and Au-Si, however, show different interfacial reactions [165,186]:

- On Cu: AuCu or Au_3Cu for Au-Ge solder, and AuSiCu for Au-Si solder.
- On ENIG: Ni_3Ge_5 or NiGe for Au-Ge solder, and AuNiSi for Au-Si solder.

After aging at 300°C for 300 hours, interfacial delamination occurs within or near the IMC layers on Cu for both Au-Ge and Au-Si [20,28], whereas failures are rarely observed on ENIG under the same conditions. Au-20Sn remains widely used due to its good electrical/thermal conductivity and strength below ~200°C, supporting applications such as LEDs, biomedical devices, and down-hole electronics. Its primary limitation is cost, which restricts widespread industrial adoption.

3.5.2.2 Bismuth-Rich Solders

Bismuth-rich alloys have been considered as potential drop-in replacements for high-lead solders [187–200] owing to their melting temperatures and thermomechanical behavior. Bi-Ag has a eutectic composition of Bi-2.6Ag with a eutectic temperature of 262°C [18], yielding a Bi matrix with dispersed Ag-rich particles. Bi-Cu has a slightly higher eutectic temperature of ~270°C at Bi-0.3Cu.

Bismuth and Bi-rich alloys are intrinsically brittle and tend to fail by cleavage, but single-crystal Bi wires show ductility strongly dependent on the angle between the loading direction and the (111) plane: ductility exceeds 100% at small angles but can fall to ~2% at larger angles [86].

To improve ductility, several strategies are used:

1. Alloying soluble elements to strengthen the Bi matrix by hindering dislocation motion.
2. Introducing fine particles in sufficient quantity to impede dislocation movement.
3. Refining the grain size to increase grain-boundary-controlled resistance to sliding.

Bi-11Ag ribbon [83], containing well-dispersed Ag particles, demonstrates a true yield strength of ~52 MPa and true strain >20% (Figure 3.43). The Bi-0.3Cu eutectic composition does not provide adequate strength or ductility due to the low Cu content. Increasing Cu to ≥2 wt% can enhance strength, provided that precipitates are refined, well-dispersed, and stabilized with additional alloying elements. However, according to the phase diagram [23], higher Cu contents significantly raise the liquidus temperature and widen the pasty range, requiring proportionally higher reflow temperatures.

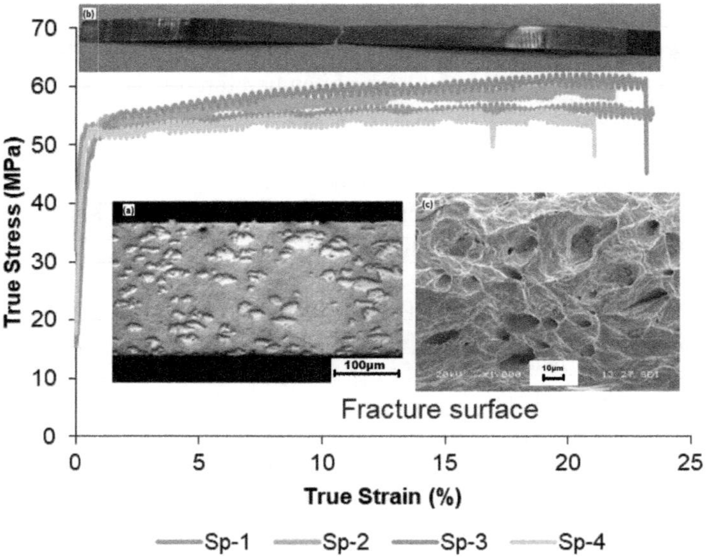

FIGURE 3.43 Mechanical property, microstructure, and fracture morphology of Bi-11Ag ribbon [15].

Both BiAg and BiCu binary alloys exhibit poor wetting on common metallization surfaces such as Cu, Ni, and Ag [190–195], which limits their direct use as solder materials. Attempts to improve wetting by alloying reactive elements such as Sn or In have shown limited benefits: small additions do not sufficiently enhance wetting, while excessive additions produce low-melting phases (BiSn or BiIn) that compromise high-temperature performance.

Wetting performance can be significantly improved through the BiAgX paste design [192–195], in which a secondary solder powder (a Sn-containing alloy) is blended with the primary Bi-Ag powder (Ag content typically 2.6–15 wt%). During soldering, the additive powder melts earlier—or concurrently with—the primary Bi-Ag powder and reacts with the metallization to promote wetting. This additive is engineered to dominate interfacial IMC formation, and after reflow, the active components of the additive powder are fully converted into IMCs. The remaining constituents dissolve into the molten primary alloy, resulting in a homogeneous solder joint.

BiAgX pastes are fully compatible with standard high-lead solder paste processes. Discrete power devices assembled with BiAgX have successfully passed preconditioning (three reflows at 260°C), moisture sensitivity level (MSL) ratings 1–3, 150°C thermal aging, and thermal cycling (−55°C to +150°C). MOSFET R_DS(on) measurements before and after reliability testing show performance comparable to high-lead solder assemblies [180]. Under more severe TCT conditions (−40°C to +200°C), BiAgX maintained strong mechanical performance for SiC/CuMo packages through 2,000 cycles [43]. These results—combined with practical materials cost and compatibility with high-lead processes—position BiAgX as a viable lead-free solution for applications up to ~200°C [172,196–200].

To further increase the remelting temperature to ~270°C, mixed-powder pastes incorporating BiCu or BiCu-X (where X is a hardening element) with appropriate additives have been developed. Early results indicate reliability comparable to BiAgX [171].

Beyond paste form, BiAgX wire and preforms have also been demonstrated [196,197]. Bismuth-silver wires and ribbons can be extruded directly from cast ingots under optimized conditions, then coated with a thin Sn or Sn-alloy layer to enhance wetting. Ribbons can be rolled to the desired thickness and punched into preforms of various geometries.

Although the BiAgX system improves mechanical processability and long-term reliability for discrete power applications, the intrinsically high electrical resistivity of Bi (112 μΩ·cm) results in inferior device performance compared with the high-Pb solder, limiting adoption in applications requiring high efficiency and low R_DS(on).

3.5.2.3 Tin–Antimony High-Temperature Solders

When the antimony (Sb) content in Sn–Sb binary alloys exceeds 43.6 wt%, Sn_3Sb_2 and β(SnSb) compounds form upon solidification at ~324°C, as shown in the phase diagram (Figure 3.44). Sn_3Sb_2 may decompose into β-Sn and β(SnSb) when cooled below 242°C. Alloying Sn–Sb with Cu and Ag not only strengthens the SnSb matrix but also reacts with any remaining β-Sn to form Cu_6Sn_5 and Ag_3Sn, eliminating low-melting Sn-rich phases.

A SnSb-based high-temperature alloy (SnSbAgCu) [15] exhibiting a melting onset of 326°C and a final melting point of 365°C has been developed. Preliminary studies revealed that this alloy can fracture thick dies (3.175 × 3.175 × 0.71 mm^3, Ti/Ni/Au finish) under bond-shear loading or during thermal shock testing (−55°C to +155°C), as shown in Figure 3.45. However, no die damage was observed in thin Si dies (2.45 × 2.75 × 0.2 mm^3) on Cu lead-frame packages under identical TST conditions.

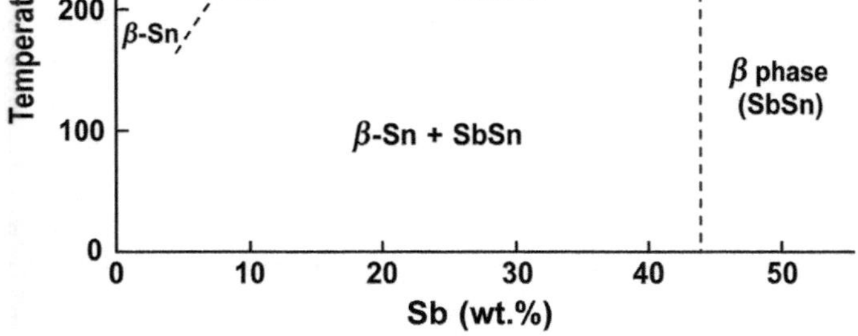

FIGURE 3.44 Sn-Sb phase diagram [123]. (Reused with permission from ASM International. All rights reserved.)

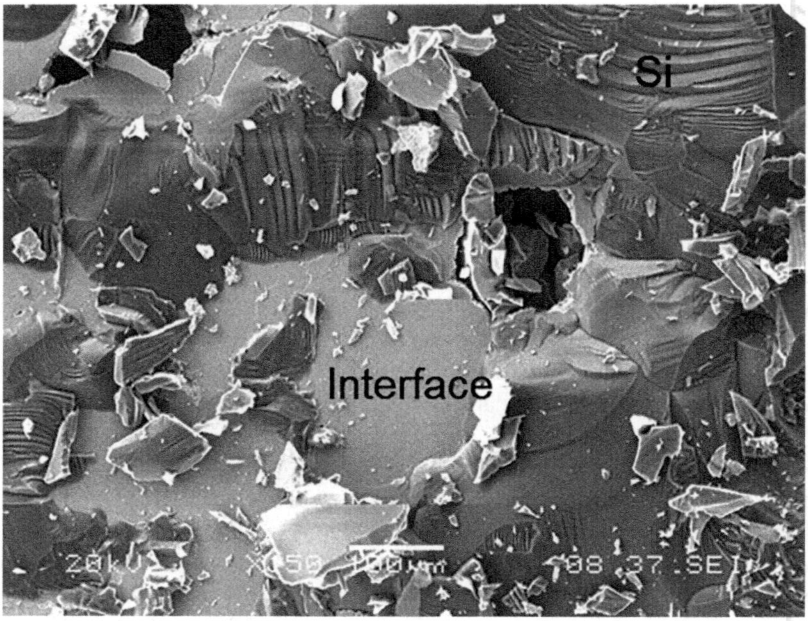

FIGURE 3.45 Die shattering after bond shear testing for Si/SiN DBC package [15]. (Reused with permission from John Wiley and Sons. All rights reserved.)

These observations motivated the development of Durafuse® HT (DFHT), a Sn-based high-temperature, lead-free solder paste employing a mixed-powder architecture designed to:

- serve as a drop-in replacement for high-lead pastes,
- improve both thermo-mechanical and electrical performance, and
- avoid the drawbacks associated with monolithic SnSbAgCu systems.

DFHT pastes are engineered by combining the advantages of a SnSbAgCu high-temperature alloy (with a melting onset of 326°C) and a Sn-rich SnAgCu solder alloy (Sn > 85 wt%, solidus > 217°C). In the final joint, the microstructure is expected to be dominated by a SnSb matrix strengthened by Ag_3Sn and Cu_6Sn_5 precipitates. Any residual Sn-rich regions are anticipated to remain isolated within the surrounding SnSb matrix. Because Sn intrinsically offers a higher thermal conductivity and lower electrical resistivity than Pb, this joint architecture is expected to outperform high-Pb solders in both thermal and electrical performance.

The SnSbAgCu high-temperature alloy [15], whose phase diagram is shown in Figure 3.45, exhibits a solidus of 326°C and a liquidus near 365°C—about 30°C–40°C higher than traditional high-Pb solders—requiring correspondingly higher reflow temperatures. Although SnSbAgCu joints provide high strength and rigidity, the elevated reflow temperature and stiffness can cause fracture of large or thick Si dies during soldering [15,83,201–205]. In contrast, Sn-rich phases are more ductile and melt at lower temperatures, enabling them to absorb residual stresses generated during solidification. Thus, combining high-melting SnSb-based alloys with lower-melting ductile Sn-rich alloys provides a means to reduce the reflow peak temperature and increase joint ductility, lowering the risk of Si die breakage.

However, the presence of Sn-rich phases in the final joint must be carefully controlled: excessive Sn-rich material may partially re-melt during subsequent SMT reflow at ~260°C, leading to solder squeeze-out or even joint delamination. Because high-Pb solders have historically performed well in die-attach and clip-bond applications—surviving multiple SMT reflows without squeeze-out—DFHT joints must be designed to deliver equal or greater mechanical robustness. After proof-of-concept trials and multiple screening iterations, two formulations—HTLF-1 and HTLF-2 out of DFHT design —were selected as drop-in paste solutions that not only withstand SMT reflow but also outperform high-Pb solders in R_DS(on) for the tested power discrete devices.

Figure 3.46 presents the DSC curves for HTLF-1 and HTLF-2 alongside the SnSbAgCu alloy. The SnSbAgCu alloy shows a single high-temperature melting peak spanning 326°C–365°C. In contrast, both HTLF-1 and HTLF-2 exhibit two distinct endothermic peaks (P1 and P2), corresponding to the melting of the low-melting Sn-rich phase and the high-melting SnSb phase, respectively. The mixed-powder formulation significantly alters the melting behavior, shifting part of the melting range to lower temperatures and enabling lower reflow peaks compatible with drop-in high-lead process conditions.

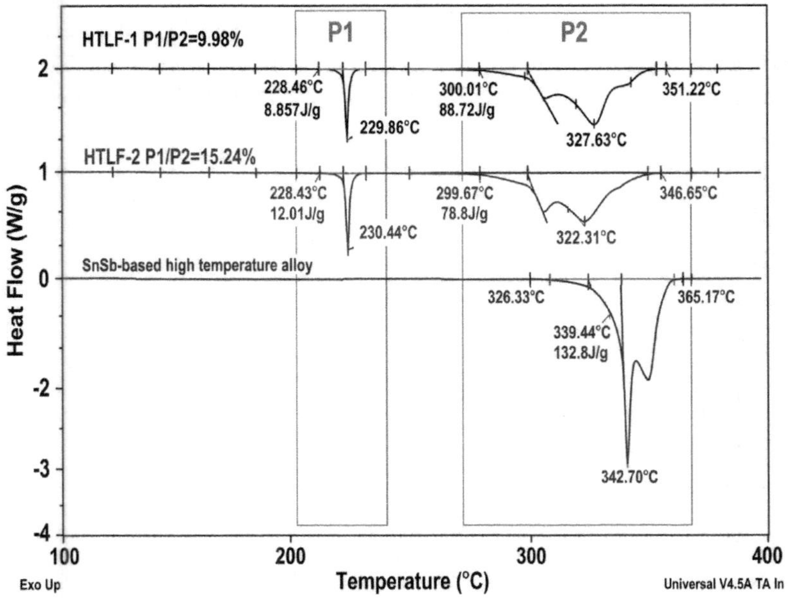

FIGURE 3.46 DSC of HTLF-1, HTLF-2, and SnSb-based high-temperature (SnSbAgCu) alloys. HTLF-2 has a lower melting Sn-rich phase than HTLF-1.

The areas under P1 and P2 reflect the relative heat absorbed by each melting process. The P1/P2 ratio is used to quantify the fraction of low-melting Sn-rich phase in the final joint. HTLF-1 shows a P1/P2 ratio of ~10%, whereas HTLF-2 exhibits a ratio of ~15%, indicating that HTLF-2 contains substantially more Sn-rich material than HTLF-1.

Bond shear testing was performed on Cu/Cu joints reflowed using a traditional high-temperature high-Pb profile (360°C peak). The commonly used high-Pb alloy Pb92.5/Sn5/Ag2.5 showed a significant reduction in shear strength, decreasing from ~30 MPa at room temperature to ~5 MPa at 290°C. In contrast, HTLF-1 consistently maintained shear strengths above 15 MPa from room temperature through 295°C, as shown in Figure 3.47—meeting the Die-Attach 5 (DA5) Consortium's minimum requirement of 15 MPa and delivering roughly three times the high-temperature strength of Pb92.5/Sn5/Ag2.5.

HTLF-2 exhibited room-temperature shear strength near 60 MPa but declined to ~5 MPaat295°C,makingitshigh-temperatureperformanceonlymarginallycomparable to Pb92.5/Sn5/Ag2.5. For reference, the monolithic SnSb-based high-temperature alloy-maintained shear strengths above 40 MPa even at ~280°C.

These results demonstrate that blending a high-melting SnSb alloy with a low-melting, ductile Sn-rich alloy reduces high-temperature strength but significantly increases ductility—an outcome confirmed by the flattened peaks observed in die shear–displacement curves. Notably, no large-die fracture was observed either after reflow or following thermal cycling testing (TCT) [202–205], further validating the role of the ductile Sn-rich phase in mitigating die-shattering risks.

Because the HTLF pastes contain a low-melting Sn-rich phase, there is a potential risk of joint degradation when power discrete devices are later soldered onto PCBs using traditional SMT reflow. The design strategy therefore relies on maintaining sufficiently high joint strength at elevated temperatures to minimize or prevent such damage. HTLF-1 and HTLF-2 were selected based on their high-temperature

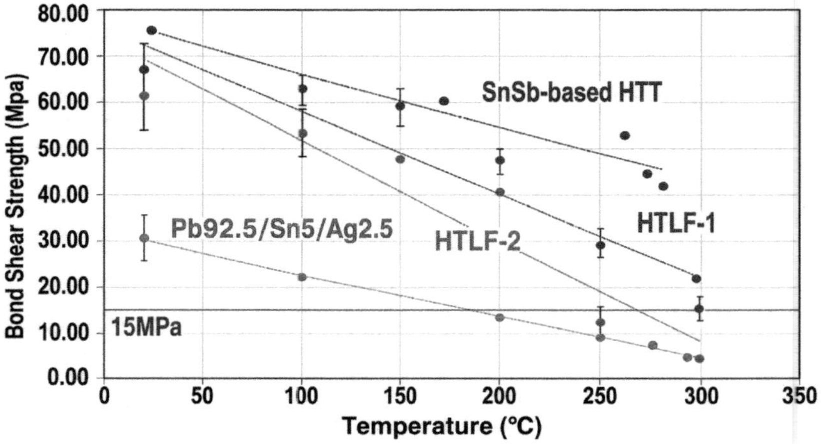

FIGURE 3.47 Bond shear strength of HTLF pastes: HTLF-1, HTLF-2, and Pb92.5/Sn5/Ag2.5.

shear performance (up to 295°C). HTLF-1 retained shear strength above 15 MPa at 295°C—significantly higher than the high-Pb solder under the same conditions—whereas HTLF-2 maintained shear strength only comparable to high-Pb at that temperature.

Two encapsulated device types were evaluated:

1. A diode-type package, consisting of two 2.5 × 1.6 mm² Ti/Ni/Ag-finished Si dies mounted on separate Cu-leadframe pads, with a Cu clip bridging the die tops to the leads.
2. A TO-leadless MOSFET package, using a 6.8 × 4.7 × 0.1 mm³ Si die with Ti/Ni/Ag backside metallization (0.1 µm Ti/0.2 µm Ni/1 µm Ag) and Ti/NiP/Au frontside metallization (0.2 µm Ti/3 µm NiP/0.05 µm Au), bonded to the bottom Cu-leadframe and two top Cu clips.

During assembly, both HTLF-1 and HTLF-2 pastes, along with Pb92.5/Sn5/Ag2.5, were processed using the same high-Pb manufacturing flow (reflow → cleaning → wire bonding → molding). The reflow peak temperature was 385°C, the long-established production condition used successfully for high-Pb soldered components in the same facility.

Voiding levels, measured via X-ray inspection (Figure 3.48), showed that both HTLF pastes and Pb92.5/Sn5/Ag2.5 exhibited comparable and acceptable voiding (<10%). The ability of HTLF-1 and HTLF-2 to meet voiding specifications under the standard high-Pb process confirms their drop-in compatibility with the existing manufacturing infrastructure.

Using a 395°C peak reflow profile—identical to the current high-Pb process and conducted under a forming gas atmosphere—HTLF-1 was successfully employed to assemble a TOLL MOSFET package, whose Si die area is roughly eight times larger than that used in the diode package. Post-reflow voiding was evaluated for both the high-Pb solder and HTLF-1, with results summarized in Figure 3.49. The high-Pb solder exhibited more favorable voiding performance, with an average voiding level of 1.1% and a maximum single void of 0.25%, compared with HTLF-1, which showed an average voiding level of 2.6% and a maximum single void of 0.4%. It is important

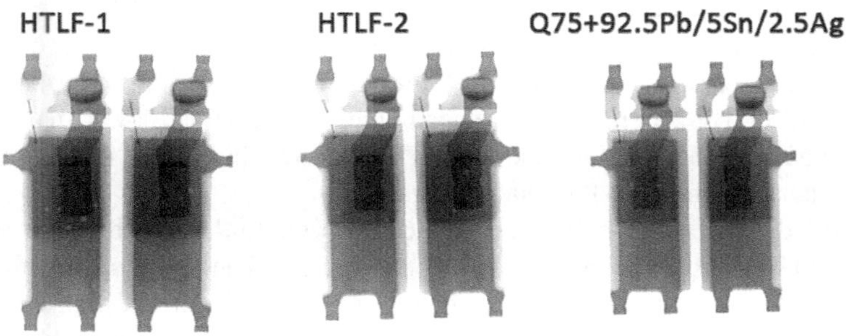

FIGURE 3.48 Voiding performance of both HTLF pastes (HTLF-1 and HTLF-2) and Pb92.5/Sn5/Ag2.5, under the same high-Pb reflow profile with a peak temperature of 385°C, for a diode package and TOLL component.

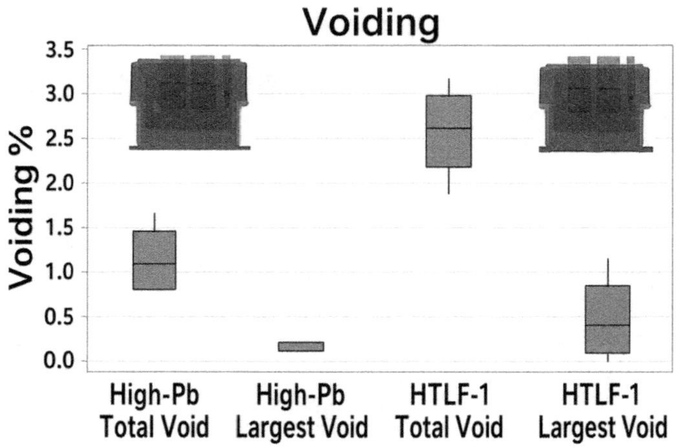

FIGURE 3.49 Voiding performance of solder samples showing the average voids and the largest single void for both Hi-Pb and HTLF-1.

to note that the reflow profile was optimized for high-Pb paste and not yet tuned for HTLF-1. Nevertheless, both materials satisfied the voiding requirement of <10%.

Moisture-sensitivity level (MSL) testing was performed in accordance with IPC/JEDEC J-STD-020D [206]. Both HTLF-1 and HTLF-2 successfully passed MSL-1 for the diode and TOLL MOSFET packages. Pre-conditioning consisted of three SMT reflow cycles at a peak temperature of 260°C. Post-test ultrasonic scanning revealed no internal delamination in any component. Measurements of R_DS(on) showed only minimal change before and after the three reflow cycles, confirming that the joints—whose high-temperature strength meets or exceeds that of high-Pb solder—maintained their integrity throughout the process. A slight increase in the interfacial IMC thickness due to partial re-melting may explain the small R_DS(on) variation observed. Overall, both HTLF pastes exhibited superior R_DS(on) performance compared with Pb92.5/Sn5/Ag2.5, as will be shown later.

Following the MSL testing, temperature cycling tests (TCT) from –55°C to 175°C were conducted on the diode package following the AEC-Q101 protocol [207]. Electrical performance was monitored throughout via R_DS(on), defined as the drain-to-source resistance under operating power. R_DS(on) was recorded both before and after cycling. As shown in Figure 3.50, the R_DS(on) distributions for HTLF-1 and HTLF-2 were similar to each other and remained lower than those of Pb92.5/Sn5/Ag2.5 both before and after 1,000 cycles, indicating that both HTLF pastes outperform the high-Pb solder electrically.

HTLF-1 displayed a distribution shape comparable to that of Pb92.5/Sn5/Ag2.5, whereas HTLF-2 exhibited a wider distribution range. The cause of this difference is not yet fully understood, though compatibility with the current high-Pb process is suspected—suggesting HTLF-2 may benefit from additional process optimization to narrow its distribution. It was further observed that the performance gap in R_DS(on) between the HTLF pastes and Pb92.5/Sn5/Ag2.5 decreased after 1,000 TCT cycles, as shown in Figure 3.50.

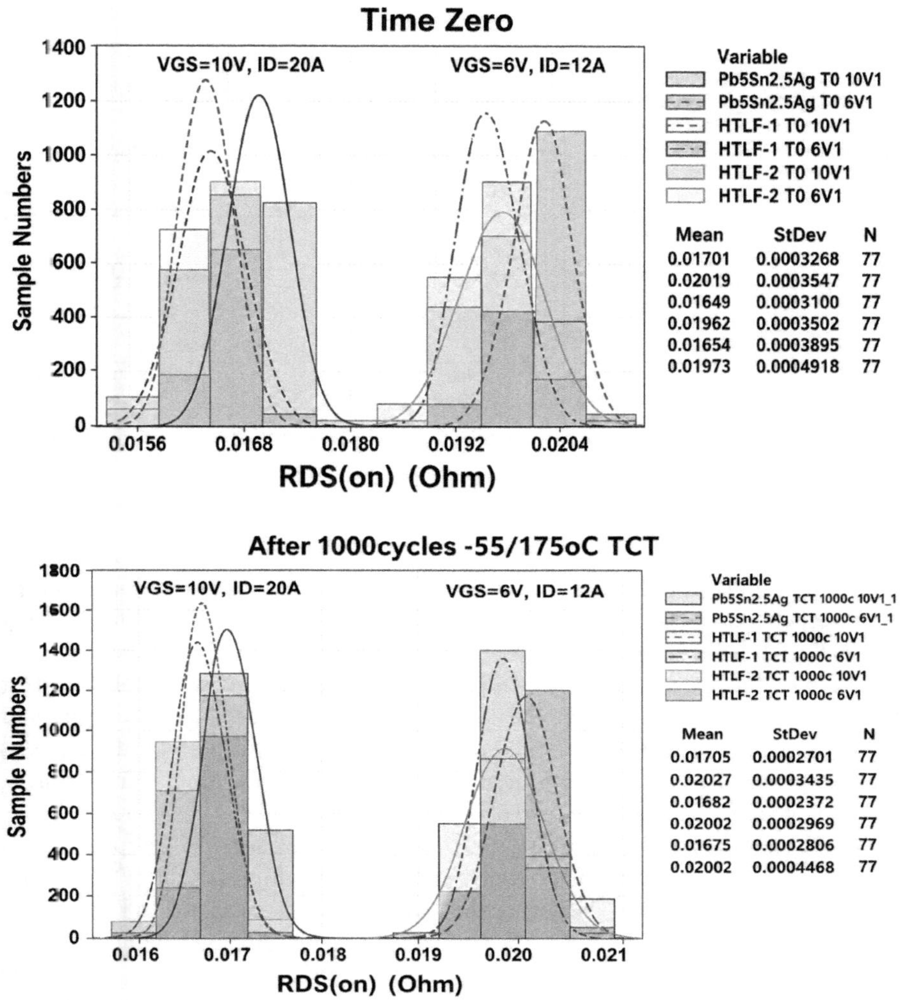

FIGURE 3.50 RDS(on) distribution before and after 1,000 cycles of TCT −55°C/175°C for the diode package built with Pb92.5/Sn5/Ag2.5, HTLF-1, and HTLF-2.

To better illustrate the trend in R_DS(on) during thermal cycling, Figure 3.51 summarizes the change in R_DS(on) at 200-cycle intervals for HTLF-1, HTLF-2, and Pb92.5/Sn5/Ag2.5. Three observations can be made. First, both HTLF pastes consistently outperformed Pb92.5/Sn5/Ag2.5 throughout the entire TCT sequence, confirming that the inherently lower electrical resistivity of Sn-based alloys provides measurable performance benefits. Second, HTLF-1 exhibited slightly better R_DS(on) stability than HTLF-2; additional process optimization may help narrow the gap between the two pastes. Third, the difference in R_DS(on) between the HTLF pastes and the high-Pb solder diminished after 1,000 TCT cycles.

Following 1,000 cycles, components were removed from the chamber, cross-sectioned, and examined microstructurally. As shown in Figure 3.52, neither Pb92.5/

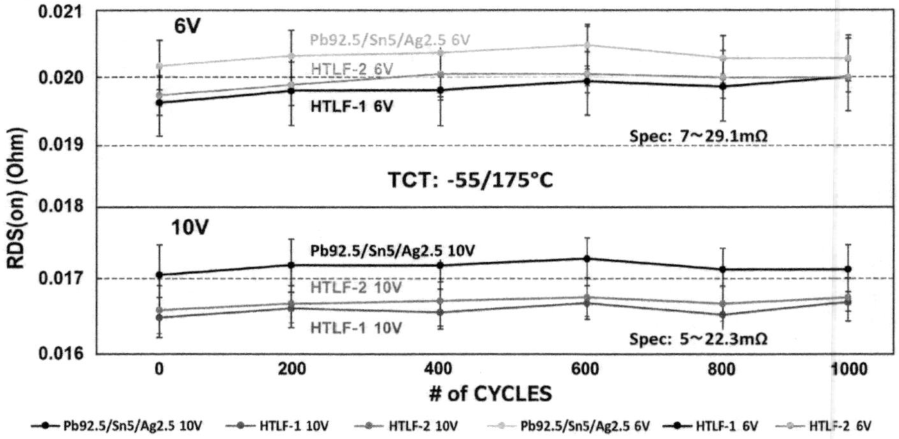

FIGURE 3.51 Evolution of R_DS(on) of the diode package built with Pb92.5/Sn5/Ag2.5, HTLF-1, and HTLF-2 during TCT.

Sn5/Ag2.5 nor the HTLF pastes exhibited corner cracks or delamination in either the die-attach (die-to-leadframe) or clip-bonded (clip-to-die-top) joints. Both HTLF-1 and HTLF-2 are Sn-rich solders, whereas Pb92.5/Sn5/Ag2.5 contains only 5 wt% Sn. Because Sn governs interfacial IMC formation, thicker IMCs were observed for the HTLF pastes compared with the high-Pb solder—an expected result given their higher Sn content. Consequently, the observed R_DS(on) shifts in all three pastes are likely linked to microstructural evolution and interfacial IMC growth during thermal cycling.

R_DS(on) data for the TOLL MOSFET were collected from 82 samples each at (i) time zero, (ii) after pre-conditioning, and (iii) through up to 2,000 TCT cycles for HTLF-1 and 1,000 TCT cycles for the high-Pb control, as shown in Figure 3.53. Three production lots of HTLF-1 were evaluated and compared with one lot of high-Pb solder. R_DS(on) was measured under two operating conditions: (i) 6 V/20 A, and (ii) 10 V/20 A. All samples from all lots completed testing successfully, and the results presented in Figure 3.53 show that every device remained within specification. The performance difference between HTLF-1 and high-Pb solder was minimal, indicating that HTLF-1 delivers comparable electrical behaviors across TCT and pre-conditioning stresses.

Figure 3.54 summarizes the average R_DS(on) values for all three HTLF-1 lots through 2,000 TCT cycles and for the high-Pb lot through 1,000 cycles. The three HTLF-1 lots exhibit substantial overlap, demonstrating strong production repeatability. Under the 6 V/20 A condition, the difference in R_DS(on) between HTLF-1 and high-Pb is approximately 1.5%. Under the 10 V/20 A condition, the difference increases slightly to just above 3%. For both HTLF-1 and high-Pb, R_DS(on) drift remained below 3% after thermal cycling under both test conditions.

Figure 3.55 shows the joint microstructures of the high-Pb and HTLF-1 assemblies. The high-Pb joints exhibit both a thinner bond-line thickness (BLT) and a thinner interfacial IMC layer compared with HTLF-1. No cracks were observed in either

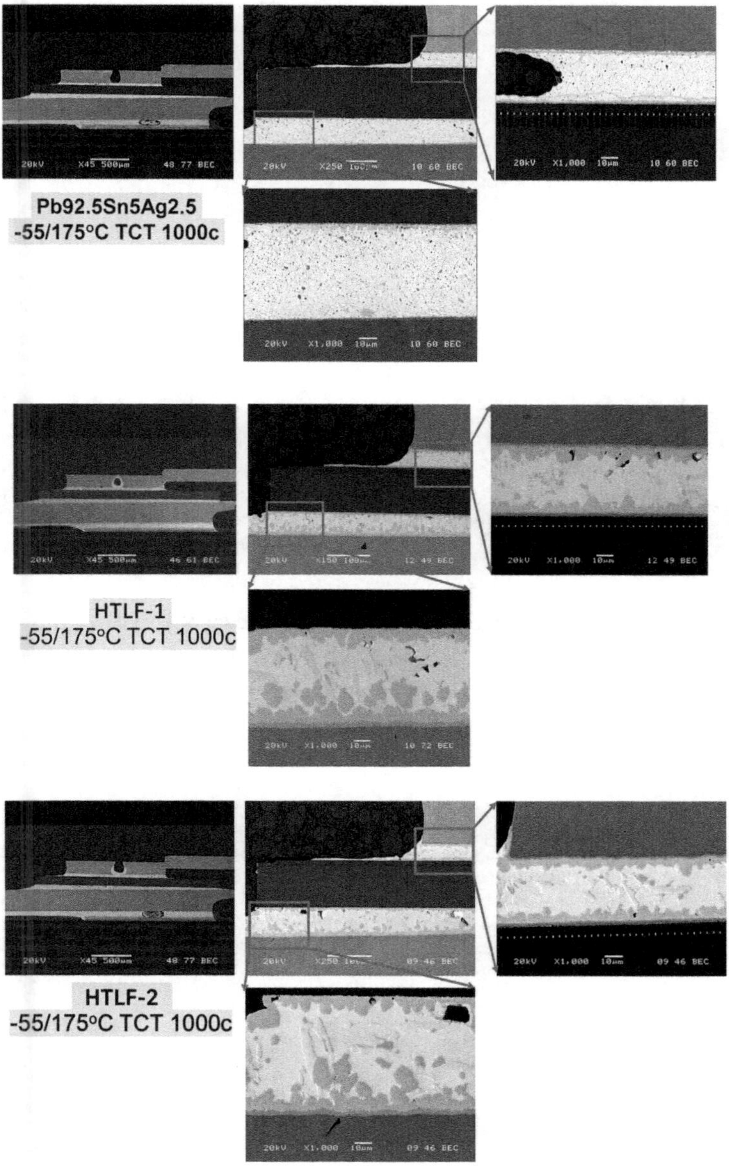

FIGURE 3.52 Joint morphology of the diode package after 1,000 cycles of TCT (–55°C/175°C) from Pb92.5/Sn5/Ag2.5, HTLF-1, and HTLF-2.

joint under the current test conditions, indicating that cracking is not a contributing factor to the observed R_DS(on) differences.

Instead, the thicker BLT and thicker interfacial IMC in the HTLF-1 joints are believed to account for their slightly inferior R_DS(on) performance relative to high-Pb, despite HTLF-1 possessing an inherently lower electrical resistivity and higher thermal conductivity. The initial BLT and IMC thickness primarily influence

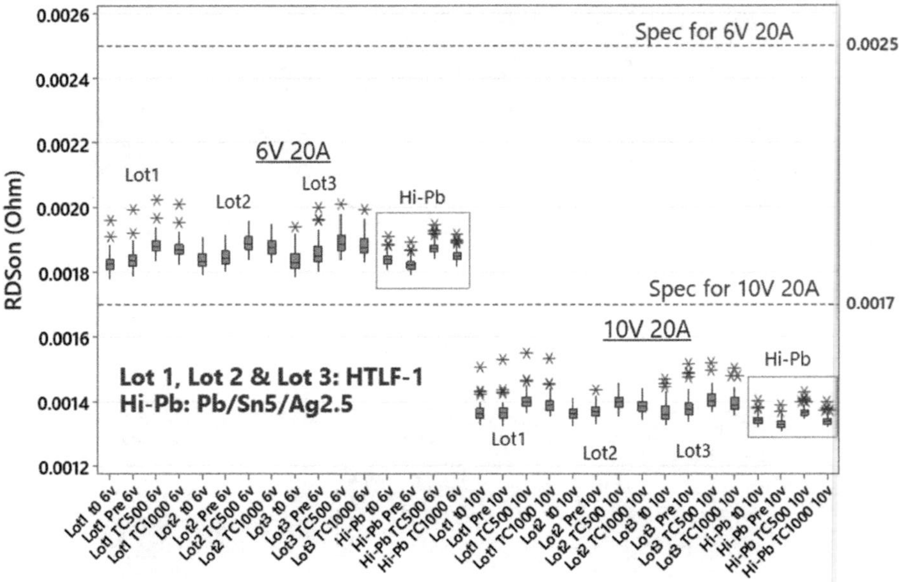

FIGURE 3.53 RDS(on) for three lots of HTLF-1 and one lot of Hi-Pb.

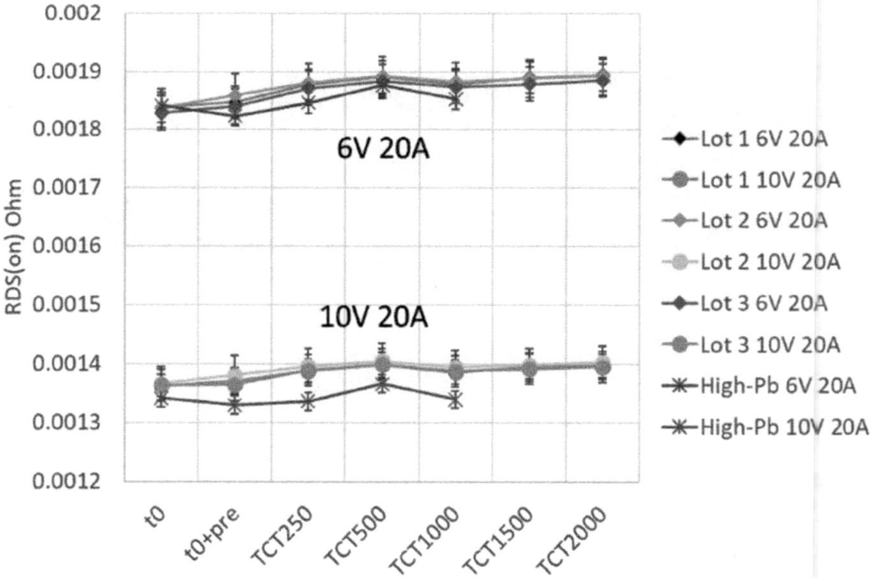

FIGURE 3.54 Summary of the average RDS(on) from time zero, pre-con, TCT up to 2,000 cycles for HTLF-1 and 1,000 cycles for Hi-Pb.

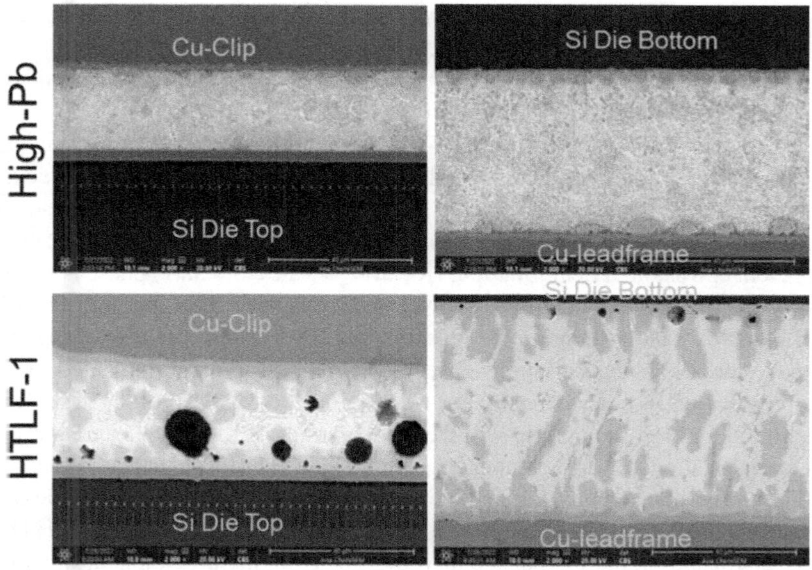

FIGURE 3.55 The joint microstructure of high-Pb and HTLF-1 for TOLL MOSFET packages.

the time-zero R_DS(on), while subsequent IMC growth during pre-conditioning and TCT contributes to the evolution of R_DS(on) over time.

To reduce the bond-line thickness (BLT) and die tilt—which directly affect R_DS(on)—multiple process variables must be optimized, including the paste volume, metal loading, flux chemistry, and reflow profile. For interfacial IMC control, a "cold" reflow profile with a lower peak temperature and a shorter time above liquidus (TAL) can suppress IMC growth during soldering. However, because HTLF-1 contains a substantial amount of Sn, IMCs continue to grow during the subsequent pre-conditioning and TCT. Notably, the R_DS(on) data for HTLF-1 stabilized after approximately 250 TCT cycles, suggesting that the IMC growth slows once a critical thickness is reached—consistent with the much slower kinetics of solid-state IMC growth compared to liquid-state formation during reflow. A more detailed investigation is required to fully understand the IMC evolution in HTLF-1 joints.

Unlike high-Pb solder, HTLF-1 contains more than 50 wt% Sn, raising a key concern regarding minimum metallization thickness—particularly the solderable layer on the Si die surface. For reliable soldering, the die metallization typically consists of an adhesion layer (Ti), a solderable layer (Ni or NiV), and a protective layer (Ag or Au). The solderable Ni/NiV layer must be thick enough to allow formation of a continuous IMC layer during soldering. Full consumption of this layer by Sn-rich solders can cause IMC spalling from the underlying Ti adhesion layer. Because Ti is poorly wettable, exposure leads to interfacial dewetting and severe voiding. This issue does not occur in high-Pb solders because Pb does not react with Ni or Ag, and the limited Sn content prevents complete Ni consumption.

A study demonstrated that the Ni/NiV solderable layer must exceed ~0.35 μm to avoid consumption during Sn-rich soldering [208]. Increasing the thickness to

~0.5 µm provides margins for reflow optimization while reducing risks of dewetting, BLT instability, and die tilt. In the TOLL package evaluated here, the Si die used a 0.2 µm Ni + 1 µm Ag stack; the thick Ag layer functioned effectively as a solderable layer and protected the Ni layer, preventing interfacial degradation.

HTLF-1, designed using the mixed-powder Durafuse® architecture, maintains high-temperature shear strength (~15 MPa at 290°C) [202–205], roughly three times higher than that of the high-Pb solder. This strength is attributed to the high-melting SnSb phases in the joint, which help resist internal stresses during pre-conditioning and prevent extrusion or delamination. However, because the SnSb phase is rigid and not very ductile, HTLF-1 exhibits a lower peel resistance than the high-Pb solder. Under complex loading conditions (tension + shear + tearing), the risk of delamination may increase. Yield variations observed in mass production indicated that further optimization is needed, including (i) refining the reflow profile, (ii) selecting fluxes suited for high-temperature Pb-free processes, and (iii) ensuring compatibility of die metallization and package design with Sn-rich HTLF soldering.

HTLF-1 also showed preliminary success in a power module (Si/AMB) die-attach application [209]. While multiple tests have demonstrated the feasibility of HTLF-1, several challenges remain—particularly the need for metallization designs compatible with Sn-rich solders. For Si die soldering, typical Ti/Ni/Ag stacks used for high-Pb soldering may be too thin for Sn-rich alloys. With the high-Pb solder, Sn from the alloy reacts with Ni and Ag to form IMCs, but because the Sn content is low, Ni and Ag layers as thin as 200 nm are sufficient and do not become fully consumed. Pb, being soft, also absorbs strain, reducing the interfacial failure risk.

In contrast, Sn-rich HTLF solders continuously react with Ni and Ag; if these layers are too thin, they can be completely consumed, causing the IMC layer to detach from the Ti adhesion layer and resulting in dewetting and void formation. For robust reliability, thicker metallization is recommended. Practical experience has shown that Si dies with 500 nm Ni + 200 nm Ag exhibit excellent compatibility with the HTLF solder. Even 350 nm Ni + 200 nm Ag can be viable when used with "cold" and/or shortened reflow profiles. Work is ongoing to evaluate these effects in detail to guide industry-wide transitions from high-Pb to high-temperature Pb-free solutions.

In summary, Sn-based HTLF pastes developed using mixed-powder (Durafuse®) technology were prepared and evaluated as drop-in replacements for high-Pb pastes in power discrete applications. These HTLF pastes were successfully processed using existing high-Pb manufacturing lines. Voiding and MSL performance were comparable to Pb92.5/Sn5/Ag2.5, and no cracks or delamination were observed in die-attach or clip joints after TCT. Electrical performance, measured through R_DS(on), was substantially better—or at least equivalent—to high-Pb solder across pre-conditioning and TCT. These results validate the Durafuse® HT design approach: both HTLF-1 and HTLF-2 function as viable drop-in replacements for high-Pb solder pastes, delivering superior performance while maintaining compatibility with existing production processes. Overall, the Durafuse® HT platform provides new, practical, Pb-free solutions for power discrete packaging, with clear advantages in reliability and electrical performance compared with traditional high-Pb solders.

3.5.2.4 Zinc-Aluminum Solders

Zinc–aluminum eutectic alloys (Zn–(4–6 wt%)Al) have a melting temperature of approximately 382°C, which is higher than that of AuSi and AuGe eutectics. At room temperature, the Zn–Al eutectic consists of two phases: an η-(Zn) matrix and α-(Al) particles. Around 275°C, the α phase undergoes a solid-state transformation into the β phase. Zn–Al and Zn–Al–Cu alloys have been used for decades as bearing-support materials, where hard, lamellar α phases are dispersed within the softer Zn matrix [15], as illustrated in Figure 3.56.

Zinc–aluminum eutectic alloys exhibit yield strengths exceeding 100 MPa—approximately three times that of traditional high-lead solders. They also offer excellent thermal conductivity (~100 W/m·K), outperforming both Au–Sn eutectic (~60 W/m·K) and high-lead solders (~26 W/m·K) [210,211]. In terms of materials cost, Zn–Al alloys are comparable to or even less expensive than high-Pb solders. However, because both Zn and Al are highly reactive metals, formulating a stable solder paste with an acceptable storage life and working life is extremely challenging. As a result, Zn–Al is generally limited to wire or preform formats when used as a solder material.

Soldering with Zn–Al requires a controlled atmosphere—vacuum, inert gas, forming gas, or formic acid—to mitigate oxidation. The oxide crust (ZnO and Al_2O_3) inhibits molten flow and severely degrades wetting and interfacial reactions. Breaking this oxide crust typically requires applied pressure during reflow [18,19]. Zinc governs the interfacial reactions, producing complex IMC structures depending on the metallization. For example, soldering Zn–4 wt% Al onto Cu forms two stacked IMC

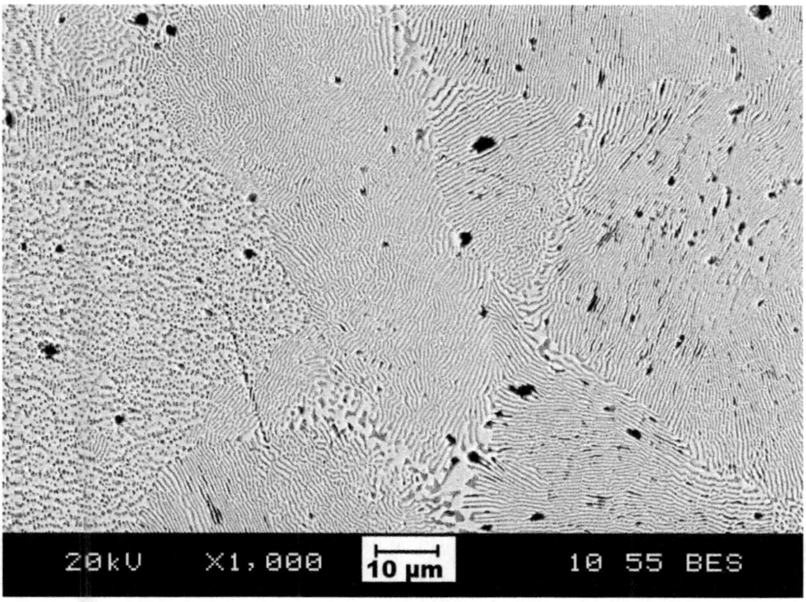

FIGURE 3.56 Microstructure of cast Zn-4Al alloy [15]. (Reused with permission from John Wiley and Sons. All rights reserved.)

layers: ε-($CuZn_4$) and γ-(Cu_5Zn_8). On Ag surfaces, three IMC layers—ε-($AgZn_3$), ζ-(AgZn), and γ-(Ag_5Zn_8)—have been observed [210–215].

Tanimoto [27] reported that Zn–Al joints in SiC/ZnAl/AMB-Si_3N_4 packages exhibited an initial bond shear strength of ~120 MPa, decreasing to ~80 MPa after 1,500 thermal cycles (–40°C to +200°C). Despite this reduction, the strength remained more than ten times higher than the IEC 60794-19 minimum requirement of 6.2 MPa. However, thermal cycling induced crack propagation into the metallization layer, ultimately leading to interfacial failure.

The high reactivity of Zn–Al alloys—coupled with stringent process requirements (vacuum and pressure)—limits their practical adoption in discrete power component manufacturing. Nonetheless, their high melting point, excellent thermal conductivity, high mechanical strength, and low material cost, make them promising candidates to replace Au–Ge or Au–Si for die-attach applications in power devices, provided that the environment is non-corrosive and processing challenges can be managed.

3.5.2.5 Silver-Sintering Materials

Silver-sintering materials have been used for die-attach in power devices for several decades due to their excellent electrical and thermal conductivities [160–162]. Traditionally, pastes formulated with micron-sized silver particles were sintered under high pressure—typically 9–40 MPa—at temperatures around 250°C, a process known as *pressure sintering*. With the introduction of nano-silver particles, the required pressure can be significantly reduced to 1–5 MPa, and the sintering temperature can be lowered to approximately 200°C while still achieving robust joints [158,159]. Because of these improvements, silver-sintering pastes have been explored as potential drop-in replacements for high-lead solders in power semiconductor applications [13,153,158].

Joint formation in silver sintering is driven by the reduction of surface energy [149]. The process begins at contact points—both particle-to-particle and particle-to-metallization—through solid-state atomic diffusion [80]. Atoms migrate toward these contact areas and form "necks" between the particles. As sintering proceeds, the necks grow, increasing the bonded area until the neck curvature radius (*r*) becomes comparable to half the diameter (*x*) of the contact area (Figure 3.57a), at which point interconnected pores remain between particles [153]. Thereafter, pore evolution is governed by reductions in pore-specific surface area and pore spheroidization, which can be monitored through changes in joint density or porosity. In pressure-less sintering, the initial compact density (related to paste metal loading), sintering shrinkage, and particle size are the primary factors governing the final density/porosity. Applying external pressures significantly increases the compact density and enhances plastic or viscous flow of Ag, accelerating neck formation and reducing porosity; under pressure, densities reaching ≥95% of bulk silver are achievable. Figure 3.57b schematically compares the porosity of strong and weak sintered connections.

Within sintered silver joints, pores serve as precursors to microcracks, which can grow and propagate across the entire bond-line. Under shear creep testing at 325°C (homologous temperature < 0.5) and 2 MPa, 3-D volumetric imaging shows pore growth into microcracks within 30 hours, with cracks coalescing rapidly during the final stages of failure (21–29.3 hours) [155]. Higher shear stresses and higher porosity

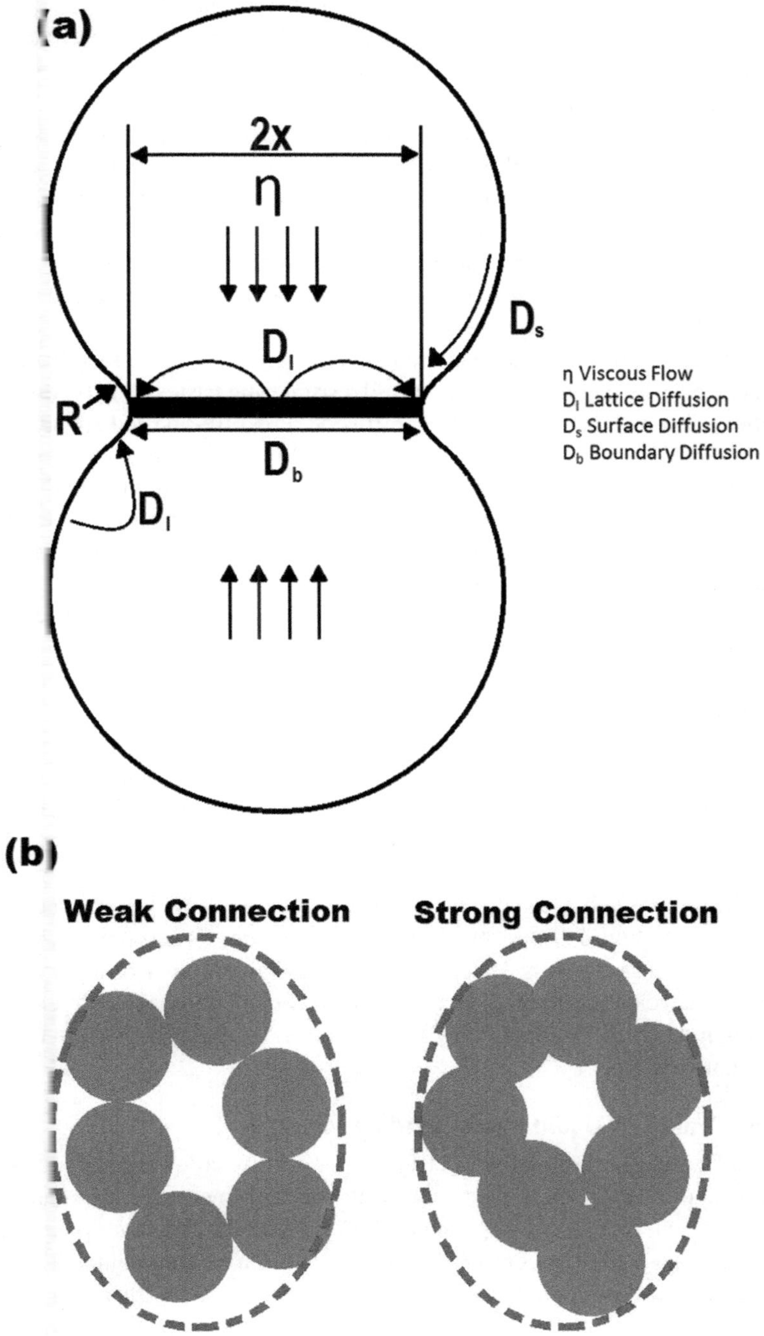

FIGURE 3.57 (a) Joining the powders from the contact area during sintering through diffusion and viscous flow, and (b) the porosity in weak connection and strong connection [15,83]. (Reused with permission from John Wiley and Sons and Spinger Nature.

accelerate failure onset [155]. During thermal cycling, both the pore size and the pore volume fraction increase substantially [153,155], and pore coalescence leads to microcrack formation and propagation—similar to the mechanism observed during shear creep.

The initial porosity, pore morphology (size and shape), and pore evolution strongly influence the reliability of silver-sintered joints. Low porosity with small, spherical pores is preferred. Pore evolution is continuous and driven by free-energy minimization, involving spheroidization, reduction of the total surface area, vacancy migration via dislocations, and pore linkage. Meanwhile, the silver matrix undergoes microstructural changes such as grain coarsening and, in some cases, recrystallization under thermomechanical stresses [155].

Interfacial bonding between sintered silver and the substrate metallization also plays a critical role in joint performance. Silver-sintering pastes bond readily to silver and other noble metals (Au, Pt, Pd) via diffusion. Pressure-assisted Ag-sintering has long been used for die-attach (Si/SiC to AMB/DBC) and for module-level bonding (AMB/DBC to heatsinks), owing to its exceptional electrical and thermal conductivity. Cu metallization on AMB or DBC, once considered challenging due to oxidation and poor bonding, has also been successfully adopted; inert or reducing atmospheres significantly improve joint formation.

Pressure-less silver sintering is highly desirable as a drop-in replacement for high-Pb solders, but several challenges remain. These include:

- high initial porosity and pore evolution within the joint,
- grain coarsening and potential Kirkendall void formation,
- reduced interfacial adhesion on non-noble metallization, and
- reliability concerns stemming from the microstructural instability of porous silver.

Although pressure-assisted sintering greatly improves joint quality, it is incompatible with the existing high-Pb mass production infrastructure for discrete power devices. Additionally, silver-sintering materials cost more than an order of magnitude more than high-lead solders, and the environmental and health impacts of nano-silver powders remain debated—leading many manufacturers to seek alternatives that do not involve nanomaterials.

3.5.2.6 Transient Liquid Phase Bonding Materials

Transient liquid-phase bonding (TLPB) operates by allowing the low-melting constituent to melt first and then react with the high-melting constituent to form high-melting intermetallic compounds (IMCs). Once the reaction is complete, the low-melting phase is fully consumed—or reduced to a minimal amount—enabling the joint to withstand high service temperatures. Figure 3.58 illustrates the TLPB mechanism, in which phase A represents the low-melting material and phase B represents the solid, high-melting phase. During bonding, B (or its constituent elements) dissolves into the molten A, and high-melting β-phase IMCs containing both A and B begin growing at the interface. This process continues until the low-melting phase A is entirely converted into the β-phase material.

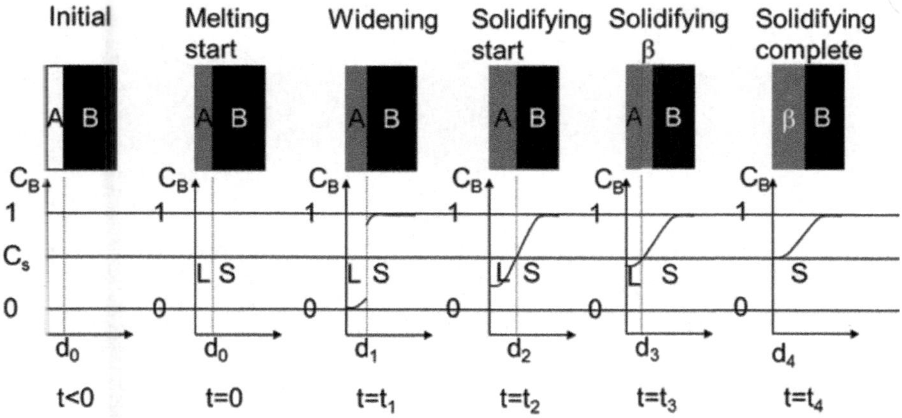

FIGURE 3.58 The schematic reaction mechanism of TLPB materials to form the high temperature joint [15,83]. (Reused with permission from John Wiley and Sons and Spinger Nature.

In practice, TLPB materials can be supplied either as pastes or as composite ribbons consisting of low-melting metal layers plated onto both sides of a high-melting metal core [182]. A typical TLPB paste—such as a blend of Cu and Sn powders—forms Cu_3Sn (adjacent to the copper particles) and Cu_6Sn_5 (above the Cu_3Sn layer) both around individual particles and on Cu-metallized substrates after reflow, as shown in Figure 3.59a. Residual Sn may remain in the joint if the liquid-phase constituent is not fully consumed. Incomplete conversion can result from an insufficient reflow time, too low a peak temperature, or the slow diffusion of atoms through increasingly thick IMC layers surrounding the Sn-rich regions.

Pores and voids formed during the TLPB paste process tend to be distributed unevenly throughout the joint (Figure 3.59a). By contrast, joints formed using composite TLPB ribbons develop IMC layers from each bonded interface that grow inward into the molten plating layers. Because these IMCs grow with a characteristic scallop morphology rather than as flat films, the layers eventually meet near the joint center, concentrating voids in the middle of the bond-line. Figure 3.59b illustrates this behavior in a Cu/Sn/Cu assembly produced from a composite preform consisting of 15 μm Sn plated onto a copper core; clusters of pores are visible at the center of the joint.

TLPB can achieve very high remelting temperatures by converting the entire joint into intermetallic phases. While these IMCs offer significantly higher strength than high-lead solders, they are also more brittle, making TLPB joints more susceptible to interfacial delamination under thermo-mechanical loading [147]. Because IMC formation is governed by diffusion kinetics, the process may require long reflow times or elevated temperatures to fully transform the low-melting constituent. Depending on the materials formulation and required processing conditions, both cost and manufacturability may vary substantially. Additionally, the brittle nature of the resulting IMC joints can result in catastrophic interfacial failures, which differ fundamentally from the fatigue-driven solder failures observed in high-lead systems.

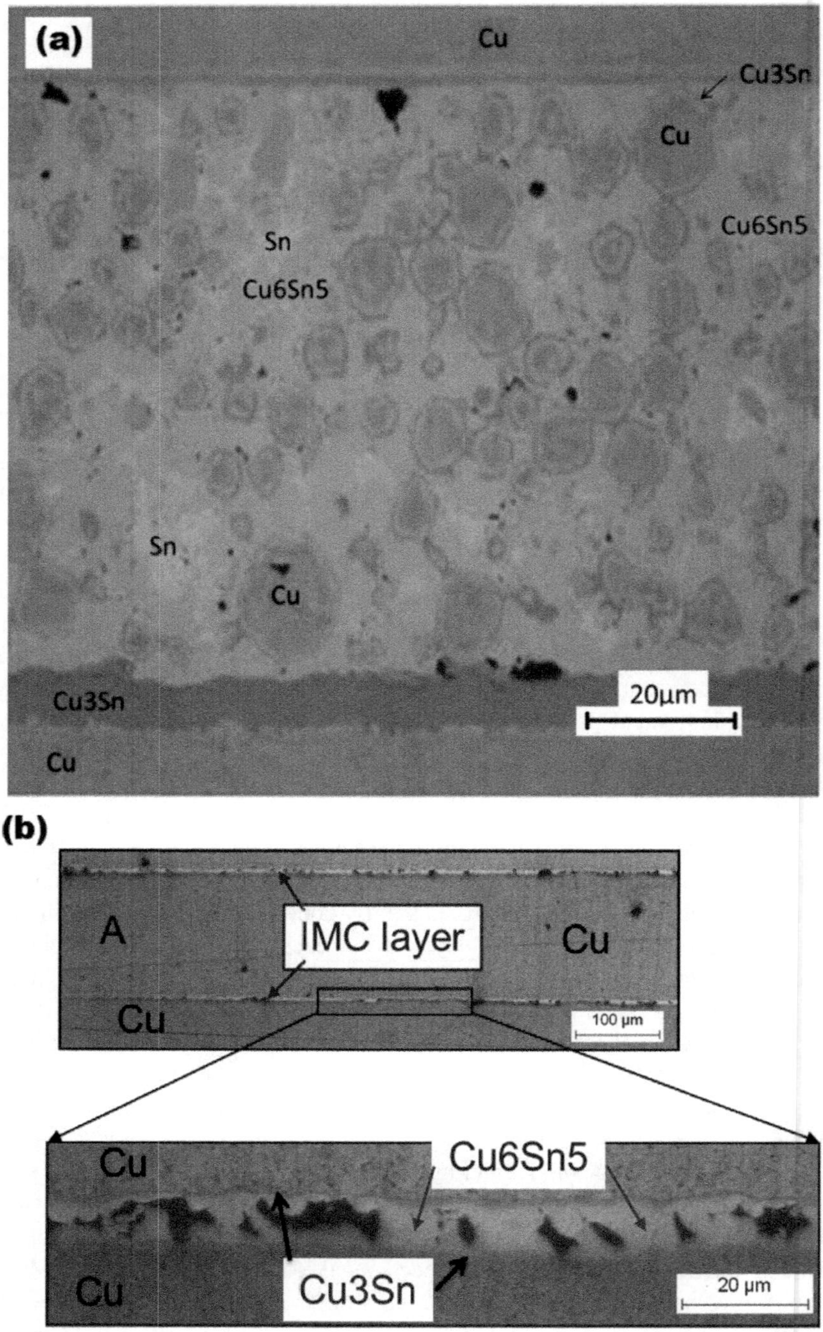

FIGURE 3.59 The morphology of the TLPB joint from (a) copper-tin paste between two copper blocks; (b) Tin/Copper/Tin composite ribbon between two copper blocks [15,83]. (Reused with permission from John Wiley and Sons and Spinger Nature. All rights reserved.)

3.5.3 Outlook

Solder alloys, silver-sintering materials, and TLPB systems represent three major categories of lead-free, high-temperature die-attach solutions. Table 3.12 summarizes their physical and mechanical properties, materials forms, and relative costs compared with Pb-5Sn-2.5Ag. As the demand for lead-free bonding materials grows, no single solution is universally suitable; instead, the best choice depends on the specific requirements of the package, device architecture, operating environment, and reliability expectations.

Each material category offers unique advantages and limitations. BiAg/BiAgX and BiCu/BiCu-X alloys, with final joint solidus temperatures around 262°C–270°C and favorable thermo-mechanical behaviors, are promising candidates for replacing high-Pb solders in smaller discrete and module-level power devices. SnSbCuAg and Durafuse® HT, which exhibit major melting transitions near 300°C, are being evaluated for higher-junction-temperature power modules, with encouraging results and ongoing production qualification.

Silver-sintering, which relies on solid-state atomic interdiffusion, remains attractive for high-power module die attach due to its exceptional thermal conductivity and mechanical robustness. However, inherent porosity and long processing times often require pressure-assisted sintering; pressure-less sintering is possible only for select package architectures.

TABLE 3.12
Physical and Mechanical Properties of Some Potential High Temperature Lead-Free Die Attach Materials [83]

	Physical and Mechanical Properties					Material Forms			Material Cost
	Melting Temp (°C)	Shear Strength (MPa) @ RT	Shear Strength (MPa) @200°C	Electrical Resistivity (μΩ*cm)	Thermal Conductivity (W/mK)	Paste	Wire	Preform	Relative to Pb5Sn 2.5Ag
Pb5Sn 2.5Ag	298	28	7.5	19	26	Yes	Yes	Yes	1
ZnAl	>360	120	70	7.5	100	No	Yes	Yes	0.2–1
BiAgX	>260	45	22	86	14	Yes	Yes	Yes	5-Feb
SnSb-CuAg	>320	>50	>30MPa	N/A	34	Yes	N/A	N/A	10-Mar
Durafuse® HT	P1: > 227; P2: >299	>50	>30	20	>34	Yes	N/A	N/A	10-Mar
AuSn	280	130	100	16.4	59	Yes	Yes	Yes	>2,000
TLP Bonding	>300	20–40	20–40	Material dependent	Material dependent	Yes	No	Yes	Material dependent
Ag Sintering	>900	20–80	20–40	5	>80	Yes	No	No	>40

TLPB materials, which convert low-melting phases into high-melting IMCs through solid–liquid interdiffusion, are also being explored for high-temperature interconnects. Challenges such as residual low-melting phases, pore formation, and process feasibility, remain under investigation.

In practice, no single die-attach material can meet all requirements across every package type, performance target, and cost constraint. However, each material system offers distinct strengths that make it suitable for specific application niches. To accelerate the industry's transition away from high-Pb solders, iNEMI has launched a collaborative effort to evaluate existing high-temperature, lead-free bonding materials for die-attach applications [136]. In parallel, the DA5 consortium has continued its high-temperature lead-free development work for nearly two decades [82,137]. Recently, the renewal period for high-lead solder exemptions was shortened from five years to just two years, increasing the urgency for industry-ready alternatives. This regulatory pressure is expected to drive faster adoption of mature lead-free solutions—or multiple solutions tailored to different device needs. Together, these initiatives aim to provide clearer guidance on the applicability, benefits, and limitations of each high-temperature lead-free bonding option.

3.6 SUMMARY

This chapter provides a comprehensive review of solder materials used across electronic and power semiconductor interconnects, covering low-, mid-, and high-temperature lead-free solder systems, their microstructural behavior, processing considerations, and reliability performance. It highlights the evolution of solder alloys driven by miniaturization, harsher operating environments, and regulatory pressures to eliminate high-lead (Hi-Pb) materials.

3.6.1 Evolution of Lead-Free Solder Technology

The transition from SnPb to lead-free solders began with first-generation Sn–Ag–Cu (SAC) alloys, such as SAC305. These materials depend on fine Ag_3Sn precipitate networks to provide creep resistance; however, they are prone to precipitate coarsening and recrystallization during thermal aging and cyclic loading. Later generations of SAC alloys incorporated micro-alloying additions (e.g., Bi, In, Sb, Ni, Mn) and reduced silver content, leading to improved drop-shock performance and enhanced microstructural stability. Third-generation high-reliability (Hi-Rel) SAC alloys further improve joint performance through tailored precipitate engineering, optimized solute additions, and intermetallic compound (IMC) stabilization. Durafuse® HR builds on these advances by delivering improvements in both voiding behavior and long-term reliability.

3.6.2 Low-Temperature Solders (LTS)

Low-temperature soldering (<200°C) helps reduce warpage, energy consumption, and thermal stress during assembly. Most LTS materials are BiSn-based, but these alloys exhibit inherent brittleness, electromigration susceptibility, Bi-phase coarsening, and poor drop-shock reliability. iNEMI studies [62,66–68,112–114, 116–118] confirm that even modified BiSn compositions remain limited for long-term use, especially above 125°C.

3.6.3 Mid-Temperature Solders (MTS)

To overcome LTS limitations, Durafuse® LT (DFLT)—a Bi-free, In-containing, Sn-rich mixed-powder solder, was developed enabling above 200°C reflow while producing solder joints remelting above 180°C–190°C. DFLT offers:

- Excellent drop-shock performance (orders of magnitude better than BiSnAg)
- Stable electromigration behavior similar to SAC305
- Strong thermal cycling reliability once joints are well-formed
- Superior warpage tolerance for large BGAs
- Good manufacturability using traditional SMT equipment

These benefits arise from its mixed-powder architecture (SnInAg + SnAgCu), enabling controlled melting, enhanced wetting, and flexible stress accommodation during reflow.

Large BGA assemblies exhibit substantial dynamic warpage, leading to defects such as HIP, NCO, NWO, hot tearing, and necking. The chapter presents detailed studies showing:

- Low reflow temperatures (185°C–200°C) are insufficient for full SAC dissolution, causing severe defects.
- Mid-range profiles (210°C–220°C) enable much better joint formation.
- DFLT significantly reduces warpage-related failures compared with BiSnAg.
- Proper paste-to-ball volume ratio (PBVR) and reflow profile optimization are essential.

When optimized, DFLT delivered robust thermal cycling performance and has already been validated in demanding applications such as large GPU modules.

3.6.4 High-Temperature Lead-Free (HTLF) Materials

Replacing Hi-Pb solders (>85 wt% Pb) remains challenging due to their high melting points (250°C–310°C) and exceptional process robustness. The chapter reviews all major HTLF candidates:

a. **Gold-based eutectics (AuSn, AuGe, AuSi)**
   - Excellent mechanical and thermal performance
   - Extremely high cost limiting broad adoption
b. **Bi-rich solders (BiAg, BiAgX, BiCu)**
   - Good high-temperature strength and reasonable cost

  - Requiring wetting enhancement strategies (e.g., BiAgX paste technology)
  - Limited by the high electrical resistivity of bismuth

c. **Sn-Sb-based systems and Durafuse® HT**
  - High melting SnSbAgCu alloys provide strong joints but are brittle
  - Durafuse® HT (HTLF-1, HTLF-2) mixes SnSb phases with Sn-rich ductile phases, with demonstrated benefits:
    - Drop-in compatibility with existing Hi-Pb processes
    - Comparable or superior R_DS(on) and thermal cycling performance
    - Strong shear strength (~15 MPa at ~290°C)
    - Requirements for thicker Ni-based die metallization

d. **Silver-sintering materials**
  - Exceptional thermal and electrical conductivity
  - Performance limited by porosity, interfacial adhesion, and need for pressure
  - High cost and concerns with nano-Ag limit mass adoption in discrete devices

e. **Transient liquid phase bonding (TLPB)**
  - Forms high-melting IMC joints through solid–liquid interdiffusion
  - High strength but brittle
  - Joints prone to interfacial delamination; long or high-temperature reflow required

A key conclusion is that no single lead-free material satisfies all die-attach and solder-interconnect requirements across the diverse semiconductor landscape, at the present time. So far, DFHT is one of the most feasible solder solutions to replace Hi-Pb, although it is still far from maturity.

Overall, this chapter provides a comprehensive framework for understanding modern solder materials development—from microstructural fundamentals to reliability mechanisms, alloy design strategies, and application-specific trade-offs. It underscores the critical role of innovative mixed-powder technologies in meeting next-generation electronics challenges, bridging the gap between traditional SAC solders, low-temperature manufacturing needs, and high-temperature power device requirements.

REFERENCES

1. Tummala, R.R., Rymaszewski, E.J., Klopfenstein, A.G., *Microelectronics Packaging Handbook*. London: Springer, 1997.
2. Harman, G., Mach, P., *Microelectronic Interconnections and Assembly*. London: Springer, 1996.
3. Laurila, T., Vuorinen, V., Mattila, T.T., et al., Materials and Interfaces in Microsystems, pp. 7-23, *Interfacial Compatibility in Microelectronics*, London: Springer, 2012.
4. Shenai, K., Khandelwal, P., Microelectronics packaging, in W.-K. Chen (eds.), *VLSI Technology*. Boca Raton, FL: CRC Press LLC, 2000.
5. Hsiung, C.K., Chen, K.N., A review on hybrid bonding interconnection and its characterization, *IEEE Nanotechnology Magazine*, vol. 18, no. 2, pp. 41–50, April 2024.

6. Lau, J., Lee, N.C., *Assembly and Reliability of Lead-Free Solder Joints*. London: Springer, 2020.
7. Datta, S., Ten nanometer CMOS logic technology, *Nature Electronics*, vol. 1, pp. 500–501, 2018.
8. Zhao, S., Tong, Y., Wang, C., Challenges and progress in packaging materials for power modules with high operation temperature: Review, *Journal of Materials Science: Materials in Electronics*, vol. 35, p. 2237, 2024.
9. Wang, H., Blaabjerg, F., Power electronics reliability: State of the art and outlook, *IEEE Journal of Emerging and Selected Topics in Power Electronics*, vol. 9, no. 6, pp. 6476–6493, 2021.
10. Calabretta, M., Renna, M., Vinciguerra, V., et al., Power packages interconnections for high reliability automotive applications, in *Proceeding of 49th European Solid-State Device Research Conference*, Cracow, Poland, 2019, pp. 35–39.
11. Agirrezabala, E., Aizpuru, I., Garrido, D., et al., Reliability of power semiconductor modules: A state-of-the-art review, *IEEE Open Journal of Power Electronics*, vol. 99, pp. 1–33, 2025.
12. Zheng, Y., Power electronic packaging interconnects and thermo-mechanical reliability, in Y. Liu (ed.), *Power Electronic Packaging*. New York: Taylor & Francis, 2025.
13. Manikam, V.R., Cheong, K.Y., Die attach materials for high temperature applications: A review, *IEEE Transactions on Components, Packaging and Manufacturing Technology*, vol. 4, pp. 457–478, 2011.
14. Siow, K.S., *High Temperature Die-Attach Materials for Microelectronics Packaging*. London: Springer, 2019.
15. Bath, J., *Lead-Free Soldering Process Development and Reliability*. New York: John Wiley & Sons, 2020.
16. Cheng, S.F., Huang, C.M., Pecht, M., A review of lead-free solders for electronics applications, *Microelectronics Reliability*, vol. 75, pp. 77–95, 2017.
17. Laurila, T., Vuorinen, V., Kivilahti, J., Materials in microelectronics interconnections, *Materials Science and Engineering R*, vol. 49, pp. 1–60, 2005.
18. Murarka, S.P., Hymes, S.W., Copper metallization for ULSL and beyond, *Critical Reviews in Solid State and Materials Sciences*, vol. 20, no. 2, pp. 87–124, 1995.
19. Lu, D., Wong, C.P., High performance conductive adhesives, *IEEE Transactions on Electronics Packaging Manufacturing*, vol. 22, no. 4, pp. 324–330, Oct. 1999.
20. Havemann, R.H., Hutchby, J.A., High-performance interconnects: An integration overview, *Proceedings of the IEEE*, vol. 89, no. 5, pp. 586–601, May 2001.
21. Lu, D., Wong, C.P., *Materials for Advanced Packageing*. New York: Springer, 2009.
22. Shamiryan, D., Abell, T., Lacopi, F., et al., Low-K dielectrics materials, *Materials Today*, vol. 7, no. 1, pp. 34–39, Jan 2004.
23. Hwang, J., *Environment-Friendly Electronics: Lead-Free Technology*. British Isles: Electrochemical Publications, 2001.
24. Suganuma, K., *Lead-Free Soldering in Electronics: Science, Technology and Environmental Impacts*. New York: Taylor & Francis Group, 2003.
25. Shangguan, D., Lead-Free Solder Interconnect Reliability. Materials Park, OH: ASM Handbook, 2005..
26. Lee, T.K., Bieler, T.R., Kim, C.U., Ma, H.T., *Fundamentlas of Lead-Free Solder Interconnect Technology from Microstructures to Reliability*. Berlin: Springer, 2015.
27. Wong, C.P., *Polymers for Electronic Packaging*. London: Springer, 2005..
28. Coyle, R., Sweatman, K., Arfaei, B., Thermal fatigue evaluation of pb-free solder joints: Results, lessons learned, and future trends, *Journal of the Minerals, Metals, and Materials Society*, vol. 67, p. 2394, 2015.
29. Coombs, C.F., *Printed Circuits Handbook*, 7th ed. New York: McGraw-Hill, 2016.

30. Siow, K.S., Are sintered silver joints ready? *Journal of Electronic Materials*, vol. 43, pp. 947–961, 2014.
31. Siow, K.S., *High Temperature Die-Attach Materials for Microelectronics Packaging*, London: Springer, 2019.
32. IPC-A-610: Acceptability of Electronic Assemblies, Institute of Printed Circuits, https://shop.electronics.org/ipc-a-610.
33. IPC J-STD-001: Requirements for Soldered Electrical and Electronic Assemblies, Institute of Printed Circuits, https://shop.electronics.org/ipc-j-std-001.
34. Thomas, O.T.K., Gopalan, P.P., *Electronics Production Defects and Analysis*. Berlin: Springer Nature Singapore, 2022.
35. Judd, M., Brindley, K., *Soldering in Electronics Assembly*. New York: Elservier/Newnes, 1999.
36. Lee, N.C., *Reflow Soldering Processes and Troubleshooting SMT, BGA, CSP and Flip Chip Technologies*. New York: Newnes, 2002.
37. Zhang, H., Lee, N.C., High-reliability, high-melting, lead-free, mixed solder paste system-BiAgX, *Review of Scientific Instruments*, vol. 92, p. 099501, 2021.
38. Stevens H., Metallic Resources, IPC Solder Products Value Council, Round robin testing and analysis of lead-free solder pastes with alloys of Tin, Silver and Copper Final Report, https://avipre.com/products/printers/Shenmao%20Solder,%20Solder%20Paste,%20Solder%20Sphere,%20Solder%20Wire%20and%20Bar//Technical%20Paper/SVPC_Final_ExecSumm.pdf.
39. Biocca, P., Rivas, C., Case study on the validation of SAC305 and SnCu based solders in SMT, wave and hand-soldering at the contract assembler level, in *The proceedings from the Boston SMTA Lead-Free Conference*, Eden Prairie, MN, June 2007.
40. Siewert, T., Liu, S., Smith, D.R., Madeni, J.C., Database for Solder Properties with Emphasis on New Lead-free Solders, NIST, 2004, https://www.msed.nist.gov/solder/NIST_LeadfreeSolder_v4.pdf.
41. Subramanian, K.N., *Lead-Free Solders: Materials Reliability for Electronics*. New York: Wiley, 2012.
42. George, E., Osterman, M., Pecht, M., Coyle, R., Parker, R., Benedetto, E., Thermal cycling reliability of alternation low solver tin-based solders, in *Proceeding of 46th International Symposium on Microelectronics (IMAPS)*, Orlando, FL, 2013, pp. 120–127.
43. Lau, J.H., State of the art of lead-free solder joint reliability, *Journal of Electronic Packaging*, vol. 143, p. 020803-1-36, 2021.
44. Henshall, G., Healey, R., Pandher, R., et al., iNEMI Pb-free alloy alternatives project report: State of the industry, *Proceedings SMTAI Proceedings*, vol. 28, pp. 109–122, 2008.
45. Cheng, F., Gao, F., Zhang, J., Jin, W., Xiao, X., Tensile properties and wettability of SAC0307 and SAC105 low Ag lead-free solder alloys, *Journal of Materials Science*, vol. 46, pp. 3424–3429, 2011.
46. Liu, W., Lee, N.-C., Porras, A., et al., Achieving high reliability low cost lead-free SAC solder joints via Mn Or Ce doping, *Proceeding of SMTAI*, vol. 134, pp. 920–934, 2009.
47. Lee, N.C., Choosing Solders for the New Era: Low Cost High Reliability solder Alloys, short course, IMAPS, 9–12 October, 2017.
48. Henshall, G., Healey, R., Pandher, R., Sweatman, K., Howell, K., Coyle, R., Sack, T., Snugovsky, P., Tisdale, S., Hua, F., O'Malley, G., Addressing opportunities and risks of Pb-free solder alloy alternatives, in *2009 European Microelectronics and Packaging Conference (EMPC)*, Rimini, Italy, 2009, pp. 1–11.
49. Shi, H., Ueda, T., Effects of edge and corner bond adhesives on the drop reliability of package-on-package bottom package assemblies, in *3rd International Conference on Computer Research and Development*, Shanghai, China, 2011, pp. 416–420.

50. Perng, S., Lee, T.K., Guirguis, C., Enhancing mechanical shock performance using edgebond technology, in *SMTA International Conference Proceedings*, Eden Prairie, MN, 2013, pp. 357–361.
51. Mishiro, K., Ishikawa, S., Abe, M., Kumai, T., Higashiguchi, Y., Tsubone, K., Effect of the drop impact on BGA/CSP package reliability, *Microelectronics Reliability*, vol. 42, pp. 77–82, 2002.
52. Plachý, Z., Pražanová, A., Dušek, K., Géczy, A., Underfill: A review of reliability improvement methods in electronics production, *Polymers*, vol. 17, p. 2206, 2025.
53. Coyle, R., Parker, P., Howell, K., et al., A collaborative industrial consortia program for characterizing thermal fatigue reliability of third generation Pb-free alloys, in *Proceedings of SMTAi*, Minneapolis, MN, 2016, pp. 188–196.
54. Henshall, G., Healy, R., Pander, R.S., et al., iNEMI Pb-free alloy alternatives project report: state of the industry, *Journal of Surface Mount Technology*, vol. 21, no. 4, pp. 11–23, 2008.
55. Snugovsky, P., Bagheri, S., Romansky, M., et al., New generation of Pb-free solder alloys: possible solution to solve current issues with main stream Pb-free soldering, *Journal of Surface Mount Technology*, vol. 25, no. 3, pp. 42–52, 2012.
56. Coyle, R., Read, P., McCormick, H., et al., The influence of alloy composition and temperature cycling dwell time on the reliability of a quad flat no lead (QFN) package, *Journal of Surface Mount Technology*, vol. 25, no. 1, pp. 28–34, 2011.
57. Coyle, R., Osenbach, J., Collins, M., et al., Phenomenological study of the effect of microstructural evolution on the thermal fatigue resistance of Pb-free solder joints, *IEEE Transactions on Components, Packaging, and Manufacturing Technology*, vol. 1, no. 10, pp. 1583–1593, 2011.
58. Coyle, R., Hillman, D., Johnson, C., et al., Alloy composition and thermal fatigue reliability of high reliability Pb-free solder alloys, in *Proceedings of SMTAI*, Rosemont, IL, October 2018.
59. Geng, J., Jilani, H., Zhang, H., Low voiding high reliability lead-free solder paste for automotive applications, in *Proceedings of SMTAI*, Minneapolis, MN, 2023.
60. Bath, J., Itoh, M., Clark, G., et al., An investigation into low temperature tin-bismuth and tin-bismuth-silver lead-free alloy solder pastes for electronics manufacturing applications, in *Proceedings of the IPC APEX Conference*, San Diego, CA, 2012.
61. Aspandiar, R., Byrd, K., Tang, K.K., et al., Investigation of low temperature solders to reduce reflow temperature, improve SMT yields and realize energy savings, in *Proceedings of the IPC APEX Expo Conference*, Hillsboro, OR, 2015.
62. Fu, H., Aspandiar, R., Chen, J., et al., iNEMI project on process development of BiSn-based low temperature solder pastes, in *Proceedings of the 2017 SMTA International Conference*, Rosemont, IL, 2017, pp. 207–220.
63. Ribas, M., Hunsinger, T., Cucu, T., et al. (2018). The Printed Circuit Assembler's Guide to…™ Low Temperature Soldering. BR Publishing, https://i-007ebooks.com/my-i-connect007/books/lts (accessed 4 February 2020)..
64. Takao, H., Yamada, A., Hasegawa, H., et al., Mechanical properties and solder joint reliability of low-melting Sn-Bi-Cu lead free solder alloy, *Journal of Japan Institute of Electronics Packaging*, vol. 5, no. 2, pp. 152–162, 2002.
65. Osgood, H., Geiger, D., Pennings, R., et al., Low temperature SMT solder evaluation, in *Proceeding of IPC APEX Conference*, Milpitas, CA, 2019, p. S14.
66. Singh, P., Palmer, L., Hamid, M., et al., Comparison of electromigration in tin-bismuth planar and bottom terminated component solder joints, in *Peoceedings of SMTA Internation*, Minneapolis, MN, 2023, pp. 361–367.
67. Singh, P., Palmer, L., Wassick, L., et al., Microstructural evolution during electromigration in eutectic tin-bismuth BTC solder joints, in *Proceedings of SMTA International*, Rosemond, IL, 2024, pp. 240–246.

68. Singh, P., Aspandiar, R.F., Coyle, R., et al., Electromigration in Sn-Bi and SAC305 solder alloys, in *Proceedings of SMTA International*, Rosemond, IL, 2025, pp. 211–216.
69. Wen, M.Y., Zhang, H., Zheng, D.Y., et al., A Bi-free in-containing lower temperature solder with superior drop shock resistance and electromigration resistance, in *Proceedings of SMTA International*, Rosemond, IL, 2025, pp. 60–64.
70. Freer, J.L., Morris, J.W. Jr., 1992, Microstructure and creep of eutectic indium/tin on copper and nickel substrates, *Journal of Electronic Materials*, vol. 21, no. 6, pp. 647–652, 1992.
71. Mei, Z., Morris, J.W. Jr, 1992, Superplastic creep of low melting point solder joints, *Journal of Electrinic Materials*, vol. 21, no. 4, pp. 401–407, 1992.
72. Freer Goldstein, J.L., Morris, J.W. Jr., 1994, Microstructural development of eutectic Bi-Sn and eutectic In-Sn during high temperature deformation, *Journal of Electronic Materials*, vol. 23, no. 5, pp. 477–486, 1994.
73. Zhang, H., Chen, F., Mutuku, F., et al., Low Temperature Melting and Mid-Temperature Melting Lead-free Solder Paste with Mixed Solder Alloy Powders, US Patent 11267080 B2, 2019.
74. Zhang, H., Chen, F., Mutuku, F., et al., Low Temperature Melting and Mid-Temperature Melting Lead-free Solder Paste with Mixed Solder Alloy Powders, US Patent 11712762 B2, 2019.
75. Zhang, H., Lytwynec, S., Lee, N.-C., et al., Novel lead-free low-temperature solder with excellent drop-shock resistance, in *Proceedings of International Microsystems, Packaging, Assembly and Circuits Technology Conference*, Taipei, 2021, pp. 224–230.
76. Zhang, H., Richmond, T., Geng, J., et al., A novel lower-temperature lead-free solder paste for wafer-level package application, in *Proceedings of SMTA International*, Minneapolis, NM, Oct, 2023, pp. 230–238.
77. Zhang, H., Richmond, T., Geng, J., et al., A bismuth-free in-containing lower-temperature lead-free solder paste for wafer-level package application that outperforms SAC305, in *Proceedings of IEEE 73rd ECTC*, Orlando, FL, 2023, pp. 661–667.
78. Zhang, H., Richmond, R., Wang, H., et al., A lower-temperature lead-free solder paste for wafer-level package application that outperforms SAC305, in *Proceedings of IPC Apex Expo Technical Conference*, Kumamoto, Japan, 2022, pp. 1130–1147.
79. Zhang, H., Richmond, R., Mutuku, F., et al., The attempt of lower temperature soldering process for large plastic ball-grid array board-level assembly, in *Proceedings of IPC APEX ECWC16*, Big Island, HI, 2024, p. S20.
80. Goppert, C., Zhang, H., Mutuku, F., et al., Mitigating solder joint defects caused by warpage of a BGA assembly with lower temperature soldering phase I: Processing optimization, *Proceeding of SMTA International*, vol. 27, no. 1, pp. 185–193, 2025.
81. Zhang, H., Zheng, D., Wen, M., et al., Mitigating solder joint defects caused by warpage of a BGA assembly with lower temperature soldering phase II: TCT reliability, *Proceeding of SMTA International*, vol. 27, no. 1, pp. 205–210.
82. Behrens, T., DA5 Consortium Report, Sept, 2025, https://www.nxp.com/docs/en/supporting-information/DA5_Customer_Presentation.pdf.
83. Zhang, H., Minter, J., Lee, N.C., A brief review on high-temperature, Pb-free die-attach materials, *Journal of Electron. Materials*, vol. 48, p. 210, 2019.
84. Zhang, H., Lee, N.C., Reliability of BiAgX solder as a drop-in solution for high temperature lead-free die-attach applications, *SMTA Journal*, vol. 26, pp. 28–32, 2013.
85. Zhang, H., Lee, N.C., High-reliability, high melting, lead-free, mixed solder paste system-BiAgX®, *Review of Scientific Instruments*, vol. 92, p. 099501, 2021.
86. Liu, W., Lee., N.C., High Reliability Lead-Free Solder Alloys for Harsh Environment Electronics Applications, US Patent 12090579 B2, 2016.
87. Geng, J., Zhang, H., Lee, N.C., High Reliability Leadfree Alloys for Harsh Service Conditions, US Patent 11752579 B2, 2016.

88. Liu, W., Lee., N.C., High Reliability Lead-Free Solder Alloys for Harsh Environment Electronics Applications, US Patent 11413709 B2, 2016.
89. Liu, W., Lee., N.C., High Reliability Lead-Free Solder Alloys for Harsh Environment Electronics Applications, US Patent 11229979 B2, 2016.
90. Imamura, Y., Ikeda, K., Piao, J.Y., Solder Alloy, Solder Paste, and Elecronic Circuit Board, US Patent 9221132 B2, 2016.
91. Yin, L., Wentlent, L., Yang, L., et al., Recrystallization and precipitate coarsening in Pb-free solder joints during thermomechanical fatigue, *Journal of Electronic Materials*, vol. 41, pp. 241–252, 2012.
92. Arfaei, B., Wentlent, L., Joshi, S., et al., Improving the thermomechanical behavior of lead free solder joints by controlling the microstructure, in *Proceedings of ITHERM*, San Diego, CA, 2012, pp. 392–398.
93. Coyle, R., McCormick, H., Osenbach, J., et al., Pb-free alloy silver content and thermal fatigue reliability of a large plastic ball grid array (PBGA) package, *Journal of Surface Mount Technology*, vol. 24, no. 1, pp. 27–33, 2011.
94. Yin, L., Recrystallization and precipitate coarsening in lead, in *AREA Consortium Meeting*, Binghamton, NY, March 2012.
95. Coyle, R., Hillman, D., Johnson, C., et al., Alloy composition and thermal fatigue reliability of high reliability Pb-free solder alloys, in *Proceedings of SMTAI*, Rosemont, IL, October 2018.
96. Coyle, R., Parker, R., Osterman, M., et al., iNEMI Pb-free alloy characterization project report: part V – the effect of dwell time on thermal fatigue reliability, in *Proceedings of SMTAI*, Ft. Worth, TX, 2013, pp. 470–489.
97. Belyakov, S.A., Nishimura, T., Akaiwa, T., et al., Role of Bi, Sb, and In in microstructural formation and properties of Sn-0.7Cu-0.05Ni-X BGA interconnections, in *IEEE International Conference on Electronic Packaging (ICEP)*, Niigata, Japan, 2019.
98. Belyakov, S.A., Coyle, R.J., Arfaei, B., et al., Microstructure and damage evolution during thermal cycling of Sn-Ag-Cu solders containing antimony, *Journal of Electribic Materials*, vol. 50, No. 3, pp. 825–841, 2021.
99. Belyakov, S.A., Sweatman, K., Akaiwa, T., et al., Precipitation of Bi and SbSn phases in next-generation Pb-free solders, in *Proceedings of SMTAI*, Rosemont, IL, 2019, pp. 484–491.
100. Belyakov, S.A., Gourlay, C.M., Coyle, R., et al., Microstructure evolution during thermal fatigue of SAC solder containing 5.5%Sb, in *Proceedings of SMTAI*, Rosemont, IL, 2020, pp. 432–440.
101. Belyakov, S.A., Arfaei, B., Johnson, C., et al., Phase formation and solid solubility in high reliability Pb-free solders containing Bi, Sb or In, in *Proceedings of SMTAI*, Rosemont, IL, 2019, pp. 492–506.
102. Belyakov, S.A., Xian, J., Zeng, G., et al., Precipitation and coarsening of bismuth plates in Sn-Ag-Cu-Bi and Sn-Cu-Ni-Bi solder joints, *Journal of Materials Science: Materials in Electronics*, vol. 30, pp. 378–390, 2019.
103. Coyle, R., Hillman, D., Parker, R., et al., The effect of bismuth, antimony, or indium on the thermal fatigue of high reliability Pb-free solder alloys, in *Proceedings of SMTAI*, Rosemont, IL, October 2018.
104. Pearson, T., Hillman, D., Mills, J., Investigation on the impact of Pb-free solder alloy mixing on solder joint integrity using ball grid arrays (BGAs), in *Proceedings of SMTAI*, Minneapolis MN, 2022, pp. 319–324.
105. Coyle, R., Johnson, C., Hillman, D., et al., Thermal cycling reliability and failure mode of two ball grid array packages with high reliability Pb-free solder alloys, in *Proceedings of SMTAI*, Rosemont, IL, 2019, pp. 439–456.
106. Li, B., Liu, S., Sun, Y., et al., The effect of indium microalloying on lead-free solders: A review, *Materials Science in Semiconductor Processing*, vol. 185, p. 108956, 2025.

107. Nguyen, T., Zhang, H., Murling, A., et al., Enhancing PCB reliability: Voiding reduction design with a mixed alloy solder paste, in *IPC Apex Expo 2025 Technical Conference Proceedings*, Bannockburn, IL, 2025.
108. Nguyen, T., Zhang, H., Murling, A., et al., Optimizing mixed alloy solder pastes for low voiding and high PCB reliability, in *Proceedings of SMTAI*, Rosemont, IL, 2025, pp. 515–523.
109. Geng, J., Zhang, H., Yue, P., et al., High Reliability Lead-Free Solder Pastes with Mixed Solder Alloy Powders, US Patent Application US2022/0395936 A1, 2016.
110. Geng, J., *Low-Voiding High-Reliability Lead-Free Solder Paste for Automotive Applications*. New York: Insider Series Webinar, 2024.
111. i NEMI, 2017 iNEMI Roadmap. Packaging and component substrates, 2017, pp. 10–12.
112. Fu, H., Aspandiar, R., Chen, J., et al., iNEMI project on process development of BiSn-based low temperature solder pastes — Part II: Characterization of mixed alloy BGA solder joints, in *Proceedings of the SMTA Pan Pacific Microelectronics Symposium*, Big Island, HI, 2018, pp. 1–17.
113. Fu, H., Radhakrishnan, J., Ribas, M., et al., iNEMI Project on process development of BiSn-based low temperature solder pastes – Part IV: Comprehensive mechanical shock tests on pop components having mixed BGA BiSn-Sac solder joints, in *Proceedings of the 2018 SMTA International Conference*, Rosemont, IL, October 2018.
114. Fu, H., Radhakrishnan, J., Ribas, M., et al., iNEMI project on process development of BiSn-based low temperature solder pastes – Part VI: Mechanical shock results on resin reinforced mixed SnAgCu-BiSn solder joints of FCBGA components, in *Proceedings of the 2019 SMTA International Conference*, Rosemont, IL, 2019.
115. Mutuku, F., Lu, D., Zhang, H., et al., Effects of dynamic warpage on solder joints of large plastic ball-grid arrays assembled with LTS, in *Proceedings of SMTA International 2024*, Rosemont, IL, 2024, pp. 285–290.
116. Singh, P., Aspandiar, R.F., Fu, H., et al., Eleactomigration in eutectic tin-bismuth bottom-terminated-component solder joints, in *Proceedings of 74th ECTC*, Orlando, FL, 2024, pp. 1433–1438.
117. Singh, P., Palmer, L., Wassick, T., et al., Electromigration in tin-bismuth planar solder joints, *Transactions of The Japan Insitute of Electronics Packaging*, vol. 17, pp. E23–008, 2024.
118. Singh, P., Palmer, L., Hamid, M., et al., Electromigration in tin-bismuth planar solder joints, in *2023 International Conference on Electronics Packaging (ICEP)*, Kumamoto, Japan, 2023, pp. 201–202.
119. Mutuku, F., Zhang, H., Wang, H., et al., A near-eutectic Sn-Bi low-temperature alloy with high thermal fatigue resistance, in *Proceedings of SMTA International 2023*, Rosemont, IL, 2023, pp. 345–352.
120. Coyle, R., Aspandiar, R., Feng, C., et al., Thermal cycling hybrid, homogeneous, and resin reinforced low temperature solder ball grid array interconnects at a high homologous temperature, in *Proceedings of SMTA International 2024*, Rosemont, IL, 2024, pp. 223–239.
121. Tang, K.K., LTS Manufacturing Challenges and Solutions, in INEMI Low Temperature Soldering Workshop, Apr. 24, 2019.
122. Coyle, R., Aspandiar, R., Hadian, F., et al., Interim thermal cycling report on hybrid, homogeneous, and resin reinforced low temperature solder joints, in *Proceedings of SMTA International 2021*, Rosemont, IL, 2021, pp. 401–426.
123. Baker, H., Okamoto, H., Henry, S.D., et al., *ASM Handbook, Volume 3: Alloy Phase Diagram, 2.260*. New York: ASM International, 1990.
124. Zhang, H., Lytwynec, S., Lee, N.-C., et al., A novel bi-free low-temperature solder paste with outstanding drop-shock resistance, in *Proceedings of IEEE 71st ECTC*, Orlando, FL, 2021, pp. 643–653.

125. Zhang, H., Lytwynec, S., Wang, H., et al., A novel mixed powder low temperature solder with superior drop-shock resistance, in *Proceedings of IPC Apex 2021*, San Diego, CA, 2021.
126 Yaeger, A., *Area Consortium Meeting*, Binghamton, NY, Oct, 2022.
127 Su, P., Ho, D., Pu, J., et al., Chiplets integrated solution with FO-EB package in HPC and networking application, in *Proceedings of 2022 IEEE 72nd Electronic Components and Technology Conference (ECTC)*, Orlando, FL, 2022, pp. 2135–2140.
128. Wong, J. H., Wu, N.Y., Lai, W.H., et al., Warpage and RDL stress analysis in large fan-out package with multi-chiplet integration, in *Proceedings of the 2022 IEEE 72nd Electronic Components and Technology Conference (ECTC)*, Orlando, FL, 2023, pp. 1074–1079.
129. Das, R.N., Plant, J., Wynn, A., et al., Extremely large area integrated circuit (ELAIC): An advanced packaging solution for chiplets, in *Proceedings of the 2023 IEEE 73rd Electronic Components and Technology Conference (ECTC)*, Orlando, FL, 2023, pp. 258–265.
130. Koide, M., Fukuzono, K., Watanabe, M., et al., Full low temperature solder BGA development for large size BGA package, in *Proceedings of 2020 IEEE 70th Electronic Components and Technology Conference (ECTC)*, Orlando, FL, 2020, pp. 1265–1269.
131. Lai, D.S., Lee, J., Huang, J., et al., Better warpage control by using low temperature solder for large FCBGA application, in *Proceedings of the 2021 International Conference on Electronics Packaging (ICEP)*, Niigata, Japan, 2021, pp. 135–136.
132. Chen, X., Wwang, H., Zhang, H., et al., Enhanced reliability of large BGA assemblies for AI server and HPC application, in *Proceedings of 2025 EPTC*, Singapore, 2025.
133. Bradley, E., Handwerker, C.A., Bath, J., et al., *Lead-Free Electronics*. Hoboken, NJ: John Wiley & Sons, 2007.
134. Suganuma, K., *Lead-Free Soldering in Electronics*. Basel, NY: *Marcel Dekker*, 2004.
135. Tegehall, E., Review of the Impact of Intermetallic Layers on the Brittleness of Tin-Lead and Lead-Free Solder Joints, IVF Project Report, 2006/2007.
136. Lim, S.P., Pan, B., Zhang, H., et al., High-temperature Pb-free die attach material project phase 1: Survey result, in *Proceeding of 2017 International Conference on Electronics Packaging*, Yamagata, Japan, TA4-4, 2017.
137. Eilken, B., Die Attach 5 Project Report, https://www.infineon.com/da5customer-presentation.
138. Zhang, H., Lee, N-C., Lead-free bonding materials for power semiconductor die attachment, in *Proceeding of 51st International Symposium on Microelectronics*, TP3-096, Pasadena, CA, 2018.
139. Zhang, H., Minter, J., Lee, N.C., Die attach soldering, in *Proceeding of the International Conference on Soldering and Reliability*, SMTA, Markham, Ontario, Canada, 2017.
140. Zeng, G., McDonald, S., Nogita, K., Development of high temperature solders: Review, *Microelectronics Reliability*, vol. 52, pp. 1306–1322, 2012.
141. Holaday, J.R., Handwerker, C.A., *High Temperature Die-Attach Materials for Microelectronics Packaging*. Berlin: Springer/K.S. Siow, 2019.
142. Bosco, N.S., Zok, F.W., Transient liquid phase bonding of Cu and Al using metallic particles interlayers, *Acta Materialia*, vol. 52, pp. 2965–2975, 2004.
143. Chu, K., Sohn, Y., Moon, C., et al., A comparative study of Cu/Sn/Cu and Ni/Sn/Ni solder joints for low temperature stable transient liquid phase bonding, *Scripta Materialia*, vol. 109, pp. 113–117, 2015.
144. Cook, G.O. III, Sorensen, C.D., Overview of transient liquid phase and partial transient liquid phase bonding, *Journal of Materials Science*, vol. 46, pp. 5305–5323, 2011.
145. Yoon, S.W., Glover, M.D., Mantooth, H.A., Shiozaki, K., Highly reliable double-sided bonding used in double-sided cooling for high temperature power electronics, in *Proceedings of the 2012 IMAPS High Temperature Electronics Conference (HiTEC)*, Albuquerque, NM, 2012.

146. Sharif, A., Gan, C.L., Chen, Z., et al., Transient liquid phase Ag-based solder technology for high-temperature packaging applications, *Journal of Alloys and Compounds*, vol. 587, pp. 365–374, 2014.
147. Dudek, R., Sommer, P., Fix, A., et al., Reliability issues for high temperature interconnections based on transient liquid phase soldering, in *Proceedings of the 14th International Conference on Thermal, Mechanical and Multi-physics Simulation and Experiments in Microelectronics and Microsystems (EuroSime 2013)*, Wroclaw, Poland, 2013.
148. Shearer, C., Shearer, B., Matijasevic, G., et al., Transient liquid phase sintering composites: Polymer adhesives with metallurgical bonds, *Journal of Electronic Materials*, vol. 28, pp. 1319–1326, 1999.
149. Mackenzie, J.K., Shuttleworth, R., A phenomenological theory of sintering, *Proceedings of the Physical Society. Section B*, vol. 62, pp. 833–852, 1949.
150. Schwarzbauer, H., Method of Securing Electronic Components to a Substrate, US Patent 4810672, 1989.
151. Schwarzbauer, H., Method and Apparatus for Fastening Electronic Components to Substrates, US Patent 4903885, 1990.
152. Schwarzbauer, H., Apparatus for Fastening Electronic Components to Substrates, US Patent 5058796, 1991.
153. Thummler, F., Oberacker, R., *Introduction to Powder Metallurgy*. London: Institute of Materials, 1993.
154. Chen, S.W., Chen, J-W, Lin, Y-C., et al., Electromigration effects upon Cu/Ni/Sn-Ag and Cu/Ni/Sn-Ag-Cu interfacial reactions, in *Proceeding of ECS Meeting*, MA2016-02, Chicago, IL, 2016.
155. Tan, Y., Li, X., Chen, G., et al., Three-dimensional visualization of the crack-growth behavior of nano-silver joints during shear creep, *Journal of Electronic Materials*, vol. 44, pp. 3553–3561, 2015.
156. Li, J.F., Agyakwa, P.A., Johnson, C.M., Kinetics of Ag_3Sn growth in Ag–Sn–Ag system during transient liquid phase soldering process, *Acta Materialia*, vol. 58, pp. 3429–3443, 2010.
157. Bai, J., Zhang, Z.Z., Calata, J.N., et al., Low-temperature sintered nanoscale silver as a novel semiconductor device-metallized substrate interconnect material, *IEEE Transactions on Components, Packaging and Manufacturing Technology*, vol. 29, p. 589, 2006.
158. Siow, K.S., Are sintered silver joints ready for use as interconnect material in microelectronic packaging?, *Journal of Electronic Materials*, vol. 43, pp. 947–961, 2014.
159. Lei, T.T., Calata, J.N., Lu, G.Q., et al., Low-temperature sintering of nanoscale silver paste for attaching large-area (>100 mm^2) chips, *IEEE Transactions on Components, Packaging and Manufacturing Technology*, vol. 33, pp. 98–104, 2010.
160. Chen, S., Lee, N.C. Nanomicrocrystallite Paste for Pressureless Sintering, US Patent 9875983 B2, 1999.
161. Chen, S., Shambach, W., LaBarbera, C., et al., Highly reliable die-attach silver joint with pressure-less sintering for high temperature application, in *Proceeding of International Conference on High Temperature Electronics 2018*, Albuquerque, NM, 2018, p. 45.
162. Chen, S., Fan, G.,Yan, X., et al., Conventional reflow oven sinter-able pressure-less silver paste for die-attach assembly mass production, in *Proceeding of International Symposium on Microelectronics*, San Diego, CA, 2014, p. 92.
163. Nikolic, N.A., Thermoset Adhesive Films, US Patent US2005/0137340A1, 2005.
164. https://www.henkel-adhesives.com/die-attach-film-adhesives-27373.htm.
165. Tanimoto, S., Matsui, K., Murakami, Y., et al., Assessment of Au-Ge Die attachment for an extended junction temperature of 300°C, in *Proceedings of the 2010 International Conference and Exhibition of High Temperature Electronics (HiTEC 2010)*, Albuquerque, NM, 2010.

166. Tanimoto, S., Matsui, K., Zushi, Y., et al., Common metal die attachment for SiC power devices operated in an extended junction temperature range, *Materials Science Forum*, vol. 717–720, pp. 853–858, 2012.
167. Tanimoto, S., Hirama, N., Watanabe, K., et al., Development of highly reliable, pressureless-sintered Ag-nanoparticle die-attachment, in *Proceedings of the 60th JSAP Spring Meeting (Annual Spring Meeting of the Japan Society of Applied Physics)*, Japan, 2013, pp. 27–30.
168. Zhang, H., Minter, J., Lee, N.C., et al., Development of BiAgX® HT solder paste for 200°C application, in *Proceedings of 2016 IMAPS High Temperature Electronics*, Albuquerque, NM, 2016.
169. Geng, J., Zhang, H., Mutuku, F., et al., Novel solder alloy with wide service temperature capability for automotive applications, *IEEE Power Electronics Magazine*, vol. 5, no 3, p. 56, Sept. 2018.
170. Bultitude, J., Jones, L., McConnell, J., et al., Base metal electrode (BME) multilayer ceramic capacitors (MLCCs) for high temperature applications, in *Proceeding of International Conference on High Temperature Electronics*, Albuquerque, NM, 2014.
171. Zhang, H., Lim, S.P., Geng, J., et al., Pb-free bonding materials for power semiconductor die-attachment, in *Presented in China Semiconductor Packaging Test Seminar*, HeFei, China, 2018.
172. Shen, Z., Johnson, R.W., Hamilton, M.C., SiC power device die attach for extreme environments, *IEEE Transactions on Electron Devices*, vol. 62, pp. 346–353, 2015.
173. Shen, Z., Reiderman, A., Johnson, R.W., Pressure-less AgNP sintering for high-power MCM assembly for extreme environment applications, in *International Symposium on Microelectronics (IMAPS)*, Pasadena, CA, 2015.
174. Shen, Z., Fang, K., Hamilton, M.C., Johnson, R.W., Snipes, E., Bozack, M., Lead-free solder attach for 200°C applications, in *Proceedings of the IMAPS Conference and Exhibition on High Temperature Electronics Network (HiTEN 2013)*, Oxford, UK, 2013.
175. Scholes, H., *AREA Consortium Spring Meeting*, Binghamton, NY, March 2014.
176. Arfaei, B., *AREA Consortium Meeting*, Binghamton, NY, March 2012.
177. Kim, B., Lee, C.-W., Lee, D., et al., Effect of Sb addition on Bi–2.6 Ag–0.1 Cu solders for high-temperature applications, *Journal of Alloys and Compounds*, vol. 592, pp. 207–214, 2014.
178. Cui, J.Z., Johnson, R.W., Hamilton, M., Reliability of AuGe die attach on DBC substrates with different Ni surface finishes, *IEEE Transactions on Components, Packaging and Manufacturing Technology*, vol. 7, pp. 1598–1606, 2017.
179. Lee, Y., Link, D., Reinforced sintered silver for high reliability die attach, in *Proceeding of 2016 IMAPS 49thInternational Symposium on Microelectronics*, Pasadena, CA, 2016.
180. Zhang, H., Mao, R.S., Lee, N.C., et al., Reliability of lead-free BiAgX pastes for high temperature die-attach application, in *TMS Annual Meeting*, San Antonio, TX, 2013.
181. Jiang, Q., Mukherjee, S., Dasgupta, A., et al., Mechanical constitutive properties of a Bi-rich high temperature solder alloy, in *Proceedings of the 15th IEEE Intersociety Conference on Thermal and Thermomechanical Phenomena in Electronic Systems (ITherm 2016)*, Las Vegas, NV, 2016, pp. 1236–1239.
182. Liu, W.P., Lee, N.C. Composite Solder Alloy Preform, US Patent 8348139 B2, 2013.
183. Teo, J.W.R., Ng, F.L., Goi, L.S.K., Microstructure of eutectic 80Au/20Sn solder joint in laser diode package, *Microelectronic Engineering*, vol. 85, pp. 512–518, 2018.
184. Ronnie, J.W., Teo, F.L. Ng, L.S. Kip Goi, Y.F. Sun, Z.F. Wang, X.Q. Shi, J. Wei, G.Y., Li, "Microstructure of Eutectic 80Au/20Sn Solder Joint in Laser Diode Package", Microelectronic Engineering, V85, 2008, 512 (2008) .

185. Liu, Y.C., Teo, J.W.R., Tung, S.K., et al., High-temperature creep and hardness of eutectic 80Au/20Sn solder, *Journal of Alloys and Compounds*, vol. 448, pp. 340–347, 2008.
186. Zheng, P., Wiggins, A., Johnson, R.W., et al., Die attach for high temperature electronics packaging, in *International Conference and Exhinition on High Temperature Electronics*, Pittsburgh PA, 2008.
187. Lalena, J.N., Dean, N.F., Weiser, M.W., Experimental investigation of Ge-doped Bi–11Ag as a new Pb-free solder alloy for power die attachment, *Journal of Electronic Materials*, vol. 31, pp. 1244–1251, 2002.
188. Shi, Y., Fang, W., Xia, Z., Lei, Y., Guo, F., Li, X., Investigation of rare earth-doped BiAg high-temperature solders, *Journal of Materials Science: Materials in Electronics*, vol. 21, pp. 875–881, 2010.
189. Yamada, Y., Takaku, Y., Yagi, Y., Nishibe, Y., Ohnuma, I., Ishida, K., Novel Bi-based high-temperature solder for mounting power semiconductor devices, *R&D Review of Toyota CRDL*, vol. 43, pp. 43–50, 2006.
190. Zhang, H., Lee, N.C., Mixed Alloy Solder Paste, US Patent 9017446 B2, 2016.
191. Zhang, H., Lee, N.C., Mixed Alloy Solder Paste, US Patent 9636784 B2, 2016.
192. Zhang, H., Lee, N.C., Hybrid Lead-Free Solder Wire, US Patent 9802274 B2, 2016.
193. Zhang, H., Lee, N.C., Mixed Solder Paste, Korean Patent No. 10-1913994, 2022.
194. Zhang, H., Lee, N.C., A drop-in die-attach solution for high temperature lead-free BiAgX solder paste system, in *Proceeding of 2012 IMAPS High Temperature Electronics*, Albuquerque, NM, 2012, pp. 58–65.
195. Zhang, H., Lee, N.C., High melting lead-free mixed biags solder paste system, in *Proceeding of ICP Apex Expo*, Albuquerque, NM, 2012, p. S16-01.
196. Zhang, H., Lee, N.C., Hybrid Lead-Free Solder Wire, US Patent 9802274 B2, 2016.
197. Zhang, H., Minter, J., Wu, J., et al., Hybrid High Temperature Lead-Free Solder Preform, US Patent 10888958 B2, 2016.
198. Zhang, H., Mao, R.S., Lee, N.C., Reliability and corrosion resistance of Pb-free BiAgX® for die-attach, in *Proceeding of 2014 IMAPS High Temperature Electronics*, Albuquerque, NM, 2014.
199. Zhang, H., Minter, J., Lee, N.C., BiAgX® a drop-in HTLF solder and the reliability for 200°C applications, in *Electronics Packaging Symposium, GE Research Center*, Albany, NY, 2016.
200. Schoeller, H., Mallampati, S., Cho, J., *Isothermal-Fatigue of High-Temperature Solder Joints Proceedings of 2015 SMTAi*, Chicago, IL, 2015.
201. Zhang, H., Lytwynec, S., Wang, H., et al., Lead-free Solder Paste with Mixed Solder Powders for High-Temperature Applications, US Patent 11738411 B2, 2016.
202. Zhang, H., DurafuseTM HT: A drop-in high-temperature Pb-free (HTLF) solution that outperforms high-Pb solders, in *APEC Proceedings S10.2*, Houston, TX, March 20–24, 2022.
203. Zhang, H., Lim, S.P., Lytwynec, S., et al., A drop-in high-temperature lead-free solder paste that outperforms high-Pb pastes in power discrete applications, in *ICEP Proceedings*, Hokkaido, Japan, May 11–14, 2022.
204. Zhang, H., Richmond, T., Aserian, K., et al., A drop-in high-temperature Pb-free solder paste that outperforms high-pb pastes in power discrete applications, *Journal of Microelectronics and Electronic Packaging*, vol. 20, pp. 82–87, 2023.
205. Zhang, H., Hu, Y.J., Hu, D., et al., The challenges of high-temperature lead-free solder paste for power discrete applications, in *Proceeding of 2024 IEEE ECTC*, Orlando, FL, 2024, pp. 1113–1118.
206. IPC/JEDEC-J-STD-020- Revision D - Moisture/Reflow Sensitivity Classification for Nonhermetic Solid State Surface Mount Devices, Institute of Printed Circuits, https://shop.electronics.org/ipcjedec-j-std-020/ipcjedec-j-std-020-standard-only?f%5B0%5D=language%3A47.

207 AEC Q101 Rev.E Failure Mechanism Based Stress Test Qualification for Discrete Semiconductors in Automotive Applications, Automotive Electronics Council, https://www.aecouncil.com/Documents/AEC_Q101_Rev_E_Base_Document.pdf.
208 Zhang, H., DA5-Indium DurafuseTM HT Status Update Report, Nov. 2023, Reutlingen Germany.
209 Morshed, M., Wang, Y., Billa, L.R., et al., High-temperature lead-free solder alloy for high power semiconductor device packaging applications, in *Proceedings of PCIM Europe*, Nuremberg, Germany, 9–11 May 2023, pp. 1389–1393.
210. Takaku, Y., Felicia, L., Ohnuma, I., et al., Interfacial reaction between Cu substrates and Zn–Al base high-temperature Pb-free solder, *Journal of Electronic Materials*, vol. 37, pp. 314–323, 2008.
211. Rettenmayr, M., Lambracht, P., Kempf, B., et al., Zn–Al based alloys as Pb-free solders for die attach, *Journal of Electronic Materials*, vol. 31, pp. 278–285, 2002.
212. Yamaguchi, T., Ikeda, O., Oda, Y., et al., Three-layer Zn/Al/Zn clad solder for die attachment, *Journal of Electronic Materials*, vol. 44, pp. 751–760, 2015.
213. Li, L., Liu, Y., Gao, H., et al., Phase formation sequence of high-temperature Zn–4Al–3Mg solder, *Journal of Materials Science: Materials in Electronics*, vol. 24, p. 336, 2013.
214. Shimizu, T., Ishikawa, H., Ohnuma, I., et al., Zn–Al–Mg–Ga alloys as Pb-free solder for die-attaching use, *Journal of Electronic Materials*, vol. 28, pp. 1172–1175, 1999.
215. Haque, A., Lim, B. H., Haseeb, A. S. N. A., et al., Die attach properties of Zn–Al–Mg–Ga based high-temperature lead-free solder on Cu lead frame, *Journal of Materials Science: Materials in Electronics*, vol. 23, pp. 115–123, 2012.

4 Materials for Flexible Electronics

Weifeng Liu

4.1 OVERVIEW OF FLEXIBLE ELECTRONICS

4.1.1 History of Flexible Electronics

Flexible electronics refers to a category of electronics which can be bent, twisted, folded, or stretched and can conform to a certain desired shape under use conditions. Flexible electronics are made by assembling electronic components on flexible substrates with electrically conductive traces to connect the components. Electrically conductive traces can be copper circuits formed through the conventional PCB (printed circuit board) fabrication process, or conductive ink through additive manufacturing processes like screen printing.

The history of flexible electronics can be traced back to the first patent filed by inventor Albert Hanson in 1903, which describes a double-sided through hole copper circuitry bonded to a thin flexible polymer substrate [1]. New concepts of circuits were constantly developed with the evolution of new electronics such as radio and wireless electronics in the first few decades of the twentieth century. A breakthrough occurred in 1960 when Echo I, the first communication satellite, was built and launched to explore the new field of space communication [1]. The Echo I system was made by vacuum coating aluminum conductor on a polyester substrate. At the same time, Dupont introduced polyimide films under the trademark Kapton®, paving the way for more advanced and innovative flexible circuits. In 1969, Philips developed the first flexible printed circuit (FPC) made of polyimide. In the 1980s, Akzo developed a two-layer adhesive flexible copper clad laminate (FCCL), and thereafter various improved FCCLs were developed with enhanced properties such as superior dimensional stability, low moisture absorption, high heat resistance and excellent flexibility [2]. Concurrently, conductive inks were developed to print on flexible substrates to create bendable electronic circuits. To enable large-scale manufacturing, the roll-to-roll (RTR) process for making flexible circuits was developed using continuous rolls of flexible substrate fed through a series of rollers to enable efficient, high-volume output.

The 21st century marks significant advancements on new technologies, new materials, new processes and new applications for flexible electronics. In the early 2000s, organic electronics were developed, which can be directly printed onto flexible substrates. Advancements in materials and design, such as the development of organic light-emitting diodes (OLEDs), have made full-color, video-capable flexible displays a practical reality. Flexible hybrid electronics (FHE) drew wide industry attention in

 DOI: 10.1201/9781003587637-4

2015 when the first FHE manufacturing innovation institute (NextFlex) was established. Emerging wearable and medical applications added more opportunities as well as challenges for the industry to develop truly flexible, conformable and stretchable electronics.

Compared to rigid PCB, flexible electronics have many advantages. Flexible electronics can bend and flex, so they can adapt to different shapes, spaces and geometries, enabling more design freedom and customization. Flexible electronics can be lighter, take up less space and add less weight to the electronics systems, making them portable. With proper design, flexible electronics are ideal for dynamic or high flex applications, withstanding significant mechanical stresses with enhanced durability. Flexible electronics can be made through additive manufacturing processes, by directly depositing conductive traces (and potentially components) on the substrate, making them more environmentally friendly, and more affordable when cheap substrate materials are used. Flexible electronics can be made through the roll-to-roll manufacturing process, enabling high-volume scalable production. With advances in materials and technologies, flexible electronics continue to push the envelope to meet the growing demand across diverse applications with greater design versatilities and capabilities.

4.1.2 Categories of Flexible Electronics

Based on how the flexible circuit is built, flexible electronics generally can be classified into two types: the conventional flexible printed circuit (FPC) and the emerging flexible hybrid electronics (FHE). FPC refers to the electronics built on a thin polymer substrate (especially polyimide) using copper circuits, whereas FHE is the class of flexible electronics built through a combination of conventional semiconductor components and printed electronics. Although the concept of fully flexible electronics (FFE) was demonstrated successfully through printing processes, their use is currently still very limited.

Another method of classifying flexible electronics is through the extent of flexibility. FPC, although its name contains the word "flexible," is only bendable and twistable to some extent; FHE, on the other hand, is more flexible, foldable and conformable by printing conductive ink on soft substrates like TPU (Thermoplastic Polyurethane). Stretchable electronics is another type of flexible electronics which can be stretched without affecting their functionality. Conformable electronics are wearables or medical patches which conform to the human body.

Flexible electronics can also be categorized by their use and making, such as flexible display, flexible battery, flexible solar cell, conformable skin patch, flexible medical electronics, etc. Flexible electronics has found countless applications, and the most prominent applications are in the following market segments: consumer, wearable, medical, robotics, automotive, military and aerospace.

4.1.3 Flexible Electronics Applications

The global flexible electronics market size was valued at USD 27.11 billion in 2023 and is projected to grow from USD 29.40 billion in 2024 to USD 70.97 billion by 2032, exhibiting a CAGR of 11.6% during the forecast period [3]. The top applications for flexible electronics are in consumer goods, wearables, medicals, robotics, automotives, military and aerospace.

Consumer goods include smartphones, computers, printers, and cameras. The characteristics of flexible electronics such as flexibility and space savings are driving factors to integrate these flexible circuits into these consumer products. The majority of flexible circuits for these applications are passive wiring structures to interconnect different electronic components and with external connectors. Another notable application is flexible display by integrating organic light-emitting diodes (OLEDs) technology onto a flexible substrate, with the first foldable cell phone rolled out to market in 2019 [4]. Since then, significant improvements have been made in design and manufacturing processes to enhance the robustness of the flexible display.

For aerospace and defense, flexible electronics offer a lightweight and durable solution, ideal for applications where weight, space, and reliability are critical. Flexible circuits are used to replace cables to make connections between different devices and systems, one example is the flex circuits used in the Mars Rovers launched in 1996 [1]. Other applications include wearable technologies for military personnel.

In the automotive field, flexible circuits are used in apparatus panels, under-hood controls, circuits to be covered within the headliner of the cabin, and in ABS systems [5]. Additionally, the automotive industry leverages flexible displays and sensors to enhance vehicle interiors with curved touchscreens and support the development of smart connected vehicles. One emerging application is in-mold electronics (IME) or in-mold structural electronic devices (IMSD), for which electronic components are mounted on printed ink traces, formed on a curved structure and then encapsulated with injection molding [5].

For healthcare and medical applications, flexible electronics are used to monitor biophysical and biochemical signals without being invasive, facilitating real-time health monitoring and personalized patient care. They are also used to integrate sensors, displays, and wireless connectivity in a slim and comfortable wristband. Thin bendable flexible circuits are used in the glucose patch, allowing for the electrical connection and data transmission from the glucose sensor to a reader device, enabling continuous monitoring of blood glucose levels without the need for frequent finger pricks. Flexible circuits are also used in electrical implants, such as Bionic Eyes which are surgically inserted into eyes to restore vision [1]. For this application, the flex circuit needs to conform to the shape of the eye's anatomy, allowing for the implantation of retinal electrodes to potentially restore vision to people with blindness. The core concept of this application is to design a thin, flexible circuit to connect with delicate retinal implants within the eye, enabling signal transmission for vision restoration.

For robotics, flexible circuits are used to provide electrical connections in areas where a rigid board is impractical to use due to rigidity, as flexible circuits can offer more flexibility, adaptability and ease of movement in robot designs, particularly in joints, limbs, or other areas requiring bending, twisting, torsion or stretching [6]. Flexible circuits are also used to connect sensors to the robotic body, joints and other electronics, allowing for seamless integration into moving parts of robotics without compromising the connection to sensors.

Wearable technology is a broad category that includes many different devices and applications, including consumer, healthcare, automotive, aerospace and military [7].

Wearable devices can provide real-time monitoring of human biomarkers and environments and feedback to users and healthcare providers. Typical examples include smartwatches, fitness trackers, VR headsets, smart glasses, and smart textiles. For smart textiles, sensors and electronics are integrated into garments to fully conform to the body curvatures and movement. For this application, printed electronics play a vital role due to their use of highly conformable and stretchable substrates and conductive inks.

4.1.4 Challenges and Opportunities

Flexible electronics face many challenges in terms of materials, design, manufacturing, reliability and applications. The most commonly used material, polyimide, in FPC, is very expensive, less flexible and conformable, and is hygroscopic by adsorbing moisture even in normal ambient conditions. Alternative materials are used to replace polyimide because they are lower in cost, more flexible, conformable, and even stretchable, but many of them cannot withstand higher processing temperatures. New materials are being developed; however, they are still in the early development or commercialization stage, and adoption by the market and users is still a long way to go.

Although flexible electronics enable design freedom, there is a lack of design tools and design rules to guide the design engineers, especially for emerging technologies like FHE. Flexible electronics face more challenges in the assembly process, as special fixtures need to be designed to hold the flexible circuit in place. Limited interconnect solutions are available to mount components on substrates which cannot withstand high processing temperature. Flexible electronics pose an increased risk of damage during handling and use, especially for the interconnect between components and circuits. This requires either fastener or encapsulation to reinforce the interconnect structure, which adds to the manufacturing complexity and cost. Although flexible technologies advance quickly, some of the emerging technologies are not mature yet and market adoption is still far away, due to lack of functionality, user experience or reliability.

Overall, the benefits of flexible electronics enable significant versatile opportunities in applications, design and prospect of low-cost manufacturing processes. With advances in materials, manufacturing processes, performance and reliability, flexible electronics will continue to find a great variety of opportunities across a variety of industries, including healthcare, automotive, space and military, supply chain logistics, robotics and automation, and consumer electronics.

4.2 FLEXIBLE PRINTED CIRCUIT (FPC)

4.2.1 History of FPC

The development of flexible printed circuits (FPC) began with DuPont's invention of polyimide films in the 1950s [2]. Since then, various innovations and products of flexible circuits have been developed and rolled out to the market with enhanced properties such as high dimensional stability, low moisture absorption, excellent heat resistance, and superior flexibility. At the same time, the roll-to-roll or reel-to-reel (RTR) process for making a flexible circuit was developed by feeding a continuous roll of flexible substrate material through a series of rollers to enable a continuous, high-volume manufacturing process. Over the last twenty years, new technologies, new materials and new processes have constantly emerged on the market [2]:

- New FPC substrate materials, such as liquid crystal polymer, PEN film, heat stabilized PET film, ultra-low dielectric constant porous polyimide film, and aramid fiber ultrathin substrate.
- New electrolytic copper foils with low profile and excellent bending performance.
- Ultra fine line and space of copper traces (line width/spacing down to 10/10 μm).
- High difficulty rigid-flex PCB with the rigid PCB area using any layer technology.
- Multi-layer FPC with embedded components.

4.2.2 FPC Construction

IPC has defined five variations of FPCs [8]. Type 1 FPCs are single-sided flexible printed boards containing one conductive layer. Type 2 FPCs are double-sided flexible printed boards containing two conductive layers with plated interconnect holes providing access to the conductive traces. Type 3 FPCs are made of three or more layers and feature an insulating material between their respective layers and plated-through-holes (PTHs) to connect the conductive layers. Type 4 FPCs are multi-layer rigid and flexible materials combinations containing three or more conductive layers with PTHs. Type 5 FPCs are similar to type 4, but without PTHs.

An FPC consists of three key components, as shown in Figure 4.1 [9]:

- **Flexible Substrate**: Thin insulating base material that provides flexibility. Common materials are polyimide and LCP.
- **Conductive Traces**: Usually copper traces that are etched or patterned on the substrate to form the circuits.
- **Coverlay**: Thin insulating film laminated over the traces for protection and isolation.

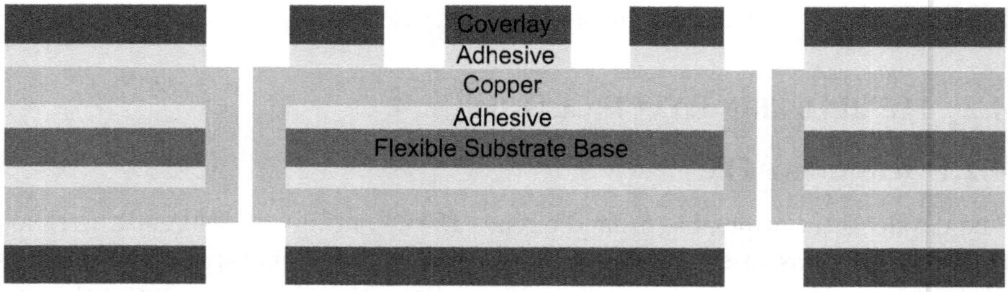

FIGURE 4.1 Schematic of FPC stack-up.

4.2.3 FPC Materials

4.2.3.1 Flexible Substrate Base

The flexible substrate base for FPC is a thin, flexible material that acts as support for the conductive traces connecting components and provides insulation between the conductors. Common substrate materials include polyimide (PI), polyester (PET), liquid crystal polymer (LCP), and polyethylene naphthalate (PEN). Although it provides exceptional high frequency performance and thermal stability, polytetrafluoroethylene (PTFE) is rarely used for FPC applications due to its high cost and manufacturing complexities.

The most common substrate material for FPCs is polyimide. Polyimide provides excellent dimensional stability, excellent electrical and chemical properties, high adhesion, and high thermal and mechanical resistance.

Another commonly used FPC base material is liquid crystal polymer (LCP), particularly when high frequency performance is needed, in millimeter wave applications for example, due to its excellent dielectric properties, high temperature resistance, and low moisture absorption capabilities.

PET film is a low-cost alternative to polyimide and LCP and is used for applications where the circuit is exposed to non-hostile environmental conditions. To enhance the thermal performance, heat stabilized PET was developed with added thermal stabilizers to enhance the material's resistance to high temperature while maintaining structural integrity without breakdown and distortion.

The polyethylene naphthalate (PEN) film is developed for cost-effectiveness with performance between that of polyethylene terephthalate (PET) and polyimide. It demonstrates excellent dimensional stability, low thermal shrinkage, chemical and solvent resistance, low moisture adsorption, great hydrolysis resistance, and high mechanical strength. It offers better performance than PET and is a cost-effective alternative to PI for demanding environmental and use conditions [10]. Heat stabilized PEN was also developed to further enhance its properties.

A comparison between these materials is listed in Table 4.1 [11–16].

4.2.3.2 Coverlay

Coverlay is the insulating layer placed over the exposed conductors. It is typically constructed of a pre-cut film laminated over the conductors using either acrylic or epoxy adhesive.

The coverlay material plays a critical role in the performance and reliability of FPC. The coverlay must exhibit excellent electrical insulation properties, thermal stability, chemical resistance, and mechanical flexibility. Key coverlay materials and their properties are listed below [17]:

Polyimide

- Excellent flexibility and durability
- High thermal stability and chemical resistance
- Can withstand temperatures up to 400°C
- Good electrical insulation

TABLE 4.1
Comparison between FPC Substrate Materials [11–16]

	PI	PET	Heat Stabilized PET	LCP	Heat Stabilized PEN
		Physical Properties			
Color	Yellowish-brown (most common)	Transparent	Transparent	Off-white or black	Transparent
Density (g/cm³)	1.37	1.36	1.38–1.45	1.7	1.35
Water absorption (24 hours)	1.4%	0.6%	0.6%	0.038%	0.1
		Thermal Properties			
Tg (°C)	366	75	78	120	120
Melting temp. (°C)	N/A	246	265	309	265
Max operating temp. (°C)	260	80	150	130	200
CTE (ppm/°C)	42	25	25	6.9	21
Shrinkage (150°C)	0.2%–1.2%	0.2%–3%	0.1%–0.2%	0.1%–0.6%	<0.1%
Flammability	UL 94-V0	UL94 HB-5VB	UL94 HB-5VB	UL94 V-0	UL94 VTM-2
		Mechanical Properties			
Tensile strength (psi)	17,500	10,674	11,500	18,564	16,389
Tensile modulus (psi)	540,000	471,372	580,151	1,725,949	725,189
Flexural strength (psi)	25,000	15,083	18,000	21,900	14,500
Flexural modulus (psi)	550,000	530,838	1,420,000	1,667,934	536,640
		Electrical Properties			
Dielectric constant	3.2, at 1 MHZ	3.26	2.9	3.81	3.2@10KHZ
Dissipation factor	0.005, at 1 MHZ	0.00855	0.005	0.0133	0.0048@10KHZ
Dielectric strength (V/mil)	395, short time, 1/8″ thickness	599	3,000	952	4,000
Surface resistivity (ohm/square)	$>10^{13}$	1.89×10^{15}	10^{17}	1.46×10^{16}	10^{14}

Polyester

- Lower cost alternative to polyimide
- Low thermal and chemical resistance, but can be improved through heat stabilizer

Liquid Crystal Polymer (LCP)

- Extremely high thermal stability (up to 280°C continuous use)
- Low moisture absorption and excellent electrical properties

Flexible Epoxies

- Low-cost coverlay option
- Limited thermal and chemical resistance

The coverlay thickness is typically in the range of 25–100 µm (1–4 mil), with thinner materials providing maximum flexibility. The adhesive to bond the coverlay to the base substrate can be acrylic, epoxy, or polyimide.

4.2.3.3 Conductive Layer

Copper is widely used as FPC conductors, which offers excellent electrical conductivity and flexibility. The copper foil thickness of flex PCBs typically is in the range of 18 µm (0.5 oz) to 35 µm (1 oz). Two types of copper are used for FPCs: ED (Electrolytic deposition) copper or RA (rolled annealed) copper. ED copper foil is made by electroplating an acid copper plating solution on stainless steel to precipitate a uniform copper film. RA copper is made from a copper ingot which is rolled and annealed repeatedly until it reaches the required thickness [18].

ED copper features a vertical grain structure and is less expensive than RA copper. It is also easier to etch, enabling a higher circuit density and more consistent trace width. However, ED copper is easier to break and does not have much bending capability; therefore, ED copper is typically used for flexible circuits which do not require repeatable bending. ED copper exhibits a very wide range of surface roughness from <1 to 4 µm. For high frequency applications, low roughness is essential to minimize signal loss. To achieve this, ED copper can be specially treated to make low-profile foils with very low roughness and at the same time with sufficient roughening for adhesion to the substrate. These copper foils are often categorized as ultra-low profile (ULP, ~0.1 to 0.3 µm roughness), very low profile (VLP, ~0.25 to 0.5 µm roughness), and simply low profile (LP, ~1 µm average roughness) [18]. ED copper is typically bonded to the substrate through the lamination process.

RA copper has an in-plane grain structure which provides greater copper elongation, flexibility and ductility, and supports more repetitive bending. RA copper uses adhesives to bond the conductor layers and the base layers.

4.2.3.4 Adhesives

Adhesives are used to bond different layers of the FPC structure together. The most common adhesives used in FPC manufacturing include polyesters (PET), polyimides (PI), acrylics, and modified epoxies, with each having their advantages and disadvantages, depending on the application [19].

4.2.3.5 Stiffener

An FPC stiffener is a relatively thick, rigid mechanical structure that is fixed to the FPC in certain areas using adhesives to prevent excessive flexing, twisting or bending of the FPC during assembly and use [20]. Stiffeners protect delicate components as well as the FPC substrate itself from damage due to mechanical stresses. They also provide mechanical reinforcement for connectors and interface contacts to external devices.

FPC stiffeners are commonly made using materials like stainless steel, aluminum, FR4, polyimide and acrylics, bonded to FPC through adhesives such as epoxies and acrylics. The stiffener often includes holes and cutouts to allow mounting of ICs, connectors, buttons and other components.

4.2.4 FPC Fabrication

Similar to rigid PCBs, the fabrication of FPCs follows a subtractive manufacturing process. It starts with copper clad laminate (CCL), and after photoresist deposition, exposure and development, a conductive pattern is formed through etching. Coverlay is then laminated using adhesives over the circuitry as an insulating layer. SMT pads are exposed through laser ablation of the coverlay, followed by plating a protective surface finish. For multi-layer FPCs, steps also involve hole drilling and plating through-holes. To optimize the materials utilization, the working materials are typically pre-cut according to the design and layout before the subtractive manufacturing process begins.

4.2.5 FPC Assembly

Components can be assembled on FPC boards through either a standard SMT process or a roll-to-roll process. For the SMT process, due to the nature of flexibility of FPC, a jig is needed to secure the FPC to a rigid pallet to facilitate baking, solder paste printing, component placement, reflow and final inspection. The typical FPC material, polyimide, is known to absorb moisture from ambient, therefore the FPC needs to be baked prior to the SMT assembly process; otherwise, the damped FPC will delaminate during the reflow process. Typical baking conditions are 105°C–115°C for 2–4 hours, depending on the layer count and environment exposure history. After baking, FPCs need to be stored in a vacuum bag, or an oven at 35°C with low humidity, prior to assembly. Besides the standard lead-free SMT process, other assembly techniques are developed to reduce the heat exposure for FPC assembly, either through reducing the peak reflow temperature by using a low temperature solder, reducing exposure duration by induction heating or photonic soldering, or using low temperature curable conductive adhesives.

The roll-to-roll or reel-to-reel manufacturing method for FPC is a continuous production process where the flexible circuit sheet is assembled with semiconductor components in a roll form. This process involves several steps: baking the unwound FPC roll, feeding the FPC roll into the SMT line, solder paste printing, component placement, reflow soldering, automated optical inspection, visual inspection, rewinding, and finally cutting into individual units. Throughout the process, tension is maintained to ensure the flatness of the FPC, eliminating the need for a jig.

4.2.6 FPC Design Rules

Many factors need to be considered when designing a flexible circuit, such as the minimal bending radius, trace design, placement of components, and use of stiffeners. A designer needs to consider specific use conditions to make sure that the design characteristics of FPC meet the requirements.

TABLE 4.2
Rule of Thumb for Bending Radius Requirement of FPC [21,22]

Number of Layers	Static Bending Radius	Dynamic Bending Radius
Single sided	FPC thickness X6	FPC thickness X100
Double sided	FPC thickness X12	FPC thickness X100
Multi-layer	FPC thickness X24	FPC thickness X100

The minimum bend radius of FPC is the minimum degree up to which the flexible area of FPC can be bent without causing damage. The bending can be static, i.e., only bent during the assembly process, or dynamic, in which case the flex circuit will see many cycles of bending during its use. The bending area and bending requirements must be properly defined during the design stage. In general, the bending radius depends on the FPC layer count and thickness as well as the required number of cycles during use. A rule of thumb has been established by industry and is shown in Table 4.2 [21,22].

From Table 4.2, reducing the overall thickness of FPC will improve flexibility. This can be achieved in multiple ways, such as reducing the copper and dielectric layer thicknesses and using an adhesiveless base material. Conductors need to be balanced with equal weight on each side of the neutral bending axis and staggered from layer to layer. Teardrop designs need to be used on traces and pads at locations where traces meet pads or vias (or where there is a change of trace width), to reduce stress concentration and minimize the risk of cracking. Curved traces (instead of sharp corners) should be employed to reduce stress at the bending area. Components should be kept out of the bending area and plated-through-holes (PTHs) should be at least 20 mils away from the bend area [21]. For the transition from flex to rigid areas, sufficient space should be allocated. Tear guards can be applied along the inside bending radius to reinforce the flex substrate and prevent tearing of the substrate; alternatively tear relief can be incorporated to minimize stress concentration to prevent initiation and propagation of tears. A reinforcement stiffener can also be bonded to the surface of FPC to provide additional stiffness and structural support, preventing FPC from excessive bending and twisting.

4.2.7 FPC Challenges and Opportunities

The main challenges with FPC include design complexity due to bending restrictions, a lower component density due to limited layer counts for routing and space to accommodate bending, assembly and test complexity due to flexibility, and a high initial cost due to use of expensive materials and manufacturing complexities. Polyimide is known for its high cost as well as moisture absorption. Flexible circuits using polyimide base material must be baked prior to the soldering process; otherwise, delamination may occur. Cheaper materials like polyester can be a good alternative to polyimide; however, polyester has limitations in terms of the maximum temperature it can withstand during the assembly process. Thermally stabilized polyester films were developed; however, it still cannot withstand the typical lead-free reflow environment. To address this challenge, low temperature soldering solutions have been developed to minimize damage to the FPC circuit materials.

With benefits in many areas, flexible circuits continue to show a promising future with opportunities in many applications such as wearable electronics, medical and healthcare, consumer electronics, automotive, military and space applications, driven by their miniaturization, compactness, flexibility and conformity to complex shapes. At the same time, flexible circuits continue to witness rapid technological advancements and innovations in materials, design and manufacturing techniques. For FPC materials, higher flexibility, higher Tg, lower dielectric constant and loss tangent, and halogen-free, are some of the FPC characteristics for future development trends. A key area of ongoing development is the use of the graphene material or graphene composites in FPC applications, as they provide improved flexibility, durability, and superior thermal and electrical performance.

4.3 FLEXIBLE HYBRID ELECTRONICS (FHE)

4.3.1 Introduction to FHE

Flexible hybrid electronics (FHE) refers to a special category of electronic assemblies made by mounting components onto circuits formed by printing electrically conductive paste on a flexible substrate. By integrating semiconductors and passive components with printed pastes and flexible substrates, FHE has the potential to revolutionize industries such as IoT and wearables by providing unique form-factors, lighter-weight assemblies and reduced manufacturing costs using high-volume ink printing technologies. With FHE, heterogeneous system integration can be achieved by forming input devices like printed sensors as well as output devices like haptics onto the same substrate as active and passive semiconductor components. These constructions can be designed to be flexible, pliable, and even stretchable, allowing them to conform to the curves of a human body or other irregular shapes, while maintaining the full functionality of traditional electronic systems [23].

Electronic assembly manufacturing technology has continued to evolve for more than 60 years. The early generations of electronic assemblies were rigid, bulky constructions with components mounted on stiff, multi-layer printed circuit boards using tin-lead based solder. Later, flexible printed circuit (FPC) boards were developed for applications requiring static or dynamic bending. Future generations of electronic assemblies may be made entirely with printing or other additive manufacturing technologies. Not only may the circuit traces be formed by printing conductive pastes, but the passive and active semiconductor devices may be printed using functional pastes. However, compared to conventional surface mount semiconductor components, the performance and functionality of the printed semiconductors are still far from meeting most application performance requirements. Conventional semiconductor components will continue to dominate the market in the foreseeable future.

Flexible hybrid electronics (FHE) technology is the "next generation" solution for making electronic assemblies that provide the benefits of using conventional semiconductor and passive component form-factors, while leveraging the capabilities of printed electronics to enable scalability, lower manufacturing costs, and system flexibility. Table 4.3 provides a relative comparison between PCB, FPC, and FHE.

TABLE 4.3
Qualitative Comparison of Typical PCB, FPC and FHE

	PCB	FPC	FHE
Circuit Formation	Subtractive (Etching)	Subtractive (Etching)	Additive (Printing)
Typical Conductor Material	Copper	Copper	Conductive paste (e.g. silver ink)
Conductivity	High	High	Low
Solderability	Yes	Yes	Possible
Typical Substrate	Epoxy-Glass	PI, PEN, LCP	PI, PET or TPU
Resin Class	Thermoset	Thermoset or Thermoplastic	Thermoset or Thermoplastic
Substrate Flexibility	Rigid	Bendable	Bendable (PI, PET) Stretchable (TPU)
Temp. Tolerance	High	Low to High	Low to High
Assembly Process	SMT	SMT	ACP, ACF or ECA

4.3.2 Structure of FHE

Flexible hybrid electronics (FHE) are made by combining printed electronics with traditional circuit boards, and are constructed using substrates, inks, components and encapsulation to form a heterogeneous structure.

- **Substrate**: A flexible substrate is used as the base for printing conductive traces. Common substrates include PI, PET, TPU, paper, or textile.
- **Ink**: Conductive and functional inks are printed onto the substrate to create conductive traces, printed sensors, and printed active and passive devices. Conductive inks can be made from a variety of materials, including silver, copper, zinc, nickel, and carbon.
- **Components and Integration**: Conventional semiconductor devices, either active or passive, can be mounted on the conductive traces through a variety of interconnecting methods, such as anisotropic conductive paste (ACP), anisotropic conductive film (ACF), or electrically conductive adhesives (ECA).
- **Encapsulation**: The FHE platform needs encapsulation to protect the circuit, the interconnect and the devices from various use conditions such as flexing, folding, and environmental exposure.

4.3.3 FHE Materials

4.3.3.1 Substrate

Flexible hybrid electronics (FHE) substrates are flexible materials that support electronic devices and circuits. Commonly used substrates for ink printing include polyethylene terephthalate (PET), polyethylene naphthalate (PEN), Polycarbonate (PC), polyimide (PI), and thermoplastic polyurethane (TPU). Among these materials, PI demonstrates exceptional heat resistance and mechanical durability; however, it is much more expensive (100× more expensive than PET), and therefore is not suitable for low-cost applications, such as disposable smart labels. PI's orange color also prevents it from optical applications like flexible display. Moreover, PI is not stretchable and to some extent is not foldable. PI is conformable; however, its conformability is not as good as TPU. PET on the other hand, is the least expensive substrate material among plastics; however, it is heat sensitive, and deformation or shrinkage may occur at elevated temperatures for conductive ink curing (typically 120°C). To enhance its thermal resistance, heat stabilized PET was developed to maintain dimensional stability even at elevated temperatures without significant shrinkage. Figure 4.2 shows a comparison of the upper operating temperatures for different substrate materials [24].

Polyethylene naphthalate (PEN) is chemically similar to PET but is more temperature resistant. To further enhance its heat resistance, heat stabilized PEN was developed to raise its shrinkage temperature above 190°C and is certified for long-term electrical use at 155°C. PEN also has better UV resistance and is more resistant to hydrolysis in aqueous alkaline conditions. However, PEN films are more expensive and have lower flex durability.

Polycarbonate (PC) is mainly used in structural electronics or in-mold electronics (IME), due to its thermoplasticity, transparency, high impact strength, high heat deflection resistance, low moisture adsorption, and high voltage insulating properties. Upon thermoforming to create the 3D structure, PC will maintain excellent mechanical and dimensional stability.

Thermoplastic polyurethane (TPU) is a melt-processable thermoplastic elastomer with the characteristics of both plastics and rubber, in terms of high durability, flexibility and stretchability. It has been widely used in the textile industry as a lamination material due to its ease of processing, strong adhesion, excellent abrasion resistance, UV and weather resistance, and breathability. TPU has been a default substrate for the development of stretchable electronics due to its high elongation and low hysteresis. There are three types of TPU films: polyester, polyether, and polycaprolactone. Polyester TPUs have excellent abrasion and chemical resistance, and superior tensile strength. Polyether TPUs have excellent hydrolysis and microbial resistance along with outstanding temperature resistance. Polycaprolactone based TPUs offer performance between that of polyester- and polyether-based TPUs.

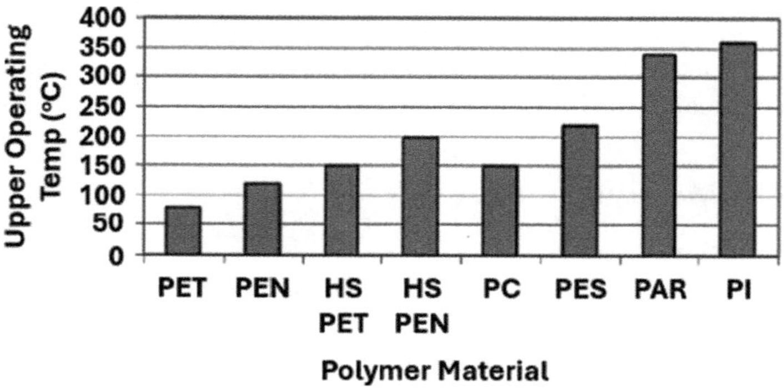

FIGURE 4.2 Upper operating temperature limits of substrate materials [24].

Key attributes to assessing the substrate materials applicability for FHE applications include:

Mechanical Performance: this may include mechanical strength and modulus, flexibility, stretchability, and conformability based on specific application needs. The material shall also demonstrate low hysteresis when the external load is removed.

Heat Resistance: the substrate material needs to demonstrate physical and chemical stability during manufacturing and use. This is particularly critical when high-precision fine line ink printing is required, as any distortion or shrinkage of the substrate at typical ink curing temperatures can compromise the accuracy of fine line printing.

Compatibility with Ink: The substrate material needs to be compatible with the conductive ink in terms of good adhesion and printability. The adhesion strength between the substrate and the conductive ink shall be strong enough to withstand intended use conditions, especially cycles of bending and stretching. For optimal printability, the ink must exhibit good wetting on the substrate to prevent broken lines; however, excessive wetting must be avoided to prevent issues such as blurred edges and line bridging. This balance is especially crucial for fine line ink printing. To assess the ink printability, the surface tension of the conductive ink is typically used to compare with the surface energy of the substrate. In ideal situations, the surface tension of an ink should be equal to or a little less than the surface energy of the substrate, as shown in Figure 4.3.

The surface energy of typical substrate materials used in FHE applications is listed Table 4.4. However, the actual value may vary significantly depending on the material type, additives, the manufacturing process, and treatment (such as coating).

Process Compatibility: Compared to polyimide, other substrate films used for FHE applications have much lower processing temperature limits. This limits their

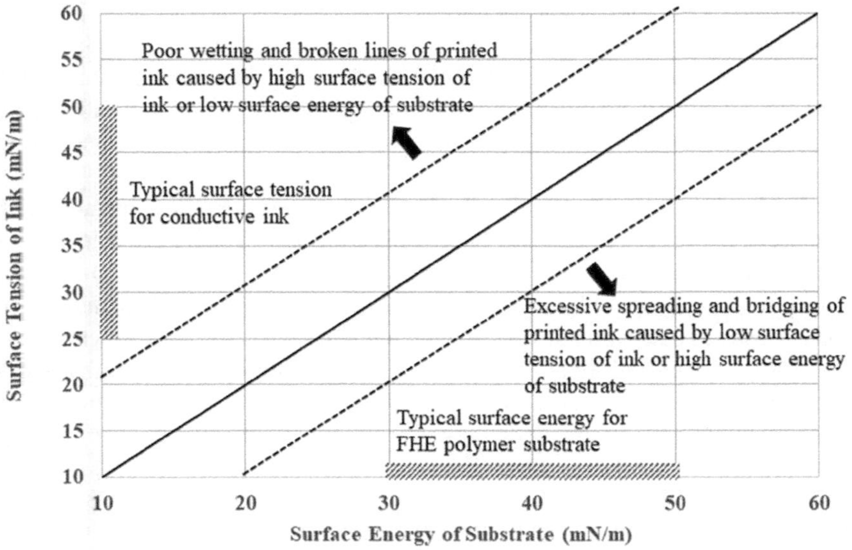

FIGURE 4.3 Compatibility between the conductive ink and the substrate.

TABLE 4.4
Surface Energy of Substrate Materials

Substrate Material	Surface Energy (mN/m)
polyethylene terephthalate (PET)	42
Polycarbonate (PC)	46
Polyimide (PI)	45–50
Polyethylene naphthalate (PEN)	40
Thermoplastic polyurethane (TPU)	30–40

TABLE 4.5
Process Temperature Limits for Substrate Materials

Substrate Film	Melting Temp.	Process Temp. Limit
Polyimide	N/A	350°C–400°C
PET	235°C–260°C	120°C
Heat stabilized PET	265°C	180°C–200°C
PEN	269°C	150°C
Polycarbonate (PC)	280°C–320°C	150°C–190°C
TPU	185°C–220°C	<150°C

application using conventional assembly processes. Thermoplastic films like PET and TPU have relatively low temperature resistance and cannot withstand the current lead-free solder reflow temperature regimens. Table 4.5 lists typical process temperature limits for FHE substrate materials.

Besides operating temperature limits, another key attribute for the substrate film is its shrinkage during high temperature manufacturing processes. A film with low shrinkage is essential during the manufacturing process to maintain the dimensional accuracy of the printed ink traces, which is especially critical for printed ink with fine features.

To address these challenges, the FHE industry requires new substrate and circuit materials that are stretchable/conformable and can withstand conventional SMT assembly temperatures, as it is considered an attractive proposition to utilize the existing high-volume SMT infrastructure and equipment for FHE manufacturing. Figure 4.4 shows a new film developed with high thermal resistance, which can withstand 260°C solder bath float, while maintaining flexibility, stretchability and low shrinkage [25]. By comparison, both PET and TPU films are observed to fully melt at such temperature.

4.3.3.2 Printed Inks

Printed inks used in printed electronics can be divided into three categories: conductive inks, dielectric inks, and functional inks. Conductive inks contain electrically conductive materials (like metal particles) embedded in a polymer matrix, enabling

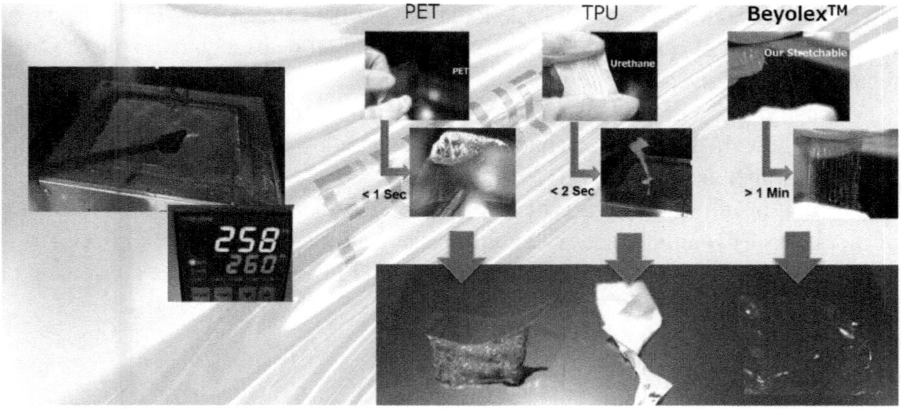

FIGURE 4.4 A new stretchable substrate material which can withstand lead-free reflow temperature [25].

them to be printed onto a substrate to form electrical connections. Dielectric inks are non-conductive inks to encapsulate conductive traces as an insulating layer, acting as an electrical barrier to prevent short circuits and protect the circuitry against the use environment. Functional inks (also called specialty inks) are formulated to perform certain functions or unique capabilities such as sensors, heaters, resistors and capacitors. Inks can also be classified based on how they are deposited on the substrate or the manufacturing process, such as inks for screen printing, inkjet ink, etc.

Key attributes for assessing the performance of conductive inks include electrical conductivity or resistivity, viscosity and rheology, adhesion, flexibility and stretchability, and durability.

Electrical Conductivity: The primary function of conductive inks is to form electrical pathways that interconnect various functional components within a system and establish connections to external devices. Conductive inks with low resistivity can minimize energy loss and heat generation while maintaining the integrity of signal transmission. A conventional way to characterize the conductivity of conductive ink is the sheet resistance. Sheet resistance is the resistance of a square piece of thin material with uniform thickness.

In a regular three-dimensional conductor, the electrical resistance is expressed as follows:

$$R = \rho \frac{L}{A} = \rho \frac{L}{WT} = \frac{\rho}{T}\frac{L}{W} = R_s \frac{L}{W} \tag{4.1}$$

Where

R_s is the sheet resistance

$$R_s = \frac{\rho}{T} \tag{4.2}$$

ρ is the material resistivity

L is length of the conductive ink trace; and

A is the cross-sectional area, which is width W times thickness T.

Sheet resistance is also commonly denoted as "ohm per square" or "Ω/sq." The reason for using the sheet resistance is that it is the actual resistance of a square sheet, regardless of the size of the square.

Viscosity: The viscosity of conductive ink is a measurement of how easily an ink flows and how thick it can be printed. An ink with a low viscosity flows more easily and gets thinner as it flows.

The unit of ink viscosity is typically expressed as centipoise (cP), which is a unit of measure of dynamic viscosity based on the centimeter-gram-second (CGS) system. The SI unit of viscosity is expressed as Pascal second (Pa.s). The conversion of the two units is as follows:

$$1 \text{ Pa.s} = 1{,}000 \text{ cP} \tag{4.3}$$

Typical viscosities for printed inks are listed in Table 4.6 [26].

There are different methods of measuring ink viscosity, such as the Zahn cup method, the vibrational viscometer, and the Elcometer flow cup method. The Zahn cup method measures the time it takes for a set volume of ink to flow through the cup, as specified by the ASTM D 4212, ASTM 1084 and ASTM D816 methods. The vibrational viscometer method determines the viscosity value by measuring the damping of a resonator immersed in the ink. In the Elcometer flow cup method, the ink flows through an orifice, and the flow time is used as a relative measurement of viscosity. A variety of cup designs are specified by the Elcometer method [27].

Rheology: Rheology of a conductive ink characterizes how the ink deforms and flows when stress is applied during the manufacturing process. It is crucial to achieve proper rheology to obtain optimal printing quality, as rheology directly impacts ink flow and deposition on the substrate and accordingly printing resolution and quality. To characterize ink rheology, a rheometer is used to measure the ink viscosity and flow behavior under different shear rates to assess the ink's viscosity, yield stress, and thixotropic and viscoelastic properties. It provides crucial information about how ink behaves during the manufacturing process. Different types of rheology test methods have been developed, including [28–30]:

Rotational Rheometry: This method uses instruments such as rotational viscometers and rotational rheometers to measure viscosity. These instruments typically apply a shear stress to the material and measure the resistance to

TABLE 4.6
Conductive Ink Viscosity [26]

Ink Printing Methods	Conductive Ink Viscosity (cP)
Screen printing	3,000–100,000
Flexography	10–1,000
Gravure	10–1,000
Inkjet	1–100

flow using geometries such as a rotating rod inside a cup filled with the liquid sample, two parallel double plates, or cone-and-plate configurations.

Oscillatory Rheometry: This method uses oscillatory rheometers to apply a controlled sinusoidal oscillating shear stress to the material, measuring the torque and the deflection angle of the measuring bob to determine both viscous and elastic properties of the material.

Capillary Rheometry: This method involves forcing the liquid sample through a capillary tube. The flow rate and pressure drop are measured as the liquid is pumped through the tube, to determine viscosity and other rheological properties.

Extensional Rheology: This method, also known as elongational rheology, is the study of how the material deforms under an extensional or stretching force. It measures the material's resistance to stretching or elongation by monitoring changes in its cross-sectional area during stretching, providing information about its extensional viscosity and other related properties.

Thixotropy Measurement: Thixotropy is a time-dependent shear thinning property of a material. It is a reversible process where a material becomes less viscous when agitated but gradually recovers to its original state when the agitation ceases. Thixotropic properties in ink are essential for fine line ink printing because they allow the ink to remain thick and stable when stationary, yet readily flow and thin out when the printing process starts, enabling precise control of the printing process to create sharp, well-defined fine lines without excessive spreading or bleeding. The thixotropic property of ink is typically expressed as viscosities at different shear rates.

Flexibility and Stretchability: FHEs are designed to be flexible or stretchable, meaning that they can bend or stretch to a certain extent without compromising the integrity of the electronic circuit, enabling them to conform to irregular surfaces and be integrated into applications which require adaptability to movement and deformation. Figure 4.5 shows the stretchability of a stretchable conductive ink in terms of change in electrical resistance at 5% and 20% stretch respectively [31]. With more

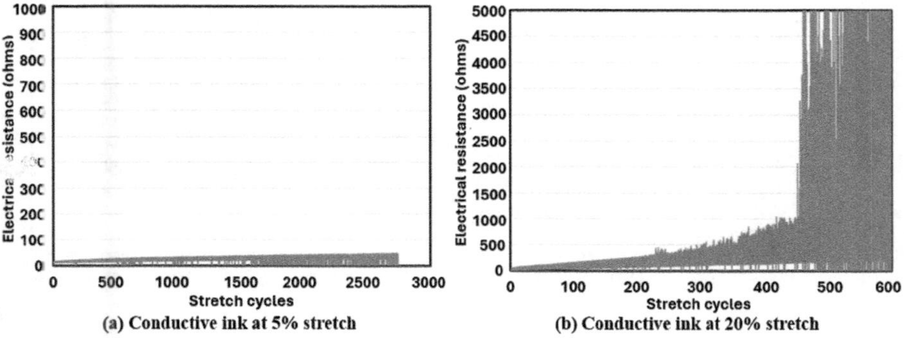

FIGURE 4.5 Stretchability of conductive ink at different stretch percentages and cycles: (a) 5%, (b) 20% [31].

stretchable cycles, the electrical resistance continues to increase, the higher the percentage of the stretch, the more pronounced increase of the electrical resistance. Based on the application requirements, the durability of the stretchable ink can be determined.

Many factors affect the stretchability of conductive ink, such as the ink formulation, printed ink width, ink circuitry design, substrate film type, substrate film thickness, and encapsulation. To enable good flexibility and stretchability, conductive ink must be formulated using conductive particles with optimal particle shape and size distribution, combined with a highly flexible and stretchable polymer binder. Additionally, the substrate film and the encapsulant must also exhibit excellent flexibility and stretchability. Strong adhesion of the conductive ink to both the substrate and the encapsulant is essential to ensure reliable and durable performance of the conductive ink. Figure 4.6 shows a comparison of conductive ink printed on two different substrate films at 20% stretch, which clearly highlights the significant difference between the two test configurations [31].

The stretchable conductive ink is typically composed of metal flakes embedded in a polymer matrix. At a low stretch percentage, metal flakes tend to slide against each other, which may cause an increase in electrical resistance; however, it still maintains the connectivity within the circuitry. When the stretch percentage is high, the polymer matrix may not be able to hold the particles in place, and cracks will form, as shown in Figure 4.7 [31]. The cracks serve as a stress relief, with the metal flakes still connected at the tips of the cracks. With these stress-relieving cracks, the printed ink can be stretched at a higher percentage without losing connectivity.

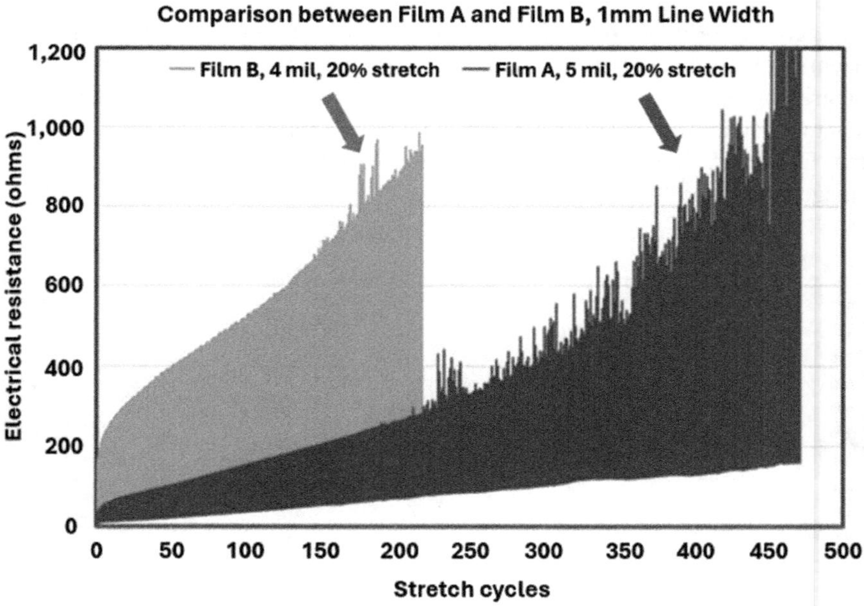

FIGURE 4.6 Comparison of stretchability (20% stretch) of conductive ink printed on different substrates [31].

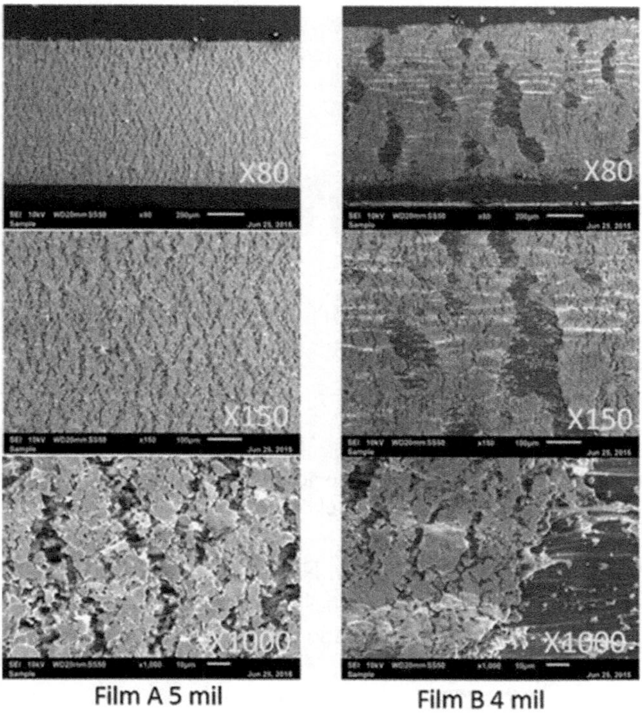

FIGURE 4.7 Cracks formation during conductive ink stretching [31].

This mechanism occurs on the condition that the bonding strength between the printed ink and the substrate is adequate to withstand the stresses during stretching. If the bonding strength is low, the metal flakes in the conductive ink may break from constraints of the substrate film, causing delamination of the ink from the substrate and accordingly electrical discontinuity. The crack mechanism as a stress relief is explained in Figure 4.8 [31].

4.3.3.3 Conductive Ink Types

The most used conductive inks are silver-based conductive inks due to silver's high electrical conductivity and stability. Copper is another highly conductive material used in conductive inks; however, due to the oxidation of copper, copper particles are usually coated with silver. Carbon-based conductive inks are used for their adaptability and cost-effectiveness; however, due to their higher electrical resistivity, they are mostly used in sensors, printed heaters, and electromagnetic shielding, or as a coating over other conductive inks for further protection.

Nano Copper Ink: Metallic nanoparticles have a large ratio of surface area to volume and correspondingly large surface energies. Compared with conventional materials, nanoparticles have many unique physical and chemical properties, and have lower melting points, and can therefore be processed at relatively low temperatures. Silver, copper, and gold are the three most frequently studied metals for nanoparticles. Among them, nano copper ink has received much attention due to its

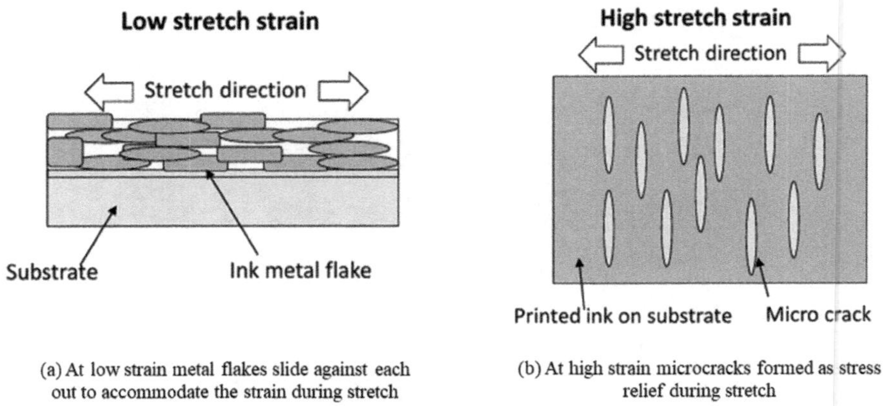

FIGURE 4.8 Mechanisms of crack formation as stress relief during ink stretching: (a) at low strain, metal flakes slide against each other to accommodate the strain during stretch, (b) at high strain, microcracks formed as stress relief during stretch [31].

low cost and high electrical conductivity. Nano copper inks have been widely used in many applications, such as photovoltaics, RFID and NFC antennas, and flexible electronics. To form conductive traces, nano copper ink requires sintering under controlled heat and pressure to coalesce the nano particles into compact blocks. Due to oxidation, sintering may be performed in an inert environment or with reducing agents. Low temperature snap sintering is a key to enable the use of nano copper inks.

Solderable Ink: Typical conductive inks are composed of silver particles, which are not solderable to typical solder materials like SAC305 or low temperature solders. During the soldering process, the silver will be quickly dissolved in the molten solder, causing breakup of the conductive traces. Solderable inks have been developed by employing non-silver metal particles (especially copper); to make the copper solderable, copper particles are typically coated with a thin layer of silver.

Liquid Metal: A liquid metal is a metal or a metal alloy which is liquid at or near room temperature. One of the commonly used liquid metals is gallium (Ga) which has a melting temperature of 30°C. Typical gallium based liquid metal alloys include eutectic gallium-indium (EGaIn), Galinstan (ternary alloy of gallium, indium and tin), and composite inks combining silver particles with liquid metal. Liquid metal conductive inks become attractive due to their high electrical conductivity, excellent stretchability and flexibility, and compatibility with direct ink writing (DIW). Figure 4.9 shows the stretchability of a liquid metal conductive ink at different strains and with long cycles (at 100% strain) [32]. Liquid metal conductive inks demonstrate superior stretchability with a stable electrical resistance at very high strains and long durability cycles as compared to conventional silver-based conductive inks. Issues with liquid metal conductive inks include challenges to interconnect with components and encapsulate thereafter to prevent oxidation of the liquid metal.

4.3.3.4 Dielectric Ink

A dielectric ink is a non-conductive ink that creates insulating layers to protect the conductive ink from the external environment and prevent electrical shorting

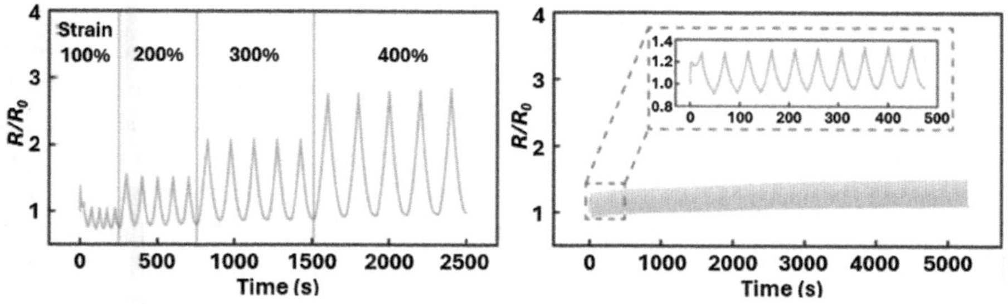

FIGURE 4.9 Stretchability of conductive ink made of liquid metal [32].

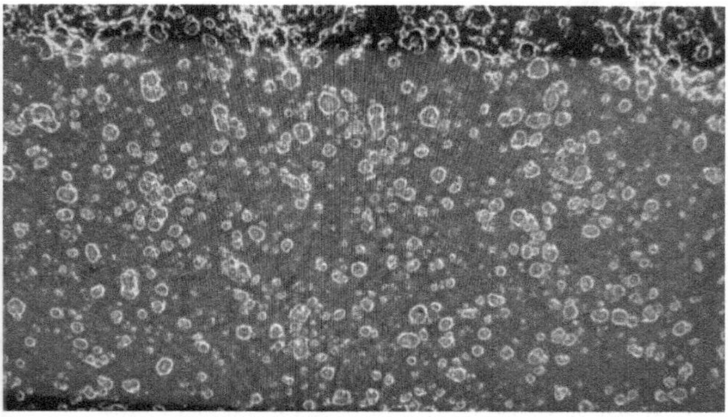

FIGURE 4.10 Pin hole formation during cure of a dielectric ink.

between neighboring conductive traces. A dielectric ink can help minimize silver electrochemical migration by acting as a protective barrier against moisture intrusion, which is a critical factor for silver ions to migrate between printed silver ink traces under an applied voltage. Dielectric inks also serve as a non-conductive divider for conductive ink trace crossovers, allowing for the construction of multi-layer complex circuitry. Dielectric inks help smooth the surface of substrates, creating a uniform layer allowing for subsequent ink printing. Furthermore, a dielectric ink provides mechanical reinforcement to the conductive ink to enhance durability against bending and stretching.

Similar to the conductive ink, a dielectric ink needs to be compatible with the substrate to have a good adhesion with the substrate as well as the conductive ink. It needs to form crosslinks with the substrate and conductive ink polymers during the curing process, either through heat or UV, to make an effective insulating protective barrier. A dielectric ink should be hydrophobic with low moisture adsorption to prevent moisture intrusion effectively.

One common issue with the dielectric ink is the so called "pin hole" defect, which refers to a microscopic hole within the printed dielectric layer. The pin hole defect breaks the insulating barrier, allowing moisture intrusion to the conductive ink and an electrical current to pass through, which can lead to electrochemical migration and circuit shorting, and ultimately compromise the functionality of the printed electronic devices. Figure 4.10 shows a dielectric layer with pin hole defects.

Many factors contribute to the pin hole defects of dielectric ink: the dielectric ink material and thickness, the ink printing process, and the curing process. Most dielectric inks, especially thermal cured ones, contain solvents which will evaporate during the curing process, leading to bubbles and pin holes. Printing dielectric inks with a higher thickness may prevent the air bubble linking together to create pin holes to the conductive layer. It is good practice to print multiple layers of dielectric inks to minimize the likelihood of open pin holes. Some UV curable dielectric inks are made up of 100% solids without solvent, which is less prone to pin hole formation; however, it is essential to validate the integrity of the dielectric ink, even for pin hole free dielectric inks as claimed by ink suppliers.

4.3.3.5 Functional Inks

Resistor Inks: Printed thick film (PTF) resistive inks are formulated through blends of silver, carbon, and non-conductive pigments to precisely tailor the resistance levels of the conductive ink. PTF resistor values can be adjusted by two means. The first is by adjusting the sheet resistance either through varying contents of the blend or changing the printing thickness. The other method is to adjust the width and length of the printed resistor based on the resistance requirement. Tolerance control is key for achieving precise resistance values of printed resistors, as these values depend on the ink formation, the ink trace dimension, and the consistency of the printing process. For some cases, post processing like laser trimming may be necessary to keep tight tolerance control [33]. Figure 4.11 shows a schematic of a resistor printed on conductor electrodes.

The electrical resistance of the resistor is calculated as follows:

$$R = \rho \frac{L}{WT} \tag{4.4}$$

Where

R is the electrical resistance of the printed resistor,
ρ is the resistivity of the printed resistor ink,
L is the length between two conductors,
W is the width of the printed resistor ink, and
T is the thickness of the printed resistor ink.

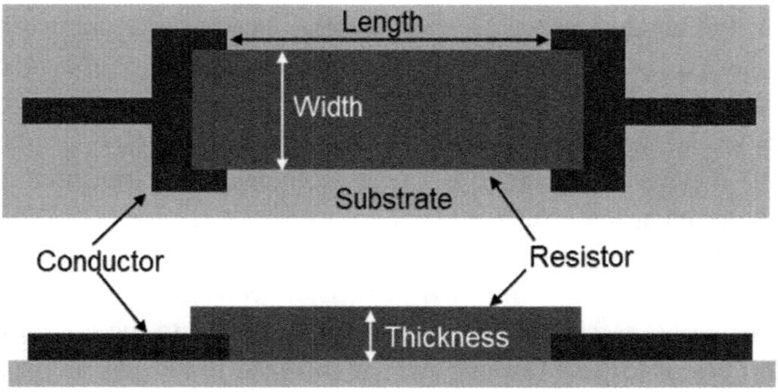

FIGURE 4.11 Schematic of a resistor printed on conductive electrodes.

One special category of resistor inks is the force sensing resistor (FSR). FSR is a printed functional ink whose resistance changes when a force is applied to it. Electrical resistance decreases under the applied force and returns to its original value when the force is removed. An FSR resistor is typically constructed by stacking two thin flexible substrates: one with conductive ink and one with piezoresistive ink, which are separated by a spacer to create a small air gap between the substrates. When pressed, the conductive ink will contact the piezoresistive ink, creating a short circuit to lower the electrical resistance.

PTC ink or "positive thermal coefficient ink" refers to a type of ink with the property of increasing its electrical resistance as its temperature rises. It acts as a self-regulating heating element where the heat output stabilizes at a specific temperature threshold due to an increase in electrical resistance [34]. It is commonly used in printed electronics for applications like flexible heating elements in automotives, furniture, and building design. Figure 4.12 shows a comparison between a PTC ink and a non-PTC ink. A PTC ink is typically made of carbon or polycrystalline ceramic materials with additional electrically conductive dopants.

Capacitor Ink: Capacitors are devices that store electrical energy in an electric field. Capacitors are made by taking two conductors and placing an insulator between them. The insulator is basically the dielectric ink with a high dielectric constant. Typical dielectric inks with high dielectric constants include barium titanate, strontium doped barium titanate, and calcium copper titanate.

Figure 4.13 shows the typical construction of the capacitor through screen printing of conductive ink and dielectric ink. The actual capacitance is determined by the dielectric constant of the dielectric ink, the cross-section area, and the gap between the two conductors. If the required capacitance is high, a multi-layer structure may be needed by alternately stacking the dielectric ink and the conductive ink, which will inevitably increase the screen printing cycle and time.

The capacitance of the printed capacitor is calculated as follows [33]:

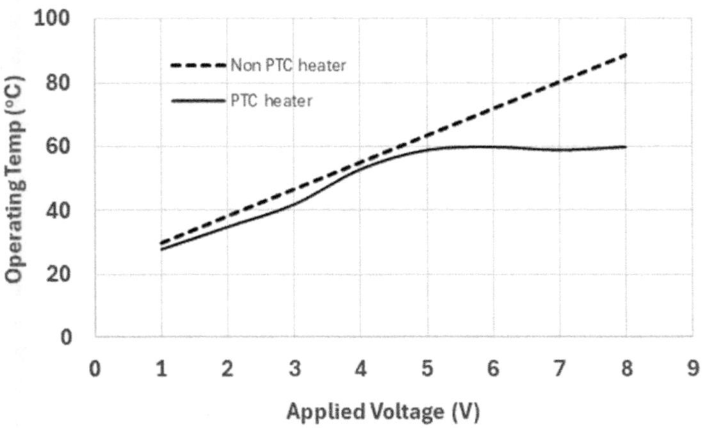

FIGURE 4.12 Comparison of temperatures between PTC ink and non PTC ink.

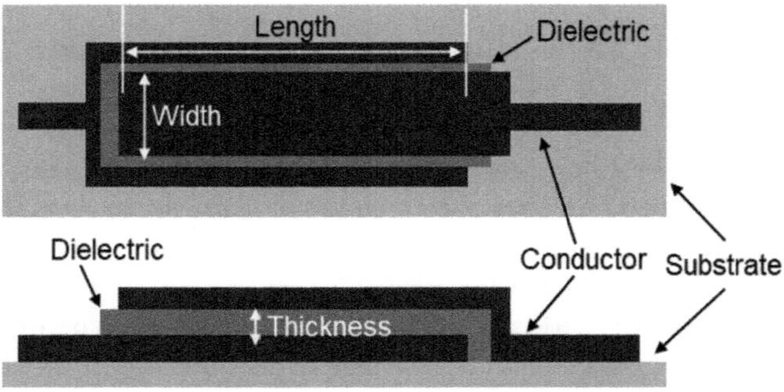

FIGURE 4.13 Construction of a capacitor through ink screen printing.

$$C = b\varepsilon LW/T \tag{4.5}$$

Where

C is the capacitance of the printed capacitor,

b is the vacuum permittivity,

ε is the dielectric constant of the printed dielectric ink,

L is the length of the printed dielectric layer (overlapped area between the top and bottom electrodes),

W is the width of the printed dielectric layer, and

T is the thickness of the printed dielectric layer.

Silver-Silver Chloride Electrode: Silver/silver chloride is a common choice of biological electrodes due to its low stable half-cell potential, reversible reaction, low impedance, good electrical conductivity, and low toxicity. It is used by many applications of biological electrode systems as biomonitoring sensors, such as electrocardiography (ECG), electroencephalography (EEG), and electromyography (EMG).

Biopotential electrodes function as an interface transducer between the biological tissue and the electronics measuring instruments, converting ion flows into an electronic signal which can be detected by the instrument. Electrodes for recording biopotentials are composed of a metal (usually silver for ECG measurement), and a salt of the metal (usually silver chloride), forming an electrolyte solution at the electrode-skin interface. Some of the silver dissolves into the solution producing Ag^+ ions [35,36]:

$$Ag \rightarrow Ag^+ + e^- \tag{4.6}$$

The operating principle of a silver/silver chloride (Ag/AgCl) electrode is based on converting the ion current at the surface of human tissues to an electrical current which can be measured by an instrument. A wet electrolyte gel is typically applied between the electrode and the tissues, which contains chloride ions such that the ion charge can be carried through the electrolyte solution. When the ion current develops, the metallic silver atoms (Ag(s)) of the electrode undergo oxidation, releasing Ag^+ cations to the solution, while the discharged electrons carry the electrical charge through the electrical wire to the measurement instrument. At the same time, the chloride anions (Cl^-) present in the electrolyte solution travel towards the anode (positively charged electrode) where they are precipitated as silver chloride (AgCl) as they bond with the silver cations (Ag^+) at the Ag(s) electrode surface. The reaction enables the ion current to transfer from the electrolyte solution to the Ag electrode while the electrical current flows through the connected wire to the measuring instrument. This process creates an electrode double layer across which a potential drop occurs, known as the half–cell potential. This reaction is a reversible reaction and is characterized by fast electrode kinetics, meaning that a sufficiently high current can be passed through the electrode with 100% efficiency of the redox reaction [35,36].

4.3.3.6 Interconnects

Electrically conductive adhesives (ECAs) are generally used to connect electronic devices to printed ink pads or traces, as most conductive inks are not solderable. ECAs are typically classified into two categories: isotropic conductive adhesives (ICAs), and anisotropic conductive adhesives (ACAs). ICAs are conductive in all directions, while ACAs are conductive only in one direction, typically along the z-axis. ICAs are typically composed of an epoxy resin, acrylic or silicone adhesives, loaded up with high contents of conductive materials such as silver, nickel, copper, carbon, or polymer spheres coated with metal. The metal fillers are loaded with sufficient concentration (up to 35% in volume, or up to 80% in weight) to ensure good contact between the metal particles and establish a continuous conductive network. The polymer matrix provides essential mechanical strength to maintain stable electrical contact between the conductive particles and ensures strong adhesion to the bonded surfaces. These conductive adhesives can be applied through dispensing or screen printing, can be thermally cured or UV cured.

ICAs offer both advantages and disadvantages compared with conventional soldering processes. ICAs typically have mild processing conditions, such as low curing temperature below 150°C, making them suitable for use with temperature-sensitive components and substrates. However, due to the nature of conductive path formation, ICAs typically have much higher electrical resistivity than solder and are more prone to the impact of use conditions, such as humidity and temperature. These disadvantages limit their use to low-cost applications with benign requirements on the current carrying capacity and reliability.

The key difference between ACAs and ICAs lies in the loading and distribution of conductive particles. By comparison, ACAs contain a much lower content of conductive particles, typically less than 10% in volume or 20% in weight. In ACAs, spherical particles are commonly used, and these conductive particles are arranged to make the conduction only in the z direction while maintaining insulation in the xy plane. As a result, ACAs provide some unique advantages over ICAs, especially for high density and fine pitch applications. Figure 4.14 shows a schematic of ACA working mechanism. Pressure is required to compress the ACAs to make contact between conductive particles to form a conductive path along the z-axis, while simultaneous heating lowers the adhesive's viscosity to fill the gap between the particles and cure the polymer resin to secure the particles in good contact between the opposing contact pads. Due to the lower concentration of conductive particles as compared

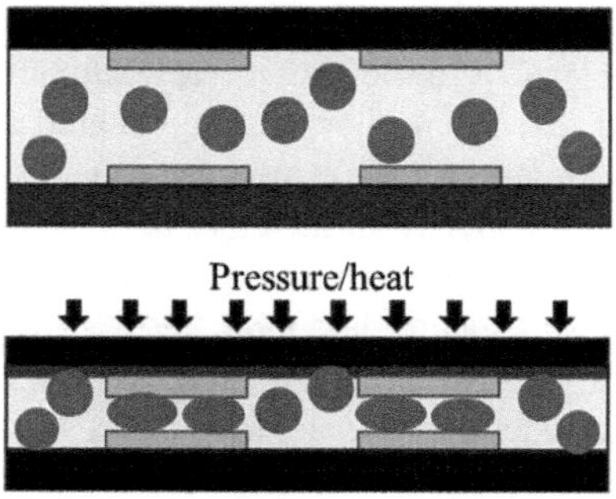

FIGURE 4.14 Schematic of ACA working mechanism.

to ICAs, ACAs have a considerably lower risk of short circuiting between adjacent conductive traces and pads.

There are two types of anisotropic conductive adhesives (ACAs): anisotropic conductive film (ACF) and anisotropic conductive paste (ACP), offering different advantages and disadvantages in terms of design requirement, process, and cost. In general, ACFs have a much higher cost, have more stringent requirements on the gaps between mating pads, and are mostly used for high density and fine pitch applications. ACPs have a lower cost but presents greater challenges in process control and are primarily used for low-end assemblies.

4.3.3.7 Encapsulation

Encapsulation plays a crucial role in enabling flexible hybrid electronics by providing flexibility, waterproofing, and other protective properties essential for the reliable performance of electronic devices.

An encapsulation material must possess the following properties to be suitable for encapsulation applications:

- **Flexibility**: Materials with a lower durometer hardness are preferred for encapsulation, as they offer greater flexibility. Typical urethane or silicone pottants have lower durometer hardness values (Shore A 30 or lower), making them the commonly used materials for flexible hybrid electronics applications.
- **Adhesion with the Substrate and Components**: The adhesion strength is critical to ensure the reliability of the FHE devices under bending and water exposure. Without strong adhesion, an encapsulant material may delaminate from the substrate during bending, allowing moisture to penetrate through the interface of delamination, causing corrosion of the encapsulated components and interconnects.
- **Process Conditions, Especially Temperature**: The FHE module may use a low-cost PET substrate with a low withstanding temperature. To minimize damage to the PET substrate and other temperature-sensitive components, a low processing temperature for encapsulation is required.

In addition to the critical material properties above, other properties must also be considered, including working life, curing duration (which affects throughput), color (dark color is preferred due to sensitivity of some components to light), and rheology.

In the following, three encapsulation methods are described: lamination, liquid coating, and low pressure molding.

Lamination: Lamination uses a hot melt film as the bonding agent which melts to the liquid phase in temperature ranges of typically 80°C–150°C, solidifying to form bonding with base substrates when they cool and at the same time to encapsulate the electronic components.

The benefits of lamination are many. The typical material, TPU film, used by lamination, has many outstanding properties, such as softness, excellent flexibility, outstanding durability, high tear resistance, resistance to oil, grease, fuel, solvent and chemicals, etc. Based on specific material composition, the TPU material can be made biocompatible, flame retardant, antistatic, anti-hydrolysis, or UV resistant. TPU is a green material and can be recycled. The material has been widely used in the textile, medical and electronics industries.

Once fully melted, the TPU material can demonstrate excellent bonding with a variety of surfaces, including PET, PC, polyimide, nylon, etc. Many TPU materials have low to medium melting temperatures, in the range 100°C–150°C, and no pressure is required during the lamination process. Due to these excellent properties, the TPU film is widely used by the textile industry as an adhesive to bond fabrics. It is commonly known as "No Sew Fusible Adhesive" or "No Sew Fusible Web".

During the lamination process, a single film can be used to encapsulate the electronics module on both sides at the same time. A cover layer can also be added to the lamination film to enable other properties, such as anti-scratch, a different color, a different surface durometer hardness, confidentiality, et al.

Furthermore, the lamination process is used widely by a variety of industries, such as textiles, graphics, electronics, and medical. Many scalable volume solutions and equipment are readily available, including roll-to-roll (R2R), roll-to-sheet (R2S), and sheet-to-sheet (S2S).

The process starts with the mechanical reinforcement of the electronic components attached to the substrate. A laser is used to cut the TPU film, removing material from designated keep-out areas of the FHE module. Then the TPU film is attached to the FHE device with the cutout window aligned to the module keep-out areas. The stack-up is placed on the conveyor of the roll laminator, where heat and pressure are applied to laminate the TPU film onto the FHE devices [37,38]. Demo samples are shown in Figure 4.15.

Liquid Coating: Liquid coating is a process for casting organic-based materials (comprising resins) to cover and seal electronics, with the materials combining solvent and water-based carriers to ensure good control of the liquid rheology with satisfactory wetting, adhesion and curing properties [37,38]. There are many variations of the liquid coating process, such as dispensing, spray coating, potting, casting, dipping, liquid jetting, etc.

Advantages of liquid coating include:

- A variety of materials options available: epoxy, acrylic, silicone, urethane, etc.
- Different curing processes: thermal, UV, moisture, or their combinations.

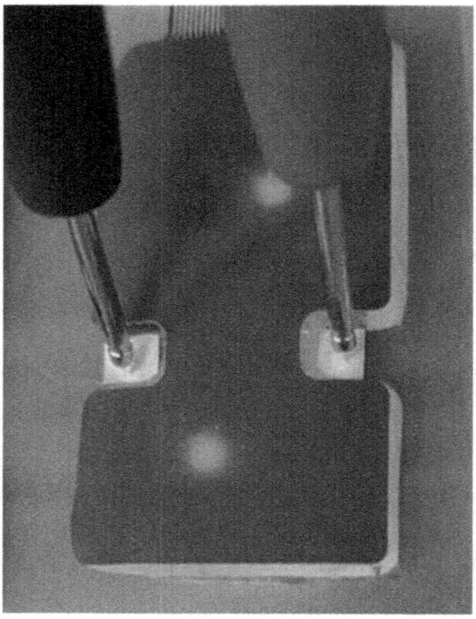

FIGURE 4.15 Encapsulated FHE devices through lamination [38].

- Liquid coatings can fill in the smallest gaps between components, due to their low viscosity.
- UV curable options available, accordingly with no heat damage to the FHE device.

Challenges or limitations with liquid coatings include:

- A double process may be required for the flexible module due to limited support to the module in the mold: one side of the module is molded at first and then the other side is molded at the second process, which inevitably affects the throughput of the manufacturing process.
- Some thermally cured materials may require prolonged thermal curing, affecting throughput.
- For UV curable materials, applying a thick coating and achieving complete curing throughout the layer is challenging. This method is generally only suitable for UV materials that are transparent or translucent.

Figure 4.16 shows a mold design with PTFE insert for liquid coating [38]. The purpose of using PTFE insert is to prevent bonding of liquid encapsulant to the aluminum mold. Figure 4.17 shows the demo samples encapsulated through the liquid coating process [38].

Low Pressure Molding: Low pressure molding is a process that molds and seals an electronic module with molten thermoplastic polyamide hotmelt. As compared to injection molding and compression molding, low pressure molding requires a lower pressure and a lower molding temperature. Typical pressures for injection molding can be in the range of 1,000–10,000 psi, whereas the pressure for low pressure molding can be as low as 10 psi. Low pressure reduces the mechanical stress on

FIGURE 4.16 Mold design for liquid coating [38].

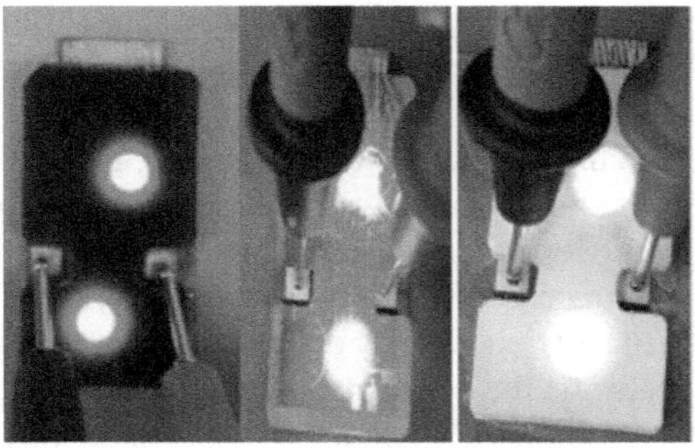

FIGURE 4.17 FHE devices encapsulated through liquid coating [38].

the FHE device and minimizes potential damage accordingly. Due to the low pressure requirement, the mold chase is considerably cheaper for the low pressure molding, as metals with a relatively lower strength can be used. As compared to potting, low pressure molding simplifies the molding process, reduces the cycle time, and increases the manufacturing throughput.

One of the main disadvantages of low pressure molding for electronics applications is the limited materials selection, with polyamide being the dominant material for this application. Polyamide is relatively more rigid compared to other encapsulation materials like TPU and silicone. Typical silicone or urethane encapsulants fall in the range of Shore A 10–50, whereas polyamide has a shore D hardness 10–40 (Shore A 70–90). The rigidity of the polyamide limits its use in FHE encapsulation to applications where only a thin layer of encapsulation is required or only certain areas of flexible FHE devices need added stiffness, such as battery holders and connectors. Furthermore, there is a limited selection of certified biocompatible polyamide materials available on the market.

4.3.4 Conductive Ink Deposition on Substrate

4.3.4.1 Conductive Ink Printing

When comparing different conductive ink printing methods, key factors to consider include the printing resolution, throughput, substrate compatibility, cost, and the method's suitability for various types of conductive inks. Techniques like screen printing offer high throughput for thick conductive traces, while inkjet printing excels at achieving high resolution and fine details but faces limitations in ink viscosity and substrate compatibility; other methods, such as flexography and gravure printing, also offer unique advantages and limitations, depending on the application requirements.

Screen printing uses a finely woven, pre-tensioned mesh, usually made of steel or nylon, to transfer conductive inks onto a substrate. A masking material (either an emulsion or a film) is applied on the mesh to define the circuit pattern where the conductive ink is deposited as the squeegee passes through the screen. Figure 4.18 shows a schematic of screen printing [39].

Compared to other ink printing techniques, screen printing is compatible with a wider range of ink viscosities, ranging from 1,000 centipoise to over 100,000 centipoises. Due to the high solid content of the ink for screen printing, screen printing can print thicker ink layers with a typical thickness range from 10 to 35 um; consequently, screen printing produces conductive traces with a much lower electrical resistance.

Gravure printing is a high-speed roll-to-roll printing process transferring the ink with a low viscosity onto a substrate. It includes a gravure cylinder (image carrier), a doctor blade, an impression roller, and an ink fountain as the main components of the typical gravure printing process, as shown in Figure 4.19 [39]. The image carrier is etched electromechanically, chemically or by laser, to form an image area based on the circuit design pattern. As the cylinder rotates, it picks up ink into the small cells and transfers it to the substrate with the impression cylinder made of rubber.

For the flexography printing, the ink is transferred first from the anilox roller to the polymer plate and next from the polymer plate to the printing substrate. As shown in Figure 4.20, an impression cylinder, a plate cylinder, an anilox roller, a doctor blade, and an inking unit are the main parts of the flexographic printing [39].

Inkjet printing works by propelling tiny droplets of conductive ink from hundreds of tiny nozzles on a print head, to form any pattern on a substrate. There are

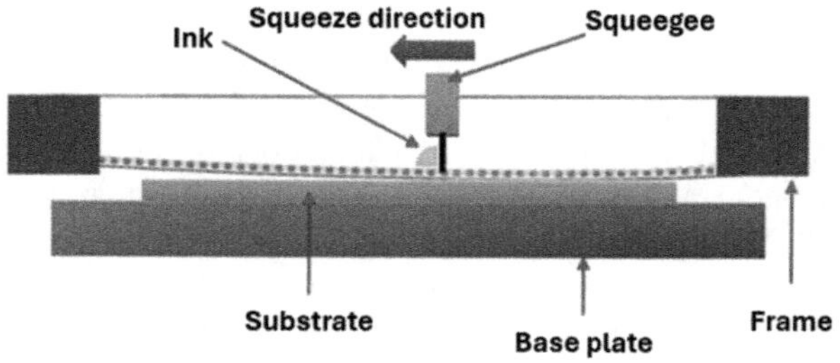

FIGURE 4.18 Schematic of screen printing [39].

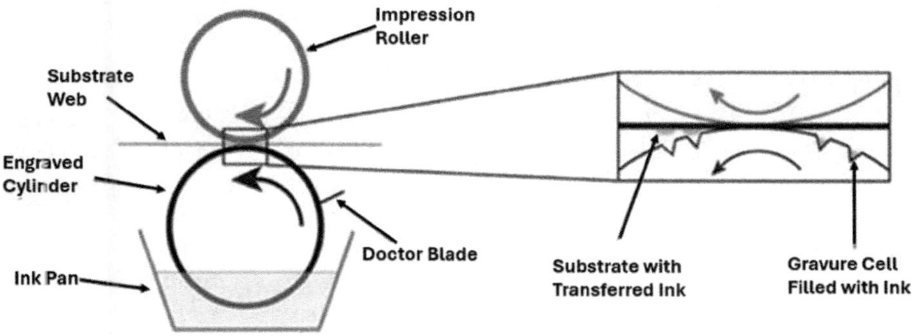

FIGURE 4.19 Schematic of gravure printing [39].

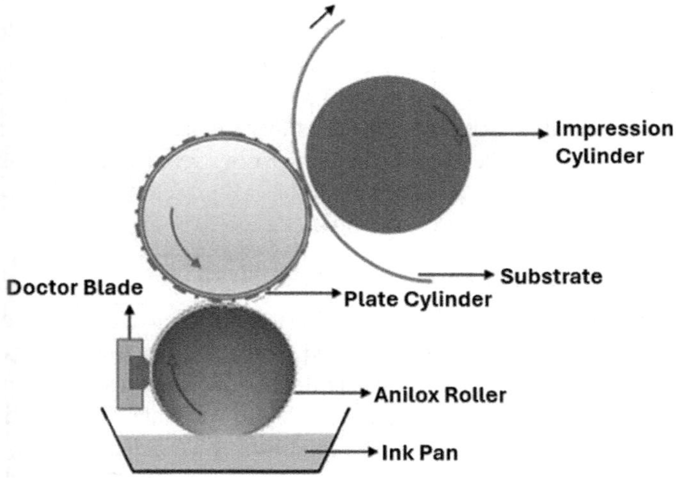

FIGURE 4.20 Schematic of flexography printing [39].

two types of inkjets: thermal inkjet and piezo inkjet. A thermal inkjet uses resistive heating to rapidly vaporize a small amount of the conductive ink, which generates pressure ejecting a small amount of the conductive ink from the nozzle. A piezo inkjet uses piezo elements to send a shockwave through the ink to propel a droplet. As compared to other printing techniques, inkjet printing does not require a physical mask or image carrier.

Aerosol Jet® technology is a fine-feature material deposition technique to directly print conductive ink onto non-planar substrates. The aerosol jet process employs aerodynamic forces to achieve high-resolution deposition of colloidal suspensions and chemical precursor solutions. An aerosol stream of the deposition material is focused, deposited, and patterned onto a planar or 3D substrate. The basic system consists of two key components: a module for atomizing liquid raw materials (mist generation), and a second module for focusing the aerosol and depositing the droplets

TABLE 4.7
Comparison between Different Printing Processes [41]

	Screen Printing	Gravure Printing	Flexography Printing	Inkjet Printing	Aerosol Jet Printing
Printing method	Contact	Contact	Contact	Contactless	Contactless
Resolution (μm)	100	2	100	30	10
Ink viscosity (mPa.s)	3,000–100,000	10–1,000	10–1,000	1–100	0.7–2,500
Ink trace thickness (μm)	3–30	0.02–12	0.17–8	0.01–0.5	0.1–10
Printing speed (m/min)	0.6–100	8–100	5–180	0.02–5	10
Material waste	Yes	Yes	Yes	No	No
3D/un-even surface	No	No	No	No	Yes

(in-flight processing) [40]. Mist generation is accomplished using an ultrasonic or pneumatic atomizer. The aerosol stream is then focused using a flow deposition head, which forms an annular, co-axial flow between the aerosol stream and a sheath gas stream. The co-axial flow exits the print head through a nozzle directed at the substrate. The aerosol jet print head can focus an aerosol stream to as small as a tenth of the size of the nozzle orifice. In summary, Table 4.7 presents a general comparison between different ink printing processes [41].

4.3.4.2 Curing Conductive Ink

After printing on substrate, conductive ink must be cured to evaporate solvents and transform the ink from liquid state into solid traces. During this process, the polymer matrix undergoes the cross-linking polymerization process to create bonding between polymer chains as well as adhesion to the substrate. This facilitates close contact or fusion between metal particles, establishing a conductive path.

Various curing techniques are available, such as thermal, UV, and photonic. Thermal curing is the most common method of curing conductive ink. It involves the application of heat to evaporate solvents and promote polymerization to hold conductive particles in place. Silver-based conductive inks are typically cured at 120°C–150°C for a duration of 10–30 minutes to achieve optimal conductivity. UV curing is a rapid photochemical process which uses high-intensity UV light to activate the photo initiators within the polymer matrix and trigger the polymer chain cross-linking process. Photonic curing uses high-intensity, pulsed visible or UV light to rapidly sinter or cure conductive ink within milliseconds.

Factors influencing the choice of curing technique include substrate material, ink chemistry, and production throughput. For the thermal curing method, substrate compatibility is essential, as the curing temperature must not exceed the maximum temperature which the substrate can withstand without degradation and shrinkage. UV curing and photonic light curing, on the other hand, offer the advantage of no heat or transient heat to minimize substrate damage, which is especially good for the heat sensitive substrates; however, both methods have application limitations, as light may not fully penetrate thick ink layers or inks with opaque or dark colors.

4.3.4.3 Fine Line Screen Printing

Standard screen processes typically print conductive line width over 100 um for production. Driven by miniaturization, fine line printing less than 50 um becomes more critical. Fine-line screen printing of conductive inks allows for the creation of high-precision electronics circuits on flexible substrates, enabling the creation of smaller, more complex, and flexible electronic components and devices.

Key elements of fine line screen printing hinges on three factors: ink formulation, screen selection, and the printing process. Besides general property requirements like strong adhesion to the substrate, the ink needs to be formulated with small metal flakes and particles to ensure good rheological properties, especially high thixotropic property. The thixotropic property defines the ink's ability to flow through the screen aperture but minimize flow once transferred to the substrate. As a rule of thumb, the screen mesh openings should be at least 4x the size of the largest particle in the paste.

For screen selection, the critical factors include the mesh wire type, wire diameter, mesh count, emulsion material, and thickness. Typically for fine line ink printing, a screen mesh with a high mesh count and a fine wire diameter is required; however, as the wire diameter gets thinner, its strength reduces, and the weaving process to fabricate screen becomes challenging. To enhance the mechanical strength, high strength stainless-steel wires are selected over low-cost nylon and polyester threads. The superior strength of the stainless-steel wire allows for use of smaller diameters (less than 18 μm), compared to a polyester mesh, which typically requires larger diameters (around 27 μm). A trade-off is necessary to balance the mechanical strength and the wire diameter. Typical fine line screens using stainless-steel mesh have a mesh count 300–500 with the wire diameter in the range 16–20 μm. The wire mesh is typically calendared to flatten the mesh by passing between rollers under high pressure to further improve the mesh performance, which will stabilize the mesh structure to improve the dimensional accuracy, lower the screen thickness, and improve mesh surface smoothness and mesh thickness consistency.

An emulsion material is applied on the stainless-steel mesh to define the ink printing pattern. Factors influencing the emulsion selection include the minimum artwork size requirements, mesh to stencil ratio, and aspect ratio of the screen. Image accuracy and clarity are critical to achieving a sharp, clean and dimensionally acceptable print. Emulsion over mesh (EOM) refers to the thickness of the emulsion on the print side of the screen mesh, relative to the mesh thickness, as shown in Figure 4.21. A good EOM ratio ensures a gastight seal between the mesh and the substrate, preventing the ink from leaking under the mesh. The EOM thickness is typically 10%–15% of the mesh thickness. To further improve the pattern accuracy and edge sharpness,

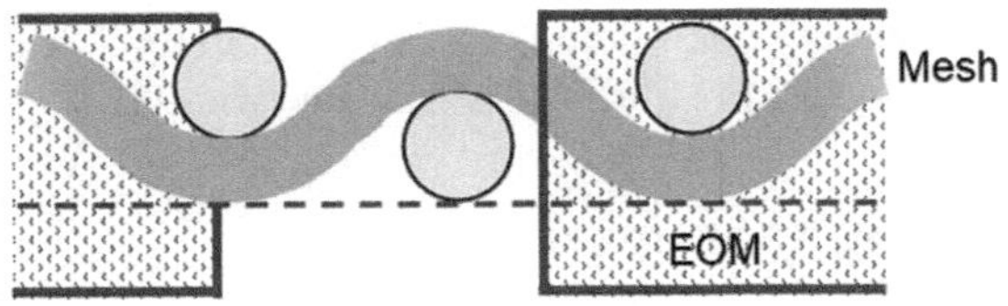

FIGURE 4.21 Concept of emulsion over mesh (EOM).

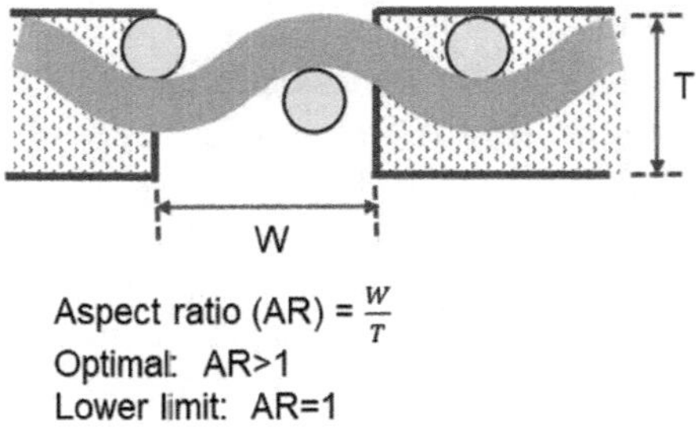

FIGURE 4.22 Schematic of the aspect ratio of a screen design.

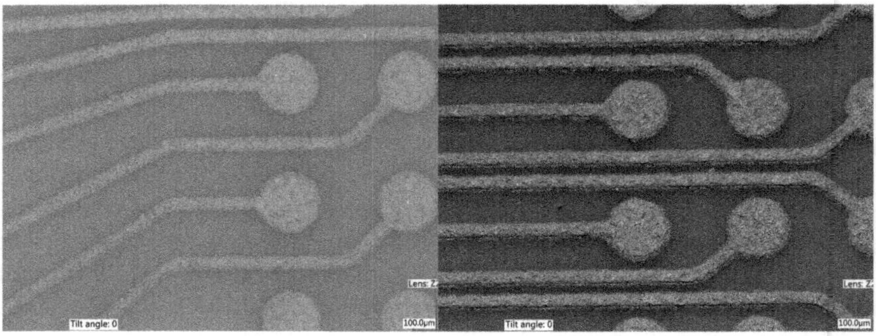

FIGURE 4.23 50 um line and space ink traces through screen printing.

capillary films were developed to replace emulsion. A capillary film provides an exceptionally uniform stencil thickness and a surface for very sharp printed edge definition.

The aspect ratio of a screen, defined as the ratio of the screen opening width over the screen thickness, is a critical parameter for fine line ink printing. Ink formation and viscosity, screen opening and printing parameters, are some of the factors determining the selection of the aspect ratio. Figure 4.22 shows a schematic of the screen aspect ratio. A higher aspect ratio is preferred to optimize the ink flow through the opening. It is challenging to print conductive inks when the aspect ratio is less than 1 (screen width is smaller than the thickness), as aspect ratio at 1 becomes the practical lower limit for ink printing. For 50 μm line/space, the 1:1 practical limit aspect ratio suggests that the overall coated screen thickness (Mesh + EOM) should be ~ 50 μm.

The screen-printing equipment and printing parameters also play a critical role for quality fine line ink printing. Key printing process parameters include the printing speed, squeegee pressure, snap off distance (gap), angle of contact, printing direction, among others. DoE experiments should be conducted to optimize the printing process for optimal printing quality and consistency. Figure 4.23 shows images of conductive ink printed on a TPU film with 50 um line width and space.

4.3.5 Challenges and Opportunities

The FHE industry is still in the early stage of development, and a variety of critical designs, materials, assembly and reliability issues remain to be addressed.

Typical silver-filled polymer based conductive pastes are not as conductive as the etched copper used to make PCBs. It is challenging to maintain performance and durability under repeated bending, torsion and stretch cycles. Additionally, typical conductive pastes are generally not solderable, so surface mount components must be mounted with electrically conductive adhesive (ECA), anisotropic conductive paste (ACP) or anisotropic conductive film (ACF). The electrical and mechanical interconnections formed with conductive adhesives are not usually as conductive and durable as conventional SMT solder joints. Additionally, commonly used thermoplastic films like PET and TPU have relatively low temperature resistance and cannot withstand the standard solder reflow processes.

To address these challenges, the FHE industry must continuously improve flexibility, stretchability and conformability of materials used in flexible hybrid electronics, while at the same time innovating new substrate and ink materials capable of withstanding high processing temperatures, as it enables the use of existing high-volume SMT infrastructure and equipment for FHE manufacturing. To accomplish this goal, the component mounting pads need to be solderable and the substrate needs to be compatible with a standard lead-free reflow process. The formation of solderable pads through printing of solderable ink onto high temperature stretchable substrates offers a promising solution for integrating FHE with conventional assembly processes.

Despite significant progress, challenges remain in developing flexible and stretchable conductive materials. Scalability remains a major roadblock, as many advanced materials, like graphene or silver nanowires, are still expensive to produce on a large scale. Reducing production costs without sacrificing performance is critical for widespread adoption. Additionally, improving the long-term durability of these materials under repeated mechanical stresses remains an ongoing challenge.

Another challenge for FHE is the lack of design rules and guidelines. Unlike the conventional electronics (rigid and FPC) with many years of extensive study to build extensive DFR/DFM libraries, FHE is still an emerging technology with limited information to guide design and manufacturing. To address this challenge, a project call was initiated by NextFlex in 2017 to establish a set of design tools and libraries to model and design FHE [42]. This effort continued over the years and is expected to continue in the foreseeable future to achieve a maturity level comparable to that of conventional electronics.

4.4 WEARABLE ELECTRONICS

4.4.1 Overview

Wearable electronics are devices designed to be worn on, around or even implanted in the body. They often include sensors, wireless transmitters and microprocessors to track body movement, collect health and fitness data, transmit data to other devices like smartphones, and provide real time feedback to the wearer. Typical examples of wearable electronics include smart watches, fitness trackers, smart glasses, smart jewelry, continuous glucose monitoring (CGM) patches, and smart clothing or e-textiles.

Wearable electronics can be generally classified into four categories based on their location and how they interact with the body: around the body, on the body, on the skin, and in the body. Around the body wearable electronics are devices directly attached to the body, such as smart watches, fitness trackers, headbands, and smart glasses. On the body wearable electronics are wrapped around the body, such as e-textiles with embedded sensors. On the skin wearable electronics are devices directly attached to the skin, such as e-tattoos and smart patches. In the body electronics are devices implanted in the body or under the skin, such as pacemakers and neuroprosthetics.

Wearables electronics can also be classified based on their function and usage into three categories: lifestyle, healthcare, and sports and fitness. Lifestyle wearables focus on enhancing everyday life, often providing entertainment, communication, or convenience. Examples include smartwatches, smart glasses, and augmented reality headsets. Healthcare wearable devices are designed to monitor and improve health, ranging from basic health tracking to advanced medical monitoring. Examples include heart rate monitors, blood pressure monitors, and glucose sensors. Sports and fitness wearables are tailored for athletes and individuals interested in fitness, often providing data on activity levels, heart rate, and other metrics. Examples include fitness trackers, smart shoes, and sports watches [43].

Flexible and stretchable electronics play a key role in enabling wearable electronics. These materials need to maintain electrical performance and flexibility under diverse wearable application conditions such as bending, torsion, stretching, compression, and crumpling, and demonstrate resistance to chemical attacks such as sweat and detergent during washing; at the same time, these materials need to demonstrate superior properties such as being compatible with the body and comfortable to wear. FPC and printed inks have been used extensively for wearable applications. New progress is being made to advance novel materials and integrate them into textiles for smart textile applications.

4.4.2 Conductive Materials for Smart Textiles Applications

There are a variety of approaches to integrate conductive materials into textiles. FPC strips can be inserted between garment layers or laminated to fabrics through adhesives. Conductive inks can be printed on TPU films and then laminated onto fabrics. Conductive yarns can be sewed to fabrics to serve as a conductive path. Conductive fabrics can be cut to strips and then laminated to garments through a TPU hotmelt film.

Due to its high electrical conductivity, copper yarns have been used for smart textile applications, which can be single copper yarn with no support, copper yarns twisted with a polymer thread or a strand of polymer microfibers, or copper wire strand. Figure 4.24 shows a hybrid conductive yarn constructed by twisting micro copper yarns with polymer fibers [44].

The hybrid approach enhances the strength of conductive yarns, improving their resistance to mechanical stress encountered during sewing, knitting, and everyday use. Another variation of the hybrid approach involves using copper wire strands to enhance flexibility and durability. These copper strands are often made by wrapping fine copper wires, with a wide range of gauges, around a central core, typically a plastic thread.

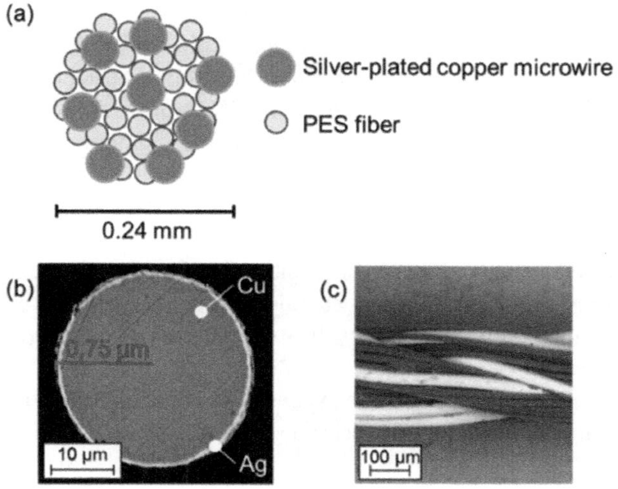

FIGURE 4.24 Hybrid construction of conductive threads by twisting micro copper fibers with polymer fibers: (a) Cross-sectional model of hybrid conductive yarn, (b) Analysis of silver-plated copper microwire by SEM, (c) Longitudinal view of hybrid yarn [44].

Similar to copper, stainless steel is formed as strands to use as conductive threads to increase strength and durability and resistance to corrosion. Due to its high strength, stainless steel can be twisted into strands without the need for a polymer core. The drawbacks of the stainless-steel thread are its high weight, high stiffness, and high electrical resistivity as compared to copper. A stainless-steel thread is not solderable, so it is typically connected to other electrical devices through mechanical fastening.

Another method to make a hybrid conductive thread is to coat a textile thread or flexible polymer thread with a thin layer of metal, such as nickel, copper or silver. This type of thread has a highly conductive outer skin, but still retains strength, flexibility, durability and light weight of conventional polymer threads, for maximum comfort and ease of construction in wearable e-textiles. The metallic skin of these threads allows them to be soldered to other metallic parts to form a good electrical connection. Durability can be a concern for these threads, as the metallic layers can de-bond and peel away from the polymer core thread. This presents a challenge for wearable textiles, as prolonged wear or repeated laundering could lead to degradation of the conductive thread.

Conductive polymers have the intrinsic ability to conduct electricity. They can be blended with various materials or dopants to optimize their properties for specific applications, enabling them to function as either semiconductors or conductors. One of the primary drawbacks of these materials is that their conductivity is typically much lower than that of metals, although sufficient alignment and doping can increase it to a level that exceeds stainless steel, as in the case of stretch-oriented polyacetylene [45].

Carbon nanotubes (CNTs) possess very high tensile strength and modulus, and are electrically and thermally conductive. Since their discovery, the industry has overcome a major challenge to develop these nano-sized materials into micro-sized yarns. In terms of material properties, CNT yarns offer excellent tensile strength, flexibility and durability, and are light weight. Additionally, they provide electrical conductivity comparable to or even better than stainless steel. While silver and nickel have higher electrical conductivity than CNT yarns, they are typically used in conductive threads as thin coatings over insulating filament cores, resulting in a lower overall conductivity. Furthermore, CNT yarns are more resistant to oxidation than stainless steel, allowing them to retain their properties and durability through laundering and daily use [45].

A layer of CNTs can also be deposited onto other fibers to create hybrid conductive threads. Alternatively, CNT yarns can also be coated with layers of metal. This process can enhance their conductivity further and make them solderable to other electrical parts or threads.

CNT yarns are currently much more expensive than the alternatives, limiting their use in e-textile applications until the price decreases.

4.4.3 Stretchable Conductive Yarns

Stretchable conductive yarns have garnered significant attention for wearable applications, where stability and durability under extensive strains are essential. However, producing high-performance conductive yarns with great stretchability and durability through cost-effective and scalable manufacturing methods remains a great challenge for wearable electronics.

Many stretchable conductive yarns have been developed and commercialized, and many more efforts are still ongoing to further improve performance of stretchable conductive yarns. Figure 4.25 shows a technique to fabricate stretchable conductive yarns by wrapping carbon nanotube yarns around a multifilament polyester yarn [46]. The spring like structure displays stable electrical conductivity even at a high tensile strain with the multifilament polyester core yarn providing mechanical strength and elasticity. Figure 4.26 shows a comparison between a carbon nanotube wrapped yarn and a pristine carbon nanotube yarn at 60% strain [46]. The carbon nanotube wrapped yarn shows great electrical stability with change of resistance less than 10% at 60% strain.

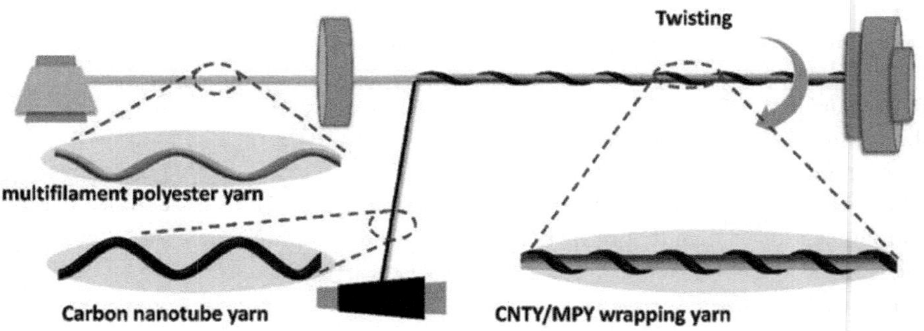

FIGURE 4.25 Schematic illustration of fabrication of stretchable conductive yarn by wrapping carbon nanotube yarn around multifilament polyester yarn [46].

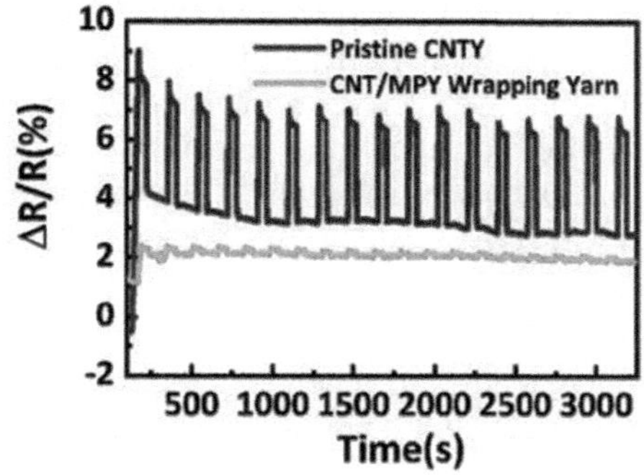

FIGURE 4.26 Relatively stable electrical resistance of carbon nanotube wrapped yarn during stretching cycles as compared to pristine carbon nano tube yarn [46].

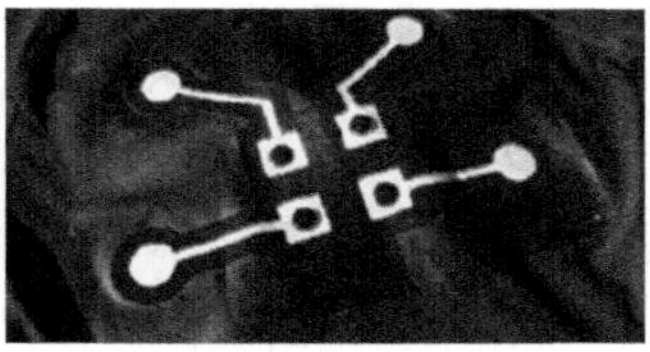

FIGURE 4.27 Conductive ink printed on TPU film and then laminated to fabric.

Another highly stretchable yarn with high performance is fabricated by electrospinning a core TPU nanoyarn with a liquid metal (LM) layer and a protective outer nanofiber layer. This yarn demonstrates ultra-robust strength and stable electrical resistance at a high strain up to 548% [47].

4.4.4 Conductive Fabrics

Conductive fabrics can be produced through various methods. Fabrics can be made electrically conductive by weaving or knitting conductive yarns. Textile metallization can also be achieved through depositing metallic layers on textile surfaces, creating metal-coated textiles. Common techniques used for this type of metallization include vacuum deposition, ion plating, electroplating, and electroless plating. Fabrics can also be made conductive by laminating conductive inks onto their surfaces, with an example shown in Figure 4.27, in which an ECG sensor is connected to an external module through conductive inks laminated to the fabric.

4.4.5 Smart Textiles Launderability Study

The launderability of e-textiles is one of the most significant reliability challenges for wearable applications. In their lifetime, e-textiles will experience many washing and drying cycles, in which they are not only attacked by chemicals such as detergent but also experience significant mechanical turbulence stresses. Conductive materials, which connect different sensors, modules and power supplies to form a body area network (BAN), may fail to deliver stable performance after multiple laundry cycles.

A launderability study was performed on four types of conductive fabrics made through weaving polyester or nylon yarns with metal coatings (Ag, Cu, Ni, NiCo, respectively), as listed in Table 4.8. Images of the samples are shown in Figure 4.28 [48]. The Ni conductive fabric is made from yarn with two metal coatings with an inner layer Cu and an outer layer Ni. The NiCo conductive fabric has more metal coating layers. Figure 4.29 is the schematic cross section of the NiCo coating on the polyester yarn [48].

The conductive fabrics were cut into stripes and then laminated through TPU film on two base fabrics: soft spandex and relatively more rigid denim. The conductive fabrics were also clamped to a rigid board after lamination on the base fabrics for

TABLE 4.8
Description of the Conductive Fabrics Used in the Launderability Study

Sample Description	Type 1	Type 2	Type 3	Type 4
Metal Coating	Cu	Ag	Ni	NiCo
Fabric	Polyester	Nylon	Polyester	Polyester
Out layer metal coating	Copper	Nylon	Nickel/Copper	Nickel-Cobalt
Sheet resistance (Ohm/sq)	0.05	<0.25	0.03	<0.1
Thickness (μm)	3.44	0.49	2.43	2.64

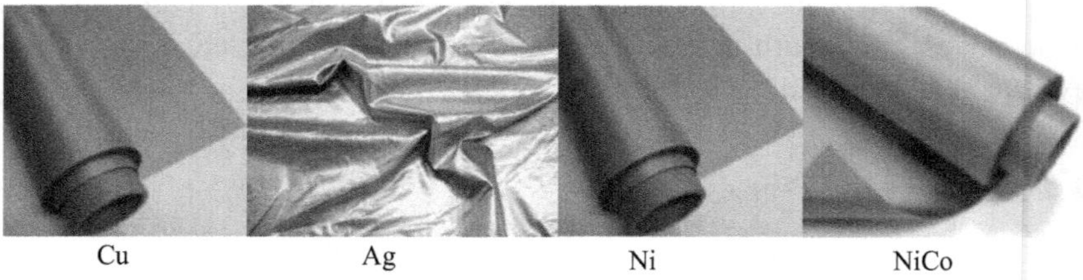

FIGURE 4.28 The conductive fabrics used in the launderability study [48].

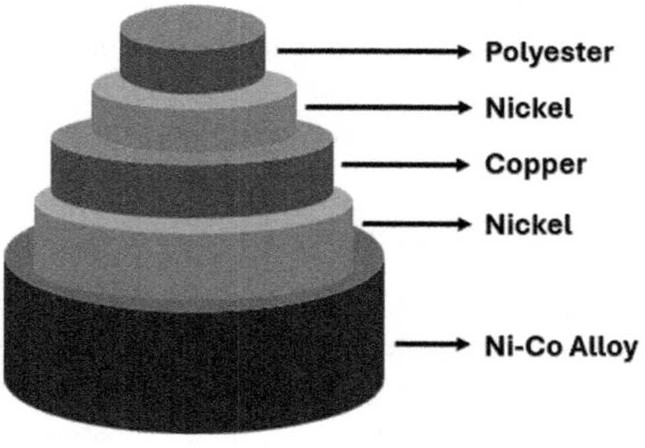

FIGURE 4.29 Metal coating layers on the polyester yarn of NiCo fabric [48].

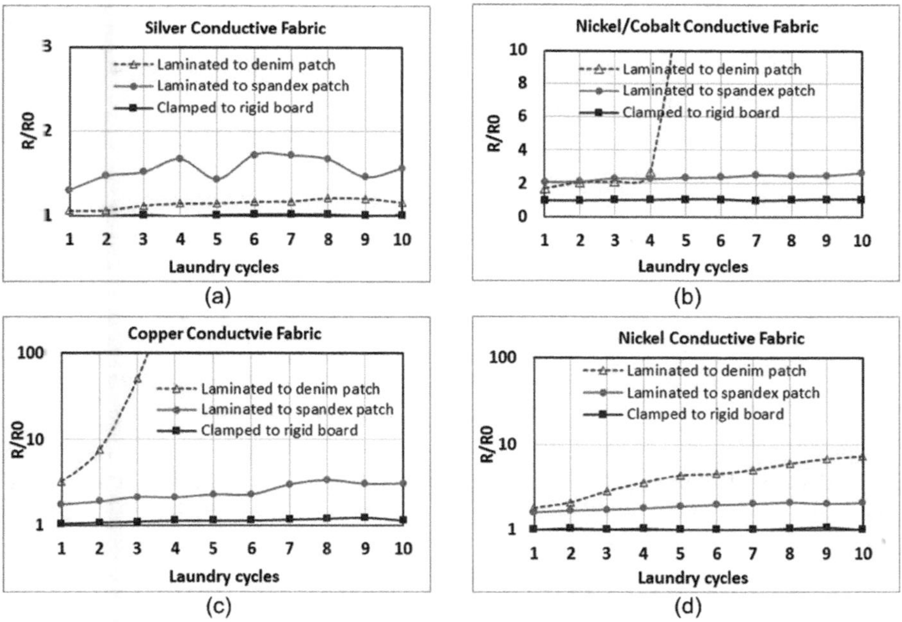

FIGURE 4.30 Increase of electrical resistance over laundering cycle for different conductive fabrics attached to different bases: (a) silver conductive fabric, (b) nickel-cobalt conductive fabric, (c) copper conductive fabric, (d) nickel conductive fabric [48].

comparison. Figure 4.30 shows the trend of the electrical resistance change over laundering cycles of conductive fabrics laminated to different base fabrics or clamped to a rigid board [48]. For all the configurations, the conductive fabrics clamped to the rigid board demonstrate very stable resistance throughout the washing and drying cycles. Except for the silver conductive fabrics, the nickel-cobalt, copper and nickel conductive fabrics laminated to flexible spandex patch show better performances than the fabrics laminated to semi-flexible denim patch. Considering that the primary difference between the bases is their rigidity, it is postulated that this mechanical property of the bases plays a major role in the observed differences in the conductive fabrics' laundering performance. The rigid board eliminates bending stress on the conductive fabrics, preventing damage to the metal coatings, consequently the electrical resistance remains stable. The conductive fabrics laminated onto a stiff but bendable base fabric may experience great bending and torsional stresses, leading to increased damage to the metal coatings, and consequently a higher increase in electrical resistance. More compliant base fabrics, such as spandex, may not be strong enough to exert sufficient force to bend and twist the conductive fabrics, therefore causing less damage to the metal layers.

Figure 4.31 presents a comparison between the metal coatings during laundry, by averaging the resistance increase in percentage of each conductive fabric laminated on base materials [48]. The resistance change of the Cu conductive fabric is the most severe among the selected conductive fabrics, while the Ag conductive fabric is the most stable, and the performances of Ni and NiCo fabrics lie in between.

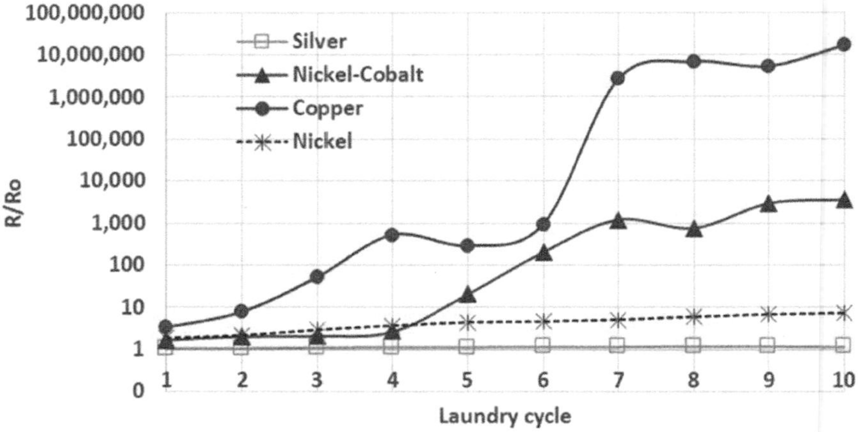

FIGURE 4.31 Increase of electrical resistance over laundry cycle for different metal coatings [48].

(a) Cu Conductive fabric (b) Ag Conductive fabric

(c) Ni conductive fabric (d) NiCo conductive fabric

FIGURE 4.32 Loss of metal coatings on the conductive fabrics laminated to denim base fabric, post 10 times laundering (SEM 500×): (a) Cu, (b) Ag, (c) Ni, (d) NiCo [48].

Ag demonstrates its robustness against mechanical stress due to its excellent mechanical properties such as stretchability, flexibility and ductility.

Figure 4.32 shows the localized area views for the conductive fabrics laminated to the denim base fabric after the laundering study [48]. All the conductive fabrics show certain metal losses; however, the failure signatures look very different. For the Cu, Ni, and NiCo conductive fabrics, the loss of metal coating exhibits itself in the form of metal fracture and peeling, likely due to weak bonding between the metal layer and the polymer fiber, which could not sustain the constant mechanical agitation

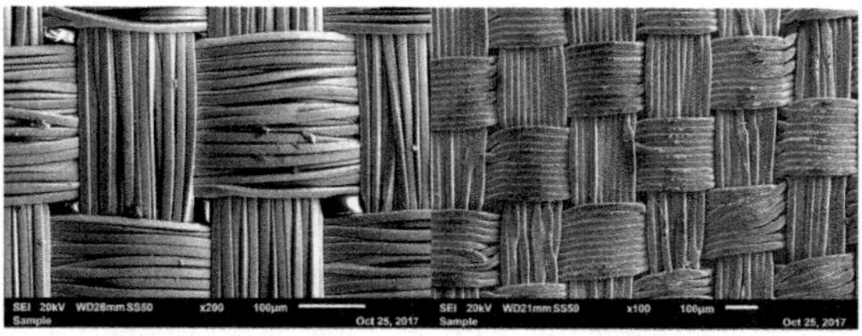

(a) Cu conductive fabric (b) Ag conductive fabric

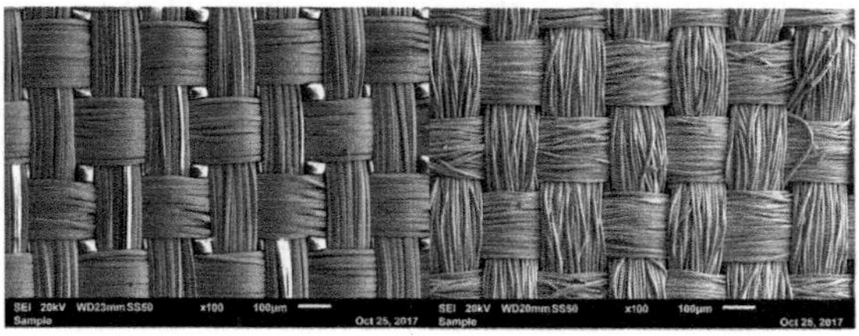

(c) Ni conductive fabric (d) NiCo conductive fabric

FIGURE 4.33 Conditions of the four conductive fabrics clamped to a rigid board after lamination to denim base fabric, after 10 cycles of laundering: (a) Cu, (b) Ag, (c) Ni, (d) NiCo [48].

during laundry. The failure mode of the Ag conductive fabric appears distinct from the others, exhibiting metal scratches or abrasion rather than fracture and peeling. This difference is likely due to its thinner thickness, strong adhesion to nylon, and superior flexibility.

Figure 4.33 shows the conditions of the four conductive fabrics clamped to a rigid board after lamination to the denim base fabric after ten laundry cycles [48]. No visible damage is observed on the metal coatings for the conductive fabrics. Consequently, the electrical resistance of the conductive fabrics remains stable throughout the laundry cycles.

Based on the observations and analyses above, it is hypothesized that mechanical stresses from laundry agitation and turbulence caused damage to the metal coating layers on the conductive fabrics, in the forms of peeling, fracture and abrasion, which led to the observed increase in electrical resistance.

4.4.6 Challenges and Opportunities

Although great progress has been made, challenges remain for materials and integration for wearable applications, especially e-textiles. Modules and sensors need to generate accurate biomarkers to the expectation of wearers without bulkiness and causing discomfort to wearers. Conductive materials and interconnects need to be integrated into fabrics seamlessly and withstand a variety of mechanical stresses without performance degradation during rigorous use conditions, such as stretching, crumpling, sweating, and laundry cycles. Furthermore, integration of electronics into textiles raises great concern for e-waste disposal and management, for which eco-friendly biocompatible and bio-degradable materials will be essential to transform the next generation wearable technologies.

Looking forward, integrating flexible and stretchable conductive materials into more complex systems, such as wearable health monitors or soft robotic systems, is a key area for future development. As e-textile technology continues to evolve, the development of more specialized methods and materials is essential to enhance the reliability of e-textile structures under rigorous usage conditions. A fundamental understanding of the materials behavior, degradation mechanism and performance is key to developing such robust materials. Industry and research laboratories must collaborate to perform extensive studies to gather firsthand data on the behavior and performances of e-textile materials, and to fully understand failure mechanisms of materials degradation. Standardized test procedures must be developed through collaborative efforts between the electronics and textile industries, via organizations such as IPC, AATCC, ASTM, and IEC, to validate components and conductive elements, including conductive threads, yarns, fabrics, and conductive inks, for compliance with end use requirements.

4.5 FLEXIBILITY TESTING OF FLEXIBLE ELECTRONICS

4.5.1 Overview

Flexibility test of flexible electronics involves subjecting the electronic components mounted on flexible substrates to various mechanical stresses such as bending, stretching, twisting, and folding, to assess their ability to maintain functionality and performance through the mechanical stress cycles. The industry has developed a variety of static and dynamic tests to simulate the actual use conditions experienced by flexible electronics, utilizing specialized setups to precisely control the applied mechanical stresses and monitor electrical parameters during the process. In general, IPC standard 9204 classifies these tests into five categories, as listed in Table 4.9 [49]:

TABLE 4.9
Categories of Flexibility Tests [49]

Category	Test Items	Applicable Standard
Stretchability testing	Tensile test under static and dynamic conditions	IPC 9204
Bending test	Variable radius bending test	IEC 62715-6-1
	Variable angle bending test	IPC 9204
	Free arc bending test	IPC 9204
	Demattia flexibility test	ASTM D813, D430, ISO 7854
	Loop bending test	ASTM D1388
	Folding endurance test	ISO 5626, IEC 62899-201
Torsion test	Twist test under different static and dynamic conditions	IPC 9204
Rolling test	Coiling flex test	IPC 9204
	Sliding plate test	IPC 9204
Crumpling test	Crease test	ASTM F2749
	Crumple flex test	ISO 7854. ASTM F392

A combination of the above tests in the table will give rise to more test options. The determination of the test to be used, test sequences and test stresses are based on the use conditions and agreement between the user and the supplier. Prior to conducting the flexibility tests, flexible electronics may be exposed to certain environment conditions as "pre-conditioning", such as temperature and humidity, to simulate the real use conditions.

4.5.2 Flexibility Testing

4.5.2.1 Stretchability Testing

Stretchability testing is used to evaluate the functionality of flexible electronics under repeated or constant uniaxial or biaxial stresses. Typical stretchability tests may include stretchability limit, stretchability under constant elongation, cyclic stretchability, and stretchability in combination with other mechanical stresses such as torsion.

The stretchability limit test is used to determine the elongation limit, in percentage, of printed electronics by uniaxial tensile testing. Samples may be prepared in the original product form or using a test coupon such as straight edge strips or "dog bone" strips as described in ASTM E-345, or other forms agreed upon between the user and the supplier. A schematic test setup for evaluating the stretchability of conductive traces for flexible electronics is shown in Figure 4.34 [49].

During the test, the stretchable circuit is held firmly between two fixtures. Cushions may be added at the interface between the sample grips and the conductive ink to prevent stress concentration which may cause premature cracks. In this test, the bottom fixture can remain stationary while the top fixture is movable. By moving the fixture(s), the specimen is stretched to a certain percentage until the circuit physically or electrically fails during the test. It is crucial to continuously monitor the functionality or resistance of the circuit during the test.

The stretchability (%) can be calculated as:

$$\text{Stretchability}\ (\%) = \frac{\text{Final circuit length} - \text{Initial circuit length}}{\text{Initial circuit length}} \tag{4.7}$$

The conductive circuit length may be obtained by measuring the distance between the two sample grips. The final length may be determined by previously established failure criterion in terms of electrical resistance increase, such as the maximum value of resistance, resistance increase percentage, etc.

Similarly, other stretchability tests may utilize the same test set up. Cyclic stretchability test is used to evaluate if a stretchable circuit can withstand repeated elongation cycles under a tensile load. The constant stretchability test is to evaluate the stretchability of a circuit at a fixed elongation for a prolonged duration. The flexible electronics can also be evaluated under combined stretch and torsion conditions, with either one of the applied stresses static or both stresses dynamic.

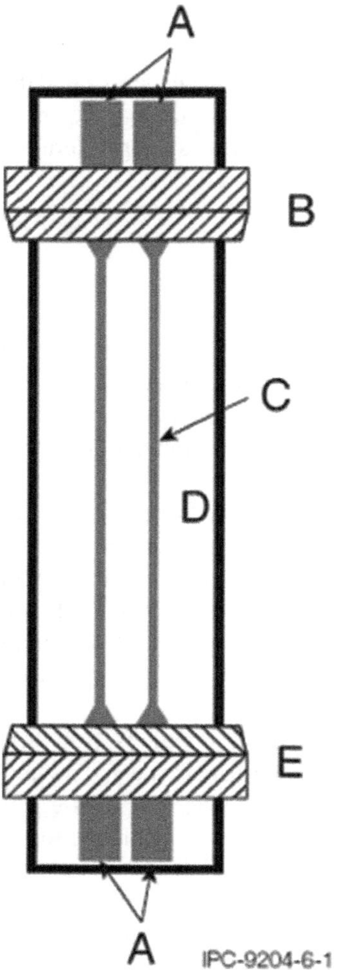

FIGURE 4.34 Test coupon design for stretchability testing (A: connection to resistance measurement, B: moving sample holder, C: stretchable ink trace, D: stretchable substrate, E: stationary sample holder) [49].

4.5.2.2 Variable Radius Bending Test

The variable radius bending test is used to evaluate the functionality of printed electronic devices under repeated bending conditions with a changing bending radius. A schematic test setup is shown in Figure 4.35 [49]. The specimen may be prepared in the original product form or by using a test coupon such as straight-edge strips or other forms agreed upon between the user and the supplier.

The specimen is bent in a U shape between two parallel plates, with each end affixed to a plate. The bottom plate is stationary, while the top plate is movable. The maximum distance between the two plates should be sufficiently larger than twice the minimal bending radius of the specimen. By moving the top plate up and down, the specimen can be bent in a constantly changing bending radius with repeating cycles.

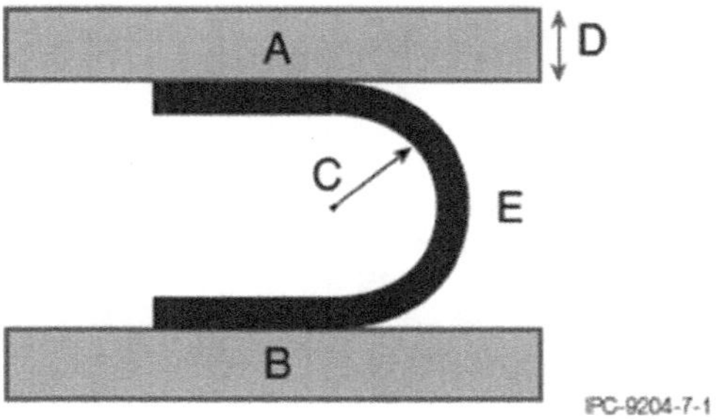

FIGURE 4.35 Variable radius bending test set up (A: movable plate, B: fixed plate, C: bending radius, D: moving direction, E: sample under bending) [49].

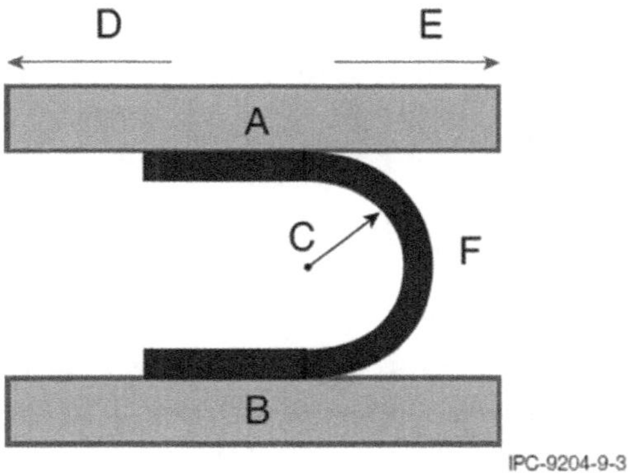

FIGURE 4.36 Sliding plate test setup (A: movable plate, B: fixed plate, C: bending radius, D: forward moving, E: backward moving, F: sample under test) [49].

4.5.2.3 Sliding Plate Test

The sliding plate test is used to evaluate the functionality of flexible electronic devices under repeated rolling conditions while subjected to a constant bending condition. The test setup is shown in Figure 4.36 [49]. The specimen is bent in a U shape with a preset bending radius between two parallel plates, with both ends fixed to either plate. The bottom plate remains stationary, while the top plate moves parallel to the bottom plate. By moving the top plate back and forth, the specimen is stressed under combined rolling and bending conditions.

4.5.2.4 Variable Angle Bending Test

The variable angle bending test is used to evaluate bending endurance of flexible electronics during their application. The specimen may be prepared in the original product form or by using a test coupon such as straight-edge strips or other forms agreed upon between users and suppliers. Figure 4.37 shows a schematic of the variable angle bending test [49]. The flex circuit is affixed to a certain weight which keeps the circuit under required tension. The circuit flexes back and forth to a certain bending angle around a mandrel. Functionality or resistance of the circuit can be monitored before and after the testing, or throughout the testing either periodically or continuously.

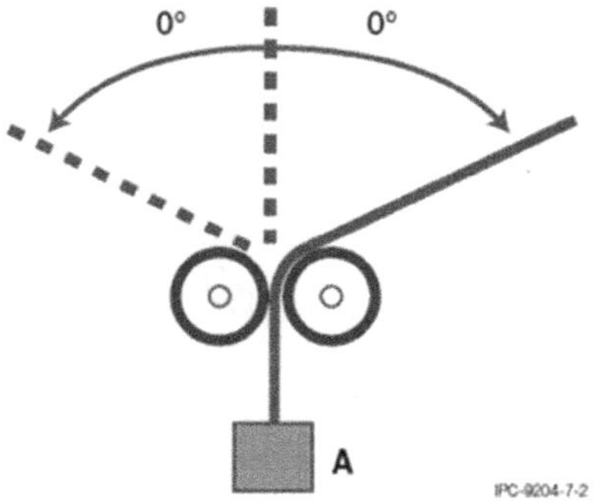

FIGURE 4.37 Variable angle bending test setup [49].

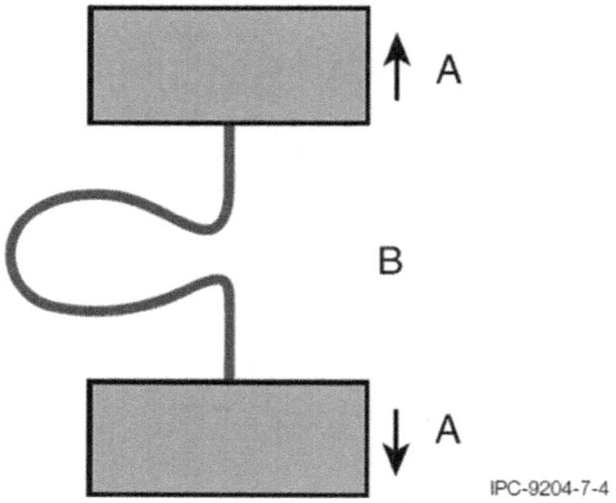

FIGURE 4.38 DeMattia flex test setup (A: grip, B: sample under test) [49].

4.5.2.5 DeMattia Flexibility Test

The Demattia flex test is used to evaluate flexible electronics and materials that undergo alternate bending and stretching during their use conditions, in reference to ASTM D813, ASTM D430 and ISO 7854. This test utilizes a DeMattia Flexometer. A test setup is shown in Figure 4.38 [49]. The two ends of the specimen are held tightly between two grips, with the specimen lying in the same vertical plane. One grip is stationary, while the other moves. The reciprocation of the grip enables alternate stretching and bending of the specimen. The functionality or resistance of the circuit before and after testing or throughout the testing needs to be monitored either periodically or continuously.

4.5.2.6 Folding Endurance Test

The folding endurance test is used to evaluate the folding endurance of flexible electronics during their application, in reference to ISO 5626 and IEC 62899-201. This test utilizes the MIT Folding Endurance Tester [50]. The equipment has a movable part which holds the specimen and bends it by 180° (90° on both sides of the vertical direction), a bending part for applying a constant curvature to the specimen, and a

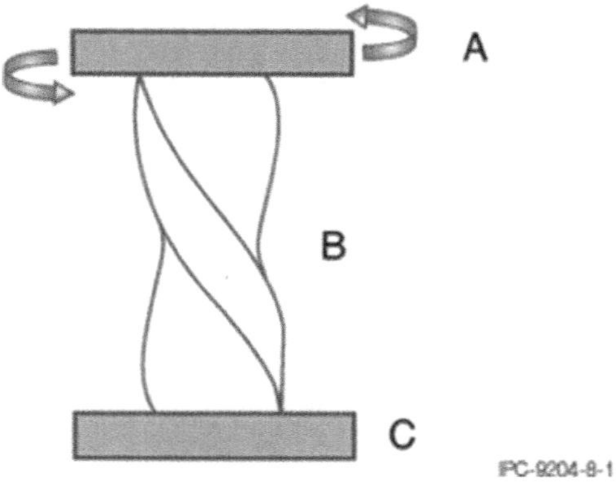

FIGURE 4.39 Torsion test setup (A: twisting sample grip, B: sample under test, C: fixed sample grip) [49].

mechanism for applying a constant tension to the specimen. In this test, the specimen is attached to the clamp connected to the plunger and to the bender, so that the specimen is kept flat and does not touch the mounting face of the bender without deflection to the specimen. Tension is applied to the specimen by loosening the locking screws on the plunger.

4.5.2.7 Torsion Test

The torsion test is used to evaluate flexible electronics and materials against torsional stresses during their application. The test schematic is shown in Figure 4.39 [49]. A flexible circuit is anchored to two fixtures. With one fixture stationary, the rotation of the second fixture will twist the flexible circuit to a certain angle, return it to its original state, and then twist it in the opposite direction. Both fixtures can also rotate in opposite directions. The functionality or resistance of the flexible circuit needs to be monitored before and after testing or continuously throughout the test.

4.5.2.8 Coiling Flex Test

The coiling flex test is used to evaluate the functionality of flexible electronic devices under cyclic and/or constant coiling conditions. A coiling test schematic is shown in Figure 4.40 [49]. In this test, a flexible printed electronic circuit is rolled up through rotation of a mandrel, while the circuit is kept under a certain tension. For cyclic coiling, the circuit is rolled and unrolled for a certain number of cycles, whereas for constant coiling, the circuit is kept under the rolled state for a prolonged duration. The functionality or resistance of the flexible electronics is to be monitored before and after the test or continuously throughout the test.

4.5.2.9 Crease Test

The crease test is used to evaluate flexible electronics and materials against creasing stresses during their application, in reference to ASTM F2749 [51]. A flexible circuit specimen is looped on a flat table with both ends of the specimen secured. A roller is rolled from the clamped ends of the specimen toward the end of the loop to create a

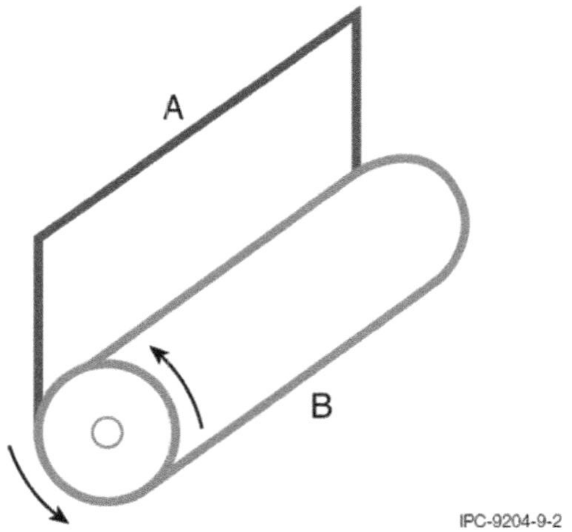

FIGURE 4.40 Coiling flex test setup (A: flexible circuit, B: rotational mandrel) [49].

crease with the conductive traces in compression. The creased specimen is unfolded and straightened. The specimen is then looped again to start the second cycle in the opposite direction (extension of the conductive traces) and then repeated for the following cycles. The circuit resistance needs to be monitored before and after each cycle or continuously throughout the cycles.

4.5.3 Challenges and Opportunities

Flexibility testing is a critical component in validating the performance and reliability of flexible electronics. This involves subjecting flexible electronics to repeated bending and flex cycles and measuring their performance and durability over time. IPC 9204 provides the first comprehensive guidelines to the industry on flexibility testing of flexible electronics; however, due to their dynamic nature and complexity in design and use conditions, it is a challenge to include all the use conditions and configurations. Flexible electronics test engineers must carefully review the designs and use conditions and select the appropriate test methods, test setups, the number of flex cycles, the bend radius, and the environmental exposure (temperature, humidity) to ensure that the tests are representative of the actual use conditions. Looking forward, with more understanding of flexible electronics use conditions and collaboration between manufacturers and users, flexibility test standards will continue to evolve to provide more precise test methods targeted for more specific use conditions and better correlation between the test conditions and the use conditions.

4.6 SUMMARY

This chapter provides a comprehensive overview of the materials for flexible electronics, covering flexible printed circuits, flexible hybrid electronics, and wearable electronics. The history of these materials, their applications and recent progress are introduced. The structure and construction of flexible electronics using flexible materials are discussed, as well as their characteristics and the manufacturing process to build the flexible electronics. Different techniques to evaluate flexibility of the flexible materials are provided. Challenges and limitations of the flexible materials are discussed, followed by opportunities for future development to satisfy the requirements of application conditions.

REFERENCES

1. Gilleo, K., The First 105 Years of Flexible Circuitry, Magazines007, 2007. https://magazines007.com/pdf/KGilleo.pdf (accessed December 4, 2024).
2. PCBSKY, A Brief History of Flexible PCB Technology, 2024. https://www.pcbsky.com/a-brief-history-of-flexible-pcb-technology.html (accessed December 4, 2024).
3. Semiconductor and Electronics, Flexible Electronics Market Size, Share, Revenue, Forecast 2032, 2024. https://www.fortunebusinessinsights.com/flexible-electronics-market-109105 (accessed December 4, 2024).
4. Savov, V., The world's first foldable phone is charmingly awful, 2019. https://www.theverge.com/2019/1/8/18174278/royole-flexpai-foldable-phone-android-ces-2019 (accessed December 4, 2024).
5. Essemtec, In-Mold Structural Electronics (IMSE™) - What's so revolutionary, 2018. https://essemtec.com/en/news/news-detail-en/-/-/in-mold-structural-electronics-imse-what-s-so-revolutionary/ (accessed December 15, 2024).
6. Flex Plus Tech, What do we know about Flex printed circuits for Humanoid Robot, 2024. https://www.flexplusfpc.com/post/what-do-we-know-about-flex-printed-circuits-for-humanoid-robot (accessed December 15, 2024).
7. Beecham Research Ltd and Wearable Technologies AG, World of Wearable Technology Applications: Function with Style, 2013.
8. IPC 6013D Standard, *Qualification and Performance Specification for Flexible/Rigid-Flexible Printed Boards*, IPC, Bannockburn, IL, 2017.
9. Flexible Circuit Technologies, Single Sided Flex Circuit, 2024. https://www.flexiblecircuit.com/all-products/single-sided-flex/ (accessed January 4, 2025).
10. Tekra, LLC, Teonex® PEN Film, 2026. https://www.tekra.com/products/brands/teonex (accessed June 13, 2026).
11. Tekra, Melinex® ST506, 2024. https://www.tekra.com/products/films/polyester-films/heat-stabilized-pet/melinex-st506 (accessed January 5, 2025).
12. Gordon, S., Substrates: Polyester Film for the Flexible Electronics Industry, *IPC Apex Conference*, Las Vegas, NV, March 25–27, 2014. https://www.ipc.org/system/files/technical_resource/E15%26S17_02.pdf (accessed January 5, 2025).
13. Upmold, Plastic Shrinkage Rates, 2017. https://upmold.com/plastic-shrinkage-rate-chart/ (accessed January 4, 2025).
14. Omnexus, Comprehensive guide on polyimide, 2025. https://omnexus.specialchem.com/selection-guide/polyimide-pi-plastic (accessed February 8, 2025).
15. Laminated Plastics, Technical Data Sheet Polyimide, 2025. https://laminatedplastics.com/polyimide.pdf (accessed February 8, 2025).
16. AZO Materials, Polyethylene Naphthalate (PEN), 2003. https://www.azom.com/article.aspx?ArticleID=1933 (accessed February 8, 2025).
17. RigidFlex PCB, Coverlay Flex PCBs: A Comprehensive Guide – Rigid Flex PCB, 2023. https://rigidflexpcb.org/coverlay-flex-pcbs-a-comprehensive-guide/ (accessed February 10, 2025).
18. Peterson, Z., Types of PCB Copper Foil for High-Frequency Design, 2024. https://resources.altium.com/p/types-pcb-copper-foil-high-frequency-design (accessed February 10, 2025).

19. Mad PCB, What's Bonding Sheet (Adhesive) in Flex PCB?, 2016. https://madpcb.com/bonding-sheet/#:~:text=Bonding%20Sheet%2C%20Sheet%20Adhesive%2C%20Pure,bond%20stiffeners%20and%20heat%20sinks (accessed February 15, 2025).
20. RAYPCB, What is the Stiffener for FPC?, 2025. https://www.raypcb.com/the-importance-of-fpc-stiffener/ (accessed February 15, 2025).
21. Yogendrappa, M., Flex PCB Design Guidelines, 2024. https://www.protoexpress.com/blog/flex-pcb-design-guidelines-for-manufacturing/ (accessed February 20, 2025).
22. Minco, Flex Circuit Design Guide, 2019. https://www.minco.com/wp-content/uploads/Minco-Flex-Circuits-Design-Guide-2019.pdf (accessed February 25, 2025).
23. CMTC, *Flexible Hybrid Electronics (FHE) is the Next Big Thing in Manufacturing: An Entirely New Platform for Electronics*, 2016. https://www.cmtc.com/blog/flexible-hybrid-electronics-in-manufacturing (accessed February 25, 2025).
24. Macdonald, W. A., Latest advances in substrates for flexible electronics, *Journal of the Society for Information Display*, vol. 15, p. 12, June 2007.
25. Behr, A., Novel Thermosetting Stretchable Material Beyolex, 2024. https://industrial.panasonic.com/ww/products/pt/beyolex-film (accessed March 1, 2025).
26. Reder, T., What Is Viscosity? Voltera, September 30, 2022. https://www.voltera.io/blog/what-is-viscosity (accessed May 31, 2025).
27. Elcometer, Viscosity Flow Cups, 2025. https://www.elcometerusa.com/Laboratory/Flow-Cups/?srsltid=AfmBOopKcC8ODZLSv3kjAxpAeWIptDmQ4W1uGAs4TfXrDiwrZ4Wo1gYd. (accessed March 1, 2025).
28. Paar, A., Basics of Rheology, 2025. https://wiki.anton-paar.com/us-en/basics-of-rheology/. (accessed March 1, 2025).
29. Stergard, A., Different types of Rheometers, January 20, 2021. https://fluidan.com/different-types-of-rheometers/. (accessed May 31, 2025).
30. Wikipedia, Rheometer, 2025. https://en.wikipedia.org/wiki/Rheometer#:~:text=There%20are%20two%20distinctively%20different,measure%20the%20resulting%20shear%20strain (accessed May 31, 2025).
31. Liu, W., *Evaluation of Stretchable Conductive Ink*, IPC Apex, Las Vegas, NV, March 13–17, 2016, (Printed with permission of Global Electronics Association).
32. Zu, W., Carranza, H., and M. D. Bartlett, Enhancing electrical conductivity of stretchable liquid metal–silver composites through direct ink writing, *ACS Applied Material Interfaces*, vol. 16, pp. 23895–23903, April 30, 2024.
33. Turunen, R. and D. Numakura, *Screen Printing Process for High Density Flexible Electronics, IPC Printed Circuits Expo, Apex and the Designer Summit*, IPC, Las Vegas, NV, April 1–3, 2008. https://www.ipc.org/system/files/technical_resource/E9%26S30_02.pdf. (accessed May 31, 2025).
34. Soulestin, T., *Accelerating Innovation with Henkel's PTC Ink Solutions*, TechBlick, June 2023. https://www.techblick.com/post/henkel-accelerating-innovation-with-henkel-s-ptc-ink-solutions (accessed June 14, 2026).
35. Wikipedia, Silver Chloride Electrode, 2026. https://en.wikipedia.org/wiki/Silver_chloride_electrode. (accessed June 14, 2026).
36. Townsend, N., ECG Instrumentation, Department of Engineering Science, University of Oxford, 2001. https://www.robots.ox.ac.uk/~neil/teaching/lectures/med_elec/notes2.pdf. (accessed May 31, 2025).
37. Liu, W. and P. Lall, *NextFlex Project Call 4.1: FHE Device Encapsulation and Overmold*, NextFlex, San Jose, CA, 2020.
38. Liu, W., *Development of Encapsulation Solutions for Flexible Hybrid Electronics*, SMTA, Minneapolis, MN, November 1–4, 2021 (As previously published in the proceedings of SMTA International 2021).
39. Avuthu, S.G. R., *An Introduction to the Process of Printed Electronics*, SMTA, Rosemont, IL, Sept 25–29, 2016 (As previously published in the proceedings of SMTA International 2016).

40. Optomec, Aerosol Jet Printed Electronics Overview, 2014. https://optomec.com/wp-content/uploads/2014/04/AJ_Printed_Electronics_Overview_whitepaper.pdf (accessed March 4, 2025).
41. Sumaiya, S., Organic solar cell by inkjet printing-an overview, *MDPI Technologies*, vol. 5, p. 53, 2017.
42. Huang, J., *NextFlex Project Call 3.0: Process Design Kit for Flexible Hybrid Electronics*, NextFlex, San Jose, CA, 2017.
43. Hanna, K. T., What is Wearable Technology? Definition, Uses and Examples, 2025. https://www.techtarget.com/searchmobilecomputing/definition/wearable-technology (accessed June 14, 2026).
44. Kalas, D., Novel SMD component and module interconnection and encapsulation technique for textile substrates using 3D printed polymer materials, *Polymers*, vol. 15, p. 2526, 2023.
45. Dexmat, The Best Conductive Threads for E-Textiles, 2020. https://dexmat.com/blog/the-best-conductive-threads-for-e-textiles/ (accessed April 4, 2025).
46. Wei, X., Highly stretchable electro-conductive yarn via wrapping carbon nanotube yarn on multifilament polyester yarn, *Journal of Industrial Textiles*, vol. 51(3S), pp. 2489S–4602S, 2022.
47. Arica, T. A., Advances in electrospun fiber-based flexible nanogenerators for wearable applications, *Macromolecular Materials and Engineering*, vol. 306, p. 2100143, 2021.
48. Liu, W., *Washability of E-Textile Materials*, IPC Apex, San Diego, CA, February 27–March 1, 2018, (Printed with permission of Global Electronics Association).
49. IPC 9204 Standard, *Guideline on Flexibility and Stretchability Testing for Printed Electronics*, IPC, Bannockburn, IL, 2017 (Printed with permission of Global Electronics Association).
50. ISO 5626:1993, *Standard, Determination of Folding Endurance*, ISO, Geneva, Switzerland, 1993.
51. ASTM F2749, *Standard Test Method for Determining the Effects of Creasing a Membrane Switch or Assembly*, ASTM, West Conshohocken, PA, 2015.

5 Materials for Thermal Management for Advanced Semiconductor Packaging

Naarendharan Meenakshi Sundaram, Min Jong Kil, Dongkai Shangguan, Subramanian Iyer, and Timothy Fisher

5.1 INTRODUCTION

The unabated increase in the power density of microelectronics demands a greater amount of cooling for these systems. The 'scale-down' and 'scale-out' approaches in the heterogeneous semiconductor packaging bring chips closer, resulting in efficient signal integrity with increased bandwidth and reduced latency at low energy per bit. It further reduces operating power at the cost of increasing power density, potentially beyond 4 W/mm^2 [1]. Cooling such heat fluxes demands high heat transfer coefficients and low thermal resistances for the heat to flow to the cooler (a cold plate or a heat sink). The goal is to maintain the chip under its operating temperature of 85°C. Advancements in cold plate design have enabled efficient liquid cooling, and to meet the need for high heat transfer coefficients, efficient two-phase cooling utilizing the latent heat of the liquid is of increasing relevance.

Designing efficient cooling modules with high heat transfer coefficients may still fail to maintain the chip temperature within safe operating limits if the contact between the chip and the cooling surface is poor, resulting in high interfacial thermal resistance and consequently high chip temperatures. Hence, designing cooling solutions for such microprocessors should employ an 'all of the above' approach - good contact, efficient cooling solutions, shorter thermal paths. This chapter presents a review of thermal interface materials (TIMs), examining various materials and their properties, as well as the underlying need for their use in thermal management applications.

The interface or contact between two objects is often never automatically smooth. All surfaces have non-zero roughness (Figure 5.1), which can trap air/gaseous medium between them. Air is a poor thermal conductor ($k \approx 0.026$ W/mK), resulting

 DOI: 10.1201/9781003587637-5

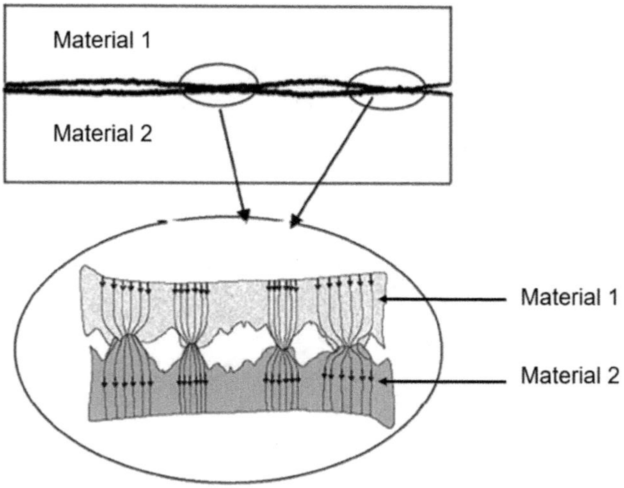

FIGURE 5.1 Roughness of two contacting surfaces. (Reprinted with permission from [2].)

in increased resistance to heat flow. The increased resistance to heat flow is also attributed to the scattering of electrons and phonons due to abrupt change of materials. This resistance results in an increased temperature difference between the hot and the cold surfaces. TIMs are interface materials with higher thermal conductivity (2–200 W/mK) that bridge the gap between such surfaces, enabling a lower overall thermal resistance and a smaller temperature drop across the two surfaces. Figure 5.2 depicts how a TIM sits between two surfaces and shows the thermal lines.

Consider an example where a chip generates 1 W/mm^2 and a microchannel cooler sits atop the chip (Figure 5.3).

Assuming that the chip generates 100 W of heat, for evaluating two different TIMs - gap pad (2 W/mK) and indium (80 W/mK), the temperature drop across the TIM (assuming 2 mm thickness and perfect contact across the interfaces) (Figure 5.4) can be estimated as:

$$P = 100 = \frac{55 - T_b}{t/(kA)} \tag{5.1}$$

Here, t represents the bondline thickness (thickness of the TIM layer), k represents the thermal conductivity of the TIM, and A represents the cross-sectional area. T_b has a value of 215°C for the gap pad and 58.7°C for indium, assuming the cooler temperature to be 55°C. Hence filling the gap between the heat source (chip) and the heat sink (microchannel cooler) with a high conductivity material helps in reducing the temperature drop from 160°C to 3.7°C. However, thermal conductivity is not the only relevant property of a TIM. It should also have favourable mechanical properties to withstand thermal cycling, high viscosity in the liquid phase to hinder pump out, the ability to withstand warpage, and many other properties to ensure long-term contact between two surfaces. This chapter explores various materials, their properties, and different testing methods for the evaluation and use of TIMs.

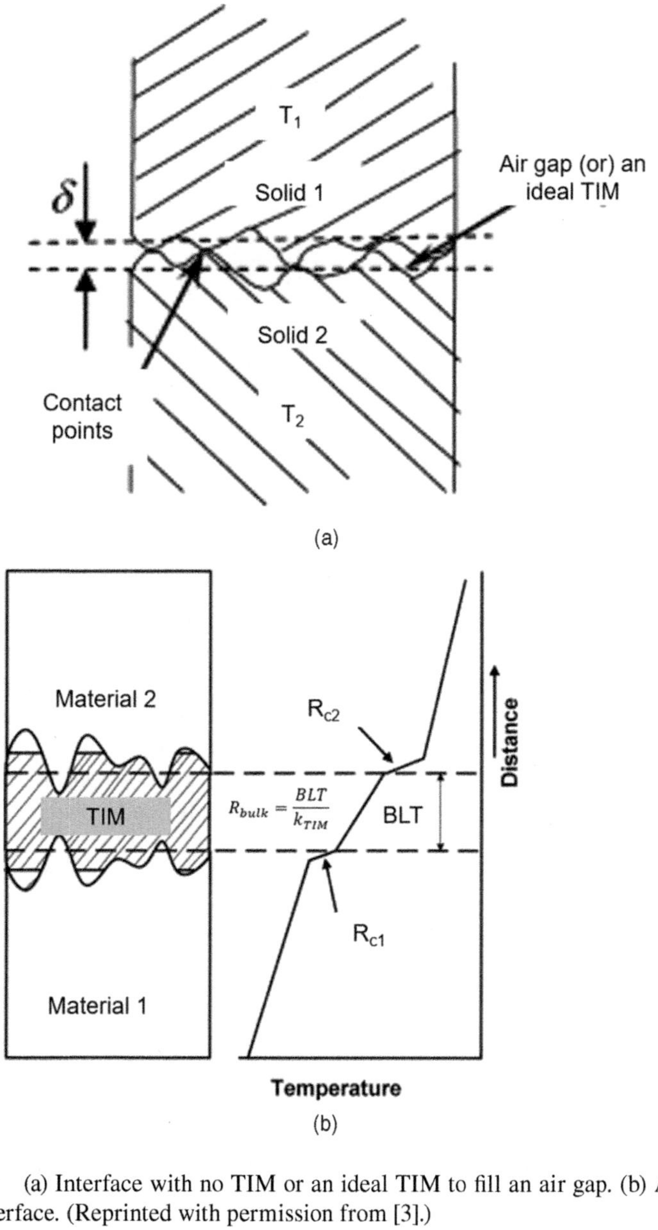

FIGURE 5.2 (a) Interface with no TIM or an ideal TIM to fill an air gap. (b) Actual TIM filling an interface. (Reprinted with permission from [3].)

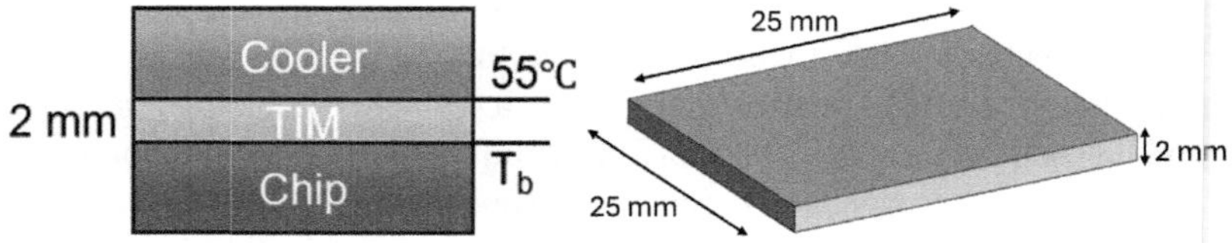

FIGURE 5.3 A chip cooler assembly with an ideal thermal interface material.

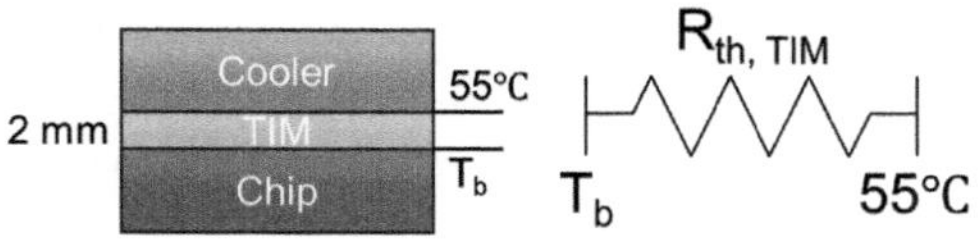

FIGURE 5.4 Thermal interface resistance across a TIM.

5.2 HEAT TRANSFER BASICS

Heat is thermal energy in transit and always flows from a body of higher temperature to a body of lower temperature. The three main modes of heat transfer are conduction, convection, and radiation. Conduction heat transfer is a mode of heat transfer that transmits heat through molecular vibrations (diffusion) or by the motion of free electrons. Convection involves the simultaneous transfer of heat through advection and diffusion. Radiation consists of the exchange of heat energy through electromagnetic waves. Thermal radiation primarily occurs in the infrared spectrum. In this section, a brief overview of conduction and convection heat transfer and their corresponding thermal resistances is provided.

5.2.1 Heat Conduction

Conduction heat transfer is a mode of heat transfer that transfers heat due to the molecular or atomic-level motion and vibrations of particles. In simple terms, conduction can be understood as the transfer from high-energy particles to low-energy particles through interaction between them. Heat conduction is depicted in Figure 5.5 (quiescent gas between two surfaces with different temperatures), showing the transfer of heat from high-temperature gas molecules to low-temperature gas molecules through molecular collisions.

Fourier's law of heat conduction governs the heat diffusion equation. In the 19th century, Joseph Fourier discovered the empirical correlation that governs the heat conduction in solids through a series of experiments on different metals.

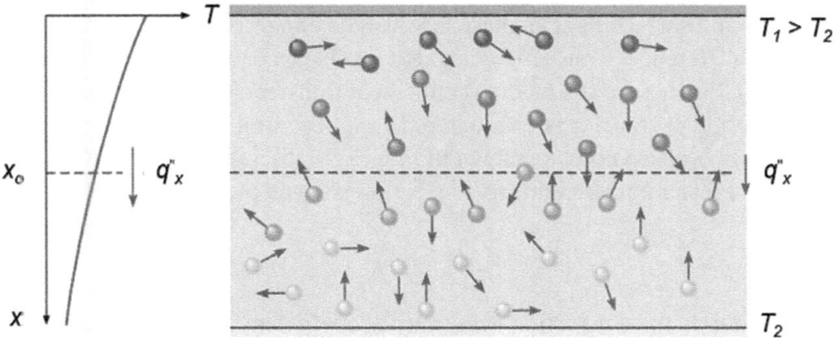

FIGURE 5.5 Conduction heat transfer showing the diffusion of energy due to molecular activity [4]. (Used with permission conveyed through Copyright Clearance Center, Inc.)

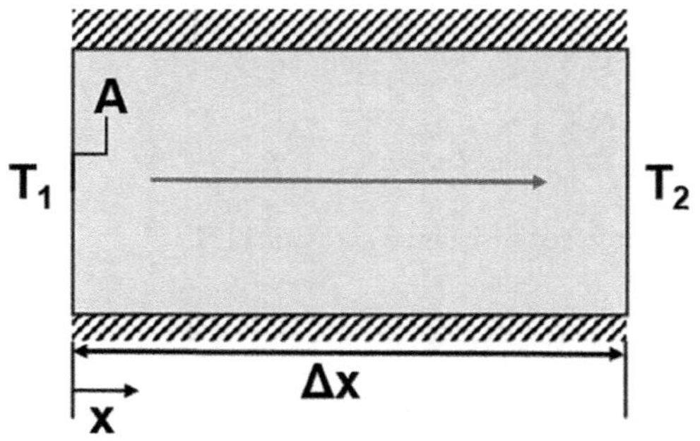

FIGURE 5.6 One-dimensional heat conduction in a rectangular slab.

In Figure 5.6, the rectangular slab is insulated on top and bottom, and the two side surfaces are maintained at T_1 and T_2 ($T_1 > T_2$). The cross-sectional area is A, and the length of the rod is Δx. The temperature difference causes the heat to flow from T_1 to T_2, and the objective is to measure the heat transfer rate q_x. It is observed that q_x is directly proportional to the temperature difference and the cross-sectional area, and is inversely proportional to the length. The constant of proportionality is the thermal conductivity, k, of the medium, as shown in the following equations,

$$q_x \propto A \frac{\Delta T}{\Delta x} \tag{5.2}$$

$$q_x = k_x A \frac{\Delta T}{\Delta x} \tag{5.3}$$

5.2.2 Convective Heat Transfer

Convection heat transfer can be decoupled into diffusion and advection, where diffusion is the energy transfer due to molecular vibrations/collisions, and advection is the energy carried by the flow (forced or natural) of the fluid. As the liquid flows over the solid surface, the friction from the solid diffuses perpendicularly into the fluid (hydrodynamic boundary layer), and likewise, the thermal energy of the solid (thermal boundary layer). Figure 5.7 shows the boundary layers for external flow of fluid over a solid surface. Single phase flow can be either forced or natural due to the temperature difference. Convection can also involve a change of phase (solid-liquid, liquid-vapor). Further details on convection heat transfer can be found in [4].

Convection heat transfer is encountered in a heat sink for air cooling or a cold plate (single/two-phase flow) as shown in Figure 5.8 [5] and Figure 5.9 [6].

The heat transfer equation due to convection is given by,

$$q = hA\Delta T \tag{5.4}$$

Here, h represents the convection heat transfer coefficient, and A is the area of contact between the fluid and the solid.

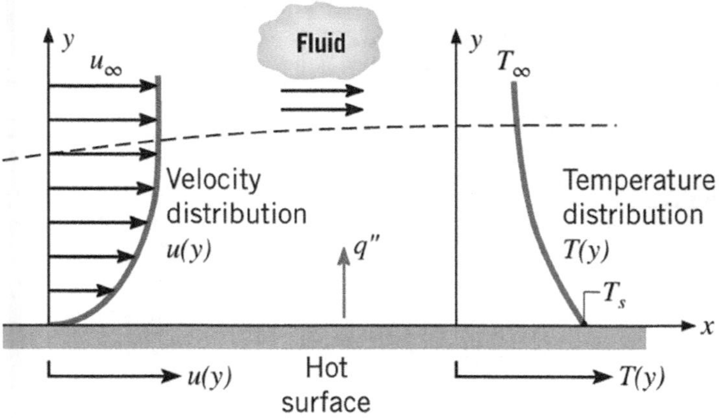

FIGURE 5.7 Convection heat transfer showing the boundary layer development [4]. (Used with permission conveyed through Copyright Clearance Center, Inc.)

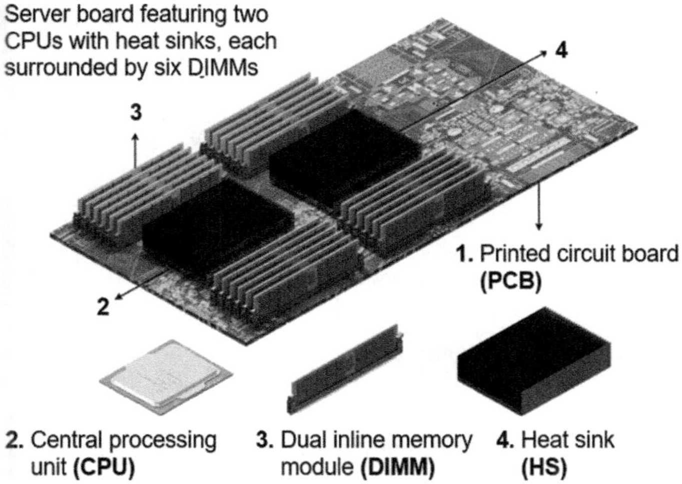

FIGURE 5.8 Schematic showing air-cooled heat sink in a 1U rack server [5]. (Reprinted from Copyright (2025), with permission from Elsevier.)

5.2.3 Thermal Resistance

Heat flow can be represented similarly to electric flow, with the temperature difference analogous to the voltage difference, the current flow analogous to the heat flow (heat transfer rate or heat flux), and the resistance to heat flow analogous to the resistance to current flow. Figure 5.10 shows the representation of thermal resistance for heat flow from temperature T_1 to temperature T_2 (any medium) and its analogy with current flow. The analogy between Ohm's law for current flow to that of heat flow can be understood from the following equations,

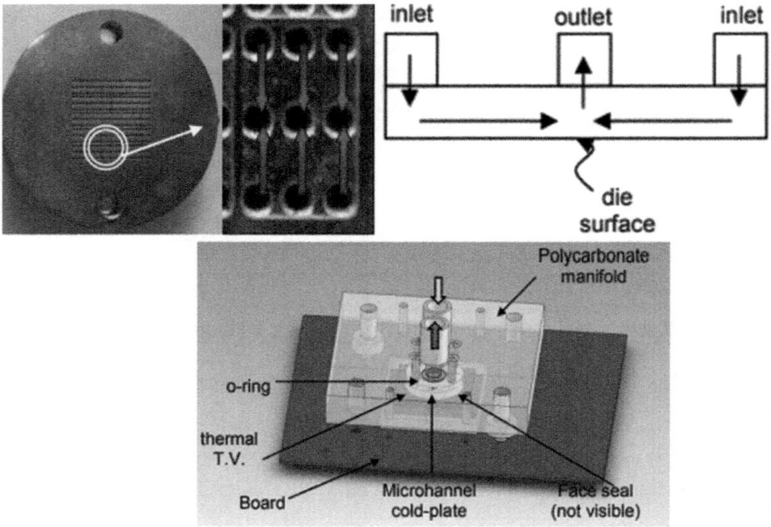

FIGURE 5.9 Illustrative schematic of a cold plate sitting on top of a chip. (Reprinted with permission from [6].)

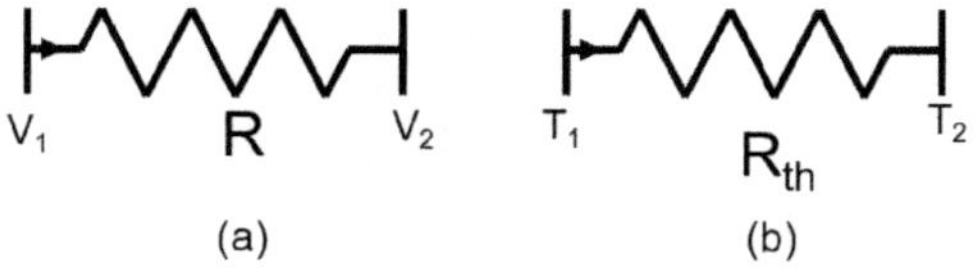

FIGURE 5.10 Resistance networks (a) electrical, (b) thermal.

$$\text{Ohm's Law}: i = \frac{V_1 - V_2}{R} \tag{5.5}$$

$$\text{Heat Transport}: q = \frac{T_1 - T_2}{R_{\text{th}}} \tag{5.6}$$

The conductive thermal resistance is given by,

$$R_{\text{cond}} = \frac{\Delta x}{kA} \tag{5.7}$$

The convective thermal resistance is given by,

$$R_{\text{conv}} = \frac{1}{hA} \tag{5.8}$$

Figure 5.11 shows an interface between two surfaces and the possible modes of heat transport between them. Heat conduction dominates in small gaps (in the range of a few hundred microns) when compared to radiation and convection.

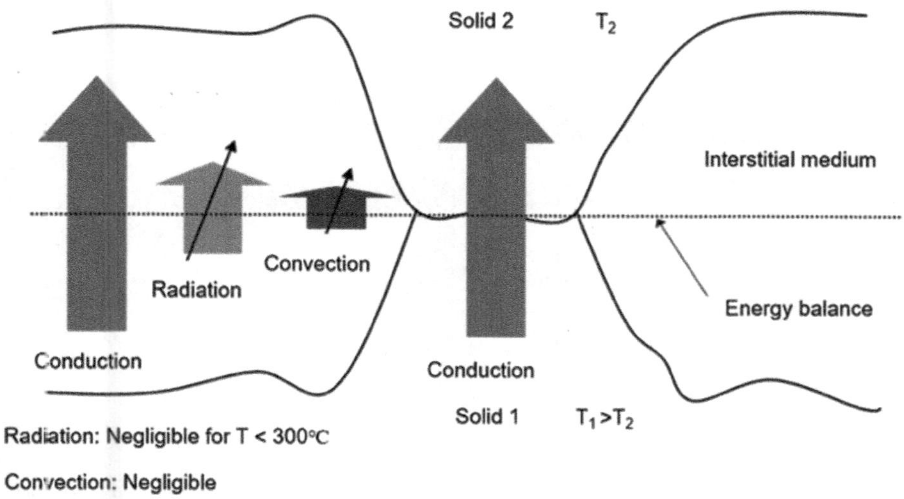

FIGURE 5.11 Modes of heat transport across an interface.

The factors affecting heat transport or the thermal resistance across an interface are:

- Contact geometry
  - Surface roughness and waviness
  - Flatness of the contact
- Thermal/physical properties
  - Contacting members
  - Gas gap (or filler material)
- Applied pressure
  - Elastic or plastic surface asperity deformation
- Interface temperature
  - Change in properties
  - Change in phonon scattering at contact spots

The thermal resistance across an interface is given by,

$$R = R_{C1} + R_{C2} + R_{\text{TIM}} \tag{5.9}$$

Here, R_{C1} and R_{C2} represent the contact resistances between the upper surface of the TIM and the heat sink, and the lower face of the TIM and the heat source, respectively, whereas R_{TIM} is the thermal resistance of the TIM itself as represented in:

$$R_{\text{TIM}} = \frac{\text{BLT}}{kA} \tag{5.10}$$

Here, k represents the thermal conductivity of the TIM while the BLT represents the bondline thickness of the TIM, and A represents the contact cross-sectional

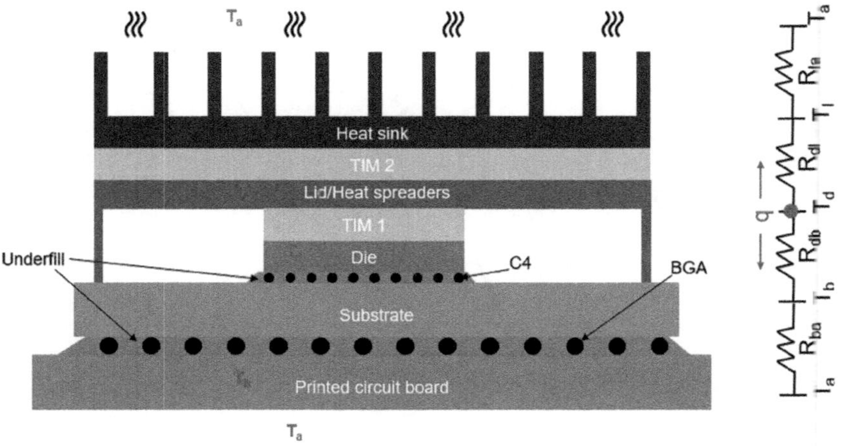

FIGURE 5.12 Typical electronics packaging stack and the associated 1D thermal resistance model.

area. Heat conduction is the main mode of heat transport across the two interfaces (Figure 5.11).

Figure 5.12 shows the thermal resistance of a traditional printed circuit board package. An air-cooled heat sink is used to cool the chip, and the package has a lid. From the thermal resistance network shown in Figure 5.12, the heat can either go to the top or bottom of the stack.

The total heat transport equation for a chip that generates heat at the rate of q (watts) is given by,

$$q = \frac{T_d - T_a}{R_{dl} + R_{la}} + \frac{T_d - T_a}{R_{db} + R_{ba}} \tag{5.11}$$

If the total heat load, the thermal resistance of the individual elements and the ambient temperature (T_a) are known, the die temperature (T_d) can be computed.

5.2.4 Heat Spreading and Constriction Resistance

A couple of other thermal resistances are quite significant when it comes to the cooling of electronics. One provides the thermal resistance encountered when heat flows from a smaller area to a larger area, where the heat flux lines spread out (spreading resistance), and the other occurs when heat is transferred from a larger area to a smaller area (constriction resistance). The isoflux lines and the pictorial depiction of the heat spreading are shown in Figure 5.13. Referencing the following [7,8] for the derivation, the corresponding thermal resistance relations are given by:

5.2.4.1 Circular Constriction in Half Space

$$\frac{\partial^2 T}{\partial r^2} + \frac{1}{r}\frac{\partial T}{\partial r} + \frac{\partial^2 T}{\partial z^2} = 0 \tag{5.12}$$

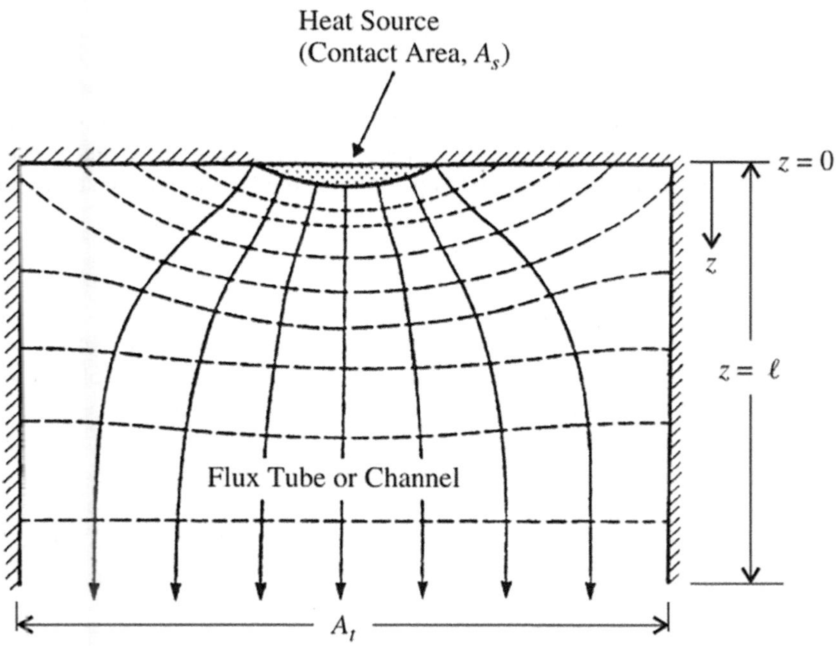

FIGURE 5.13 Heat spreading from a small area (chip) to a larger area (heat spreader) [7]. (Reproduced from Ref. [7] with permission.)

$$R_{cd,t} = \frac{1}{4\text{ ka}}(\text{Uniform temperature}) \tag{5.13}$$

$$qR_{cd,q} = \frac{8}{3\pi^2\text{ka}} = \frac{1}{3.7\text{ ka}}(\text{Uniform heat flux}) \tag{5.14}$$

5.2.4.2 Spreading Resistance

$$qR_{\text{Total}} = \overline{T}_S - \overline{T}_{z=r} \tag{5.15}$$

$$qR_{1D} = \overline{T}_{z=0} - \overline{T}_{z=\ell} \tag{5.16}$$

$$R_{\text{Total}} = R_S + R_{1D} \tag{5.17}$$

$$R_S = \frac{\overline{T}_S - \overline{T}_{z=0}}{q} \tag{5.18}$$

$$q = \iint -k\frac{\partial T}{\partial z}\Big|_{z=0}\, dA_s \tag{5.19}$$

Here, $\bar{T}$ represents the area averaged temperature.

More often, as covered in Section 5.3, heat spreaders are used in electronics packages to reduce the power density of cooling. But as we scale towards advanced packaging and heterogeneously integrated systems [9], where many high-power processors are brought in near proximity of each other, the heat spreading resistance becomes significant as the operating temperature of the chip might exceed its limit before the heat can spread to a larger area.

5.2.5 Heat Transport by Phonons

In conduction heat transfer, the thermal energy is mainly carried by two carriers - phonons and electrons. The free electrons in materials help in transport of both electricity and heat. Phonons are quasiparticles that are governed by the Bose-Einstein statistics (Bosons). Phonons are the quantized lattice vibrations that exist in materials and occupy discrete energy levels ($\hbar\omega$). There are two types of phonons (Figure 5.14) - acoustic and optical phonons [10]. Acoustic phonons are characterized by lower frequency, in-phase vibrations that are primarily responsible for the propagation of sound waves in a medium, and optical phonons are characterized by higher frequency, out of phase vibrations that interact with light (electro-magnetic waves).

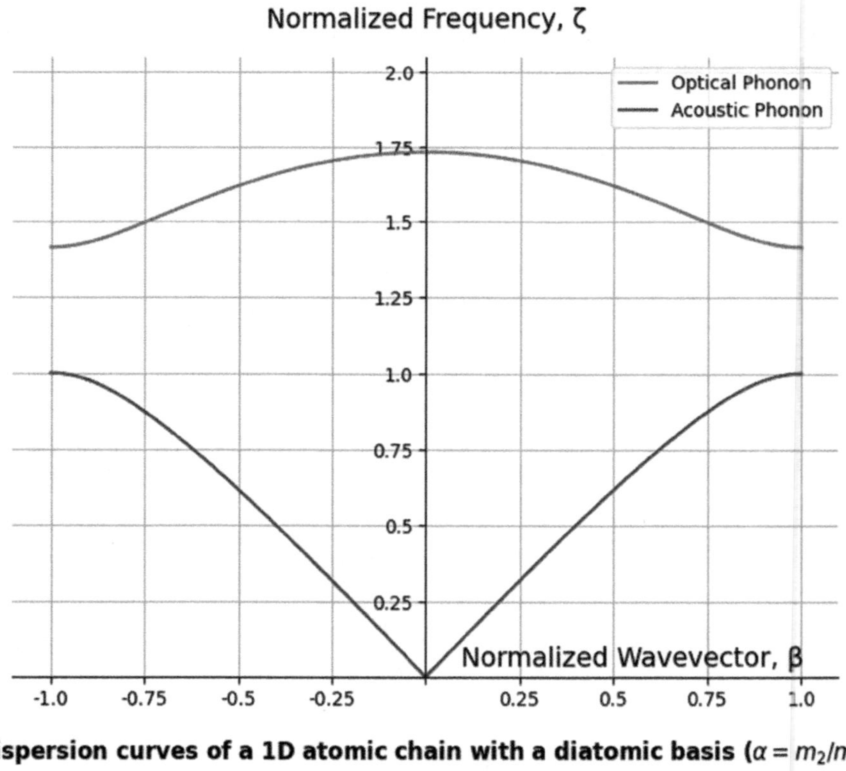

FIGURE 5.14 Phonon band dispersion curves with a 1D atomic chain with a diatomic basis.

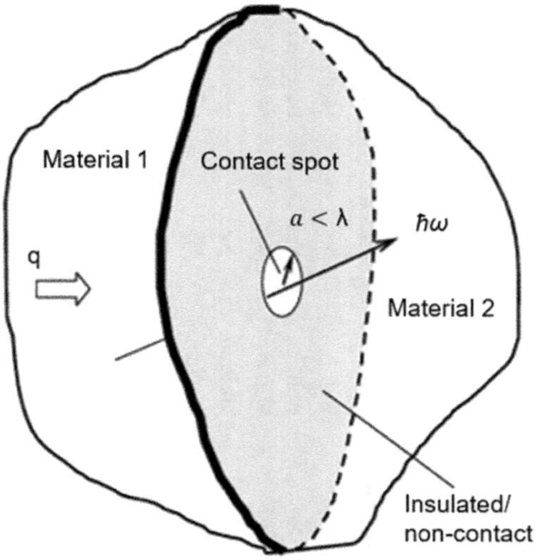

FIGURE 5.15 Phonon tunneling across two materials.

In the context of TIM, when we have two surfaces in close proximity such that the distance between them is less than the phonon wavelength, heat can still transport from one medium to another, and the transport rate is given by [10],

$$q''_{12} = \frac{v_g}{4\pi}\sum_p \int_0^{\omega_D} \frac{\hbar\omega . D(\omega)}{\exp\left(\frac{\hbar\omega}{k_B T_1}\right)-1} \int_0^{2\pi}\int_0^{\pi/2} T(\omega,\theta)\cos(\theta)\sin(\theta)d\theta d\Phi d\omega \quad (5.20)$$

Equation (5.20) is valid for the following criteria (Figure 5.15),

- Contact size $a < \lambda$, phonon ballistic transport
  - λ, phonon mean free path
  - v_g, frequency-independent phonon group velocity
  - $D(\omega)$, phonon density of states
  - T, transmissivity
- Heat flows through the a-contact

5.3 THERMAL INTERFACE MATERIALS

Different categories of thermal interface materials (TIMs) are utilized across a wide range of applications, with each type exhibiting unique advantages and constraints depending on the specific thermal management requirements. Table 5.1 summarizes the general classification of TIMs [11].

This section begins with a discussion of the fundamental properties required of an effective thermal interface material (TIM), followed by a detailed examination of the composition, advantages, and limitations of each category of TIM introduced earlier.

TABLE 5.1
Broad Classification of Thermal Interface Materials

Type	Thermal Conductivity (W/mK)	BLT (μm)	Thermal Interface Resistance (mm²K/W)	Pump-Out	Absorbs Stress	Reusable	Replaceability
Thermal grease [12]	0.4–4	20–150	10–200	Yes	Well	No	Medium
Thermal pad [12]	0.8–3	200–1,000	100–300	No	Well	Yes	Excellent
Phase change material [12]	0.7–1.5	20–150	30–70	Yes	Well	No	Medium
Thermal gel [12]	2–5	75–250	40–80	Yes	Medium	No	Medium
Thermal adhesive [12]	1–2	50–200	15–100	No	Medium	No	Poor
Solder [12]	20–80	25–200	<5	No	Poor	No	Poor
Carbon-based [13,17,18]	1–36	20–300	3–250	No	Medium	Yes	Medium
Low melting alloys (including liquid metals) [12,14–16]	10–80	<50	<10	Yes	Well	Yes	Medium

5.3.1 Essential Properties of a Thermal Interface Material

The two key parameters for selecting a TIM, as derived from Equation (5.10), are: the bondline thickness and the thermal conductivity of the TIM. A thinner bondline thickness and a higher thermal conductivity jointly reduce the overall thermal resistance across the TIM layer, thereby minimizing the temperature drop between the mating surfaces. However, as indicated in Equation (5.9), the total interfacial thermal resistance also includes the contact resistances (R_{C1} and R_{C2}) at the interfaces between the TIM and the adjoining substrates. These interfacial resistances are strongly influenced by parameters such as surface roughness, clamping pressure, compressive modulus, and surface flatness [2]. Consequently, the conformability of a TIM—its ability to fill microscopic voids and asperities under pressure—becomes a critical design consideration.

Beyond thermal and mechanical characteristics, several secondary but application-specific properties also guide TIM selection. These include biocompatibility, non-toxicity, a wide operational temperature range, chemical stability, appropriate viscosity (particularly for liquid or semi-liquid TIMs), electrical conductivity or insulation as required, contact pressure requirements, and cost-effectiveness. A comprehensive overview of the current status and emerging challenges in TIM development is provided by [2,13,19–22].

5.3.2 Thermal Greases/Pastes

Thermal grease, also referred to as thermal paste, is a polymer-based TIM designed to enhance heat transfer between contacting surfaces [2]. It primarily consists of two components: a thermally conductive filler dispersed within a polymer matrix. Common filler materials include aluminum oxide, zinc, silver, and various carbon-based allotropes, which provide high thermal conductivity. The polymer matrix typically comprises silicone oil, hydrocarbon oil, polyethylene glycol (PEG), or sodium silicate-based binders [23]. Among these, silicone oil is most widely used because of its low moisture absorption, excellent wettability, thermal stability, high dielectric strength, and ionic purity [24,25].

Based on the type of filler material, thermal greases are generally categorized as metal-based, ceramic-based, or carbon-based [2,26–28]. The filler characteristics such as thermal conductivity, coefficient of thermal expansion, geometry, particle size distribution, and concentration strongly influence the effective thermal performance of the compound [29]. A summary of commonly used filler materials is provided in Table 5.2 [22,30]. A commercially available thermal grease (OmegaTherm201) is shown in Figure 5.16.

The thermal conductivity of some of the common filler materials is provided in Table 5.3 [11]. Even though the thermal conductivity of the filler material is high, the effective (bulk) thermal conductivity of the composite is dominated by the binder (the polymer matrix). The thermal conductivity of the polymer matrices (Table 5.4) is significantly lower than that of the filler material leading to a low thermal conductivity of the grease (0.4–4 W/mK). The trade-off lies between the lower viscosity and wetting characteristics provided by the polymer base and the improved thermal

TABLE 5.2
Filler Material Choices

Metal-based	Silver, aluminum, copper
Ceramic-based	Alumina, aluminum nitride (AlN), silicon carbide (SiC), zinc oxide (ZnO)
Carbon-based [30]	CNT, graphene, graphite nanoparticles, few-layered graphene sheets (FLG), carbon black dispersions

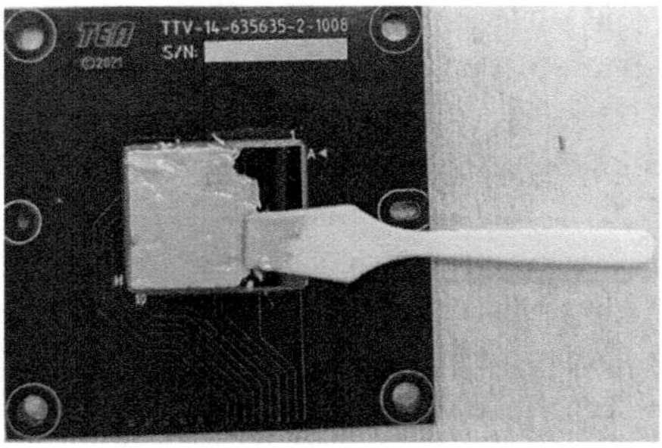

FIGURE 5.16 Thermal grease on a silicon chip.

TABLE 5.3
Thermal Conductivity of Filler Materials

Material	Thermal Conductivity (W/mK)
Carbon nanotubes [31]	6,600
Aluminum nitride (AlN) [4]	150
Silver (Ag) [4]	430
Copper (Cu) [4]	400
Gold (Au) [4]	315
Aluminum (Al) [4]	235
Boron nitride (BN) [11]	250–300
Alumina (Al_2O_3) [4]	35
Zinc oxide (ZnO) [11]	21

TABLE 5.4
Thermal Conductivity of Commonly Used Polymer Matrices

Material	Thermal Conductivity (W/mK)
Silicone oil	0.165 [32]
Epoxy resin	0.1–0.3 [3]
Hydrocarbon oil	0.1–0.18 [33]
Polyethylene glycol	0.19–0.32 [34]

conductivity provided by the filler material. It is found that a lower viscosity can be achieved by mixing particles of different sizes for the same volume fraction [2]. Models to evaluate the effective thermal conductivity of composite TIMs are discussed in Section 5.7.1.

Some notable benefits of thermal grease include [11]:

- Good thermal conductivity (0.4–4 W/mK)
- Small bondline thickness (20–150 μm) and minimal contact pressure
- The low viscosity of thermal grease enables filling of surface crevices
- Ease of application – no curing is required
- Delamination of the TIM is not a concern

Some notable limitations of thermal grease include [11]:

- Susceptible to pump-out and phase separation
    - The frequent power cycling results in the pump out of the grease, hence making it difficult to be used for long-term applications [21,35,36]
- The low viscous nature means that it is easy to migrate and becomes messy while applying the TIM to the component
- The thermal performance of grease decreases with temperature and time [37]

A study by [38] employed finite element analysis (FEA) to investigate the influence of voiding within thermal grease layers. The results indicate that central voids have a significantly greater impact on the overall thermal resistance than peripheral voids. The effect of voiding area on the thermal resistance for different grease thicknesses is illustrated in Figure 5.17.

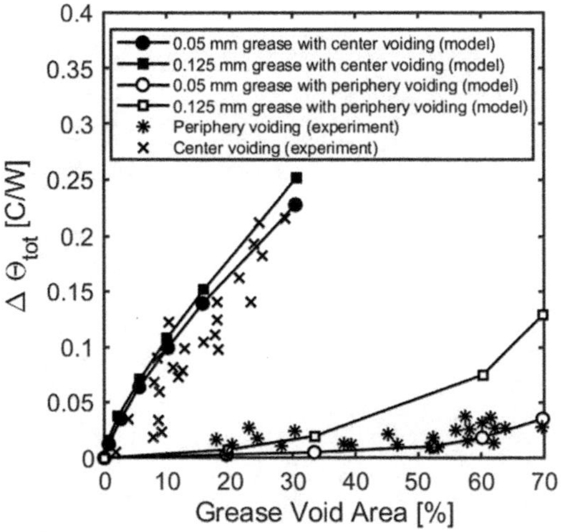

FIGURE 5.17 Plot showing the effect of voiding in thermal grease on the thermal interface resistance. (Reprinted with permission from [38].)

5.3.3 Thermal Pads and Elastomers

Thermal pads, also referred to as elastomeric pads or gap-filler pads, are composed of polymeric elastomers embedded with ceramic fillers to enhance thermal conductivity. The typical thickness of thermal pads ranges from 0.2 mm to 1.0 mm, and they are primarily used as a TIM in low-power electronic devices such as chipsets, memory modules, and mobile processors [36]. Since the thickness of the thermal pads is typically larger than that of other TIMs, they must exhibit sufficient conformability and are generally required to be compressed to at least 25% of their original thickness to achieve optimal thermal coupling between the mating surfaces [36]. This compressibility allows them to compensate for assembly tolerances and surface non-uniformities, making them suitable for a variety of electronics packaging configurations.

The polymer matrix is usually composed of silicone, epoxy, or rubber-based elastomers [39], while the thermally conductive fillers commonly include aluminum oxide, zinc oxide, boron nitride, or aluminum nitride. In some formulations, a woven fiberglass mesh or a dielectric film is incorporated as reinforcement to improve mechanical integrity and handling during assembly [40]. A representative commercial thermal gap pad is shown in Figure 5.18.

Some notable benefits of the thermal pads are [27,41,42]:

- Not messy, and handling is easy, thus reducing assembly time
- Pump-out issue is eliminated
- Added benefit of acting as a vibration damper and protecting the electronic components
- Better stability when compared to phase change TIMs
- Does not conduct electricity

FIGURE 5.18 Thermal pad on a silicon chip.

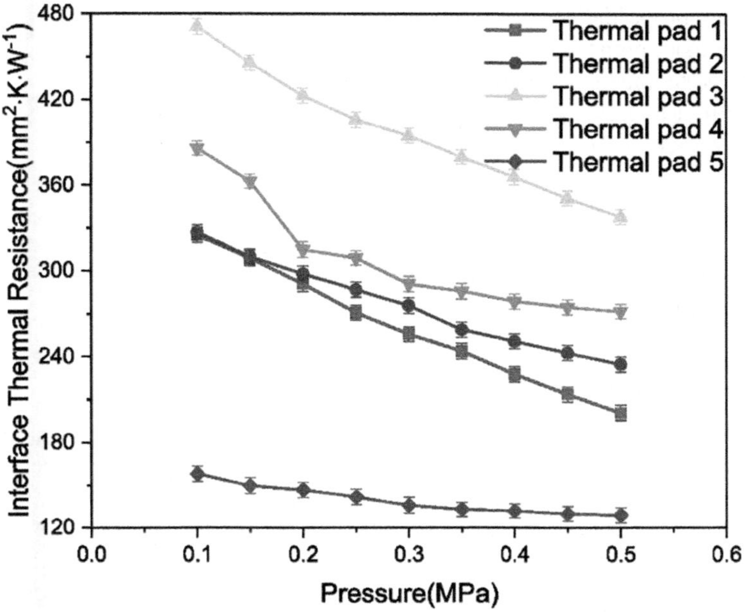

FIGURE 5.19 Effect of mounting pressure on the thermal resistance of thermal pads (for boron nitride, polymer, three carbon-based materials) [43]. (Reprinted from Copyright (2019), with permission from Elsevier.)

Some notable limitations of the thermal pads are [41,42]:

- Does not flow easily like grease. Needs to be compressed with sufficient pressure to ensure that voids are filled. So, it demands consistent mounting pressure to ensure lower thermal resistance (Figure 5.19)
- A trade-off between the addition of filler particles (improved conductivity) and conformability (compressibility) must be considered

5.3.4 Gels

Thermal gels are closely related to thermal greases, consisting of metallic or ceramic thermally conductive fillers dispersed within a silicone polymer matrix. However, unlike greases, gels undergo an additional curing or cross-linking step, which increases the interconnect density of the polymer network. As a result, the silicone base in gels exhibits a higher cross-link density than that of grease but remains less rigid than elastomeric thermal pads [36]. This intermediate level of cross-linking imparts gels with the combined characteristics of both greases and thermal pads: they can flow to conform to surface asperities while maintaining structural stability after application.

Some notable benefits are [28]:

- Good shape retention
- Can fill voids and crevices on rough surfaces

- Offer properties of both solid and liquid
- No major pump-out and migration concerns
- Good temperature stability
- Similar to grease, gels can withstand thermo-mechanical stresses
- Reworkable

Several notable limitations are [28]:

- Lower thermal conductivity than grease
- Delamination concerns

5.3.5 Metal-Based Thermal Interface Materials

Metals inherently exhibit much higher thermal conductivity than polymers or liquids due to the combined contributions of free-electron transport and efficient lattice vibrations facilitated by their densely packed atomic structure. However, when used in solid form (such as a preform) as a TIM, metals tend to exhibit limited surface conformity, which leads to poor interfacial contact and consequently high contact resistance at the mating surfaces. This limitation reduces the overall effectiveness of solid metals as TIMs. To overcome this issue, it is desirable for the metal-based TIM to remain in the liquid phase under the operating temperature range of electronic devices (typically below 100°C). This requirement has driven the development of low melting alloys (LMAs) that remain liquid during normal operation and can thus serve as efficient liquid metal TIMs [44]. Metal alloys in their liquid form can easily flow into the microscopic crevices between contacting surfaces and fill interfacial voids, which greatly reduces the thermal contact resistance [2].

Table 5.5 lists various metals and liquid metals with low melting points along with their key thermophysical properties, and Table 5.6 presents the composition of

TABLE 5.5
Thermal Properties of Typical Metals with Low Melting Points [44]

Liquid Metal	Melting Point (°C)	Evaporation Point (°C)	Evaporation Pressure (mmHg)	Specific Heat (kJ/kgK)	Density (kg/m³)	Thermal Conductivity (W/mK)	Surface Tension (N/m)
Mercury	−38.87	356.65	1.68×10^{-3a}	0.139^b	$13{,}546^a$	8.34^b	0.455^a
Cesium	28.65	2,023.84	10^{-6c}	0.236^c	$1{,}796^c$	17.4^c	0.248^d
Gallium	29.8	2,204.8	10^{-12}	0.37^b	$5{,}907^b$	29.4^b	0.707^b
Rubidium	38.85	685.73	6×10^{-6}	0.363^m	$1{,}470^m$	29.3^m	0.081
Potassium	63.2	756.5	6×10^{-7}	0.78^m	664^m	54.0^m	0.103^a
Sodium	97.83	881.4	10^{-7}	1.23^m	926.9^m	86.9^m	0.194^d
Indium	156.8	2,023.8	$<10^{-10}$	0.27^m	$7{,}030^m$	36.4^e	0.55^m
Lithium	186	1,342.3	10^{-10}	4.389^b	515^b	41.3^b	0.405^b
Tin	232	2,622.8	$<10^{-10}$	0.257	$6{,}940^e$	15.08^b	0.531^m

Note: ᵃ25°C; ᵇ200°C; ᶜ160°C; ᵈ100°C; ᵐmelting point.

TABLE 5.6
LMA Alloys with Melting Temperature between 10°C and 81°C [47]

Liquidus (°C)	Solidus (°C)	Composition (%)
10.7E	10.7	62.5Ga, 21.5In, 16Sn
15.7E	15.7	75.5Ga, 24.5In
25	15.7	95Ga, 5In
29.8E	29.8	100Ga
47.2E	47.2	44.7Bi, 19.1In, 6.3Sn, 22.6Pb, 5.3Cd
58E	58	49Bi, 21In, 12Sn, 18Pb
60E	60	32.5Bi, 51In, 16.5Sn
69	58	49Bi, 18In, 5Sn, 18Pb
72E	72	33.7Bi, 66.3In
79E	79	57Bi, 26In, 17Sn
81E	80	54.02Bi, 29.68In, 16.3Sn

metal alloys that melt within the temperature range of 10°C and 81°C, including alloy compositions of gallium, indium and tin. It is important to note that another widely used variant of gallium-indium-tin alloy system is Galinstan (R) (68.5% Ga, 21.5% In, 10% Sn) [45]. Owing to its lower tin content, Galinstan exhibits a lower melting temperature of −19°C compared to the eutectic EGaInSn alloy which has a melting temperature of 10.7°C. Furthermore, Galinstan has a surface tension of 534 mN/m which is lower than that of the eutectic counterpart (605 mN/m). Galinstan is one among LMAs that is gaining traction as a TIM [46].

Among the commonly used low-melting alloys are those based on gallium and bismuth. Because of toxicity and environmental concerns, alloys containing cadmium or lead are seldom employed in TIM applications. From Table 5.6 (E stands for eutectic), indium stands out due to its relatively high thermal conductivity, making it a promising candidate for alloying with low-melting metals or liquid metals such as gallium. In addition, indium's soft and ductile nature, combined with its excellent wettability, enhances its ability to form efficient thermal interfaces [48]. Gallium-based liquid-metal-embedded elastomers have also been reported as suitable materials for low-temperature wearable and flexible electronic devices [49]. For optimal reliability, LMA based TIMs must remain mechanically compliant and sufficiently soft to accommodate stresses that arise from the mismatch in coefficients of thermal expansion and from package warpage during power or temperature cycling.

Some notable benefits of LMA are [2]:

- High thermal conductivity
- Good wettability on surfaces, which helps ensure filling of crevices avoiding airgaps
- Can be used between dissimilar materials that have different coefficients of thermal expansion [47]
- No curing cycle, and reworkability

FIGURE 5.20 (a–c) Depiction of the effect of corrosion on LMA TIM post the exposure to an environment consisting of 85°C and 85% humidity (85/85) [47]. (With permission.)

Some notable limitations of LMA are [47]:

- Corrosion is a significant issue in LMAs. Gallium is highly corrosive to metals with the exception of tungsten and tantalum [50]. Exposure of iron, chromium and nickel specimens to gallium at 400°C forms $FeGa_3$, $CrGa_4$, and Ni_2Ga_3 (with traces of $NiGa_4$) intermetallics. The corresponding corrosion rates of iron, chromium and, nickel are 1.9, 0.04, and 0.21 mm/day, respectively. The corrosion is caused primarily by oxygen, heat, and moisture [47]. Figure 5.20 shows the corrosion of LMA [47]. In industrial practice, proper measures can be taken to prevent the occurrence of corrosion in electronics packaging applications.
- Proper considerations need to be taken to avoid formation of intermetallics. Intermetallics have lower thermal conductivity and affect the performance of the LMA as a TIM

Solder-based TIMs have also been investigated for high-performance cooling applications [51]. However, a major challenge associated with melting and subsequently solidifying a metal between two surfaces is the formation of intermetallic compounds at the interfaces. These intermetallics tend to reduce the effective thermal conductivity, increase the bondline thickness, and introduce brittleness into the joint, thereby compromising mechanical reliability. Therefore, proper selection of the solder alloy is important for TIM applications. Of particular interest are indium and indium alloys which have been widely used as TIM in the semiconductor industry. A recent study [52] employed molten indium as a solder TIM between a test chip and a cold plate. They reported that at a heat flux of 1 W/mm^2, the temperature drops across the TIM decreased from 85°C for thermal grease to about 10°C for the indium TIM. A commercially available indium solder TIM is shown in Figure 5.21.

Further, under metal-based TIMs, sintered nano-Ag and nano-Cu are gaining traction [54,55]. In this approach, a metallic bonding layer between the die and the cooler/heat spreader, or between the package and the cooler, is formed by sintering nanoscale silver or copper particles between the mating surfaces. The process, as described [54] for Ag-based TIM, involves applying Ag paste on the bond substrate (heat spreader/cooler—direct-bonded copper, or DBC), followed by drying, die placement, and heating under applied pressures. External pressure is required to prevent the formation of porous Ag, and this can limit process throughput. The sintered paste, when

FIGURE 5.21 Commercial solder metal TIM [53]. (With permission.)

FIGURE 5.22 (a) Printed nano-Cu paste sample. (b) Dispensed nano-Cu paste sample. (Reprinted with permission from [56].)

used as a TIM, avoids the formation of intermetallic compounds and voids (when sufficient pressure is applied) compared to solder-based TIMs. Researchers have explored sintered copper [56] as a TIM as well. Zhao et al. [56] developed a sintered Cu paste as an alternative to Ag paste, as the latter was sensitive to surface finishes. Figure 5.22 shows the sintered Cu application process. Recent advances have significantly improved the application processes for sintered Ag and sintered Cu, in terms of throughput and process robustness.

5.3.6 Phase Change Materials as Thermal Interface Materials

Phase change thermal interface materials (PCTIMs) are a class of TIMs that utilize a reversible solid–liquid phase transition to enhance heat transfer between heat-generating components and heat spreaders or heat sinks [57]. PCTIM reacts to a certain temperature range to reversibly change phase through endothermic and exothermic processes, reducing thermal resistance by filling gaps between two contacting surfaces. A key characteristic of PCTIMs is their ability to function as latent heat storage media, absorbing and releasing heat during phase transitions, which helps to buffer sudden temperature fluctuations and improve thermal stability. The transition temperature can be tailored so that the material changes phase near the operating temperature of the device. PCTIM can have different base materials such as organic materials (paraffin, polyethylene glycol, fatty acid, sugar alcohol PCM), inorganic materials (hydrated salt, metals), or hybrid composition that combine the advantage of both the systems.

Some notable benefits of PCTIMs are:

- Ability to buffer temperature spikes or thermal shock
- Low thermal contact resistance after phase change
- Could be engineered to have different phase change temperatures
- Durability over thermal cycles [28]

Some notable limitations of PCTIMs are:

- Intrinsic low thermal conductivity [58], especially those that are non-metal based
- Liquid leakage issue during phase transformation [59]
- Weak mechanical properties such as tensile strength and stiffness [60]

The intrinsically low thermal conductivity of PCTIMs can be improved by incorporating thermally conductive fillers or inorganic additives. Liquid leakage and weak mechanical properties can be mitigated by using appropriate encapsulation materials, which help control leakage, improve mechanical strength, and maintain stability during thermal cycles.

To expand the application range of PCTIMs, both mechanical and chemical modification approaches have been explored. Mechanical approaches include the introduction of porous structured fillers [61,62] and the use of encapsulation techniques [63,64]. A representative chemical approach involves incorporating thermoplastic polymers that induce crosslinking reactions, forming a three-dimensional network that enhances mechanical strength and leakage resistance [65]. A typical PCM TIM is shown in Figure 5.23.

5.3.7 Adhesive-Based Thermal Interface Materials

Adhesive TIMs are a type of thermal interface material designed to provide both thermal conduction and mechanical bonding between two components. Adhesive TIMs require a curing process, either physical or chemical, to form semi-permanent or permanent bonding between two contacting surfaces and help conduct heat. This mechanical bonding capability eliminates the need for external pressure and minimizes the risk of detachment caused by dimensional changes in the TIM during operation [12]. Adhesive TIMs are widely commercialized and are available in several forms, including tapes, films, and liquid adhesives.

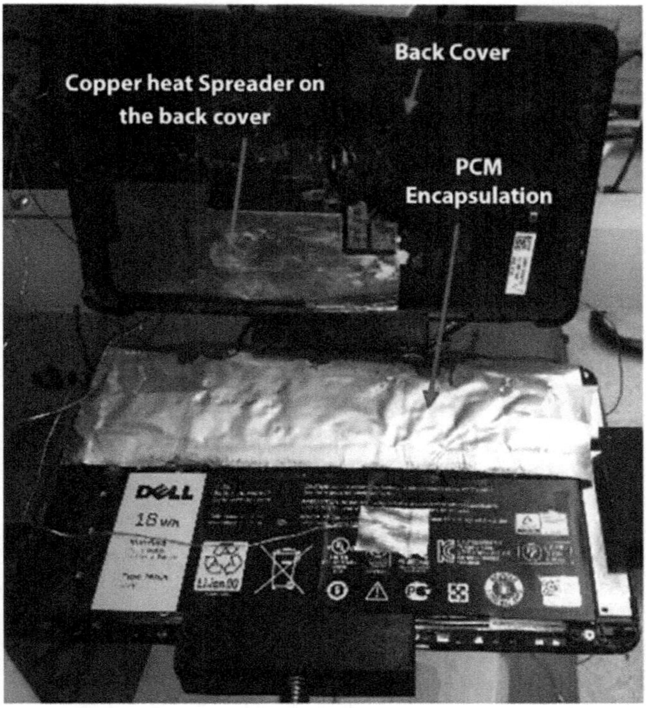

FIGURE 5.23 Experimental photo showing PCM encapsulated TIM used on the display of a PC [66].

Some notable benefits of an adhesive TIM [28] are:

- Do not require additional fasteners which could make the assembly lighter
- Maintains consistent bondline thickness after curing
- Mechanical stability due to bonding after curing

Some notable limitations of an adhesive TIM are:

- Difficult to rework or replace due to strong bonding
- Low thermal conductivity without fillers
- Additional curing process required

To improve the heat-spreading capability of adhesive TIMs, research has focused on incorporating thermally conductive fillers. Epoxy resin is typically used as the polymer matrix, while fillers such as boron nitride and graphene are introduced to enhance thermal performance [67,68]. The addition of boron nitride fillers has been shown to increase the thermal conductivity to 3.6 W/mK, representing a twenty-fold improvement over the base polymer. Similarly, incorporating three-dimensional graphene fillers, cured at 110°C for one hour, raised the thermal conductivity from 0.2 to 4.9 W/mK at a loading of 30 wt%. In summary, adhesive TIMs offer advantages such

FIGURE 5.24 Thermal tape on a silicon chip.

as mechanical strength and the elimination of external pressure components due to their curing process but face challenges in achieving high thermal conductivity, which can be addressed by optimizing the filler type and loading. A commercially available adhesive TIM is shown in Figure 5.24.

5.3.8 Carbon-Based Thermal Interface Materials

Carbon-based TIMs have gained significant attention as promising candidates for high-performance thermal management applications. Carbon exhibits high thermal conductivity and its versatility in structural forms allows for a broad range of design possibilities. Carbon exists in different allotropes, primarily graphite (sp^2 hybridization) and diamond (sp^3 hybridization), as well as amorphous forms. For TIM applications, carbon materials, such as graphite, graphene, carbon nanotubes (CNTs), carbon fibers, carbon nanofibers (CNFs), and carbon black (CB), are utilized either individually or as hybrid composites to tailor the desired thermal and mechanical properties.

Some notable benefits of carbon based TIMs are:

- High intrinsic thermal conductivity of certain carbon allotropes such as graphite, graphene, and carbon nanotubes
- Wide range of forms: powders, fibers, arrays, and composites
- Potential for lightweight, mechanically robust structures
- Tunable anisotropy for specific applications (in-plane vs. through-plane conduction)

Some notable limitations of carbon based TIMs are:

- Thermal conductivity often limited by interfacial resistance between fillers and matrix

- Anisotropic heat transport in many carbon structures such as graphene and CNTs
- Presence of air voids in porous or vertically aligned structures, reducing effective conductivity
- Higher manufacturing cost for highly engineered carbon nanomaterials

From a structural standpoint, carbon-based TIMs can be broadly categorized into four main types: carbon paste TIMs, graphene-based TIMs, carbon nanotube (CNT)-based TIMs, and three-dimensional structured TIMs.

Carbon paste TIMs are composed of carbon fillers dispersed within a liquid medium, allowing them to conform to surface irregularities and effectively fill interfacial gaps [69] The fillers, such as carbon black or graphene, provide pathways for thermal conduction, while the liquid carrier—often oil-based rather than polymeric—maintains conformability. Carbon paste TIMs typically have lower thermal conductivity than solid composites, but they are capable of achieving a low bondline thickness due to their flowability [70]. For application, carbon based TIMs are appropriate where high conformability outweighs the need for very high thermal conductivity.

Graphene, consisting of a single layer or a few layers of sp^2-bonded carbon atoms arranged in a two-dimensional lattice, exhibits exceptionally high in-plane thermal conductivity but relatively low through-plane conductivity. To improve cross-plane heat transport, several strategies have been proposed. One approach involves infiltrating graphene papers with a mixture of large and small graphene flakes to enhance out-of-plane heat conduction [71]. Another effective method employs horizontally oriented metal-capped graphene structures, which reduce the interfacial resistance at the mating surfaces and improve overall thermal transfer performance [66–69].

Carbon nanotubes (CNTs) are cylindrical graphene structures that can either be single-walled (SWCNT) or multi-walled (MWCNT). Their high thermal conductivity makes them suitable for TIMs in both composite and structured forms. CNTs can be used in paste forms dispersed in oils or polymers, often combined with other fillers to improve percolation and conductivity [70,72]. CNTs can be in structured forms like CNT buckypaper and vertically aligned CNT arrays, which can be oriented to maximize heat conduction in the desired direction. CNT arrays, in particular, can achieve very low thermal resistance when well-aligned and adequately interfaced [73–80]. A commercially available CNT based TIM is shown in Figure 5.25.

5.4 INTERFACES IN SEMICONDUCTOR PACKAGING

The preceding sections introduced the fundamentals of heat transfer, the nature of interfaces, and various thermal interface materials used to bridge surface gaps. In this section, an elaborate overview of electronics packages, the interfaces encountered in them, and the requirement of TIMs, will be covered.

5.4.1 Packaging Levels

Thermal management in electronics packaging can be broadly classified into chip-level, board-level, and system-level (Figure 5.26) [81]. Chip-level thermal

FIGURE 5.25 A commercially available CNT TIM on a silicon chip.

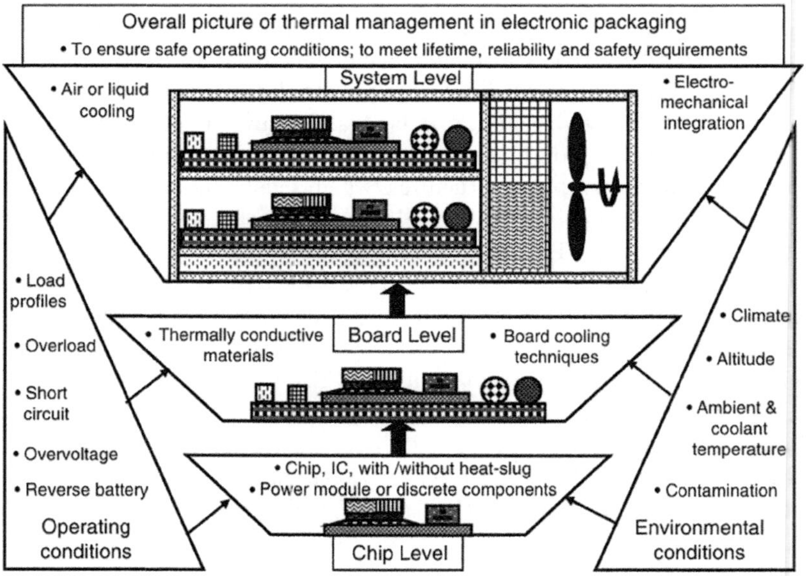

FIGURE 5.26 Overview of thermal management approaches across multiple electronic packaging levels [81]. (Used with permission conveyed through Copyright Clearance Center, Inc.)

management involves the use of various heat-spreading technologies, using metals such as copper, copper-tungsten, and copper-molybdenum mixtures, as well as micro heat pipes and thermoelectric coolers. It mainly deals with the dissipation of heat within the semiconductor device. Board-level thermal management involves distributing heat from the chip to the chassis or the electronic system; mostly passive heat spreaders like copper or aluminum, are employed here. Finally, system-level thermal management involves removing heat from the package using heat exchangers, heat sinks, and/or cold plates. Some advanced semiconductors use direct chip cooling, where the coolant directly contacts the chip, eliminating the need for any interface materials. Details on the heat spreaders, heat sinks, and cold plates are covered in Section 5.5 of the chapter.

5.4.2 Thermal Interface Materials in Packaging

Broadly, the TIM in an electronics package can be classified into TIM 1, TIM 1.5, and TIM 2, based on the mating surfaces. TIM 1 and TIM 2 work together within the packaging hierarchy, where TIM 1 forms the interface between the bare die and the package lid, while TIM 2 provides the interface between the lid and the external heat exchanger, such as a heat sink or cold plate (Figure 5.27a). It is generally preferred that TIM 1 possesses a low bondline thickness and high thermal conductivity to minimize thermal resistance and prevent overheating of the chip. Common materials used for TIM 1 include reflow solders, phase-change metal alloys, thermal gels, and liquid metal TIMs, all of which offer good thermal performance and mechanical reliability at the chip-to-lid interface [82]. Due to reliability concerns such as pump-out and void formation, thermal grease is generally avoided as TIM 1. However, because of its ease of application, reworkability, and good surface conformity, it remains a viable option for use as TIM 2, where it interfaces between the package lid and the external heat exchanger [82]. Indium as a TIM 1 material was extensively studied [83]. The backside of the die, where heat dissipation occurs, plays a critical role in the effectiveness of TIM applications. Proper wetting behaviors and backside metallization are essential to enable the use of solder-based TIMs as TIM 1 materials. TIM 1.5 is typically employed in lidless package configurations, where it interfaces directly between the bare die and the external heat exchanger (Figure 5.27b). Lidless packaging eliminates the thermal resistance introduced by the heat spreader (lid) and removes additional contact resistances associated with multiple interfacial layers (Figure 5.28).

5.4.3 Thermal Pathways in Chip Assembly

A traditional printed circuit board (PCB) package is shown in Figure 5.12. The figure illustrates a flip chip package [84] employing controlled collapse chip connection

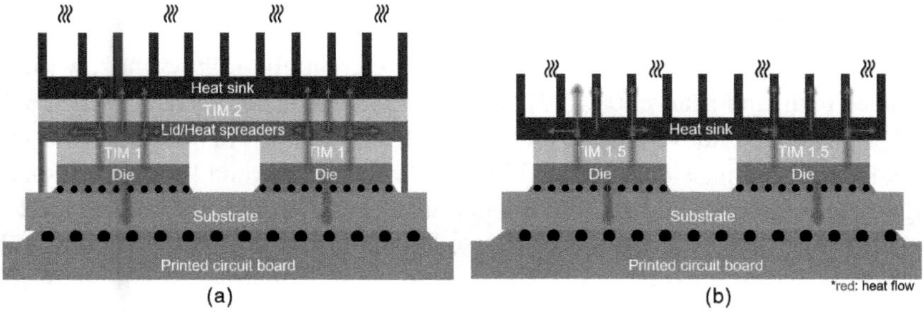

FIGURE 5.27 (a) Package with a lid/heat spreader. (b) Package without a lid/heat spreader.

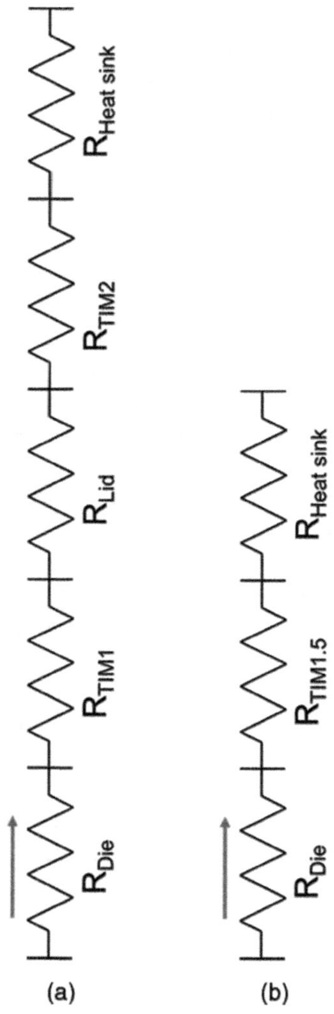

FIGURE 5.28 Heat flow path to the heat sink from the dies in (a) lidded and (b) lidless packages.

(C4) bumps [85]; an alternate way of connecting the die is through wire bonding [86]. A package with semiconductor chips is often referred to as a module [87]. As shown in Figure 5.12, heat generated within the die can flow upward to the cooling solution or downwards to the substrate/PCB. Heat preferentially follows the low thermal resistance path, and nearly 90% [87] of the generated heat is removed by the cooling solution positioned above the die (covered in Section 5.5). This occurs because the combined thermal resistance of the TIM, the heat spreader and the cooling solution is significantly lower than that of the solder bumps, underfills and the substrate/PCB. The remaining 10% of the heat is conducted downwards through the chip assembly, represented by the resistance R_{db}. The resistance R_{db} comprises the thermal resistance of the solder bumps (C4), solder balls (ball grid array, or BGA), underfills, and the substrate/PCB.

TABLE 5.7
Thermal Conductivity of Commonly Used Materials in Chip Assembly

Material	Thermal Conductivity (W/mK)
UBM and solder [88]	Cu: 398
	Sn: 67
	Ni: 90.5
	Cu_6Sn_5: 34.1
	Cu_3Sn: 70.4
	Ni_3Sn_4: 19.6
Underfill (epoxy resin) [89]	0.2–1.5
Organic substrates [90]	0.2–0.5
Ceramic substrates [90]	20–180

The most common materials for solder bumps and balls in C4 and BGA are lead-free solders such as tin-silver and tin-silver-copper (SAC) alloys. Underfill materials protect the solder bumps from thermally induced mechanical stresses arising from package warpage or coefficient of thermal expansion (CTE) mismatch, thereby improving solder joint reliability. Typical underfills are thermosetting polymers, such as epoxy resins. Under bump metallization (UBM) is required to enable electrical and mechanical interconnection between the solder bumps and the die. The UBM stack typically consists of three functional layers: adhesion layer (e.g., Ti, TiW, Cr, Al); solder barrier layer (e.g., Ni, Co, NiX, Cu pillar or Cr-Cu); and wetting layer (e.g., Cu, Au, Pd). Substrates may be organic, such as flame retardant 4 (FR-4) or Ajinomoto build-up film (ABF), or inorganic, including ceramics (alumina (Al_2O_3), aluminum nitride (AIN)) as well as glass or silicon. Table 5.7 summarizes the typical thermal conductivities of the materials discussed.

With the advent of heterogeneous integration and advanced packaging, the use of silicon substrates has gained significant traction. One such platform is the silicon interconnect fabric (Si-IF) in which advances in silicon processing and packaging technology are leveraged to develop heterogeneously integrated, wafer scale systems [87]. This approach scales down the interconnect pitch and height, enabling reduced latency and improved bandwidth of communication. In addition, the use of silicon substrates can enhance heat dissipation as silicon has a relatively high thermal conductivity of approximately 150 W/mK.

5.5 THERMAL MANAGEMENT SYSTEMS

This section provides an overview of key cooling solutions (Figure 5.29) for electronics systems, categorized by their fluid configuration and heat transfer mechanisms. The primary modes of cooling are air cooling (natural and forced convection), single-phase liquid cooling (forced convection), two-phase liquid cooling (flow with

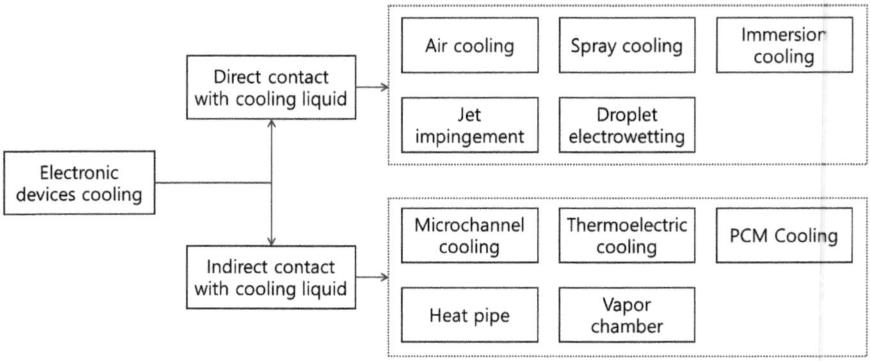

FIGURE 5.29 Broad classification of thermal management techniques.

phase change), and immersion cooling. Although radiation heat transfer is another mode in which heat can be dissipated, conduction and convection dominate in removing heat from electronics systems due to their much higher heat transfer rates. Overall, cooling methods can be grouped into two categories: direct cooling and indirect cooling. In direct cooling, the working fluid directly contacts the heat-generating component (e.g., a chip), eliminating intermediate interfaces and associated thermal resistances. In indirect cooling [91], the cooling fluid does not come into contact with the chip but flows through a secondary surface. Zhang et al. [91] presents a comprehensive review of different techniques used in cooling electronic equipment. The following subsections introduce major cooling techniques within these two categories.

5.5.1 Direct Cooling

Direct cooling allows coolants to directly contact a chip or other heat-generating surfaces, eliminating the need for a TIM. By removing solid-solid interfaces, this approach minimizes the total thermal resistance for the heat flow.

5.5.1.1 Air-Cooling

Air-cooling uses either natural or forced airflow ($k \approx 0.026$ W/mK) over the heated surface of a chip. Air cooling is the simplest method for achieving thermal management of electronics systems, particularly when the power density is moderate. The heat transfer coefficient ranges from 5 to 25 W/m^{2}K for natural convection and to about 300 W/m^{2}K for forced convection [4]. As shown in Figure 5.30, radiation and free convection are generally applicable for heat-flux densities below approximately 1,550 W/m^2 [92]. Air-cooling often employs heat sinks attached to the actual chip to enlarge the effective heat-transfer area. The use of vapor chambers and heat pipes as heat spreaders has increased the potential for air-cooling to be used for higher heat flux thermal management.

5.5.1.2 Spray-Cooling

Spray cooling involves using a nozzle to atomize the cooling fluid into smaller droplets which contact the chip directly, as shown in Figure 5.31. Its high heat flux

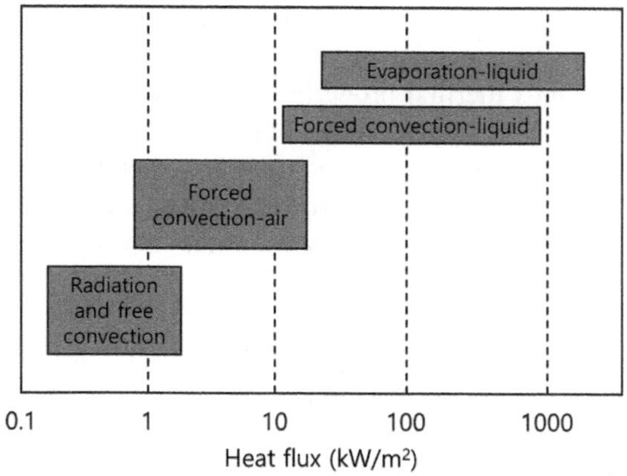

FIGURE 5.30 Typical heat flux range for conventional approaches [93]. (Reprinted from with permission from Elsevier.)

FIGURE 5.31 Visualization of spray-cooling droplets pore size of 1.2 mm for a flow rate of 40 *L/h* [100]. (Reprinted from with permission from Elsevier.)

dissipation ability makes it a suitable candidate for cooling high power generating electronics [94]. Depending on the working fluid and operating conditions, spray cooling can operate in either single-phase or two-phase mode, where the latter benefits from latent-heat absorption during vaporization. The positioning of the nozzle is significant in spray cooling [95], and considerable research, such as [96], has found that atomized droplets will be smaller if the pressure differential across the nozzle is increased. Furthermore, multiple nozzles can be used in conjunction to improve the heat transfer rate.

As the liquid coolant's thermal properties are critical in the single-phase spray cooling approach, researchers have explored the effectiveness of binary mixtures of alcohols [97], surfactant solutions [98], and two-phase low-boiling-point fluids, such as R134a [99]. Despite its excellent performance, spray-cooling faces practical challenges such as high-pressure operating conditions and nozzle clogging.

5.5.1.3 Jet Impingement Cooling

Jet impingement cooling is similar to spray-cooling, where a nozzle is used to create continuous liquid jets, rather than atomizing the cooling fluid. These jets strike the heated surface directly, forming different regions of the jet impingement flow (Figure 5.32). Identical to spray cooling, significant research in jet impingement cooling has been done on the design, placement of the nozzle, and the choice of working fluid. Heat transfer coefficient can be as high as 15,200 W/m^{2}K for liquid synthetic jet impingement cooling [101].

5.5.1.4 Immersion Cooling

Immersion cooling (Figure 5.33) is an effective cooling method where the electronic equipment is entirely immersed in a dielectric fluid that removes the heat from the system. Dielectric fluids are preferred for their good thermal conductivity and electrical insulation. This approach offers the highest heat transfer capability among direct cooling methods and is a suitable option for cooling data centers [103] and servers [104]. The coolant may operate in either a single-phase regime or a two-phase

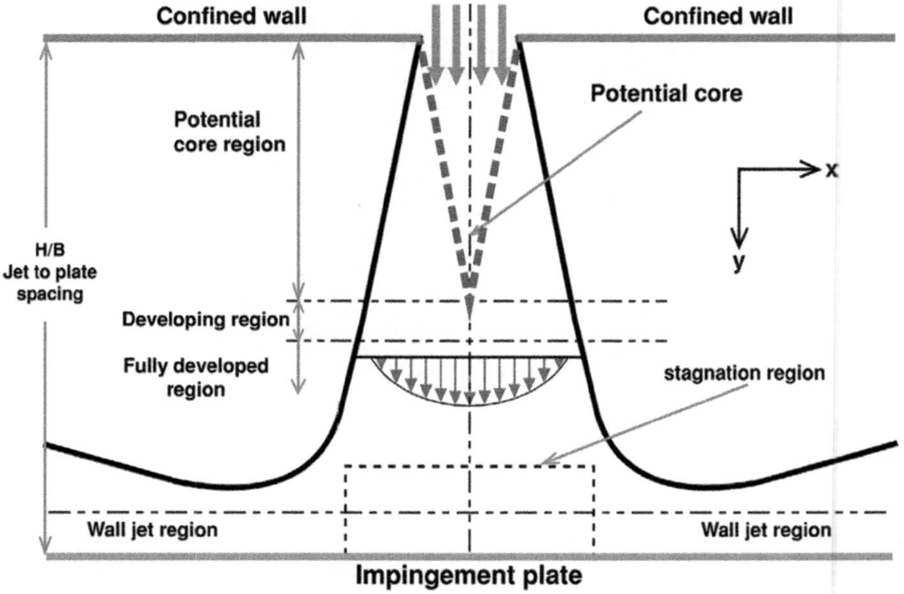

FIGURE 5.32 Different flow regimes in jet impingement cooling [102]. (Reprinted from with permission from Elsevier.)

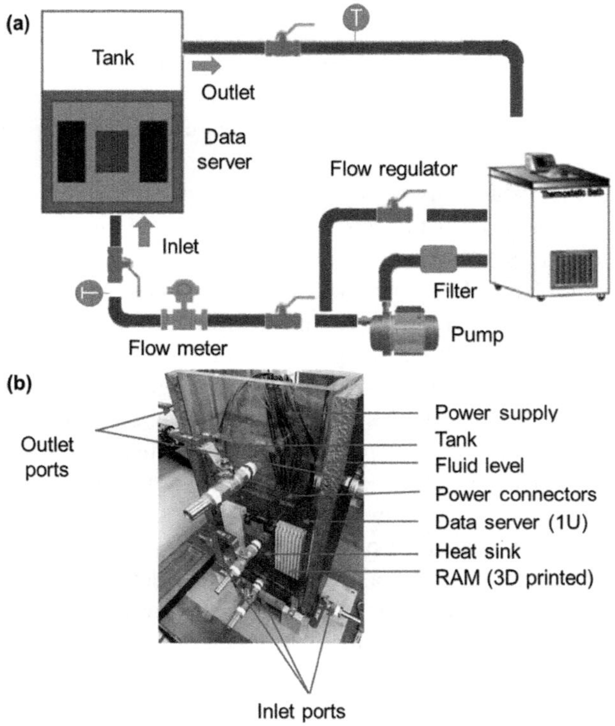

FIGURE 5.33 (a) A schematic of an immersion cooled data server. (b) Example of data server immersed in FC-40 [105]. (Reprinted from with permission from Elsevier.)

TABLE 5.8
Commonly Used Coolants in Immersion Cooling

Fluids	Thermal Conductivity (W/mK)	Cooling Mechanism	Boiling Point (°C)	Ref.
Ethanol	0.167	Two-phase	78	[106]
Water–ethylene glycol (1:1)	0.37	Single- or two-phase	107.3	[107]
Mineral oil	0.13	Single-phase	–	[108]
Novec 649	0.059	Two-phase	49	[109]
Novec 7000	0.075	Two-phase	34	[110]
FC-3283	0.066	Single-phase	128	[111]
FC-43	0.065	Single-phase	178	[111]
FC-72	0.057	Two-phase	56	[112]

in the boiling regime with natural or pumped circulation. The commonly used fluids are listed in Table 5.8 [105]. Significant research is still needed for long-term aging and compatibility of TIM and electronic components immersed in different fluids. Recent work in the industry on the compatibility of a common commercial cooling fluid and an indium alloy solid preform has shown promising results.

5.5.2 Indirect Cooling

Although direct cooling offers low thermal resistance by eliminating the TIM, the reliability of direct contact cooling (with the exclusion of air cooling) warrants further studies. Several concerns such as contamination, dielectric breakdown, and long-term reliability need to be addressed.

Indirect cooling, widely used in commercial applications, typically employs an intermediate conduction path such as a heat sink, cold plate, or vapor chamber between the chip and the coolant. Typical systems include air-cooled heat sinks or liquid-cooled cold plates. Air-cooled heat sinks have been already commercialized and used in various forms in industry (covered in Section 5.5.1.1). The liquid cooling may or may not involve the phase change of the liquid inside the cold plate. The cold plates are usually made of copper and aluminum, through which a coolant circulates to remove heat. Water remains the most effective single-phase coolant owing to its high thermal conductivity and specific heat. In practical applications, mixtures such as propylene glycol in water (PG25, 25 vol.% propylene glycol) are used to suppress microbial growth and prevent freezing [113]. The higher boiling point of water (100°C at 1 atm) limits its usefulness in two-phase applications. Researchers have tried other refrigerants such as R134a and its green alternatives – R1234yf, R1234ze(E), R515B – for data center cooling involving phase change of the liquid [114]. Recent experiments using methanol [52,115] or water/methanol mixtures [116] demonstrated cooling capabilities of up to 1 W/mm^2 through flash boiling. Figure 5.34 shows a copper cold plate sitting on top of a thermal test chip with a layer of TIM in between them.

5.5.3 Heat Spreaders

Heat spreaders are an integral part of electronics cooling in redistributing localized heat from small active areas to larger and more manageable areas in the planar direction. These passive devices, ranging from traditional metals, phase-change heat spreaders, to graphitic heat spreaders, enhance the temperature uniformity across the chip and the cooling system. Recent studies demonstrated development of

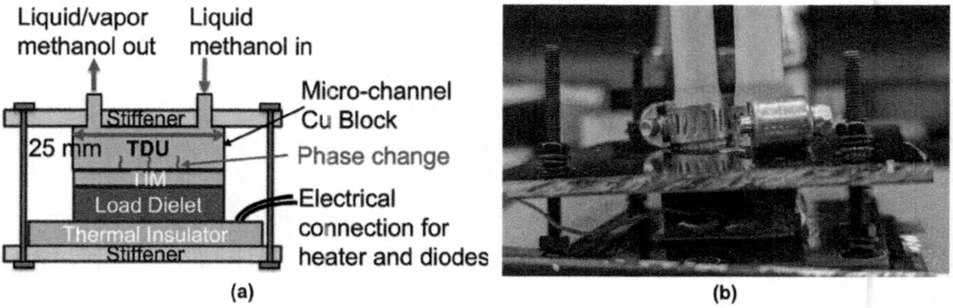

FIGURE 5.34 (a) Schematic of the cold plate and thermal test chip assembly. (b) Experimental setup of the cold plate and thermal test chip assembly. (Reprinted with permission from [52].)

approximately 200 μm thick graphitic heat spreaders with continuous and uniform structures capable of effectively dissipating device-level heat [117,118].

5.5.3.1 Metallic Heat Spreaders

The common heat spreader materials are high-conductivity metals such as aluminum ($k \approx 200$ W/mK) and copper ($k \approx 400$ W/mK). The high thermal conductivity of these metals allows rapid lateral heat transport. Figure 5.35 shows copper and graphitic heat spreaders being used for heat spreading enhancement on a high-power amplifier [117]. Researchers have also explored the use of diamond as heat spreading layer owing to its high thermal conductivity [119].

5.5.3.2 Phase Change Heat Spreaders

Phase change heat spreaders can be heat pipes (Figure 5.36) [120], vapor chambers (Figure 5.37) [121], or oscillating heat pipes (OHPs) (Figure 5.38) [122]. The primary operation of heat pipes and vapor chambers is the evaporation of liquid at the evaporator (chip) region and condensation of the vapor at the condenser (cooler) end,

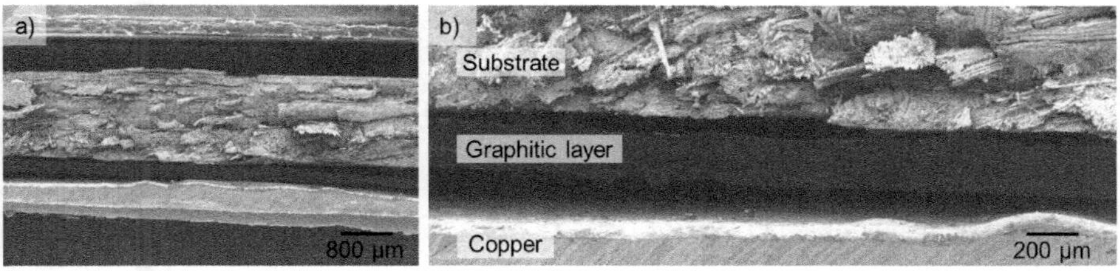

FIGURE 5.35 Copper and graphitic layers being used as heat spreaders for a power amplifier. (Reprinted with permission from [117].)

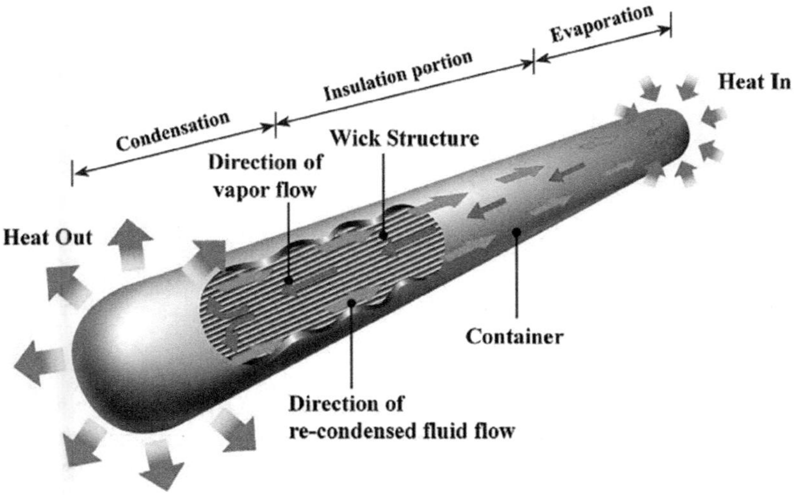

FIGURE 5.36 Schematic of operation of a heat pipe [120]. (Reprinted from with permission from Elsevier.)

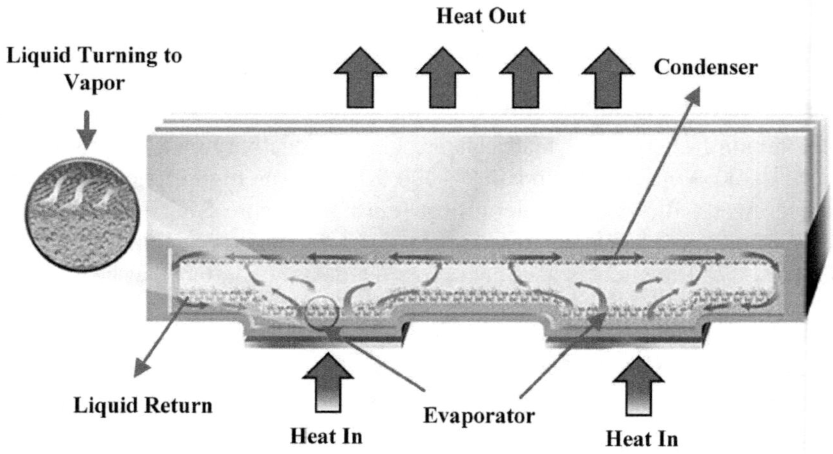

FIGURE 5.37 Schematic of operation of a vapor chamber [120]. (Reprinted from with permission from Elsevier.)

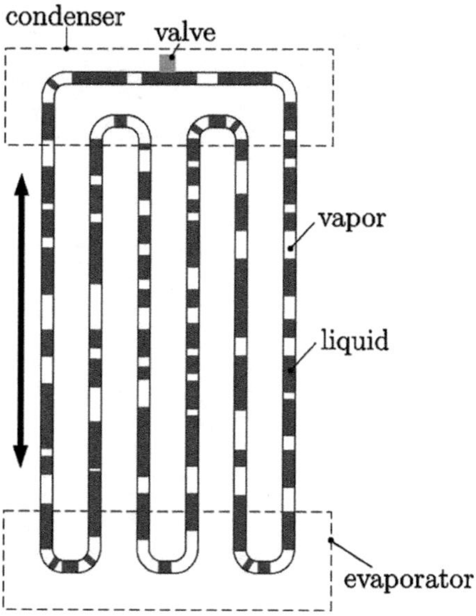

FIGURE 5.38 Schematic of operation of an oscillating heat pipe [124]. (Reprinted from with permission from Elsevier.)

while capillary action within a wick returns the liquid. OHPs operate without wicking structures, but instead, oscillatory motion in a network of channels circulates the fluid. Recent developments include 3D vapor chambers (Figure 5.39) [123] enabling efficient transfer of heat when embedded in air-cooled heat sinks. These devices can achieve effective conductance 100–200 times higher than that of metallic heat spreaders, offering substantial improvements in spreading performance.

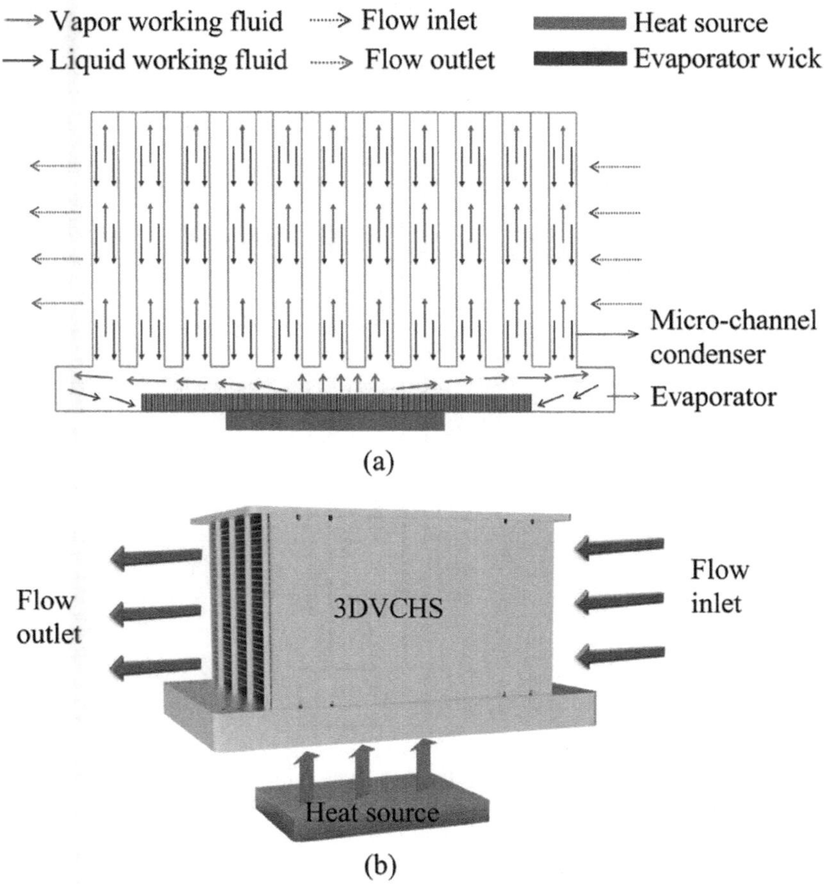

FIGURE 5.39 Schematic of 3D vapor heat sink: (a) two-dimensional view, (b) three-dimensional view [123]. (Reprinted from with permission from Elsevier.)

5.6 TESTING AND CHARACTERIZATION OF THERMAL INTERFACE MATERIALS

Accurate testing and characterization of TIMs are essential to evaluate their heat transfer capability, structural integrity, and reliability. This section introduces thermal performance measurements in steady-state and transient techniques, materials characterization, and reliability testing protocols.

5.6.1 Thermal Performance

5.6.1.1 Steady-State Measurements

Steady-state measurements for TIMs determine the heat flux and temperature gradients based on Fourier's law of heat conduction when thermodynamic properties become independent of time. Common steady-state approaches include the reference

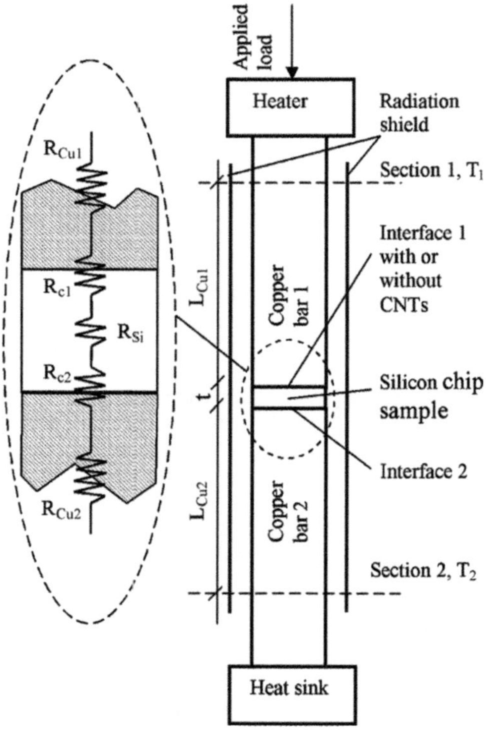

FIGURE 5.40 ASTM D5470 standard measurement configuration. (Reprinted with permission from [128].)

bar method, microbridge method, and Raman thermometry [125]. Among these, the reference bar method is the most widely adopted.

Based on the ASTM D5470 standard (Figure 5.40), the one-dimensional steady-state temperature reference bar method [126,127] is widely used to measure the total thermal resistance (two contact resistances and bulk thermal resistance of TIM). The reference bar measurement assumes one-dimensional, uniform heat conduction through vertically aligned copper reference bars and TIM specimen. Each copper bar is designed to have multiple thermocouples; by extrapolating their temperature gradients, the temperatures at the top and bottom of the specimen are determined. Using Fourier's law, the effective thermal conductivity can be derived from the heat flux, the temperature difference between two surfaces of the specimen, and the BLT.

Proper experimental configuration is critical to minimize measurement uncertainty. Key considerations include:

- Precise vertical alignment of bars and specimen with identical cross-sectional areas; a ball-joint alignment ensures a point pressure and parallelism
- Uniform surface roughness and thickness of specimen and bars to reduce contact resistance variability
- Vacuum or insulation control to suppress convective heat losses and ensure conductive heat flow through aligned setup

- Calibrated sensors on temperature, BLT, and data acquisition to limit uncertainty
- Operation within a temperature range to avoid phase transition within the TIM
- Stable mechanical loading using parallel rigid supports to maintain consistent pressure and thickness

5.6.1.2 Transient Temperature Measurements

Transient temperature techniques analyze the time-dependent temperature response of a specimen to an applied heat pulse or periodic heating. Common approaches include the hot-wire method, laser flash analysis, the 3-omega method, photoacoustic techniques, Ångström's method, and ultrafast optical pump–probe measurements [125]. In this chapter, we focus on Ångström's method.

Ångström's method, first introduced by Ångstrom, in 1862 [129], employs sinusoidal heating to characterize how rapidly temperature waves propagate through a material. A heat source generates a periodic temperature oscillation at a single angular frequency, ω. By analyzing the temperature amplitude ratio, T_1/T_2, and the phase lag, $\Delta\varphi$, between two points separated by a distance, the thermal diffusivity, α, can be derived as:

$$\alpha = \frac{\omega \Delta x^2}{2\Delta\varphi \ln(T_1/T_2)} \tag{5.21}$$

With known specific heat, c_p, and density, ρ, the thermal conductivity, k, can then be determined via:

$$k = \alpha \rho c_p \tag{5.22}$$

This method offers several benefits. Unlike the 3-omega or laser flash methods, Ångström's approach requires relatively simple sample preparation and is less sensitive to uncertainties from surface emissivity or coating materials. Moreover, it reduces the need for extensive optimization against surrounding heat losses. However, accurate interpretation demands careful tracking of the amplitude and phase across different frequencies. At high frequencies, temperature oscillations become difficult to resolve due to reduced thermal penetration depths [130]. To mitigate environmental influences, forced convection around the setup is often employed. In practice, this measurement is usually suggested for thin film specimens.

5.6.1.3 Thermal Test Vehicles

Standard thermal performance characterization and testing, as discussed above, are very useful in characterizing the fundamental properties of TIMs. At the same time, it is highly desirable to replicate the complex thermal environment of real semiconductor devices, and to emulate dynamic power profiles, non-uniform heating, complex interfacial thermal behaviors, and packaging-induced mechanical constraints that influence TIM performance in actual systems. For heterogeneous integration with multiple chips of various functionalities, sizes, and power distributions all in the same package, characterization of the thermal behavior becomes even more challenging.

Silicon-based Thermal Test Chips (TTCs) have been developed to emulate realistic semiconductor chips, including non-uniform power distribution (down to 1 mm × 1 mm spatial resolution), with embedded in situ sensors enabling high-resolution (down to 1 mm × 1 mm spatial resolution) temperature mapping, for accurate thermal performance characterization [52]. By performing thermal measurements and testing (steady state or transient), in realistic thermal test vehicles (TTVs) which emulate actual packaging configurations (with single or multiple TTCs), the thermal performance of TIMs can be accurately assessed [131]. More details can be found in Chapter 8 of this book.

5.6.2 Materials Characterization

A variety of characterization techniques are available to examine TIMs and their interactions with contacting surfaces in terms of optical, mechanical, and chemical aspects. This subsection introduces key methods that provide insights into the microstructure, surface quality, bonding effectiveness, and materials composition of TIMs.

5.6.2.1 Optical Microscopy

Optical microscopy is a widely used inspection tool for rapid, qualitative evaluation of TIM samples. Using a visible light source and objective lenses ranging from 5× to 1,000× magnification, this technique provides direct imaging of the sample surface, with typical spatial resolution on the order of a few micrometers. For proper measurement, the specimen must be aligned perpendicular to the illumination source, and careful adjustment of the objective lens is necessary to prevent contact with uneven samples.

5.6.2.2 Surface Profilometry

Surface profilometry characterizes the surface topography and roughness of TIMs at scales from hundreds of nanometers to several millimeters. Mechanical contact profilometers use a stylus to trace the surface, while optical non-contact methods employ interferometry or confocal techniques. Although both approaches provide quantitative roughness information in the *x*- and *y*-directions, optical methods are generally faster and more suitable for capturing large-area morphology.

5.6.2.3 Bondline Thickness

The bondline thickness is a critical parameter in evaluating TIM performance, as it directly affects thermal resistance and measurement uncertainty, particularly in thin samples (< 100 μm). BLT can be determined using mechanical spacers, displacement gauges (e.g., micrometers with fixtures), or optical and microscopic methods. Selection of an appropriate measurement approach depends on the geometry, mechanical stiffness, and mechanical stability of the sample.

5.6.2.4 Scanning Acoustic Microscopy

Voids, such as air gaps, bubbles, or unfilled regions between a TIM and contacting surfaces, act as strong thermal insulators due to the low conductivity of air ($k \approx 0.026$ W/mK). Accurate void detection is therefore essential. Scanning acoustic microscopy is a non-destructive method that uses ultrasound waves to detect voids, cracks, and delamination with micrometer-scale resolution. X-ray imaging serves as a complementary non-destructive technique to scanning acoustic microscopy, providing 2D and 3D visualization of internal structures and identifying density variations or voids based on differential X-ray absorption.

5.6.2.5 X-Ray Diffraction

X-ray diffraction (XRD) is a non-destructive technique used to probe the crystalline structure, phase composition, lattice parameters, and crystallite size. The method relies on measuring diffraction patterns as the incident X-ray angle is varied, with typical depth analysis of nanometers to tens of micrometers. Samples may be measured in powder form or as flat bulk specimens, as long as they are properly aligned on the sample stage.

5.6.2.6 Scanning Electron Microscopy

Scanning electron microscopy (SEM) provides high-resolution imaging of TIMs, with resolutions down to 1–50 nm. A focused electron beam interacts with the sample surface, generating secondary and backscattered electrons that reveal the morphology, topography, and microstructural features. In addition, SEM combined with energy-dispersive X-ray spectroscopy (EDS) enables elemental composition analysis. The effective probing depth ranges from nanometers to micrometers, depending on the beam energy and sample properties.

5.6.2.7 Transmission Electron Microscopy

Transmission electron microscopy (TEM) offers imaging at near-atomic resolution, down to a few angstroms, by transmitting an electron beam through an ultrathin specimen. It provides detailed information on the morphology, nanostructure, crystallography, phase distribution, and chemical composition. Although the probing depth is limited to several hundred nanometers, the technique delivers unparalleled insights into the nanoscale features and crystal structure of TIMs.

More details can be found in Chapter 8 of this book.

5.6.3 Reliability

Reliability is a critical characteristic of TIMs, as TIMs' performance must be sustained over long operating lifetimes under thermal, mechanical, and environmental stresses. Even when a TIM initially exhibits a high thermal conductivity and low thermal resistance, degradation mechanisms such as thermal cycling, pump-out, or humidity exposure can increase the thermal resistance and raise device junction temperatures, leading to potential failures. Therefore, accelerated reliability testing provides essential insights into the long-term durability of TIMs and their suitability for practical applications.

5.6.3.1 Thermal Stability

The thermal stability of TIMs reflects their ability to withstand high temperatures and repetitive thermal fluctuations without deterioration in thermal properties. Several accelerated thermal tests are standardized for this purpose:

- **Temperature Cycling Test**: Alternating heating and cooling between −40°C and +125°C, following JESD22-A104E [132], is applied to evaluate the ability of TIMs and solder interconnections to withstand mechanical stresses caused by thermal cycles.
- **Power and Temperature Cycling Test**: Alternating exposures of high and low temperatures while turning power on and off periodically simulate worst-case operating conditions, following JESD22-A105C [133].
- **High Temperature Storage Life Test**: This test determines the effect of time and temperature for thermally activated failure mechanisms of solid-state electronic devices, following JESD22-A103B [134].
- **Bake Test**: This test employs a TIM in a temperature-controlled oven for an extended period. The bake test conditions are determined by the TIMs considering phase change, rate of degradation, and reliability along with other components in the device. These tests target the temperature range of 95°C–150°C with hundreds and thousands of hours [135,136].
- **Thermal Aging Analysis**: Thermal aging examines materials changes, such as mass loss at elevated temperatures by thermo-gravimetric analysis (TGA), or thermodynamic transitions or chemical reactions (curing, phase change) by differential scanning calorimetry (DSC), after long-term exposure to elevated temperatures.

5.6.3.2 Mechanical Reliability

Mechanical reliability is to check the ability to maintain TIMs' structural and functional integrity under mechanical cycles.

- **Pump-Out**: Particularly relevant for grease-based TIMs, pump-out occurs when TIMs are forced out of the interface due to coefficient of thermal expansion (CTE) mismatches between contacting surfaces, combined with a cyclic warpage. Since the displaced material does not fully return during contraction, the TIM's volume gradually decreases, increasing interfacial thermal resistance.
- **Dry-Out**: Grease TIM degradation under elevated temperatures or repeated cycling causes filler separation from the polymer matrix. Volatile loss or polymer migration cause void formation and cracks, resulting in an increased thermal resistance and consequently higher junction temperatures.

5.6.3.3 Environmental Reliability

Environmental reliability tests assess the degradation of TIMs under humidity and combined temperature-humidity stress.

- **Temperature Humidity Test (THT)**: THT runs at 85°C and 85% relative humidity for 1,000–2,000 hours to simulate long-term storage or operation in humid environments.
- **Highly Accelerated Stress Test (HAST)**: HAST is designed to have similar mechanisms as THT but run at higher temperatures to exclude the undesirable effects such as phase change (condensation from vapor to water) and curing. HAST runs at 130°C and 85% relative humidity, following JESD22-A118.

5.7 MODELLING AND SIMULATION

In this section, attention is given to the methods used for modelling and evaluating the interfacial thermal resistance of a TIM. Analytical models for estimation of the thermal conductivity, interfacial thermal resistance, and bondline thickness are discussed, along with the use of commercial numerical solvers to assess the overall thermal performance of TIMs.

5.7.1 Analytical Models for Characterizing a Thermal Interface Material

The thermal resistance across a TIM as covered in Section 5.2.3 is illustrated in Figure 5.41 and expressed as

$$R_{\text{total}} = \frac{\text{BLT}}{k_{\text{TIM}}} + R_{C1} + R_{C2} \tag{5.23}$$

The thermal conductivity of a TIM is a key parameter that directly influences its ability to reduce the overall thermal resistance. Estimating the thermal conductivity of composite materials such as thermal grease, gels, and thermal pads is important in assessing their performance. Table 5.9 summarizes various models used to evaluate the effective thermal conductivity of composite TIMs containing thermally conductive filler particles [137].

For estimating the thermal contact resistance between the surfaces, the following model [138,139] can be used,

$$\frac{1}{R_c} = \frac{2.5k_1k_2}{k_1 + k_2}\left(\frac{m}{\sqrt{\sigma_1 + \sigma_2}}\right)\left(\frac{P}{H}\right)^{0.95} \tag{5.24}$$

$$m = 0.125\left(\sqrt{\sigma_1 + \sigma_2} \cdot 10^6\right)^{0.402} \tag{5.25}$$

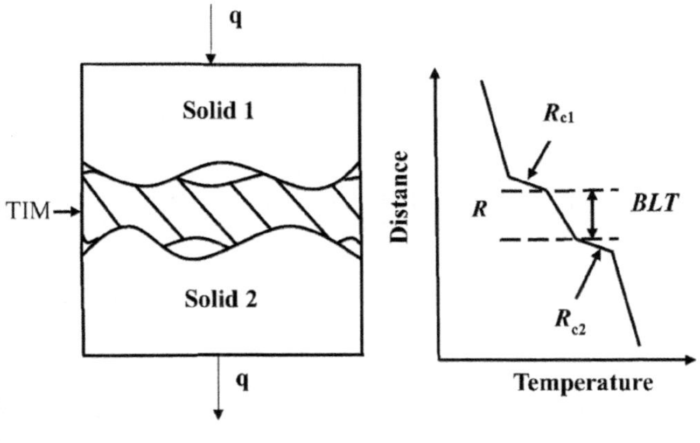

FIGURE 5.41 Thermal resistance of a thermal interface material [43]. (Reprinted from with permission from Elsevier.)

TABLE 5.9
Thermal Conductivity Models

Model	Formula	Application	References
Maxwell	$k_{\text{eff}} = \dfrac{2k_f + k_s - 2\phi_s(k_s - k_f)}{2k_f + k_s + \phi_s(k_s - k_f)} \cdot k_f$	Fillers with low volume fractions	[138]
Bruggeman	$1 - V_f = \dfrac{\lambda_f - \lambda_c}{\lambda_f - \lambda_p}\left(\dfrac{\lambda_c}{\lambda_f}\right)^{\frac{1}{3}}$	Fillers with high volume fractions	[139]
Hasselman	$k_{eff} = k_m \left[\dfrac{\left(\dfrac{k_f}{k_m} + \dfrac{2h}{ak_m} - \dfrac{k_f}{k_m} + \dfrac{2h}{ak_f}\right)V_f + \dfrac{k_f}{k_m} + \dfrac{2h}{ak_f}}{\left(1 - \dfrac{k_f}{k_m} + \dfrac{2h}{ak_m}\right)V_f + \dfrac{k_f}{k_m} + \dfrac{2h}{ak_f}}\right]$	Metal matrix composites	[140]
Lewis and Nielsen	$\dfrac{k_e}{k_m} = \dfrac{1 + AB\phi_2}{1 - B\psi\phi_2}$	Two-phase system	[141]
EMA	$k = k_m \dfrac{3 + f(\beta_{\parallel} + 2\beta_{\perp})}{3 - f\beta_{\perp}}$	Fiber fillers	[142,143]

Here, k_1 and k_2 are the thermal conductivities, P is the contact pressure, H is the surface microhardness of the softer of the two contacting surfaces, σ_1 and σ_2 are the surface roughness of the two contacting surfaces.

Another model [144] to evaluate the contact resistance is,

$$R_C = R_{c1} + R_{c2} = \left(\frac{\sigma_1 + \sigma_2}{2k_{\text{TIM}}}\right)\left(\frac{A_{\text{nominal}}}{A_{\text{micro}}}\right) \tag{5.26}$$

Here, A_{nominal} and A_{micro} are the nominal and actual microscopic areas of the two contacting surfaces. For solid TIMs, these areas can be calculated by plastic deformation, and the new total contact resistance can be calculated as provided in Equation (5.24). For liquid-like TIMs, since plastic deformation is not applicable anymore, A_{nominal} and A_{micro} are calculated by the difference in surface energy resulting from adhesion of the fluidic TIM to the contacting surface.

Modelling the bondline thickness involves the calculation of the yield stress in polymer based TIMs [145]. The relationship governing this behavior is given by Prasher [3],

$$\text{BLT} = \frac{2}{3}r\left(\frac{\tau_y}{P}\right) + \left(\frac{cr}{1.5}\right)^{0.188} d^{0.811}\left(\frac{\tau_y}{P}\right)^{0.188} \tag{5.27}$$

Here, r is the substrate radius, τ_y is the yield stress of the polymer matrix, P is the applied pressure, c is a constant, and d is the particle diameter [146].

5.7.2 Numerical Methods for Evaluating a Thermal Interface Material

Computational tools can be used to simulate the thermal performance of a TIM within an electronic package. Once the material properties of the TIM are defined, it can be modeled as an additional layer in the overall heat flow path. Sponagle and Groulx [147] conducted a steady-state thermal conductance characterization of TIMs using a finite element analysis (FEA) model. Their approach replicated the 1D copper reference bar method (ASTM D5470-06) without relying on the one-dimensional assumptions typically made in analytical models. A primary limitation of such simulations lies in the very small bondline thickness of TIMs, which makes the computational model highly demanding in terms of the time and memory usage for running the simulation. To address this, the authors introduced an approximation that incorporated the bondline thickness information into the overall thermal conductance of the TIM.

The thermal resistance of TIMs can be evaluated experimentally, and its impact on the overall device performance can be further studied through three-dimensional numerical simulations. Narumanchi et al. [148] investigated the influence of TIM thermal resistance on the predicted maximum junction temperature of an insulated gate bipolar transistor (IGBT). Similar finite element analysis has been reported, all of which highlight the importance of accurately characterizing and modelling the thermal resistance of TIMs for reliable thermal management design [149–151].

5.8 EXAMPLE APPLICATIONS OF THERMAL INTERFACE MATERIALS

The computing power requirements of modern electrical and electronic devices have increased significantly over the years and are projected to continue rising in the future. As a result, high-performance devices demand more efficient and reliable cooling solutions. With the advent of higher computing power devices, appropriate TIM ensuring efficient heat transport between surfaces becomes more significant. This section presents three detailed case studies illustrating the use of TIMs in advanced thermal management applications: high-performance computing and data center systems, batteries in electric vehicles, and high-power light-emitting diodes (LEDs).

5.8.1 High-Performance Computing and Data Center Applications

High-performance computing (HPC) demands efficient operation of chips along with good signal integrity. One approach to achieving this is by placing processors in proximity, which minimizes communication latency and improves bandwidth. A notable technique for accomplishing this is three-dimensional (3D) stacking of wafers or dies [152]. While this approach enables good signal integrity, stacking chips means that there is a high concentration of heat loads on a smaller area which requires an efficient heat dissipation pathway. The cooling solution for such devices not just involves highly efficient design of the cooling architecture but also involves reducing the thermal resistance of the stack. Lowering the thermal resistance across the thermal interface materials (TIMs) is particularly important, as it directly contributes to minimizing the overall thermal resistance of the package. For such applications, researchers have investigated high thermal conductivity TIMs, including low-melting-point metal alloys based on indium [153] for such applications.

At the data center level, the use of efficient and reliable TIMs becomes critical for reducing the overall energy consumption and improving the PUE (power usage

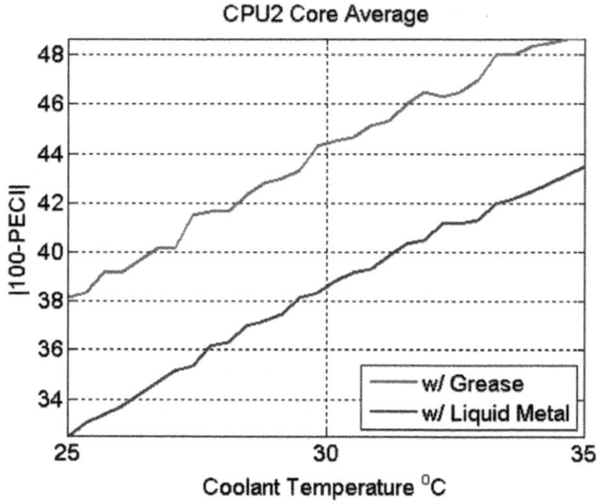

FIGURE 5.42 Thermal performance results for grease and liquid metal TIM as TIM 2 in a lidded module. (Reprinted with permission from [154].)

effectiveness). Implementing liquid metal TIMs in data center racks allows liquid-cooled systems to operate at higher coolant inlet temperatures while maintaining the same chip temperature compared to conventional TIMs [154]. This capability enables the use of cooling water at near-atmospheric conditions or even industrial waste coolants, simplifying thermal management and improving energy efficiency.

A comparison of CPU cooling using thermal grease and liquid metal TIMs is shown in Figure 5.42. The figure uses platform environment control interface (PECI) readings to represent the CPU core temperature, indicating that, for a given inlet coolant temperature, liquid metal TIMs can reduce the CPU operating temperature by approximately 5°C–6°C. Additionally, power consumption was reduced by about 15%–20% when using liquid metal TIMs, primarily due to lower outdoor heat exchanger fan speeds. These findings highlight that the selection of an appropriate TIM can have a substantial impact on both the thermal performance and the overall energy efficiency in data center design.

5.8.2 Electric Vehicle Battery Modules

TIMs are finding increasing use in the automotive industry, particularly in electric and hybrid vehicle battery systems where effective thermal management is essential. At the cell level, a study by Lee et al. [155] demonstrated that integrating graphite-based TIMs with cylindrical 21700-format Li-ion cells (21 mm in diameter and 70 mm in length) enhances heat spreading. With an in-plane thermal conductivity of approximately 220 W/mK and a through-plane conductivity of about 5 W/mK, graphite TIMs placed on both electrodes significantly reduced the interfacial temperature gradients. The application of graphite TIMs decreased the temperature difference at the solid–solid interface by nearly 90%, resulting in a variation of less than 0.3 K compared with the case without TIMs.

At the module level, TIMs such as pastes or adhesive layers are recommended to fill voids between prismatic cells and the base frame, ensuring reliable thermal coupling and uniform heat dissipation [156]. For automotive batteries, where service lifetimes typically span 10–15 years, the long-term reliability of TIMs is critical. Degradation mechanisms may include mechanical failures (pump-out, delamination) as well as chemical changes (dry-out, oxidation). Therefore, durability testing combining high-temperature storage, temperature and power cycling, vibration testing, and mixed-stress protocols that simulate real-life operating conditions, is necessary to evaluate the lifetime performance at the module level.

5.8.3 High-Power Light-Emitting Diodes

TIMs are used to dissipate heat in high-power light-emitting diode (LED) devices, where efficient thermal management is essential for maintaining performance and durability. One notable example is a flexible TIM based on self-assembled boron arsenide (s-BAs) embedded in a polymer composite [157]. This structure exhibits a thermal conductivity of up to 21 W/mK and an elastic compliance as low as 100 kPa. When applied to a 5 cm diameter power LED module, the s-BAs TIM lowered the operating temperature by approximately 45°C compared with conventional TIMs such as thermal epoxy and silicone thermal pads.

Graphene-based TIMs have also shown promising results in LED cooling. Vertically aligned graphene nanowalls (GNWs), synthesized via mesoplasma chemical vapor deposition and infiltrated with silicone, achieved a thermal conductivity of 20.4 W/mK at a graphene content of only 5.6 wt% [158]. When applied to power LEDs, this GNW-based TIM reduced the operating temperature by 22°C compared with no TIM, and by 8°C relative to a state-of-the-art commercial thermal pad with a thermal conductivity of 5.0 W/mK. These results highlight the growing opportunity for advanced TIMs to improve thermal management in high-power LED applications.

5.9 SUMMARY, FUTURE TRENDS, AND CONCLUSION

5.9.1 Summary

A comprehensive review on materials for thermal management for advanced semiconductor packaging, from fundamental principles to current applications, was provided in this chapter. Sections 5.1 and 5.2 introduced the fundamentals of heat transfer, including conduction, convection, and thermal resistance, to establish a solid basis for understanding TIMs. Section 5.3 summarized the various types of TIMs: greases, pads, gels, PCMs, metallic alloys, and carbon-based composites, highlighting their compositions, properties, benefits, limitations, and representative applications. Section 5.4 described thermal management strategies at the chip, board, and system levels, with particular emphasis on classifications such as TIM 1.0, 1.5, and 2.0. Section 5.5 reviewed cooling solutions both in direct cooling (air, spray, jet impingement, and immersion) and in indirect cooling (cold plate-based solutions), and discussed the materials used for heat spreading. Section 5.6 focused on the testing and characterization, covering steady-state and transient thermal measurements, along with optical, mechanical, and chemical characterization tools, followed by reliability assessments under thermal, mechanical, and environmental stresses. Section 5.7 introduced analytical and numerical approaches to predict thermal conductivity, thermal interface resistance, and BLT. Section 5.8 presented several applications in data centers and high-performance computing (HPC), electric vehicle (EV) battery systems, and high-power LEDs.

5.9.2 Future Trends in Thermal Interface Material Development

The future development of TIMs must continue to address materials and design requirements to manage high heat flux and ensure long-term reliability under diverse environmental conditions. CNT-based TIMs are expected to gain further interest due to their intrinsically high thermal conductivity, mechanical flexibility, and structural robustness. In addition, the ability to vertically align CNT structures offers the potential to reduce the thermal resistance. However, achieving controlled and consistent CNT alignment at a mass-production scale remains a challenge and requires further investigation.

Low melting alloys (LMAs) also show promise for future TIM development because of their high thermal conductivity and ability to fill air gaps at relatively low melting temperatures (10°C–85°C). Despite these advantages, certain LMAs face challenges such as corrosion and process robustness. Their varying melting points require careful selection to match specific application needs.

Beyond these advancements, novel fabrication methods and bio-inspired structural designs are expected to expand the range of TIM materials and architectures suitable for emerging applications, including data centers, high-performance computing, electric vehicles, flexible electronics, and space systems.

The increasingly stringent thermal requirements in advanced semiconductor packaging continue to demand innovative package-level thermal solutions. As package integration is further scaled in power density and architectural complexity, conventional thermal pathways become increasingly constrained, necessitating alternative approaches for heat removal. One promising direction is the elimination of the TIM through direct bonding between the cooling solution or heat spreader and the chip. This can be achieved through several approaches. One method involves incorporating an additional copper layer on the backside of the chip and directly attaching the copper cold plate using thermal compression bonding (TCB) [159]. Another approach is to fabricate the cold plate from silicon and either directly bond it to the chip or perform backside silicon etching to create integrated microchannels for fluid flow on the chip. Reliability studies need to be conducted on such approaches as the fluid flow can cause mechanical stresses on the chip, and the longevity of the bond needs to be studied further.

5.9.3 Conclusion

This chapter presents the materials essential for thermal management in electronic systems, along with a detailed understanding of TIMs, their classification, constituent materials, and significance. In conclusion, TIMs remain one of the most critical and challenging components in advanced semiconductor packaging. Alongside ongoing research, commercial market trends indicate a continued rise in global TIM market value and demand, driven by the increasing need for efficient thermal solutions in next-generation electronics. Following this trend, substantial progress has been made in understanding the fundamental mechanisms of heat transfer through TIMs, developing advanced materials and structural designs, and expanding their application space. However, several challenges remain. Further improvements are required in enhancing the thermal conductivity, ensuring long-term stability and reliability, and bridging the gap between laboratory-scale demonstrations and high volume manufacturing. Looking ahead, the convergence of materials innovations with optimized TIM designs, along with careful selection of TIM types tailored to specific application needs, will be essential. Such advancements will continue to reinforce the importance of TIMs in next-generation advanced electronics packaging.

ACKNOWLEDGEMENTS

We acknowledge the contributions of undergraduate students Sean Maguire and Arionna Smith for their assistance in preparing several tables and schematics in this chapter.

REFERENCES

1. CHIPS NAPMP NOFO 2, n.d. https://www.nist.gov/system/files/documents/2024/10/18/CHIPS%20NAPMP%20NOFO%202.pdf (accessed January 1, 2026).
2. Sarvar F, Whalley D, Conway P. Thermal interface materials - A review of the state of the art. *2006 1st Electronic Systemintegration Technology Conference*, San Diego, CA, 2006, pp. 1292–302. https://doi.org/10.1109/ESTC.2006.280178.
3. Prasher R. Thermal interface materials: Historical perspective, status, and future directions. *Proceedings of the IEEE* 2006;94:1571–1586. https://doi.org/10.1109/JPROC.2006.879796.
4. Bergman TL, Lavine AS. *Fundamentals of Heat and Mass Transfer*, 8th ed. New York: John Wiley & Sons, Inc., 2017.
5. Khoshvaght-Aliabadi M, Ghodrati P, Nasrolahzadeh A, Kang YT. Optimization of heat sinks for data center server CPUs cooled via single-phase immersion cooling. *Appl Therm Eng* 2025;280:127790. https://doi.org/10.1016/j.applthermaleng.2025.127790.
6. Acikalin T, Schroeder C. Direct liquid cooling of bare die packages using a microchannel cold plate. *Fourteenth Intersociety Conference on Thermal and Thermomechanical Phenomena in Electronic Systems (ITherm)*, San Diego, CA, 2014, pp. 673–679. https://doi.org/10.1109/ITHERM.2014.6892346.
7. Carslaw HS, Jaeger JC. *Conduction of Heat in Solids*. Oxford: Oxford University Press, 1959.
8. Bejan A, Kraus AD. *Heat Transfer Handbook*, vol. 1. New York: John Wiley & Sons, 2003.
9. Iyer SS, Jangam S, Vaisband B. Silicon interconnect fabric: A versatile heterogeneous integration platform for AI systems. *IBM J Res Dev* 2019;63:5:1–5:16. https://doi.org/10.1147/JRD.2019.2940427.
10. Fisher TS. *Thermal Energy at the Nanoscale*, vol. 3. New York: World Scientific Publishing Company, 2013.

11. Zhou Y, Wu S, Long Y, Zhu P, Wu F, Liu F, et al. Recent advances in thermal interface materials. *ES Mater Manuf* 2020;7:4–24. https://doi.org/10.30919/esmm5f717.
12. Chung DDL. Performance of thermal interface materials. *Small* 2022;18:693. https://doi.org/10.1002/smll.202200693.
13. Chung DDL. Thermal interface materials. *J Mater Eng Perform* 2001;10:56–59. https://doi.org/10.1361/105994901770345358.
14. Roy CK, Bhavnani S, Hamilton MC, Johnson RW, Nguyen JL, Knight RW, et al. Investigation into the application of low melting temperature alloys as wet thermal interface materials. *Int J Heat Mass Transf* 2015;85:996–1002. https://doi.org/10.1016/j.ijheatmasstransfer.2015.02.029.
15. Roy CK, Bhavnani S, Hamilton MC, Wayne Johnson R, Knight RW, Harris DK. Durability of low melt alloys as thermal interface materials. *J Electron Packag* 2016;138:1. https://doi.org/10.1115/1.4032462.
16. Webb RL, Gwinn JP. Low melting point thermal interface material. *ITherm 2002. Eighth Intersociety Conference on Thermal and Thermomechanical Phenomena in Electronic Systems (Cat. No.02CH37258)*, San Diego, CA, 2002, pp. 671–676. https://doi.org/10.1109/ITHERM.2002.1012519.
17. Hayashi K, Nakano T, Inoue Y. Enhancing thermal conduction properties of vertically aligned CNT forests by reducing interfacial thermal resistance using an aluminum interlayer. *Carbon N Y* 2025;238:120256. https://doi.org/10.1016/j.carbon.2025.120256.
18. Bai R, Wei Y, Xu J, Li X, Li M, Zou Z, et al. Encapsulated carbon nanotube array as a thermal interface material compatible with standard electronics packaging. *Nano Res* 2023;16:11389–11400. https://doi.org/10.1007/s12274-023-5872-y.
19. Martin Y, Van Kessel T. High Performance Liquid Metal Thermal Interface for Large Volume Production, 2007. https://api.semanticscholar.org/CorpusID:202706316
20. Yu H, Li L, Kido T, Xi G, Xu G, Guo F. Thermal and insulating properties of epoxy/aluminum nitride composites used for thermal interface material. *J Appl Polym Sci* 2012;124:669–677. https://doi.org/10.1002/app.35016.
21. Fletcher LS. A review of thermal enhancement techniques for electronic systems. *IEEE Trans Compon Hybrids Manuf Technol* 1990;13:1012–1021. https://doi.org/10.1109/33.62543.
22. Pathumudy RD, Prabhu KN. Thermal interface materials for cooling microelectronic systems: present status and future challenges. *J Mater Sci Mater Electron* 2021;32:11339–11366. https://doi.org/10.1007/s10854-021-05635-w.
23. Roy CK, Bhavnani S, Hamilton MC, Johnson RW, Knight RW, Harris DK. Thermal performance of low melting temperature alloys at the interface between dissimilar materials. *Appl Therm Eng* 2016;99:72–79. https://doi.org/10.1016/j.applthermaleng.2016.01.036.
24. Becker G, Lee G, Lin Z. Thermal conductivity in advanced chips: emerging generation of thermal greases offers advantages. *Adv Packag* 2005;14:14–16.
25. Peterson AL. Silicones with improved thermal conductivity for thermal management in electronic packaging. *40th Conference Proceedings on Electronic Components and Technology*, Las Vegas, Nevada, 1990, pp. 613–619. https://doi.org/10.1109/ECTC.1990.122251.
26. Kusuma WJ, Fadarina, Hasan A. Sodium silicate composite filled by zinc oxide as low resistance thermal grease. *J Phys Conf Ser* 2019;1167:012045. https://doi.org/10.1088/1742-6596/1167/1/012045.
27. Mallik S, Ekere N, Best C, Bhatti R. Investigation of thermal management materials for automotive electronic control units. *Appl Therm Eng* 2011;31:355–362. https://doi.org/10.1016/j.applthermaleng.2010.09.023.
28. Otiaba KC, Ekere NN, Bhatti RS, Mallik S, Alam MO, Amalu EH. Thermal interface materials for automotive electronic control unit: Trends, technology and R&D challenges. *Microelectron Reliab* 2011;51:2031–2043. https://doi.org/10.1016/j.microrel.2011.05.001.

29. Lewis JS, Perrier T, Mohammadzadeh A, Kargar F, Balandin AA. Power cycling and reliability testing of epoxy-based graphene thermal interface materials. *J Carbon Res* 2020;6:26. https://doi.org/10.3390/c6020026.
30. Raza MA, Westwood A. Thermal contact resistance of various carbon nanomaterial-based epoxy composites developed for thermal interface applications. *J Mater Sci Mater Electron* 2019;30:10630–10638. https://doi.org/10.1007/s10854-019-01408-8.
31. Kwon Y-K, Kim P. *Unusually High Thermal Conductivity in Carbon Nanotubes. High Thermal Conductivity Materials.* New York: Springer-Verlag, n.d., pp. 227–265. https://doi.org/10.1007/0-387-25100-6_8.
32. Adhikari A. Polymer matrix composite: Thermally conductive grease preparation and characterization. A Thesis, 2019.
33. Ahmed J, Wang QJ, Balogun O, Ren N, England R, Lockwood F. Molecular dynamics modeling of thermal conductivity of several hydrocarbon base oils. *Tribol Lett* 2023;71:70. https://doi.org/10.1007/s11249-023-01738-z.
34. Lo WT, Ma K, Abeykoon C. Enhancing the thermal performance of polyethylene glycol phase change material with carbon-based fillers. *Int J Heat Mass Transf* 2024; 220:124919. https://doi.org/10.1016/j.ijheatmasstransfer.2023.124919.
35. Gwinn JP, Webb RL. Performance and testing of thermal interface materials. *Microelectron J* 2003;34:215–222. https://doi.org/10.1016/S0026-2692(02)00191-X.
36. Viswanath R, Wakharkar V, Watwe A, Lebonheur V. Thermal performance challenges from silicon to systems. *Intel Technol J* 2000;Q3:15.
37. Chen H, Wei H, Chen M, Meng F, Li H, Li Q. Enhancing the effectiveness of silicone thermal grease by the addition of functionalized carbon nanotubes. *Appl Surf Sci* 2013;283:525–531. https://doi.org/10.1016/j.apsusc.2013.06.139.
38. Chiu C-P, Solbrekken GL, Chung YD. Thermal modeling of grease-type interface material in PPGA application. *Thirteenth Annual IEEE. Semiconductor Thermal Measurement and Management Symposium*, Austin, TX, 1997, pp. 57–63. https://doi.org/10.1109/STHERM.1997.566783.
39. Uetani K, Ata S, Tomonoh S, Yamada T, Yumura M, Hata K. Elastomeric thermal interface materials with high through-plane thermal conductivity from carbon fiber fillers vertically aligned by electrostatic flocking. *Adv Mater* 2014;26:5857–5862. https://doi.org/10.1002/adma.201401736.
40. Daniel Blazej. Thermal Interface Materials. Electronics Cooling, 2003. https://www.electronics-cooling.com/2003/11/thermal-interface-materials/ (accessed October 8, 2025).
41. Patrick Zemlianoy. Thermal control of space electronics|Electronics Cooling, 1996. https://www.electronics-cooling.com/1996/09/thermal-control-of-space-electronics/ (accessed October 1, 2025).
42. Miksa deSorgo. Thermal Interface Materials. Electronics Cooling, 1996. https://www.electronics-cooling.com/1996/09/thermal-interface-materials-2/ (accessed October 8, 2025).
43. Zhao J-W, Zhao R, Huo Y-K, Cheng W-L. Effects of surface roughness, temperature and pressure on interface thermal resistance of thermal interface materials. *Int J Heat Mass Transf* 2019;140:705–716. https://doi.org/10.1016/j.ijheatmasstransfer.2019.06.045.
44. Kunquan M, Jing L. Liquid metal cooling in thermal management of computer chips. *Front Energy Power Eng China* 2007;1:384–402. https://doi.org/10.1007/s11708-007-0057-3.
45. Lourenço MJV., Alves M, Serra JM, Nieto de Castro CA, Buschmann MH. The thermal conductivity of near-eutectic galinstan (Ga68.4In21.5Sn10) molten alloy. *Int J Thermophys* 2024;45:6. https://doi.org/10.1007/s10765-023-03293-0.
46. Fan G, Wells J, Beam M. Thermal performance and reliability of liquid metal alloys as thermal interface materials for computing electronics devices. *2024 IEEE 74th*

Electronic Components and Technology Conference (ECTC), Denver, CO, 2024, pp. 1705–1711. https://doi.org/10.1109/ECTC51529.2024.00282.
47. Macris CG, Sanderson TR, Ebel RG, Leyerle CB. Performance, reliability, and approaches using a low melt alloy as a thermal interface material. *Proceed Int Microelectron Packag Soc (IMAPS)* 2004;2:4.
48. Kim D-G, Yoon J-W, Lee C-Y, Jung S-B. Reaction diffusion and formation of Cu11/In9 and In27Ni10 phases in the couple of indium-substrates. *Mater Trans* 2003;44:72–77. https://doi.org/10.2320/matertrans.44.72.
49. Malakooti MH, Kazem N, Yan J, Pan C, Markvicka EJ, Matyjaszewski K, et al. Liquid metal supercooling for low-temperature thermoelectric wearables. *Adv Funct Mater* 2019;29:45. https://doi.org/10.1002/adfm.201906098.
50. Barbier F, Blanc J. Corrosion of martensitic and austenitic steels in liquid gallium. *J Mater Res* 1999;14:737–744. https://doi.org/10.1557/JMR.1999.0099.
51. Vijay K. Solder-TIMs (thermal interface materials) for superior thermal management in power electronics. *18th European Microelectronics and Packaging Conference*, Brighton, 2011.
52. Ren H, Yang B, Meenakshi Sundaram N, Shangguan D, Fisher TS, Iyer SS. Architecting the thermal dissipation and power delivery for chiplet-based wafer-scale systems and experimental demonstration of segmented vertical cooling for silicon dielets at 1 W/mm^2 power density. *2025 IEEE 75th Electronic Components and Technology Conference (ECTC)*, Lake Buena Vista, FL, 2025, pp. 1512–1517. https://doi.org/10.1109/ECTC51687.2025.00258.
53. Indium Corporation. Solder TIMs|Thermal Interface Materials, n.d. https://www.indium.com/products/thermal-interface-materials/solder-tims/ (accessed October 2, 2025).
54. Chen S, Fan G, Yan X, LaBarbera C, Kresge L, Lee N-C. Pressureless sintering of nano-Ag paste with low porosity for die attach. *IMAPS* 2014;2014:000092–000098. https://doi.org/10.4071/isom-TA35.
55. Chen S, Fan G, Yan X, Labarbera C, Kresge L, Lee N-C. Porosity evolution of Ag-sintering at die attach. *IMAPS* 2015;2015:000833–000841. https://doi.org/10.4071/isom-2015-THP41.
56. Zhao J, Yao M, Lee N-C. Nano-Cu sintering paste for high power devices die attach applications. *2018 IEEE 68th Electronic Components and Technology Conference (ECTC)*, vol. 2018, Lake Buena Vista, FL, 2018, pp. 557–563. https://doi.org/10.1109/ECTC.2018.00088.
57. Lee W, Kim J. A comprehensive review of research on organic-based phase-change thermal interface materials for thermal management of electric devices: Methods, performance, and applications. *Polym Test* 2025;142:108677. https://doi.org/10.1016/j.polymertesting.2024.108677.
58. Imran Khan M, Asfand F, Al-Ghamdi SG. Progress in research and development of phase change materials for thermal energy storage in concentrated solar power. *Appl Therm Eng* 2023;219:119546. https://doi.org/10.1016/j.applthermaleng.2022.119546.
59. Aiswarya V, Das S. Magnetized graphene oxide – Modified microencapsulated phase change material for enhanced heat transfer performance with reduced leakage. *Therm Sci Eng Prog* 2023;41:101807. https://doi.org/10.1016/j.tsep.2023.101807.
60. Li C, Wen X, Cai W, Yu H, Liu D. Phase change material for passive cooling in building envelopes: A comprehensive review. *J Build Eng* 2023;65:105763. https://doi.org/10.1016/j.jobe.2022.105763.
61. Hamidi E, Ganesan PB, Sharma RK, Yong KW. Computational study of heat transfer enhancement using porous foams with phase change materials: A comparative review. *Renew Sustain Energy Rev* 2023;176:113196. https://doi.org/10.1016/j.rser.2023.113196.

62. Sheikholeslami M. Efficacy of porous foam on discharging of phase change material with inclusion of hybrid nanomaterial. *J Energy Storage* 2023;62:106925. https://doi.org/10.1016/j.est.2023.106925.
63. Jiang L, Chen G, Zhao L, Li M, Zhang W, Ma X. Research on the low-temperature waste heat storage performance of a novel microencapsulated phase change material with a double-shell encapsulation of hexadecylamine. *J Energy Storage* 2024;80:110253. https://doi.org/10.1016/j.est.2023.110253.
64. Zou L, Luo Y, Zhang J, Sheng X, Chen Y, Lin P. Phase change material gel particles with suitable size and superior thermophysical properties towards highly efficient thermal management of miniature electronic components. *J Energy Storage* 2023;60:106590. https://doi.org/10.1016/j.est.2022.106590.
65 Lee W, Lee J, Yang W, Kim J. Fabrication of biobased advanced phase change material and multifunctional composites for efficient thermal management. *ACS Sustain Chem Eng* 2023;11:1178–1189. https://doi.org/10.1021/acssuschemeng.2c06598.
66 Rahman IU, Nardini S, Buonomo B, Manca O, Khan H, Siviero B. Thermal interface materials: A promising solution for passive heat dissipation in electronic appliances. *Therm Sci Eng Prog* 2025;62:103673. https://doi.org/10.1016/j.tsep.2025.103673.
67. Zhi C, Bando Y, Terao T, Tang C, Kuwahara H, Golberg D. Towards thermoconductive, electrically insulating polymeric composites with boron nitride nanotubes as fillers. *Adv Funct Mater* 2009;19:1857–1862. https://doi.org/10.1002/adfm.200801435.
68. Tang B, Hu G, Gao H, Hai L. Application of graphene as filler to improve thermal transport property of epoxy resin for thermal interface materials. *Int J Heat Mass Transf* 2015;85:420–429. https://doi.org/10.1016/j.ijheatmasstransfer.2015.01.141.
69. Leong C-K, Chung DDL. Carbon black dispersions as thermal pastes that surpass solder in providing high thermal contact conductance. *Carbon N Y* 2003;41:2459–2469. https://doi.org/10.1016/S0008-6223(03)00247-1.
70. Lin C, Chung DDL. Nanostructured fumed metal oxides for thermal interface pastes. *J Mater Sci* 2007;42:9245–9255. https://doi.org/10.1007/s10853-007-1911-4.
71. Gao J, Yan Q, Lv L, Tan X, Ying J, Yang K, et al. Lightweight thermal interface materials based on hierarchically structured graphene paper with superior through-plane thermal conductivity. *Chem Eng J* 2021;419:129609. https://doi.org/10.1016/j.cej.2021.129609.
72. Dai W, Ma T, Yan Q, Gao J, Tan X, Lv L, et al. Metal-level thermally conductive yet soft graphene thermal interface materials. *ACS Nano* 2019;13:11561–11571. https://doi.org/10.1021/acsnano.9b05163.
73. Chen H, Chen M, Di J, Xu G, Li H, Li Q. Architecting three-dimensional networks in carbon nanotube buckypapers for thermal interface materials. *J Phys Chem C* 2012;116:3903–3909. https://doi.org/10.1021/jp2086158.
74. Hao M, Kumar A, Hodson SL, Zemlyanov D, He P, Fisher TS. Brazed carbon nanotube arrays: Decoupling thermal conductance and mechanical rigidity. *Adv Mater Interfaces* 2017;4:1601042. https://doi.org/10.1002/admi.201601042.
75. Xu J, Fisher TS. Enhancement of thermal interface materials with carbon nanotube arrays. *Int J Heat Mass Transf* 2006;49:1658–1666. https://doi.org/10.1016/j.ijheatmasstransfer.2005.09.039.
76. Cola BA, Xu X, Fisher TS. Increased real contact in thermal interfaces: A carbon nanotube/foil material. *Appl Phys Lett* 2007;90:093513. https://doi.org/10.1063/1.2644018.
77. Hodson SL, Bhuvana T, Cola BA, Xu X, Kulkarni GU, Fisher TS. Palladium thiolate bonding of carbon nanotube thermal interfaces. *J Electron Packag* 2011;133:020907. https://doi.org/10.1115/1.4004094.
78. Cola BA, Xu J, Cheng C, Xu X, Fisher TS, Hu H. Photoacoustic characterization of carbon nanotube array thermal interfaces. *J Appl Phys* 2007;101:054313. https://doi.org/10.1063/1.2510998.

79. Cross R, Cola BA, Fisher T, Xu X, Gall K, Graham S. A metallization and bonding approach for high performance carbon nanotube thermal interface materials. *Nanotechnology* 2010;21:445705. https://doi.org/10.1088/0957-4484/21/44/445705.
80. Hu X, Padilla AA, Xu J, Fisher TS, Goodson KE. Thermal characterization of vertically-oriented carbon nanotubes on silicon. *Semiconductor Thermal Measurement and Management IEEE Twenty First Annual IEEE Symposium*, San Jose, CA, 2005. https://doi.org/10.1109/STHERM.2005.1412194.
81. Tong XC. *Advanced Materials for Thermal Management of Electronic Packaging*. New York: Springer, 2011.
82. Thanu DPR, Liu B, Cartas MA. Thermal interface materials and cooling technologies in microelectronic packaging—A critical review. *J Microelectron Electron Packag* 2018;15:63–74. https://doi.org/10.4071/IMAPS.654289.
83. Deppisch C, Fitzgerald T, Raman A, Hua F, Zhang C, Liu P, et al. The material optimization and reliability characterization of an indium-solder thermal interface material for CPU packaging. *Min Met Mater Soc* 2006;58:67–74. https://doi.org/10.1007/s11837-006-0186-6.
84. Lau JH. *Flip Chip Technologies*. New York: McGraw-Hill Professional, 1996.
85. Miller LF. Controlled collapse reflow chip joining. *IBM J Res Dev* 1969;13:239–250. https://doi.org/10.1147/rd.133.0239.
86. Prasad SK. *Advanced Wirebond Interconnection Technology*. New York: Springer US, 2006.
87. Iyer SS. Heterogeneous integration for performance and scaling. *IEEE Trans Compon Packaging Manuf Technol* 2016;6:973–982. https://doi.org/10.1109/TCPMT.2015.2511626.
88. Siewert T, Liu S, Smith DR, Juan M, Madeni C. Database for Solder Properties with Emphasis on New Lead-free Solders. 2002.
89. Lee WS, Han IY, Yu J, Kim SJ, Byun KY. Thermal characterization of thermally conductive underfill for a flip-chip package using novel temperature sensing technique. *Thermochim Acta* 2007;455:148–155. https://doi.org/10.1016/j.tca.2006.11.034.
90. Shearer A, Eltantawy ME, Montazerian M, Lanagan MT, Mauro JC. Glass-ceramic substrates for electronics packaging. *Adv Electron Mater* 2025;11,e00331. https://doi.org/10.1002/aelm.202500331.
91. Zhang Z, Wang X, Yan Y. A review of the state-of-the-art in electronic cooling. *Adv Electr Eng Electron Energy* 2021;1:100009. https://doi.org/10.1016/j.prime.2021.100009.
92. Scott AW. *Cooling of Electronic Equipment*. New York: Wiley, 1974.
93. Sohel Murshed SM, Nieto de Castro CA. A critical review of traditional and emerging techniques and fluids for electronics cooling. *Renew Sustain Energy Rev* 2017;78:821–833. https://doi.org/10.1016/j.rser.2017.04.112.
94. Liang G, Mudawar I. Review of spray cooling – Part 1: Single-phase and nucleate boiling regimes, and critical heat flux. *Int J Heat Mass Transf* 2017;115:1174–1205. https://doi.org/10.1016/j.ijheatmasstransfer.2017.06.029.
95. Zhao X, Yin Z, Zhang B, Yang Z. Experimental investigation of surface temperature non-uniformity in spray cooling. *Int J Heat Mass Transf* 2020;146:118819. https://doi.org/10.1016/j.ijheatmasstransfer.2019.118819.
96. Liu X, Xue R, Ruan Y, Chen L, Zhang X, Hou Y. Effects of injection pressure difference on droplet size distribution and spray cone angle in spray cooling of liquid nitrogen. *Cryogenics (Guildf)* 2017;83:57–63. https://doi.org/10.1016/j.cryogenics.2017.01.011.
97. Cai C, Liu H, Chen H, Zhao C, Gao J, Jia M, et al. Experimental investigation on heat transfer of water spray cooling by addition of n-butanol. *ASME 2020 Heat Transfer Summer Conference*, New York: American Society of Mechanical Engineers, 2020. https://doi.org/10.1115/HT2020-9136.

98. Liu N, Yu Z, Zhu T, Yin X, Zhang H. Effects of microstructured surface and mixed surfactants on the heat transfer performance of pulsed spray cooling. *Int J Therm Sci* 2020;158:106530. https://doi.org/10.1016/j.ijthermalsci.2020.106530.
99. Abdulrazzaq NM, Anumbe NC, Khan JA. Experimental investigation of the performance of refrigerant R134A working in a spray cooling. *Proceeding of 3rd Thermal and Fluids Engineering Conference (TFEC)*, Connecticut: Begellhouse, 2018, pp. 1355–1368. https://doi.org/10.1615/TFEC2018.hte.021658.
100. Bao J, Wang Y, Xu X, Niu X, Liu J, Qiu L. Analysis on the influences of atomization characteristics on heat transfer characteristics of spray cooling. *Sustain Cities Soc* 2019;51:101799. https://doi.org/10.1016/j.scs.2019.101799.
101. Azarifar M, Arik M. Liquid synthetic jets for high flux electronics cooling. *Appl Therm Eng* 2025;261:125007. https://doi.org/10.1016/j.applthermaleng.2024.125007.
102. Uysal A, Keçebaş A, Kayfeci M. Futuristic methods of electronics cooling. In Fethi Aloui, Edwin Geo Varuvel, Ankit Sonthalia (Eds.). *Handbook of Thermal Management Systems*, Amsterdam, Netherlands: Elsevier, 2023, pp. 687–702. https://doi.org/10.1016/B978-0-443-19017-9.00026-X.
103. Kuncoro IW, Pambudi NA, Biddinika MK, Widiastuti I, Hijriawan M, Wibowo KM. Immersion cooling as the next technology for data center cooling: A review. *J Phys Conf Ser* 2019;1402:044057. https://doi.org/10.1088/1742-6596/1402/4/044057.
104. Wei J. Liquid cooling, opportunity and challenges toward effective and efficient scalabilities. *2019 IEEE CPMT Symposium Japan (ICSJ)*, Kyoto, Japan, 2019, pp. 83–84. https://doi.org/10.1109/ICSJ47124.2019.8998723.
105. Muneeshwaran M, Lin Y-C, Wang C-C. Performance analysis of single-phase immersion cooling system of data center using FC-40 dielectric fluid. *Int Commun Heat Mass Transfer* 2023;145:106843. https://doi.org/10.1016/j.icheatmasstransfer.2023.106843.
106. Kang X, Wang Y, Huang Q, Cui Y, Wang C, Wen C, et al. Phase-change immersion cooling high power light emitting diodes and heat transfer improvement. *Microelectron Reliab* 2017;79:257–264. https://doi.org/10.1016/j.microrel.2017.05.033.
107. Birbarah P, Gebrael T, Foulkes T, Stillwell A, Moore A, Pilawa-Podgurski R, et al. Water immersion cooling of high power density electronics. *Int J Heat Mass Transf* 2020;147:118918. https://doi.org/10.1016/j.ijheatmasstransfer.2019.118918.
108. Suresh Patil M, Seo J-H, Lee M-Y. A novel dielectric fluid immersion cooling technology for Li-ion battery thermal management. *Energy Convers Manag* 2021;229:113715. https://doi.org/10.1016/j.enconman.2020.113715.
109. Hsu Y-T, Li J-X, Lu M-C. Enhanced immersion cooling using two-tier micro- and nano-structures. *Appl Therm Eng* 2018;131:864–873. https://doi.org/10.1016/j.applthermaleng.2017.12.067.
110. An X, Arora M, Huang W, Brantley WC, Greathouse JL. 3D numerical analysis of two-phase immersion cooling for electronic components. *2018 17th IEEE Intersociety Conference on Thermal and Thermomechanical Phenomena in Electronic Systems (ITherm)*, Orlando, FL, 2018, pp. 609–614. https://doi.org/10.1109/ITHERM.2018.8419528.
111. Matsuoka M, Matsuda K, Kubo H. Liquid immersion cooling technology with natural convection in data center. *2017 IEEE 6th International Conference on Cloud Networking (CloudNet)*, Prague, Czech Republic, 2017, pp. 1–7. https://doi.org/10.1109/CloudNet.2017.8071539.
112. Niazmand A, Chauhan T, Saini S, Shahi P, Bansode PV, Agonafer D. CFD simulation of two-phase immersion cooling using FC-72 dielectric fluid. *ASME 2020 International Technical Conference and Exhibition on Packaging and Integration of Electronic and Photonic Microsystems*, New York: American Society of Mechanical Engineers, 2020. https://doi.org/10.1115/IPACK2020-2595.

113. Shahi P, Heydari A, Modi H, Reddy Chinthaparthy LS, Barigala A, Eslami B, et al. Maximizing cooling potential: A step-by-step guide to commissioning liquid-cooled data centers. *2024 23rd IEEE Intersociety Conference on Thermal and Thermomechanical Phenomena in Electronic Systems (ITherm)*, Orlando, FL, 2024, pp. 1–12. https://doi.org/10.1109/ITherm55375.2024.10709522.
114. Heydari A, Al-Zu'bi O, Manaserh Y, Mehrabi M, Hosseini F, Sammakia B. System-level assessment of green refrigerant replacements for direct-to-chip two-phase cooling. *2024 23rd IEEE Intersociety Conference on Thermal and Thermomechanical Phenomena in Electronic Systems (ITherm)*, Orlando, FL, 2024, pp. 1–10. https://doi.org/10.1109/ITherm55375.2024.10709484.
115. Ren H, Guo Z, Yang B, Sahoo K, Sundaram NM, Chen Z, et al. Power delivery for scaled-out chiplet-based wafer-scale systems with 8 μm Cu-Cu bond pitch on active SI-interconnect fabric substrate. *2025 Symposium on VLSI Technology and Circuits (VLSI Technology and Circuits)*, Kyoto, Japan, 2025, pp. 1–3. https://doi.org/10.23919/VLSITechnologyandCir65189.2025.11074799.
116. Meenakshi SN, Pugazhendhi R, Iyer SS, Fisher TS. Flash cooling with methanol/water mixtures for 1 W/mm^2 fluxes without lateral heat spreading. *2025 24th IEEE Intersociety Conference on Thermal and Thermomechanical Phenomena in Electronic Systems (ITherm)*, Orlando, FL, 2025, pp. 1–8. https://doi.org/10.1109/ITherm55376.2025.11235664.
117. Kil MJ, Lam E, Abuseada M, Buckwalter JF, Fisher TS. Direct solar-thermal formation of graphitic heat spreaders on organic substrates. *2023 22nd IEEE Intersociety Conference on Thermal and Thermomechanical Phenomena in Electronic Systems (ITherm)*, vol. 2023, Orlando, FL, 2023, pp. 1–7. https://doi.org/10.1109/ITherm55368.2023.10177667.
118. Kil MJ, Lam E, Abuseada M, Buckwalter JF, Fisher TS. Solar-thermal synthesized graphitic heat spreader for thermally enhanced organic substrates. *IEEE Trans Compon Packaging Manuf Technol* 2024;14:903–911. https://doi.org/10.1109/TCPMT.2024.3374599.
119. Bonhaus J, Borchert D, Denisenko A, Fahrner WR. Diamond heat spreaders for high power devices with integrated temperature sensors. *Fourteenth Annual IEEE Semiconductor Thermal Measurement and Management Symposium (Cat. No.98CH36195)*, San Diego, CA, 1998, pp. 139–146. https://doi.org/10.1109/STHERM.1998.660399.
120. Tang H, Tang Y, Wan Z, Li J, Yuan W, Lu L, et al. Review of applications and developments of ultra-thin micro heat pipes for electronic cooling. *Appl Energy* 2018;223:383–400. https://doi.org/10.1016/j.apenergy.2018.04.072.
121. Weibel JA, Garimella SV. Recent advances in vapor chamber transport characterization for high-heat-flux applications. *Adv Heat Transf* 2013;45:209–301. https://doi.org/10.1016/B978-0-12-407819-2.00004-9.
122. Li Y, Eldredge JD, Lavine AS, Fisher TS, Drolen BL. A conjugate heat transfer model of oscillating heat pipe dynamics, performance, and dryout. *Int J Heat Mass Transf* 2024;227:125530. https://doi.org/10.1016/j.ijheatmasstransfer.2024.125530.
123. Liao Y, Gan Y, Liu F, Li Y. Experimental study on the thermal performance of aluminum three-dimensional vapor chamber heat sink with a louvered-fin stacked evaporator wick for data center servers. *Energy* 2024;304:132037. https://doi.org/10.1016/j.energy.2024.132037.
124. Zhao X, Su L, Jiang J, Deng W, Zhao D. A review of working fluids and flow state effects on thermal performance of micro-channel oscillating heat pipe for aerospace heat dissipation. *Aerospace* 2023;10:179. https://doi.org/10.3390/aerospace10020179.
125. Cui Y, Li M, Hu Y. Emerging interface materials for electronics thermal management: experiments, modeling, and new opportunities. *J Mater Chem C Mater* 2020;8:10568–10586. https://doi.org/10.1039/C9TC05415D.

126. Hao M, Saviers KR, Fisher TS. Design and validation of a high-temperature thermal interface resistance measurement system. *J Therm Sci Eng Appl* 2016;8:031008. https://doi.org/10.1115/1.4033011.
127. ASTM. *Test Method for Thermal Transmission Properties of Thermally Conductive Electrical Insulation Materials.* West Conshohocken, PA: ASTM, 2017. https://doi.org/10.1520/D5470-17.
128. Xu J, Fisher TS. Enhanced thermal contact conductance using carbon nanotube array interfaces. *IEEE Trans Compon Packag Technol* 2006;29:261–267. https://doi.org/10.1109/TCAPT.2006.875876.
129. Ångstrom AJ. Neue methode, das Wärmeleitungsvermögen der Körper zu bestimmen. *Ann Phys* 1862;190:513–530. https://doi.org/10.1002/andp.18621901202.
130. Hu Y, Fisher TS. Accurate thermal diffusivity measurements using a modified ångström's method with bayesian statistics. *J Heat Transfer* 2020;142:071401. https://doi.org/10.1115/1.4047145.
131. Stever J, Yang C, Hang Y, El KP, Xu C, Wang C, et al. Silicon TTV for advanced thermal investigations of high powered lidless package AI silicon. *2025 24th IEEE Intersociety Conference on Thermal and Thermomechanical Phenomena in Electronic Systems (ITherm)*, Dallas, TX, 2025, pp. 1–7. https://doi.org/10.1109/ITherm55376.2025.11235656.
132. JESD22-A104E. Temperature Cycling. 2009.
133. JESD22-A105C. Power and Temperature Cycling. 2011.
134. JESD22-A103B. High Temperature Storage Life. 2001.
135. Goel N, Anoop TK, Bhattacharya A, Cervantes JA, Mongia RK, Machiroutu SV, et al. Technical review of characterization methods for thermal interface materials (TIM). *2008 11th Intersociety Conference on Thermal and Thermomechanical Phenomena in Electronic Systems*, Orlando, FL, 2008, pp. 248–258. https://doi.org/10.1109/ITHERM.2008.4544277.
136. Bharatham L, Fong WS, Torresola J, Koang CC. Qualification of phase change thermal interface material for wave solder heat sink on FCBGA package. *2005 7th Electronic Packaging Technology Conference*, vol. 2, Singapore, 2005, pp. 537–542. https://doi.org/10.1109/EPTC.2005.1614462.
137. Xing W, Xu Y, Song C, Deng T. Recent advances in thermal interface materials for thermal management of high-power electronics. *Nanomaterials* 2022;12:365. https://doi.org/10.3390/nano12193365.
138. Xu JZ, Gao BZ, Kang FY. A reconstruction of Maxwell model for effective thermal conductivity of composite materials. *Appl Therm Eng* 2016;102:972–979. https://doi.org/10.1016/j.applthermaleng.2016.03.155.
139. Bruggeman DAG. The calculation of various physical constants of heterogeneous substances. I. The dielectric constants and conductivities of mixtures composed of isotropic substances. *Ann Phys* 1935;416:636–791.
140. Hasselman DPH, Johnson LF. Effective thermal conductivity of composites with interfacial thermal barrier resistance. *J Compos Mater* 1987;21:508–515. https://doi.org/10.1177/002199838702100602.
141. Lewis TB, Nielsen LE. Dynamic mechanical properties of particulate-filled composites. *J Appl Polym Sci* 1970;14:1449–1471. https://doi.org/10.1002/app.1970.070140604.
142. Nan C-W, Birringer R, Clarke DR, Gleiter H. Effective thermal conductivity of particulate composites with interfacial thermal resistance. *J Appl Phys* 1997;81:6692–6699. https://doi.org/10.1063/1.365209.
143. Nan C-W, Liu G, Lin Y, Li M. Interface effect on thermal conductivity of carbon nanotube composites. *Appl Phys Lett* 2004;85:3549–3551. https://doi.org/10.1063/1.1808874.
144. Prasher RS. Surface chemistry and characteristics based model for the thermal contact resistance of fluidic interstitial thermal interface materials. *J Heat Transfer* 2001;123:969–975. https://doi.org/10.1115/1.1388301.

145. Prasher RS, Matayabas JC. Thermal contact resistance of cured gel polymeric thermal interface material. *IEEE Trans Comp Packaging Technol* 2004;27:702–709. https://doi.org/10.1109/TCAPT.2004.838883.
146. Hu X, Govindasamy S, Goodson KE. Two-medium model for the bond line thickness of particle filled thermal interface materials. *Heat Transfer* 2004;3:337–342. https://doi.org/10.1115/IMECE2004-62027.
147. Sponagle BN, Groulx D. Improvement of a steady state method of thermal interface material characterization by use of three dimensional FEA simulation in COMSOL multiphysics, Halifax, NS, Canada, 2012, pp. 4.
148. Narumanchi S, Mihalic M, Kelly K, Eesley G. Thermal interface materials for power electronics applications. *2008 11th Intersociety Conference on Thermal and Thermomechanical Phenomena in Electronic Systems*, Orlando, FL, 2008, pp. 395–404. https://doi.org/10.1109/ITHERM.2008.4544297.
149. Pour Shahid Saeed Abadi P, Chung DDL. Numerical modeling of the performance of thermal interface materials in the form of paste-coated sheets. *J Electron Mater* 2011;40:1490–1500. https://doi.org/10.1007/s11664-011-1630-0.
150. Raja B, Praveenkumar V, Leelaprasad M, Manigandan P. Thermal simulations of an electronic system using ANSYS Icepak. *Int J Eng Res Appl* 2015;5:57–68.
151. Thompson MK, Thompson JM. Considerations for predicting thermal contact resistance in ANSYS. *17th Korea ANSYS User's Conference*, Pisa, Italy, 2007.
152. Barua A, Hossain Md S, Masood KI, Subrina S. Thermal management in 3-D integrated circuits with graphene heat spreaders. *Phys Procedia* 2012;25:311–316. https://doi.org/10.1016/j.phpro.2012.03.089.
153. De Bock H, Labhart J, Chauhan S, Kirk G, Kim JH. Thermal Interface Devices. US9615486B2, 2017.
154. Schultz M, Gaynes M, Parida P, Chainer T. Experimental investigation of direct attach microprocessors in a liquid-cooled chiller-less data center. *Fourteenth Intersociety Conference on Thermal and Thermomechanical Phenomena in Electronic Systems (ITherm)*, Orlando, FL, 2014, pp. 729–735. https://doi.org/10.1109/ITHERM.2014.6892353.
155. Lee N, Choi J, Lee J, Shin D, Um S. Prognostic analysis of thermal interface material effects on anisotropic heat transfer characteristics and state of health of a 21,700 cylindrical lithium-ion battery module. *J Energy Storage* 2023;72:108594. https://doi.org/10.1016/j.est.2023.108594.
156. Stadler R, Maurer A. Methods for durability testing and lifetime estimation of thermal interface materials in batteries. *Batteries* 2019;5:34. https://doi.org/10.3390/batteries5010034.
157. Cui Y, Qin Z, Wu H, Li M, Hu Y. Flexible thermal interface based on self-assembled boron arsenide for high-performance thermal management. *Nat Commun* 2021;12:1284. https://doi.org/10.1038/s41467-021-21531-7.
158. Yan Q, Alam FE, Gao J, Dai W, Tan X, Lv L, et al. Soft and self-adhesive thermal interface materials based on vertically aligned, covalently bonded graphene nanowalls for efficient microelectronic cooling. *Adv Funct Mater* 2021;31:2104062. https://doi.org/10.1002/adfm.202104062.
159. Jangam S, Bajwa AA, Mogera U, Ambhore P, Colosimo T, Chylak B, et al. Fine-pitch (≤10 μm) direct Cu-Cu interconnects using in-situ formic acid vapor treatment. *2019 IEEE 69th Electronic Components and Technology Conference (ECTC)*, Las Vegas, NV, 2019, pp. 620–627. https://doi.org/10.1109/ECTC.2019.00099.

6 Encapsulation Materials for Advanced Semiconductor Packaging and Heterogeneous Integration

Zihao Lin and Ching-Ping Wong

6.1 OVERVIEW OF POLYMERIC MATERIALS FOR MICROELECTRONICS PACKAGING

6.1.1 Historical Evolution of Packaging Materials

The use of polymers in microelectronics packaging has evolved in parallel with the relentless scaling and functional diversification of integrated circuits (ICs). In the earliest generations of electronics (1950s–1970s), hermetic ceramic and metal packages dominated, because of their robustness, gas impermeability, and compatibility with military and aerospace reliability requirements. These packages, however, were bulky, expensive, and limited in terms of manufacturability at high volumes.

The turning point came with the expansion of consumer electronics in the 1970s and 1980s. Plastic encapsulated microcircuits (PEMs) began to replace ceramic dual in-line packages (DIPs), enabled by the development of epoxy molding compounds (EMCs). These thermosetting resins offered low-cost, high-throughput encapsulation while still providing sufficient mechanical strength and moisture resistance for commercial applications. The 1980s also saw the emergence of polymer adhesives for die attach and silicone gels for passivation of sensitive devices.

With the introduction of flip-chip technology in the 1990s, a new reliability challenge emerged: the coefficient of thermal expansion (CTE) mismatch between silicon dies and organic substrates. This mismatch caused solder fatigue under thermal cycling. The invention of underfill materials—capillary-flow epoxies filled with silica—addressed this issue by redistributing stresses and improving mechanical coupling. Underfills quickly became indispensable in flip-chip packaging, laying the groundwork for the miniaturization trend that followed.

DOI: 10.1201/9781003587637-6

The 2000s marked the widespread adoption of wafer-level packaging (WLP) and fan-out wafer-level packaging (FOWLP), both of which required photosensitive polymer dielectrics (e.g., polyimide, benzocyclobutene, polybenzoxazole) to define redistribution layers (RDLs). Concurrently, the industry sought low-dielectric-constant polymers to minimize parasitic delay in high-speed devices.

Today, polymers not only provide structural and protective functions but also actively contribute to electrical, thermal, and even optical performance. Their historical trajectory shows a consistent pattern: moving from passive encapsulants toward active functional materials that enable new device architectures, such as co-packaged optics (CPO), 2.5D/3D integration, and system-in-package (SiP) platforms.

6.1.2 Categories of Packaging Polymers

The landscape of packaging polymers has gradually diversified as device architectures have become more complex. At the foundation are epoxy molding compounds, which remain the dominant encapsulants in wire-bonded and flip-chip packages. Their formulations typically combine epoxy resin, curing agents, silica fillers, and flame retardants, allowing manufacturers to tailor thermal expansion, moisture resistance, and flow behavior to different package designs. Complementing EMCs are underfill systems, which occupy the narrow gaps beneath flip-chip dies and distribute stresses across thousands of solder joints. These materials evolved from early capillary-flow underfills into no-flow and molded varieties, each addressing the need for higher throughput, finer pitch, and improved reliability.

Another important class is thermal interface materials (TIM), designed to reduce interfacial thermal resistance between chips and heat spreaders. By embedding thermally conductive fillers such as alumina, boron nitride, or silver particles into compliant polymer matrices, TIMs provide a critical pathway for heat dissipation in high-power devices.

Wafer-level packaging has introduced yet another category: photosensitive polymers such as polyimides [1], benzocyclobutene [2], and polybenzoxazoles [3], which act as dielectric layers in redistribution structures. Their ability to be patterned with photolithography enables fine-line interconnects essential for fan-out and 3D integration.

Beyond these core families, specialized polymers are also used for through-silicon via liners, adhesive bonding in stacked dies, and conformal coatings that shield devices from corrosive or humid environments. Taken together, these categories demonstrate the adaptability of polymer chemistry to perform structural, electrical, thermal, and protective functions across a wide spectrum of packaging technologies.

6.1.3 Applications across Packaging Platforms

Polymers are deeply embedded in nearly every packaging platform used in microelectronics. In traditional wire-bonded plastic packages, epoxy molding compounds encapsulate entire assemblies, protecting fragile wire bonds and silicon dies from mechanical damage and environmental stresses. As the industry shifted toward flip-chip interconnections, polymers assumed a more specialized role: underfills became indispensable to counteract thermal expansion mismatches and to extend solder joint lifetimes. In fan-out wafer-level packaging, polymers provide both the structural body of the package and the dielectric framework for redistribution layers, enabling unprecedented input/output density without reliance on costly silicon interposers. Panel-level packaging extends these principles to larger substrates, leveraging polymer processability to achieve cost-effective scaling.

In advanced system architectures, polymers enable vertical integration. Adhesives and stress-buffering layers in 2.5D and 3D assemblies ensure the mechanical reliability of stacked dies and interposers. Through-silicon vias rely on polymer liners for electrical isolation and stress relief, while redistribution layers once again depend on photo-patternable polymers for fine-line definition. Polymers also extend into emerging domains such as co-packaged optics, where optical-grade resins with a controlled refractive index and low optical loss are required to integrate photonic and electronic functions. At the same time, thermal interface polymers bridge the gap between high-power processors and external cooling solutions, helping to manage escalating heat densities in data centers and AI accelerators. The breadth of these applications illustrates how polymers have moved beyond their traditional role as protective encapsulants to become integral design elements in virtually every packaging strategy.

6.1.4 Challenges and Future Opportunities

Despite their versatility, polymers in packaging remain constrained by several persistent limitations. Chief among these is thermal conductivity. Even with heavy filler loading, most polymer systems struggle to dissipate the heat generated in modern high-power devices, and achieving the necessary balance between thermal performance and processability continues to challenge material engineers. Electrical performance is another concern: as frequencies move into the millimeter-wave and terahertz domains, polymers must exhibit an ultra-low dielectric constant and loss tangent, while still maintaining adhesion and mechanical toughness. Reliability also poses ongoing risks, as polymers are exposed to harsh environments involving coupled stressors such as temperature cycling, humidity, mechanical strain, and electrical bias. Long-term resistance to delamination, cracking, and chemical degradation remains a key determinant of package lifetime.

These challenges, however, are matched by promising opportunities. Advances in nanostructured fillers, including boron nitride nanosheets, alumina nanowires, and graphene derivatives, are opening pathways to create percolative thermal networks without sacrificing viscosity or flow. Hybrid organic–inorganic systems are being explored to achieve property combinations unattainable by single-phase polymers. Sustainability has become another driver, pushing the development of halogen-free, recyclable, and even bio-derived polymers to meet environmental regulations.

Looking ahead, the role of polymers will only expand as heterogeneous integration, system-in-package platforms, and co-packaged optics demand multifunctional materials with unprecedented combinations of thermal, electrical, optical, and mechanical properties. Rather than passive protective layers, polymers are increasingly viewed as active enablers of future electronic systems.

6.2 EPOXY MOLDING COMPOUNDS

6.2.1 Functional Roles in Packaging

EMCs serve as the primary encapsulation materials in a wide range of semiconductor packaging technologies. As thermosetting polymer composites, EMCs are specifically engineered to fulfill multiple roles simultaneously, including mechanical protection, environmental sealing, electrical insulation, and thermal stress buffering. With the continuing advancement of integrated circuits and the emergence of complex multi-die and heterogeneous integration schemes, the performance demands placed on EMCs have grown dramatically. This section explores the core functional roles of EMCs in modern packaging, and highlights how their design influences overall package reliability and performance.

One of the most fundamental roles of EMCs is to provide mechanical protection to delicate internal components such as semiconductor dies, interconnect wires, micro-bumps, and redistribution layers, as shown in Figure 6.1. These elements are highly sensitive to external mechanical shock, vibration, and thermal expansion mismatches. The cured EMC forms a rigid protective shell that distributes and absorbs external stress, thereby preventing die cracking, wire breakage, or substrate warpage. The mechanical robustness of EMCs is especially important in automotive, aerospace, and industrial electronics, where long-term exposure to mechanical fatigue is unavoidable.

Equally important is the environmental shielding capability of EMCs. The polymer matrix and densely packed fillers together form an effective barrier against moisture ingress, ionic contaminants, corrosive gases, and dust. This is critical for maintaining electrical insulation and preventing corrosion of metal contacts and interconnects. In harsh environments, such as under-hood automotive modules or avionics systems, EMCs must remain stable under temperature extremes, high humidity, and chemical exposure. Many commercial EMCs are tested under accelerated aging conditions—such as 85°C/85%RH for 1,000 hours or more—to evaluate long-term reliability.

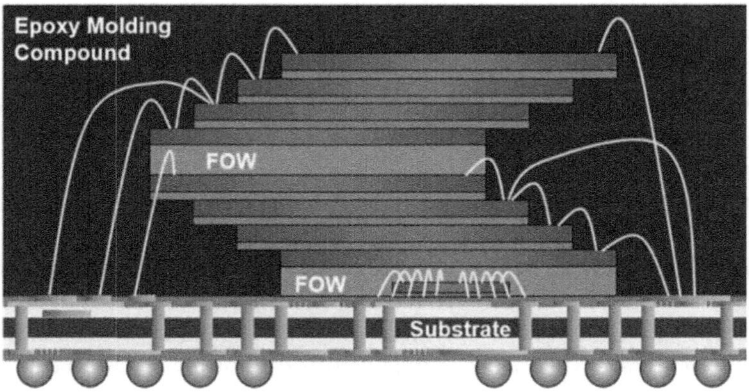

FIGURE 6.1 Schematic diagram of a wire bonded memory package with EMC as encapsulant. (Reproduced from Ref. [4] with permission from Springer Nature.)

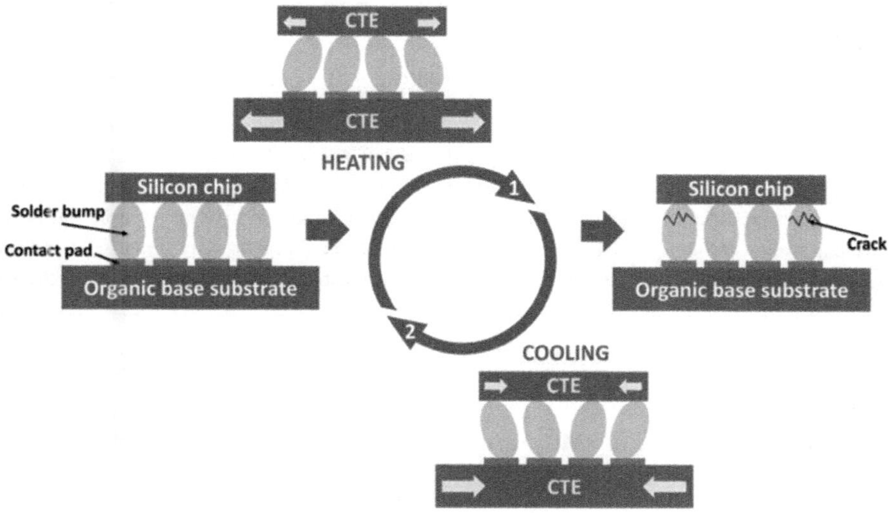

FIGURE 6.2 Schematic representation of the failure mechanism of a flip chip, which is driven by the CTE mismatch between the chip and the organic substrate under thermal cycling. (Reproduced from Ref. [5], published by MDPI under the CC BY license.)

Thermal stress modulation is another key function. In a typical electronics package, components with different coefficients of thermal expansion (CTE) are assembled together—silicon dies (CTE ~2.5 to 3 ppm/°C), organic substrates (~15 to 20 ppm/°C), leadframes, and various polymer layers. During temperature cycling, differential expansion can cause delamination or cracking at interfaces. Figure 6.2 shows the mechanism. By adjusting the EMC formulation, particularly the filler content and resin network design, the CTE can be tailored to intermediate values (~10 to 15 ppm/°C), which helps buffer interfacial stresses and improve the thermal fatigue resistance. This is particularly important in flip-chip and WLP configurations where the EMC interfaces directly with redistribution layers and underfill materials.

EMCs also play a role in dimensional stability and structural reinforcement. In advanced wafer-level and panel-level packages, EMCs are molded over the entire die or wafer, forming the structural body of the package. In these applications, planarity, warpage control, and die shift suppression are critical. The material must exhibit minimal flow-induced deformation, low cure shrinkage, and uniform filler distribution across large areas. Compression molding and vacuum lamination methods are commonly used to achieve these characteristics in fan-out and panel-level EMCs.

To summarize, the primary functions of EMCs in packaging are listed in Table 6.1.

Furthermore, EMCs are utilized in a variety of packaging technologies, each with specific functional integration (Table 6.2):

In each scenario, EMCs interact synergistically with other packaging materials—such as underfills, die-attach adhesives, and thermal interface materials—to collectively ensure the mechanical, thermal, and electrical integrity of the entire package. For instance, in flip-chip packaging, underfill provides solder joint support, while the EMC protects the die surface and improves warpage control. In 3D integrated

TABLE 6.1
The Primary Functions of EMCs in Packaging

Functional Role	Description
Mechanical protection	Prevents die crack, bump shift, and wire break under shock and vibration
Environmental shielding	Acts as moisture, ion, and gas barrier for harsh environments
Thermal stress management	Buffers CTE mismatches between die, substrate, and encapsulant
Structural reinforcement	Maintains flatness, planarity, and physical integrity of molded structures
Electrical insulation	Ensures long-term dielectric reliability and suppression of leakage paths

TABLE 6.2
Functional Integration of EMC Across Advanced Packaging Technologies

Packaging Type	EMC Role and Functional Integration
Wire-bonded packages (QFP)	Bulk encapsulation of wire-bonded die and leadframe
Flip-chip packages (FC-CSP)	Works in combination with underfill for top-side protection
Fan-out WLP (FOWLP)	Forms molded wafer body through compression molding
Panel-level packaging (PLP)	Encapsulates dies across large-area substrates for cost-effective scaling
2.5D/3D Integration	Fills cavities around TSVs, interposers, or stacked dies

systems, EMCs often encapsulate and stabilize vertically stacked chips, where they must offer precise flow behavior and minimal voiding. Their role is thus central to achieving long-term reliability in high-performance electronics.

In the subsequent sections, we will delve deeper into the materials composition of EMCs, their property-performance relationships, molding techniques, reliability concerns, and future materials development trends aligned with next-generation packaging architectures.

6.2.2 Chemistry and Filler Systems

The formulation of Epoxy Molding Compounds is a sophisticated exercise in materials engineering, involving the optimization of the chemical composition, particle morphology, rheological behavior, and network structure. Each constituent in the EMC formulation plays a pivotal role in determining the compound's processing behavior, cured-state properties, and long-term reliability in electronic packaging applications. This section outlines the key components of EMCs, their respective functions, and the rationale behind their selection and ratio balancing.

At the heart of EMCs lies the epoxy resin matrix, which serves as the primary binder and crosslinkable phase. The most widely used resins in EMCs are novolac-type epoxies and bisphenol-type epoxies (such as Bisphenol-A or Bisphenol-F [6,7]). Novolac epoxies [8,9] are favored for their higher functionality and ability to form dense, thermally stable crosslinked networks. This makes them ideal for high-temperature applications or automotive modules exposed to thermal extremes.

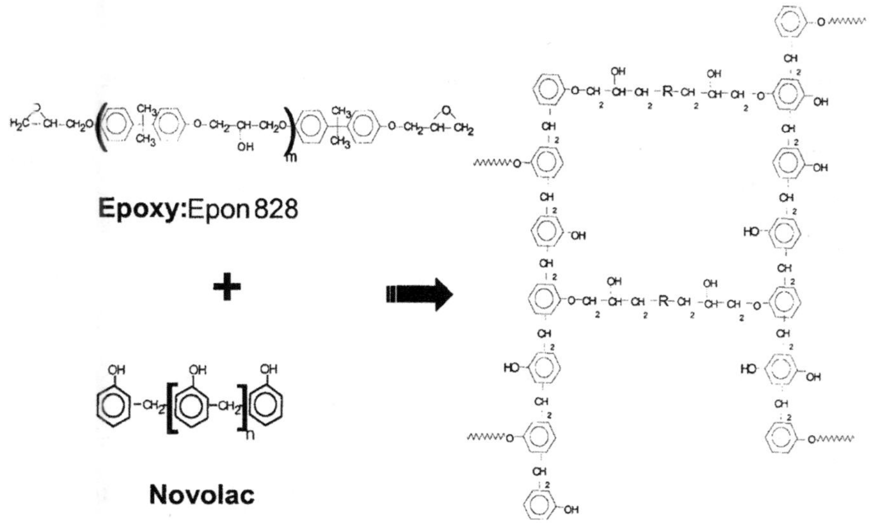

FIGURE 6.3 Schematic of the Epoxy–phenolic synthesis reaction. (Reproduced from Ref. [10] with permission from John Wiley and Sons.)

Bisphenol-type epoxies, in contrast, offer lower viscosity, which improves flow during molding and helps fill intricate mold geometries. They are preferred in fine-pitch or ultra-thin packages where complete cavity fill is critical.

Curing agents are essential for initiating the crosslinking reaction during thermal processing. Phenolic novolac curing agents are most commonly used due to their ability to create highly crosslinked, rigid networks, as shown in Figure 6.3, leading to excellent thermal and mechanical properties. In some EMCs targeting low-stress or flexible applications, anhydride curing agents may be used instead, offering reduced crosslink density and improved compliance. The cure kinetics and final properties are further modulated by the addition of catalysts such as imidazoles, tertiary amines, and phosphonium salts, typically in concentrations below 1 wt%. These accelerators control the gel time and ensure complete cure within industrial molding cycles.

Fillers, often comprising over 80% of the compound's mass, dictate much of the final EMC's performance. Fused silica (SiO_2) is the standard filler in most EMCs due to its low thermal expansion, high electrical insulation, and chemical inertness. However, in applications where thermal conductivity is critical—such as in power devices or RF modules—thermal fillers like aluminum oxide (Al_2O_3), boron nitride (BN), or aluminum nitride (AlN) are incorporated. These materials allow EMCs to achieve thermal conductivities ranging from 1.5 to 5 W/m·K depending on loading level and filler morphology.

In optical or imaging modules, specialized fillers are used to tailor optical properties. Titanium dioxide or carbon black may be added for light-blocking in CMOS image sensors, while optical clarity is preserved in transparent packages using nano-silica or refractive index-matched systems. In all cases, filler surface treatment with silane coupling agents is critical to improve dispersion, matrix adhesion, and moisture resistance.

TABLE 6.3
Overview of Typical EMC Formulation Components and Their Functions

Component	Materials	Function	Typical Loading (%)
Epoxy resin	Novolac epoxy, Bisphenol-A/F	Matrix phase, crosslinkable network	10–20
Curing agent	Phenolic novolac, anhydride	Network formation and thermal/mechanical tuning	5–10
Filler	Silica, Al_2O_3, BN, AlN	CTE control, stiffness, thermal conduction	70–85
Catalyst	Imidazole, phosphine, tertiary amine	Cure acceleration	<1
Additives	Silane agents, flow modifiers, mold release	Interfacial adhesion, rheology, demolding	<5

A simplified table summarizing typical EMC formulation components and their functions is shown in Table 6.3.

Each formulation must be finely balanced. Increasing the filler content lowers the CTE and improves the modulus, but also raises the melt viscosity and can lead to poor flow or voids in molding. Likewise, altering the resin-to-curing agent ratio affects the gel time, cure shrinkage, and final T_g. Additives such as flow modifiers or release agents improve cavity fill and mold demolding but may degrade electrical or moisture properties if not well optimized.

In the next section, we will explore how these formulation variables influence the key performance properties of EMCs—including thermal conductivity, glass transition temperature, fracture toughness, and dielectric reliability—and how trade-offs among them are managed to meet the diverse demands of advanced packaging applications.

6.2.3 Thermomechanical and Electrical Performance

The performance of EMCs in advanced electronic packaging is intimately tied to their internal structure, filler systems, and the intricate interplay between thermoset matrix design and particulate reinforcement. These materials must simultaneously meet stringent thermal, mechanical, electrical, and environmental requirements under increasingly aggressive operating conditions. A deep understanding of structure–property relationships is thus essential for tailoring EMCs to specific packaging platforms such as QFPs, FC-CSP, FOWLP, or 2.5D/3D architectures.

Thermal performance is a primary concern in most EMC formulations. For typical silica-filled systems, thermal conductivity values between 0.6 and 1.0 W/m·K are common, sufficient for low-power consumer electronics. However, power modules and RF devices demand higher dissipation capabilities, often exceeding 2 W/m·K. This is typically achieved through the incorporation of thermally conductive fillers like Al_2O_3, BN, or AlN. The geometry of these fillers—whether spherical, platelet-like, or rod-shaped—profoundly influences the formation of percolating heat conductive networks, as shown in Figure 6.4. The orientation and packing efficiency

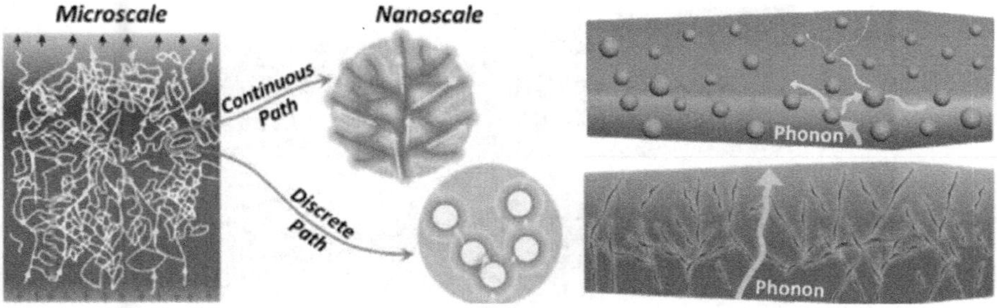

FIGURE 6.4 Schematic of the heat path through epoxy resin. (Reproduced from Ref. [11] with permission from Elsevier.)

further affect thermal transport; some advanced formulations use hybrid morphologies or magnetic alignment to guide filler arrangement during molding.

The CTE is another fundamental thermal parameter. A key design goal is to reduce the CTE mismatch between the EMC and the silicon die (CTE ~3 ppm/°C), which otherwise results in interfacial stresses during thermal excursions. By elevating the silica content to >80 wt%, CTE values as low as 10–15 ppm/°C can be achieved. Such formulations suppress warpage, enhance package flatness, and reduce mechanical strains on interconnects and solder joints.

Closely related to thermal integrity is the glass transition temperature (T_g), which marks the boundary between rigid and rubbery mechanical behaviors. Above T_g, the modulus of the EMC drops precipitously, leading to softening and dimensional instability. Therefore, T_g values exceeding 150°C are standard for high-performance EMCs, ensuring stability during Pb-free solder reflow processes (~260°C) and prolonged thermal aging. T_g can be adjusted by manipulating the epoxy backbone (e.g., novolac vs. bisphenol), curing agent functionality, and crosslink density. High-functionality novolac resins and phenolic hardeners form tightly crosslinked networks, contributing to both high T_g and dimensional stability.

In terms of mechanical behaviors, EMCs are required to offer a high elastic modulus, typically between 8 and 20 GPa, to provide structural support and resist package deformation. However, overly stiff matrices risk interfacial delamination and die cracking, particularly in thin-die or large-area packages. To mitigate this, the inclusion of rubbery modifiers or submicron silica particles is employed to boost the fracture toughness. Toughness is a critical metric under drop tests, temperature cycling, or board-level reliability assessments. Additionally, cure-induced shrinkage must be minimized to prevent void formation and residual stress buildup; this is often managed by balancing resin-to-filler ratios and controlling polymer conversion kinetics.

Electrical performance is also vital, especially in high-speed and high-frequency applications. EMCs typically exhibit high volume resistivity (>10^{14} Ω·cm), ensuring insulation between closely spaced features. Dielectric constant (D_k) values of 3.5–4.0 at 1 MHz are common and sufficient for most packages. However, for 5G, mmWave, and RF modules, materials with $D_k < 3.0$ and dissipation factor (D_f) < 0.005 are desirable. Achieving these targets often involves trade-offs with mechanical robustness, as low-D_k materials may involve air voids, porous structures, or low-polarity fluorinated systems that can weaken the matrix.

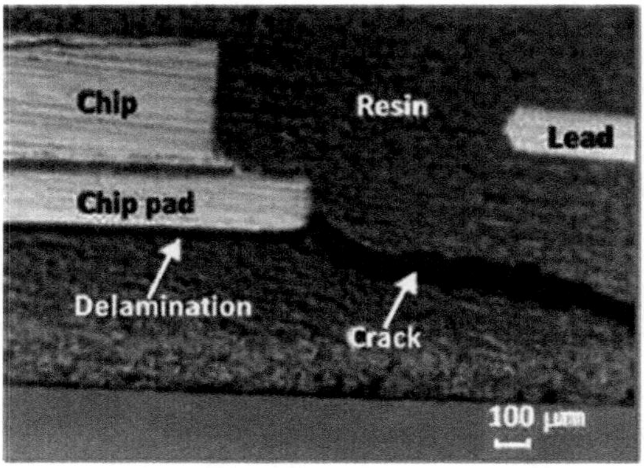

FIGURE 6.5 Delamination of chip pad from encapsulating epoxy resin. (Adapted from Ref. [12], © ASME. Used with permission.)

Moisture absorption and water vapor transmission rate (WVTR) directly affect both electrical insulation and mechanical reliability. In humid or condensing environments, absorbed moisture may cause ionic conduction, dielectric breakdown, or delamination during reflow, shown in Figure 6.5. As such, EMCs are engineered to exhibit water uptake below 0.2 wt%, using hydrophobic epoxy chemistries and silane-treated fillers to form water-resistant microstructures.

EMC design optimization involves managing the interdependencies among these properties. For instance, increasing the filler content enhances the thermal conductivity and lowers the CTE but simultaneously raises the viscosity, complicating flow during molding. Similarly, increasing the T_g through tighter crosslinking improves thermal and dimensional stability but may elevate residual stresses or reduce flexibility. Balancing these competing effects demands a system-level formulation approach, often supported by design of experiments (DoE) or finite element simulation.

As we transition into the next section, we will examine how these properties influence processing strategies, including molding techniques, flow control, and defect mitigation in high-density encapsulation schemes.

6.2.4 Molding and Processing Strategies

The successful implementation of EMCs in advanced packaging depends not only on materials formulation but also on the precise control of molding and processing methods. As device geometries shrink and package complexity grows, molding techniques must evolve to address challenges such as void minimization, warpage suppression, filler homogeneity, and fine-feature replication. The two principal molding techniques—transfer molding and compression molding—remain central to modern EMC processing, but their roles have shifted in response to new application domains and performance requirements.

As shown in Figure 6.6, transfer molding has been the workhorse technique for traditional leadframe-based packages, such as QFPs, SOPs, and BGAs. In this process, pre-weighed EMC pellets are heated in a transfer pot and then injected under

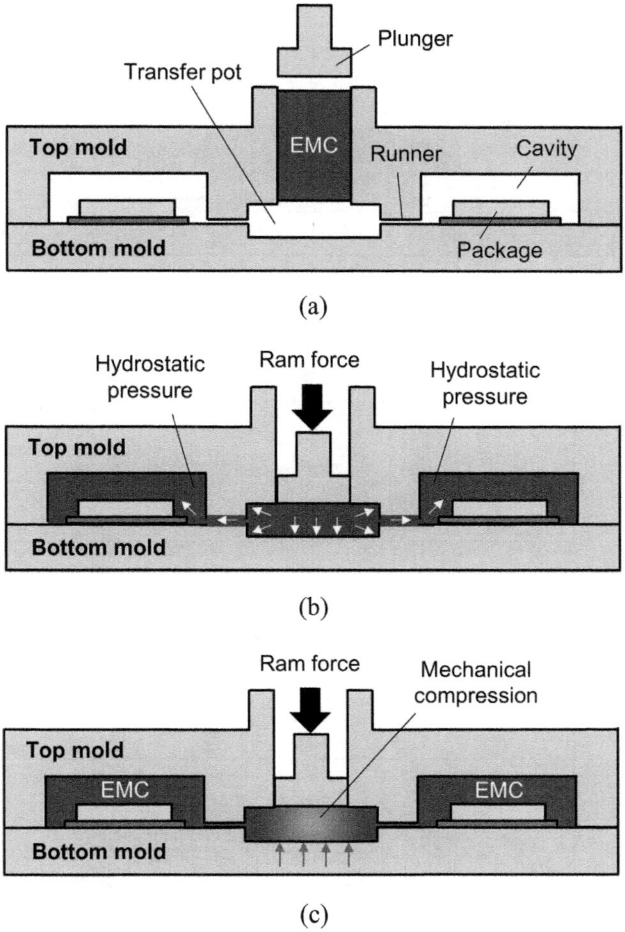

FIGURE 6.6 Schematic illustration of transfer molding process: (a) EMC is charged, and the mold is heated to the mold temperature; (b) EMC remains as a viscous liquid during EMC transfer (before EMC passes its gel point); and (c) EMC becomes a deformable solid as soon as EMC passes its gel point. (Reproduced from Ref. [13] with permission from Elsevier.)

high pressure (typically 7–15 MPa) into mold cavities containing the die, wire bonds, and substrate. The EMC must exhibit fast flow, low viscosity, and rapid cure kinetics to completely fill the cavity without entrapping air or causing die shift. Transfer molding is well-suited to high-throughput manufacturing, offering short cycle times and excellent dimensional consistency. However, the aggressive flow conditions can exert considerable mechanical stresses on delicate structures, especially for ultra-thin dies or packages with fragile interconnects. Additionally, the high-pressure flow may induce filler segregation or internal voids if rheology is not carefully tuned.

In contrast, compression molding has gained prominence in advanced wafer-level and panel-level packaging processes, including FOWLP (Fan-Out Wafer-Level Packaging), FO-CSP (Fan-Out Chip Scale Package), and PLP (Panel-Level

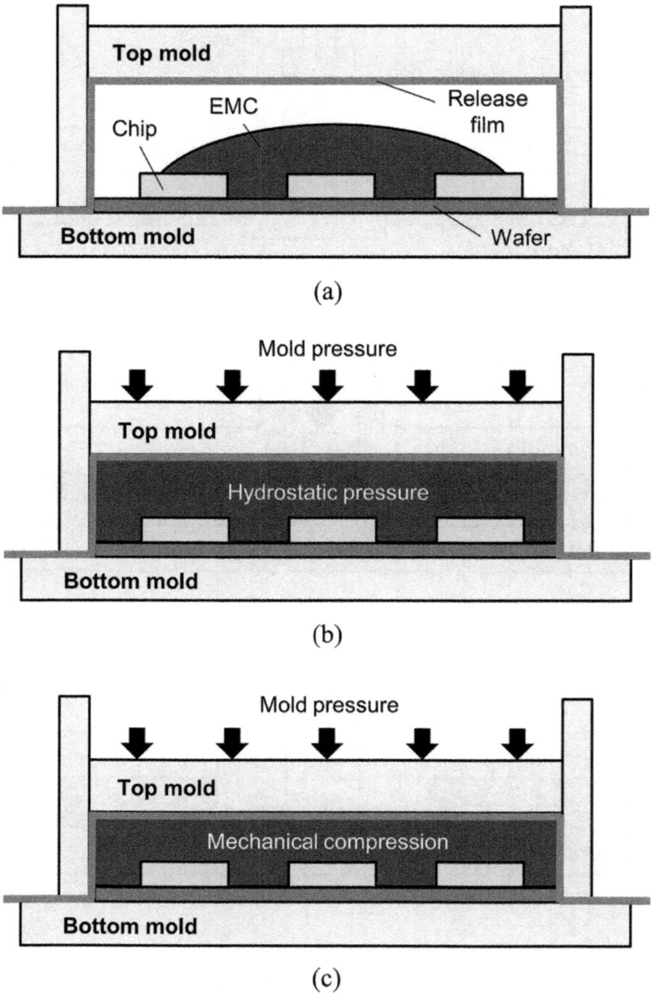

FIGURE 6.7 Schematic illustration of compression molding process: (a) EMC is charged, and the mold is heated to the mold temperature; (b) EMC remains as a viscous liquid during EMC transfer (before EMC passes its gel point); (c) EMC becomes a deformable solid as soon as EMC passes its gel point, and the pressure continuously deforms EMC. (Reproduced from Ref. [13] with permission from Elsevier.)

Packaging). The process is shown in Figure 6.7. In this approach, the EMC is introduced into the mold cavity in powder, paste, or sheet form and then compressed over the device using low-pressure lamination. This results in more uniform flow and reduced risk of void formation or bump damage. Compression molding is particularly advantageous for large-area encapsulation, where planarity and filler alignment are critical. Because it eliminates the plunger–runner–gate system of transfer molding, compression molding also allows for more flexible mold designs, better materials utilization, and simpler debonding from wafers or panels.

The choice of molding techniques influences the design of EMC flow characteristics. For transfer molding, the compound must remain sufficiently low in viscosity at processing temperatures (usually 160°C–180°C) to avoid premature gelation before full cavity fill. Therefore, formulators often optimize the balance between low-molecular-weight epoxy monomers for flow and high-functionality resins for post-cure rigidity. The use of flow modifiers, such as reactive diluents or surface-treated fillers, plays a critical role in achieving desired rheological profiles. For compression molding, the focus shifts toward uniform filler dispersion and compaction behavior under pressure. Here, spherical fillers with narrow particle size distributions are preferred to minimize sedimentation and enable predictable shrinkage during cure.

To accommodate increasingly complex packaging architectures, B-stage EMCs have emerged as a transformative solution. These materials are partially cured (pre-gelled) to a semi-solid or tacky state, enabling them to be pre-applied onto substrates, dies, or redistribution layers before final compression or reflow. This two-stage cure allows for greater alignment accuracy in chiplet integration, reduced bleed or flow-out during molding, and improved co-cure compatibility with underfills, die attach adhesives, or redistribution layers. B-stage EMCs are particularly useful in advanced 2.5D and 3D integration schemes, such as EMIB (Embedded Multi-die Interconnect Bridge) and chip-on-substrate modules, where sequential encapsulation is required.

Tooling and mold design also exert significant influence on EMC processability and final package quality. The mold geometry, runner length, gate location, and venting must be optimized to ensure complete filling, symmetric flow fronts, and minimal air entrapment. Advanced simulation tools such as finite element mold flow analysis (e.g., Autodesk Moldflow, Moldex3D) are commonly employed to evaluate cavity fill, pressure gradients, filler movement, and thermal gradients. These tools assist in predicting defects such as weld lines, voids, and residual stress concentrations, and help engineers fine-tune parameters such as mold temperature, plunger speed, and cure ramp.

Despite advances in materials design and simulation, practical processing challenges persist. Voids and delamination remain prevalent failure modes if the EMC flow path is poorly designed or trapped moisture is not properly vented. Filler sedimentation can lead to non-uniform mechanical or thermal properties across the package, particularly in horizontal mold orientations or during long cure cycles. To mitigate these risks, process engineers often employ vacuum-assisted molding, post-mold cure profiles, and filler surface treatments to maintain dispersion stability.

Table 6.4 compares key aspects of the three dominant EMC molding approaches.

In summary, the transition from conventional transfer molding to advanced compression and B-stage molding reflects the broader shift toward miniaturized, heterogeneous, and multi-functional packaging platforms. The synergy between EMC materials design and molding process development will continue to define the manufacturability, reliability, and scalability of next-generation semiconductor packaging.

In the next section, we will explore the reliability requirements, testing protocols, and failure mechanisms that must be addressed to ensure long-term performance of molded electronic packages.

TABLE 6.4
Overview of Key Aspects of Dominant EMC Molding Approaches

Molding Technique	Pressure and Flow	Applications	Advantages	Limitations
Transfer molding	High pressure, fast	Leadframe packages (QFP, SOP, BGA)	High throughput, consistent dimensions	Mechanical stress, filler segregation, voids
Compression molding	Low pressure, slow	WLP, PLP, FO-CSP, SiP	Low warpage, uniform filler, large-area coverage	Slower cycle time, mold cleaning requirements
B-stage EMC processing	Stepwise, partially cured	Chiplet, 2.5D/3D integration, EMIB	Precise alignment, reduced bleeding, multilayer processing	Complex formulation, handling sensitivity, cost

6.2.5 Reliability Risks and Mitigation

The reliability of EMCs is a critical determinant of overall package performance, particularly in advanced packaging architectures that demand high interconnect density, reduced form factors, and robust environmental endurance. Failures originating from the encapsulation layer not only threaten the mechanical integrity of the package but also compromise electrical function and thermal management. As such, a comprehensive understanding of failure modes, their root causes, and mitigation strategies are essential for both materials scientists and process engineers.

One of the most frequently observed failure mechanisms in EMC-encapsulated systems is interfacial delamination. This phenomenon typically occurs at the die-EMC, EMC-substrate, or EMC-metal trace interfaces due to cyclic thermal stresses induced by mismatched CTE. During power cycling or environmental stress tests such as thermal shock and temperature cycling, differential expansion between EMC (CTE ~10 to 15 ppm/°C) and silicon dies (CTE ~3 ppm/°C) accumulates stresses at the interface. Over time, especially under conditions of poor adhesion or moisture ingress, these stresses can lead to crack initiation and progressive delamination.

To combat delamination, both materials design and surface preparation are key. Silane coupling agents are widely employed to enhance the chemical bonding between inorganic fillers and the epoxy matrix, as well as to promote adhesion to metal traces and substrates [14]. Plasma treatments and surface roughening can also increase mechanical interlocking. For sensitive interfaces, low-modulus interlayers or compliant buffer zones may be introduced to relieve stress concentrations. The choice of the filler particle morphology and surface chemistry further influences the stress transfer efficiency across the interface.

Moisture-induced failures represent another major reliability concern. EMCs inherently absorb small amounts of moisture from the environment due to their polymeric nature. When a moisture-laden package is subjected to rapid temperature elevation, as during solder reflow, the absorbed moisture vaporizes, creating internal pressure. This can lead to the so-called "popcorn effect," where the package cracks or delaminates explosively. To mitigate this, modern EMCs are formulated with hydrophobic matrix systems and surface-treated fillers that reduce water uptake, typically targeting moisture absorption below 0.2 wt%. Moreover, packages are often subjected to dry baking prior to reflow to ensure moisture removal. Water vapor transmission rate is another critical metric for long-term reliability in humid environments, particularly in automotive or outdoor electronics.

Cracking is also a pervasive failure mode, especially in thin-die or large-area packaging. Cracks may originate from intrinsic material shrinkage during curing, rapid cooling after mold release, or mechanical handling. Formulations with a high-modulus or brittleness are more susceptible, particularly if the filler dispersion is poor or residual voids exist. Materials developers often adjust the crosslink density, add rubbery modifiers, or optimize filler morphology to improve the fracture toughness. Crack propagation paths can be predicted using finite element analysis to identify stress hotspots.

Warpage poses a subtler but equally damaging reliability challenge. Post-mold cure warpage may result in die tilt, underfill flow asymmetry, or solder joint fatigue during board-level assembly. Warpage is exacerbated by CTE mismatch between the EMC and the adjacent layers, as well as uneven cure kinetics and filler settlement. Advanced EMCs address this by adopting low-shrinkage formulations, symmetric layer stacking, and controlled filler size distribution. During process development, warpage is evaluated using laser profilometry, moiré interferometry, or warpage simulators.

Voids and filler segregation are additional contributors to reliability degradation. Entrapped air during molding or void formation from solvent outgassing can act as stress concentrators and sites for moisture accumulation. Filler sedimentation during molding leads to property gradients and localized weakening. These issues are typically addressed by using vacuum-assisted compression molding, optimizing the thixotropic behavior of EMCs, and ensuring uniform preheat profiles.

A suite of reliability evaluation protocols has been established to quantify EMC performance under various environmental and mechanical stressors:

- **Temperature Cycling Test (TCT)**: Subjects packages to thermal excursions (e.g., –55°C to 125°C) to evaluate fatigue and interfacial integrity.
- **Highly Accelerated Stress Test (HAST)**: Accelerates moisture-related failure under high temperature and humidity with applied pressure.
- **Test**: Similar to HAST but conducted at atmospheric pressure to assess hydrothermal stability.
- **Moisture-Induced Reliability Test (MIR)**: Combines moisture soak and solder reflow to assess susceptibility to popcorn cracking.
- **Acoustic Microscopy (SAM)**: Non-destructive imaging used to detect delamination, voids, and filler dispersion.
- **Finite Element Modeling (FEM)**: Used to simulate warpage, stress distribution, and potential crack propagation paths.

In summary, achieving high reliability in EMC-encapsulated packages requires an integrated approach encompassing formulation science, process control, surface engineering, and predictive simulation. As packaging technologies scale down to finer pitches and more complex interconnect schemes, the margin for error narrows, and the reliability performance of EMCs becomes even more critical. The next section will explore how emerging materials and sustainable formulation strategies are pushing the boundaries of EMC performance in response to new challenges in heterogeneous integration, high-frequency operation, and environmental regulation.

6.2.6 Advanced EMC Directions

As the semiconductor industry transitions toward heterogeneous integration, chiplet architectures, and increasingly demanding performance requirements across thermal, mechanical, and electrical domains, EMCs are undergoing a fundamental transformation. Traditional silica-filled epoxies, although mature and cost-effective, are often inadequate for emerging applications that demand ultra-low warpage, high thermal conductivity, fine-pitch compatibility, and environmental compliance. The next generation of EMCs must therefore embody multifunctional performance while maintaining processability, reliability, and cost-effectiveness.

One of the most urgent materials frontiers is the development of high thermal conductivity EMCs. In conventional EMCs, thermal conductivities typically lie between 0.6 and 1.0 W/m·K, sufficient for low-power logic and memory devices. However, in power devices, high-frequency RF components, and 3D-stacked packages, heat dissipation becomes a limiting factor for system stability and longevity. To address this, thermally conductive ceramic fillers such as Al_2O_3, BN, and AlN are incorporated in high volume fractions. These fillers not only boost the bulk conductivity (to values >3 W/m·K) but also help reduce local hot spots within the molding compound. Surface treatment of these particles with silanes or phosphates enhances matrix compatibility, reducing the interfacial thermal resistance. Hybrid filler systems—where a mix of particle sizes and aspect ratios (e.g., spherical Al_2O_3 and plate-like BN) is used—have shown promise in forming percolated thermal pathways without significantly increasing the viscosity.

Simultaneously, low-modulus or stress-relieved EMCs are gaining traction, particularly in fan-out wafer-level packaging (FOWLP), system-in-package (SiP), and thin-die assemblies where mechanical stresses must be minimized. These formulations typically introduce flexible epoxy linkages, internal plasticizers, or rubbery segments such as CTBN (carboxyl-terminated butadiene-acrylonitrile) to the resin network. By reducing the Young's modulus below 5 GPa, such EMCs help absorb thermal cycling strains and suppress die cracking or interfacial delamination. However, care must be taken to balance softness with dimensional stability, ensuring that the cured compound retains adequate shape fidelity and resistance to warpage or flow-induced distortion during post-mold processes.

Another major innovation lies in B-stage EMCs. These materials are designed to be partially cured to a tacky, semi-solid state that facilitates sequential processing in multi-die or chiplet packaging. In 2.5D/3D stack-ups or EMIB-like hybrid structures, B-stage EMCs enable intermediate die placement and alignment without premature flow or cure. The final full-cure step can then be triggered by a global thermal cycle. B-stage materials demand precise control over rheology, cure kinetics, and thermal behavior. Tailoring the onset of gelation, and achieving uniform tack strength across the material, are essential for alignment fidelity and void-free encapsulation. These EMCs are often used in conjunction with pre-applied underfills, wafer-level die-attach adhesives, or even TIMs, and thus must exhibit excellent chemical compatibility and minimal bleeding.

Wafer-level and panel-level packaging platforms have also driven the demand for ultra-flat, low-void EMC systems. Particularly in FOPLP and eWLB (embedded wafer-level ball grid array) technologies, where molding occurs over large surfaces and redistribution layers are formed atop the molded surface, planarity and filler uniformity are critical. Advanced EMCs for these applications incorporate spherical, tightly sieved fillers and thixotropy control agents that prevent sedimentation during compression molding. Moreover, vacuum-assisted or dual-stage compression processes are optimized in tandem with the EMC formulation to suppress void formation and ensure complete cavity fill. For high-resolution RDL patterning on the EMC surface, materials must also exhibit controlled surface roughness and good adhesion to subsequent polymeric dielectric layers.

From an electrical standpoint, the rise of 5G, mmWave, and high-speed computing applications has placed new emphasis on the dielectric properties of EMCs. Traditional EMCs possess D_k of 3.5–4.0 and D_f of around 0.01, which are marginally acceptable for sub-10 GHz operation. However, for millimeter-wave frequencies and radar modules, these values must be significantly lowered. Emerging formulations are exploring fluorinated epoxy matrices, porous hybrid structures, or nano-silica additives that can reduce D_k to below 3.0 and D_f to 0.003–0.005, while maintaining mechanical integrity. These low-D_k EMCs are being actively evaluated in base station components, high-speed optical transceivers, and antenna-integrated packages.

In parallel, the industry is responding to global environmental regulations and consumer pressure with the development of “green” EMCs. These include halogen-free, antimony-free, and low-VOC (volatile organic compound) formulations. The shift away from brominated epoxy systems has pushed materials scientists toward phosphorus- or nitrogen-based flame retardants. Furthermore, the adoption of water-based dispersants, recyclable packaging formats, and even bio-based fillers (e.g., cellulose-derived silica or starch-silicate hybrids) is beginning to emerge. Although these materials are not yet mainstream in high-performance electronics, they are being adopted in consumer-grade devices and automotive modules where sustainability metrics play a larger role.

As electronic packaging becomes an increasingly strategic differentiator in semiconductor performance, the materials that comprise it—especially EMCs—must evolve in tandem. Future success will depend not only on meeting thermal and mechanical benchmarks but also on enabling process innovation, environmental resilience, and functional integration. The convergence of polymer chemistry, filler engineering, and simulation-driven formulation is poised to deliver the next wave of encapsulation solutions for the semiconductor industry.

6.3 UNDERFILL MATERIALS

6.3.1 Role and Classification of Underfill Materials

Underfill materials are indispensable in modern flip-chip and heterogeneous integration packaging technologies. Their primary function is to fill the narrow gap between the semiconductor die and the substrate or interposer, shown in Figure 6.8, reinforcing the mechanical integrity of solder joints or micro-bumps while simultaneously enhancing thermal and moisture reliability. In high-density assemblies, where the pitch between interconnects can be less than 50 μm and the die sizes increasingly large and thin, the role of underfill becomes even more critical.

The mechanical reinforcement provided by underfill reduces strain concentrations during thermal cycling, especially at the interface between dissimilar materials with mismatched CTE. For example, in a typical FC-CSP structure, the CTE mismatch between the silicon die (~2.5 ppm/°C) and the BT substrate (~15 to 18 ppm/°C) can induce stresses that exceed the solder joint strength. The capillary underfill mitigates this by distributing stresses into the polymer matrix. Moreover, the underfill acts as a diffusion barrier, impeding the ingress of moisture and contaminants into vulnerable interfacial regions, which is particularly vital for mobile and automotive reliability.

Underfill materials are extensively used in various application scenarios: Flip-Chip Chip Scale Packages (FC-CSP), where they protect solder bump interconnects and improve board-level reliability. 2.5D integration structures, such as CoWoS and EMIB, where the underfill must simultaneously address both silicon-to-silicon and silicon-to-substrate interfaces, managing more complex CTE mismatches and planar variations. 3D stacked dies, where the underfill is used in die-to-die interfaces, filling narrow gaps between dies with fine-pitch micro-bumps or TSVs. In such configurations, the material must offer excellent flow through confined geometries while preventing void formation.

Based on their processing approaches (Figure 6.9) and performance characteristics, underfill materials can be broadly categorized as follows.

Capillary Flow Underfill (CUF): Applied after chip placement and solder reflow. It flows into the narrow gap through capillary action and cures thermally. It is widely used in conventional flip-chip assemblies due to its simplicity and effectiveness.

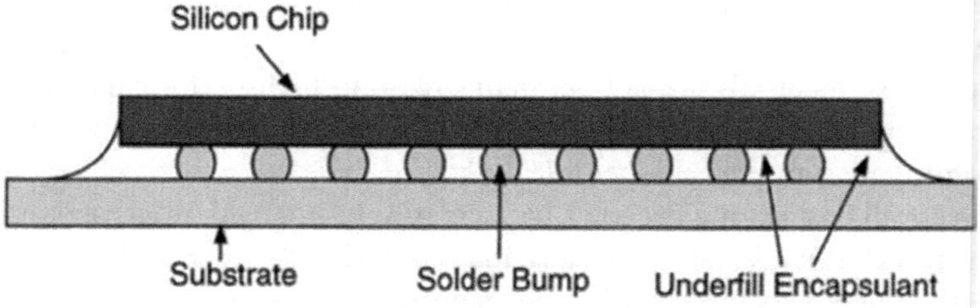

FIGURE 6.8 Illustration of flip chip packaging with underfill materials as encapsulant. (Reproduced from Ref. [15] with permission from Elsevier.)

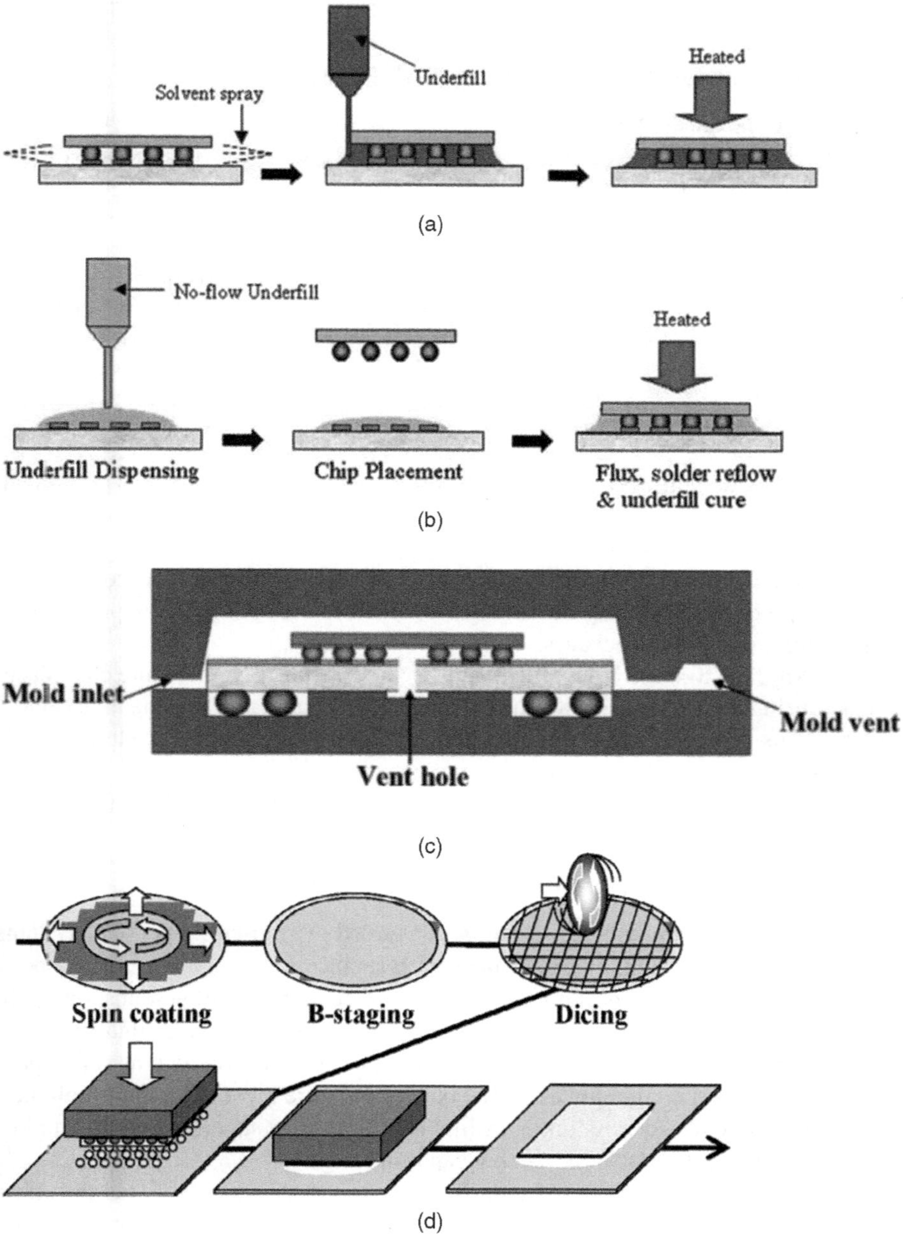

FIGURE 6.9 (a) Flip-chip process using capillary flow underfill. (b) Flip-chip process using no-flow underfill. (c) Design of flip-chip BGA with molded underfill. (Reproduced from Ref. [16] with permission from Springer Nature.) (d) Typical process flow of pre-applied underfill (wafer-level underfill) application. (Adapted from Ref. [17], © IEEE. Used with permission.)

TABLE 6.5
Overview of Key Aspects of Underfill Materials

Underfill Type	Process Flow	Advantages	Challenges
CUF	Post-reflow capillary flow + thermal cure	Good coverage, standard process	Voids in fine pitch, long cycle
NUF	Pre-applied + co-reflow	Simplified process, fast throughput	Limited flow, reflow compatibility
MUF	Mold encapsulation with filler flow	Low voids, high throughput	High filler sedimentation risk
Pre-applied/ B-stage	Pre-cured + post-assembly final cure	High alignment accuracy, fine pitch ready	Shelf life, process window control

No-Flow Underfill (NUF): Pre-applied to the substrate or die before placement, and co-cures with the solder paste during reflow. This streamlines assembly but requires precise viscosity and thermal compatibility.

Molded Underfill (MUF): Incorporated during the transfer or compression molding process, combining the roles of underfill and EMC. MUF is prevalent in wafer-level and fan-out packages.

Pre-Applied/B-stage Underfill: Deposited and partially cured before chip placement. During final curing, it transitions from solid to liquid and fills the under-bump region. This type is suitable for high *I/O* and multi-chip modules, including 2.5D interposer and 3D stacked structures.

An overview of different underfill types is shown in the following table (Table 6.5).

6.3.2 Materials Systems and Formulation Strategies

Underfill materials are typically formulated as filled thermosetting polymer systems designed for processability, mechanical performance, and reliability. The epoxy resin family remains the backbone of most commercial underfill systems, owing to its strong adhesion, dimensional stability, chemical resistance, and tunable curing behavior. Among the epoxy types, novolac and DGEBA (diglycidyl ether of bisphenol A) are the most commonly used. Novolac epoxies are favored for their high thermal stability and dense crosslinking, while DGEBA offers superior wetting and low viscosity, making it suitable for narrow-gap infiltration.

The curing agents employed in underfill systems include phenolic resins and anhydrides, often used in combination with accelerators such as imidazoles or substituted triazines. These choices influence both the cure profile and the final network properties such as T_g, modulus, and chemical resistance. In advanced underfill formulations, dual-curing strategies (e.g., UV/thermal, B-stage/thermal) are implemented to tailor the interface development and reduce residual stresses.

Fillers play a vital role in tuning the CTE and mechanical properties of underfills. SiO_2 is the predominant filler due to its low CTE, good dielectric properties, and inertness. However, in high-power or thermally demanding applications, alternative

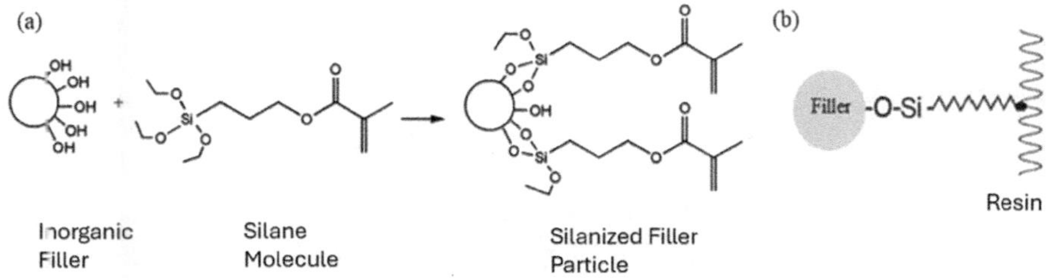

FIGURE 6.10 (a) Schematic of surface modification of inorganic filler with silane molecule. (b) silanes treatment mechanism illustrated one site of reactions. (Adapted from Ref. [18] with permission from SAGE Publications.)

fillers like Al_2O_3, BN, and AlN are used to enhance thermal conductivity. For example, BN platelets can be aligned to enable in-plane heat dissipation while maintaining electrical insulation. To balance flowability with thermal performance, mixed micro/nano-filler strategies are adopted, where nanoparticles improve packing and filler dispersion while reducing viscosity spikes.

Dispersant and interface design is equally important. Coupling agents such as (3-mercaptopropyl)trimethoxysilane (MPTS) (Figure 6.10) enhance filler–matrix adhesion, improving mechanical performance and reducing filler sedimentation during flow. Surface-treated fillers or organically modified particles further contribute to rheological control and interfacial energy optimization. Additives including wetting enhancers, flow modifiers, and adhesion promoters are tailored to application-specific needs—such as minimizing void formation in capillary underfill or improving co-curability in NUF systems.

A typical underfill formulation may contain 10–25 wt% resin matrix, 70–85 wt% filler, and <5 wt% additive/catalyst system, though the exact ratios are adjusted based on flow requirements, reliability targets, and processing methods (CUF, NUF, etc.).

6.3.3 Critical Properties and Structure–Performance Relationships

The performance of underfill materials in flip-chip and advanced packaging critically depends on a range of thermomechanical, rheological, interfacial, and dielectric properties. These properties are not only dictated by materials composition but are also highly influenced by the filler morphology, matrix crosslinking density, and interface design strategies. A systematic understanding of how the structure governs the function is essential for tailoring underfill formulations that can meet the stringent demands of fine-pitch, high-density, and thermally demanding packaging architectures.

Flow Behavior and Wettability: Capillary flow underfills must exhibit low initial viscosity and excellent wetting behavior to completely fill narrow gaps (typically <50 μm) between bumps without forming voids or air entrapment. The rheological behavior is influenced by the resin viscosity, filler volume fraction, particle shape, and surface interactions. Wettability is often characterized by contact angle measurements; underfills with contact angles below 30° on metallization surfaces

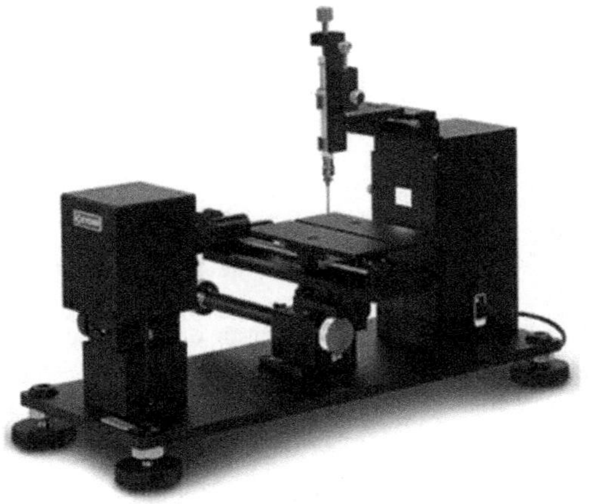

FIGURE 6.11 The contact angle instrument and representative image. (Reproduced from Ref. [19] with permission from Elsevier.)

(e.g., Cu, UBM) tend to show better gap-filling efficiency. Figure 6.11 depicts a device used to measure the contact angle.

Surface tension also plays a critical role, as it influences the capillary pressure driving the flow. Enhancing capillary action requires tuning the surface chemistry and controlling filler dispersion to avoid yield stresses that hinder flow.

Cure Behavior and Shrinkage Stress: During curing, epoxy networks undergo volumetric shrinkage, typically between 1.5% and 3%. This can generate residual stresses, particularly in systems with high crosslink density or significant thermal gradients during cure. Differential scanning calorimetry (DSC) and dynamic mechanical analysis (DMA) are commonly used to monitor the curing kinetics. For advanced packaging applications, formulations are generally designed to withstand elevated operating temperatures; consequently, they often require a cure or post-cure schedule above ~150°C, a temperature region at which many aromatic epoxy/phenolic systems begin to show accelerated reaction kinetics and can achieve the crosslink density needed for high-T_g performance. However, increasing the cure temperature also accelerates modulus build-up and amplifies internal stress, requiring careful optimization of the cure profile.

To mitigate shrinkage-induced stress, several strategies are adopted:

- Use of toughened epoxy systems with flexible segments.
- Dual-cure mechanisms (e.g., B-stage + thermal) to control gelation front progression.
- Fillers that act as stress-dissipating scaffolds.

Thermo-Mechanical Performance: Mechanical properties such as the storage modulus, toughness, and shear strength are critical to withstand thermal cycling and mechanical shock. A well-designed underfill must offer a high modulus at room temperature to support mechanical loads, but a reduced modulus at high temperatures to

TABLE 6.6
Key Target Properties of Underfill Materials for Flip-Chip Packaging

Property	Target Value/Range	Function
Viscosity (25°C)	<5,000 mPa·s (CUF)	Enables capillary flow
Contact angle (Cu)	<30°	Promotes wetting and complete gap filling
Glass transition (T_g)	>130°C–150°C	Stability during thermal cycling
CTE (post-cure)	20–30 ppm/°C	Minimizes thermal mismatch with silicon
Shear strength	>30 MPa	Maintains mechanical integrity during handling
Shrinkage	<2.5%	Reduces residual stress after curing

absorb stresses and prevent delamination. This thermomechanical balance is often analyzed using DMA over a wide temperature range.

The mismatch in CTE between the underfill and the adjacent materials (Si, Cu, substrate) drives interfacial stresses under temperature variations. The target CTE of underfill is typically in the range of 20–30 ppm/°C. Adding high-loadings of low-CTE fillers (e.g., fused silica) is the most direct route to reduce the CTE, but it also raises the viscosity.

Table 6.6 summarizes the critical materials properties and performance targets that underfill formulations must achieve to ensure reliable assembly and long-term stability in flip-chip packages.

Dielectric and Moisture Performance: As packaging moves toward high-frequency and high-speed signal environments, the electrical properties of underfill materials gain increased importance. The D_k should be kept low (typically <4.0 at 1 MHz), and the D_f should be minimized to ensure signal integrity. The use of low-polarity resin systems and surface-treated fillers can suppress dielectric losses, as shown in Figure 6.12.

Moisture resistance is equally important. Underfill materials are often the first line of defense against moisture ingress through the gap region. Excess moisture can lead to corrosion, delamination, or popcorning during reflow. Therefore, a low water absorption (<0.5 wt%) and a low water vapor transmission rate are essential. Hydrophobic resins, silane-treated fillers, and dense crosslink networks all contribute to moisture resistance.

Interfacial Integrity and Adhesion Control: Underfill must adhere strongly to multiple surfaces, including Si, Cu, UBM, passivation layers, and substrates such as BT or glass. Poor adhesion leads to interfacial delamination, which is a dominant failure mechanism under thermal cycling or drop impact. Strategies to enhance adhesion include using silane coupling agents (e.g., APTES and MPTS) tailored to surface chemistry, plasma or UV treatment of bonding surfaces prior to dispensing and interfacial nanofillers to bridge materials mismatches via mechanical interlocking or chemical compatibility.

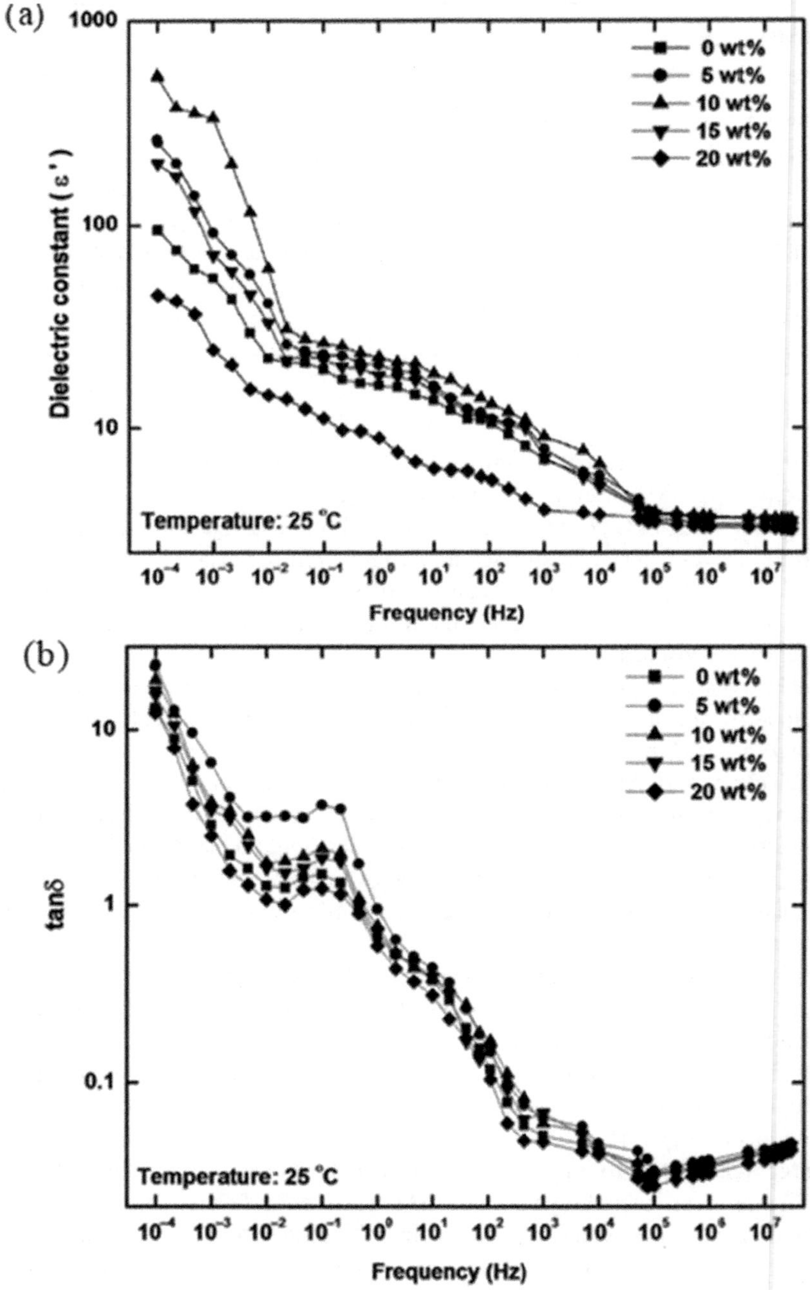

FIGURE 6.12 (a) Effect of frequency on dielectric constant for epoxy-silica nanocomposites. (b) Effect of frequency on tan δ for epoxy-silica nanocomposites. (Reproduced from Ref. [20] with permission from Springer Nature.)

Atomic force microscopy (AFM) and XPS are frequently used to evaluate surface energy and chemical bonding at interfaces. FEM simulations are also employed to identify stress hotspots and design interfacial grading strategies.

In summary, underfill performance is a complex function of flow, cure, mechanics, adhesion, and environmental durability. A formulation must be optimized holistically to meet the multi-physics challenges of flip-chip, 2.5D, and 3D packaging.

6.3.4 Processing Considerations

The processing of underfill materials is as critical as their formulation. The success of an underfill system is not only determined by its bulk properties but also by its behavior during dispensing, flow, curing, and post-process reliability.

In CUF applications, the underfill is dispensed after the flip-chip assembly and solder reflow. The capillary-driven flow fills the gap between the die and the substrate. Achieving uniform filling without voids is highly dependent on the following: Dispensing parameters: Flow rate, needle diameter, and travel speed must be optimized. Surface energy matching: Clean and chemically treated surfaces promote better flow. Die size and bump pitch: Narrower gaps increase flow resistance, necessitating low viscosity (< 5,000 mPa·s).

The fill time is governed by Washburn's equation, where x_f is the axial flow length, h is the flow parameter; σ is the liquid surface tension; θ_0 and θ_s are the contact angles at the lower and upper channel surfaces; s is the channel spacing; and μ_{eff} is the effective viscosity, with flow rate inversely proportional to the viscosity and bump gap width. In fine-pitch packages (<40 μm), filler sedimentation or incomplete wetting can lead to voids. Vacuum assist or post-fill underfill flow simulations (e.g., CFD modeling) are often used to optimize fill paths (Figure 6.13).

NUF and B-stage underfills are applied before the die placement. In NUF, the resin is screen-printed or coated on the substrate, and the die is then mounted before reflow. The material must maintain dimensional stability during placement and co-cure during reflow without interfering with solder joint formation. Key considerations include the following: Co-curability with solder paste: Resin must not hinder solder joint wetting or introduce voids. Thermal profile alignment: Cure kinetics of the underfill must synchronize with the solder reflow peaks (e.g., 220°C–250°C). Shelf life and storage stability: Partially cured B-stage materials must avoid premature gelation or moisture uptake.

$$\frac{\mathrm{d}}{dt}(x_f)^2 = b = \frac{\sigma}{6}(cos\theta_0 + cos\theta_s)\frac{s}{\mu_{eff}}$$

FIGURE 6.13 Washburn's equation. (Adapted from Ref. [21], © IEEE. Used with permission.)

For B-stage underfills, partial curing (gelation) is performed before chip placement. Final curing then completes adhesion and stress relaxation. These systems are highly attractive for high-volume manufacturing and fan-out panels but require stringent process control.

In multichip modules or interposer-based 2.5D/3D structures, underfill must flow through complex geometries, including die-to-substrate and die-to-die gaps. Major challenges include the following: Fill path obstruction: Routing congestion and TSV structures increase resistance to flow. Sedimentation: Larger fillers settle during prolonged flow, leading to non-uniform CTE. Void entrapment in die corners or cavity pockets: Design of die spacing and venting paths is critical.

Dam and fill methods, wherein a high-viscosity material (dam) is dispensed around the periphery before a lower-viscosity fill is applied, can help contain flow and improve reliability in such complex structures.

To ensure strong adhesion and uniform coverage, underfills must be compatible with a variety of surfaces, including copper, UBM, dielectric layers, organic substrates, and increasingly, glass interposers. Effective surface preparation may include the following: Plasma activation or UV ozone cleaning to increase surface energy. Silane primer application to promote chemical bonding. Thermal pre-baking to eliminate surface moisture and reduce voids.

For example, glass interposers require tailored silane treatments due to the low surface roughness and chemical inertness of glass, while copper requires oxide control to prevent delamination under stress.

The following table outlines the key processing steps, critical parameters, and practical notes for dispensing, curing, and surface preparation of underfill systems used in flip-chip packaging (Table 6.7).

Altogether, processing of underfills must be tightly integrated into the assembly flow. Yield, performance, and long-term reliability are highly sensitive to deviations in viscosity, flow path control, interface conditions, and curing uniformity. As package architectures become increasingly dense and heterogeneous, predictive process simulation and AI-driven control systems are expected to play growing roles in underfill process optimization.

TABLE 6.7
Representative Processing Steps and Parameters for Underfill Materials

Processing Step	Key Parameters	Notes
Dispensing (CUF)	Flow rate, path, standoff gap	Optimized for uniform capillary action
Reflow profile (NUF)	Ramp rate, peak temperature, time above liquidus	Aligned with joint formation and cure
B-stage curing	Partial gel temperature/time (e.g., 80°C–120°C)	Controls shelf life and placement behavior
Final cure	130°C–160°C typical, 30–60 minutes	Ensures full crosslinking and stress relief
Surface prep	Plasma, UV, silane treatment	Tailored to UBM, Si, glass, or BT substrates

6.3.5 Reliability and Failure Mechanisms

Reliability of underfill materials is a central concern in flip-chip, 2.5D, and 3D integration technologies. The underfill must maintain structural and interfacial integrity over extended lifetimes, often under harsh thermal, mechanical, and environmental stresses. Failure to do so can lead to cracking, delamination, or electrical failure, compromising the entire package.

Among the most frequently encountered failure modes, interfacial delamination remains dominant. Poor adhesion between the underfill and the adjacent surfaces—such as die passivation layers, UBM, or organic substrates—can result in progressive separation under thermal or mechanical cycling. This process is exacerbated by mismatched moduli between the materials, or by moisture uptake at the interface, which weakens bonding. Closely related is the issue of underfill cracking, which is common in brittle, high-modulus systems that lack sufficient fracture toughness. In scenarios involving low-k dielectric layers or ultra-thin dies, cracks originating from the underfill can readily propagate into the functional die stack.

Void formation is another critical concern. During underfill dispensing and capillary flow, entrapped air or insufficient wetting can generate voids within the material matrix. These voids act as localized stress concentrators and contribute to early failures during mechanical shock or thermal cycling. Figure 6.14 shows a simulation of a failure due to a void. In packages involving large dies or complex 3D structures, filler sedimentation during the underfill flow can also create local inhomogeneity, resulting in mechanical weaknesses and localized CTE mismatches. If unchecked, such filler migration can contribute to stress concentrations that result in bump fracture or warping of the entire package.

To assess and mitigate these risks, a wide range of reliability testing protocols are employed. Thermal Cycling Tests assess fatigue resistance under repeated thermal excursions, while Highly Accelerated Stress Tests (uHAST) and Autoclave Tests

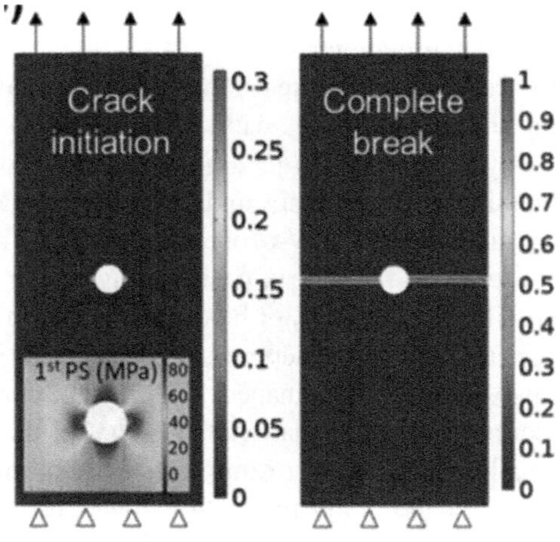

FIGURE 6.14 Simulation results of a failure due to a void. (Reproduced from Ref. [22] with permission from Elsevier.)

TABLE 6.8
Reliability and Characterization Tests to Evaluate the Performance of Underfill Materials

Test Type	Purpose	Conditions
Thermal Cycling Test (TCT)	Evaluates fatigue resistance to temperature swings	−55°C to 125°C, 1,000–2,000 cycles
Highly Accelerated Stress Test (uHAST)	Tests moisture resistance under pressure	130°C, 85% RH, 2 atm, 96–168 hours
Autoclave Test	Evaluates moisture ingress and popcorning	121°C, 100% RH, 2 atm
Drop Test	Tests mechanical shock and adhesion integrity	1,500 G, 0.5 ms, JEDEC standard drop tests
Die Shear Test	Assess interfacial bond strength	Measured in MPa, performed post-aging
X-ray/μ-CT Imaging	Detects voids, delamination, and filler agglomeration	Non-destructive, high-resolution failure analysis

focus on moisture ingress and environmental degradation under elevated humidity and temperatures. Mechanical robustness is evaluated using JEDEC-standard drop tests, while the interfacial bonding strength is quantified through post-aging die shear testing. Non-destructive inspection methods such as X-ray and micro-CT imaging are increasingly adopted to detect voids, delamination fronts, and filler agglomeration with high resolution. The following table summarizes representative reliability and characterization tests used to evaluate the performance of underfill materials, covering thermal, moisture, mechanical, and interfacial failure modes (Table 6.8).

Complementing physical testing, finite element modeling (FEM) serves as a powerful tool to predict stress accumulation and guide materials design. By simulating packaging structures under thermal, mechanical, or humidity loading, engineers can identify potential stress hotspots and optimize the filler content, resin modulus, or cure profiles to minimize delamination risk.

Enhancing underfill reliability requires a multidimensional design approach. Interfacial engineering plays a vital role, with silane coupling agents, such as 3-mercaptopropyltrimethoxysilane (MPTS) or aminopropyltriethoxysilane (APTES), providing chemical bonding between resin and inorganic surfaces. Surface treatments including plasma activation or UV-ozone exposure can improve wettability and bonding quality. In terms of the matrix design, toughening agents like rubber-modified epoxies, nano-silica dispersions, or block copolymer micelles are incorporated to enhance fracture toughness and suppress crack propagation.

Cure profiles must also be carefully managed. A dual-stage cure strategy, in which partial crosslinking occurs before full gelation, helps control the volumetric shrinkage and alleviates internal stresses. Slower ramp rates and optimized peak temperatures reduce the potential for thermally induced warpage, especially in fan-out or interposer-based packages. Meanwhile, intelligent filler design—including surface modification, size control, and nano/micro hybridization—enables improved dispersion, reduced sedimentation, and better mechanical balance.

From a manufacturing standpoint, package layout and flow path design must minimize obstacles and enable uniform underfill distribution. In 2.5D packages with silicon or glass interposers, flow resistance is often higher due to TSVs or die alignment complexity. Introducing auxiliary fill paths, vent channels, or vacuum-assisted dispensing methods can significantly improve yield and reliability.

Overall, the underfill is far more than a passive structural adhesive; it is a performance-critical material that must be engineered for multiphysics robustness. Reliable flip-chip and 3D packaging outcomes depend on tight control of interfacial bonding, stress management, thermal and moisture protection, and filler stability. Future innovations will continue to push the envelope, leveraging integrated simulation, smart materials, and adaptive processing to ensure consistent performance across diverse packaging architectures.

6.3.6 Special Topics: Dam and Fill, and High-Performance Underfill

With the continued advancement of heterogeneous integration, there has been a growing demand for underfill systems tailored to highly customized process flows and application-specific constraints. Two of the most significant evolutions in this context are the emergence of dam-and-fill underfill strategies (Figure 6.15) and the development of high-performance, multifunctional underfills suitable for extreme operating environments or sub-40 μm pitch applications.

The dam-and-fill method, originally developed to overcome challenges in large die and high bump-density packaging, relies on the sequential dispensing of two different underfill materials. A high-viscosity "dam" material is first deposited around the die perimeter to act as a containment barrier. This is followed by the dispensing of a lower-viscosity "fill" material that flows within the bounded region, effectively filling the gap between the die and the substrate. This method is especially advantageous in high-bandwidth memory (HBM) packages, 2.5D silicon interposer designs, and large-chiplet assemblies where the risk of overflow, voiding, or filler segregation is elevated. The dam material must possess sufficient thixotropy and dimensional stability, while the fill material must retain a low viscosity and high wetting capability, even in narrow gaps. Both components must be chemically compatible to co-cure without phase separation or interfacial delamination.

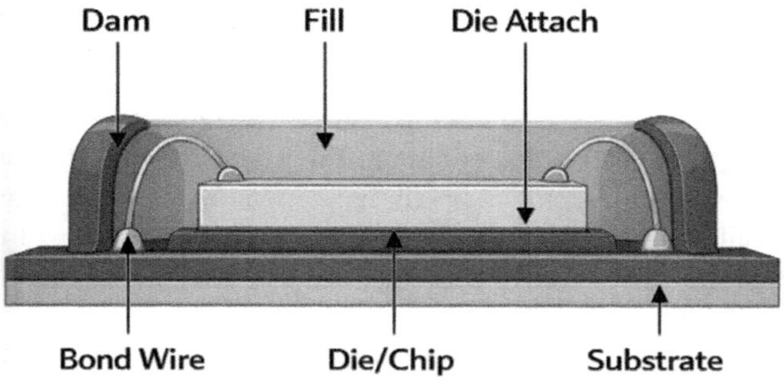

FIGURE 6.15 Schematic of dam-and-fill underfill strategies.

In parallel, the field has seen a surge in high-performance underfills designed to address increasingly stringent thermal, mechanical, and processing constraints. Among these, thermally conductive underfills have drawn significant interest for use in power devices, high-performance computing, and optoelectronic modules. These systems integrate fillers such as h-BN, AlN, or thermally modified silica to achieve high in-plane or through-thickness thermal conductivities. However, the trade-off between filler loading and flowability is a critical design challenge. Excessive filler volume fraction increases viscosity and risks sedimentation, especially during long flow paths. To mitigate this, hybrid strategies involving bimodal particle size distributions, surface-treated fillers with steric stabilization, or incorporation of anisotropic alignment methods have been explored.

Another frontier lies in the miniaturization of underfills for sub-40 μm bump pitch applications. As the die interconnect density increases, the standoff height between the die and the substrate decreases, necessitating ultralow-viscosity underfill materials capable of rapid, void-free flow without bump damage. Nanofilled epoxy systems, fluorinated or siloxane-modified matrices, and new rheology modifiers, have been developed to maintain Newtonian or shear-thinning behavior during dispensing. In such applications, precise control over capillary action, wetting dynamics, and air escape pathways become essential. Spray coating and inkjet-based dispensing systems, which allow for high-resolution and programmable materials deposition, are also being considered as alternatives to conventional needle dispensing.

These emerging trends also intersect with broader industry drivers such as environmental sustainability and functional integration. Green underfill systems with low volatile organic compounds (VOC) and halogen-free chemistries are being actively developed to comply with global regulatory and environmental mandates. In parallel, multifunctional underfills that go beyond mechanical and thermal roles are being researched. These include materials capable of self-healing microcracks, sensing interface stresses, or even adapting thermal conductivity based on local temperature gradients [23].

The convergence of these special topics signals a paradigm shift in underfill development from passive, support-oriented material to an active contributor in system performance, reliability, and intelligence. As fan-out panel-level packaging (FOPLP), glass-core substrates, and chiplet-based architectures become more prevalent, underfill materials will need to be increasingly customizable, process-compatible, and data-driven in their optimization.

6.3.7 Outlook and Future Trends

The future of underfill materials is being reshaped by the convergence of advanced packaging demands, evolving device architectures, and growing environmental and functional expectations. As chiplet-based integration, system-in-package (SiP), and 3D stacking gain momentum in the semiconductor industry, underfill technologies are facing a pivotal transition from standardized formulations to highly tailored, multifunctional systems.

One of the most pressing challenges is the increasing complexity and density of heterogeneous integration. In advanced 2.5D packages like CoWoS and EMIB, as well as 3D structures incorporating hybrid bonding and TSV-based interconnects, the number of chiplets and interconnect interfaces is rapidly rising. This trend results in smaller bump pitches, tighter standoff heights, and reduced underfill volume per interface. Under these conditions, traditional capillary flow dynamics become increasingly inadequate, prompting a shift toward B-stageable or pre-applied underfill strategies. These materials must support automated high-throughput assembly while ensuring uniform interface coverage and minimal void formation.

Alongside structural complexity, the thermal management requirements in modern electronics are becoming significantly more demanding. With power densities in high-performance computing and AI accelerators reaching several hundred W/cm^2, underfill materials must contribute to heat dissipation without compromising mechanical reliability. This calls for a new generation of thermally conductive underfills that achieve a high conductivity at low filler loadings, often through nano-engineered or anisotropically aligned filler networks. These materials must also maintain low dielectric loss for signal integrity, particularly in radio-frequency (RF) or optical interconnect environments.

In terms of reliability, the push toward thinner dies, organic interposers, and glass substrates introduce novel interfacial and mechanical challenges. The modulus mismatch between the underfill and emerging materials (such as glass or polyimide) necessitates a more nuanced approach to stress distribution. As a result, adaptive cure kinetics, gradient modulus designs, and local toughening mechanisms are being investigated. Finite element simulations integrated with real-time thermal-mechanical characterization are expected to play a critical role in co-optimizing materials and assembly parameters, as shown in Figure 6.16.

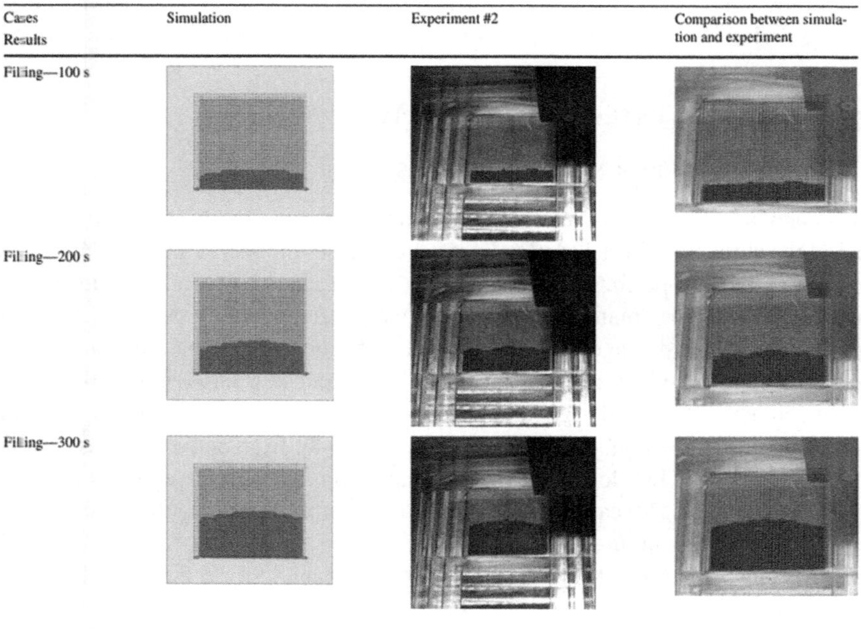

FIGURE 6.16 Comparison between simulation and experimental results of the capillary underfill process. (Reproduced from Ref. [24] with permission from Springer Nature.)

Digitalization and intelligent process control will also shape the future landscape of underfill design. AI-assisted flow path simulation, machine vision–guided dispensing, and machine learning models trained on historical reliability data, are enabling more predictive and efficient underfill selection and placement strategies. In this context, the definition of "high-performance underfill" is shifting from static material properties to dynamic, system-aware adaptability.

Environmental and sustainability concerns are accelerating the transition to green chemistries in electronic materials. Underfill systems free from halogenated components, low in VOCs, and compatible with recycling workflows are becoming increasingly attractive for global compliance and circular economy considerations. At the same time, functional integration into underfill materials is gaining traction. Early-stage research into self-healing polymers, thermochromic indicators, and stress-sensing additives is paving the way for underfills that can monitor, adapt to, or even respond to degradation over time.

Finally, as advanced packaging enters the panel-level and glass-core era, scalability becomes critical. Underfills must be compatible with larger form factors, lower-cost processing, and thinner substrate technologies. This will require materials that cure faster, exhibit minimal shrinkage, and resist warpage during high-throughput panel processing, all without sacrificing interface reliability and reworkability.

In summary, the next generation of underfill materials will need to strike a balance between processing agility, multi-functional performance, and predictive reliability. As packages become denser, more thermally stressed, and structurally heterogeneous, underfills will evolve from passive gap fillers into active enablers of system performance, lifetime extension, and real-time diagnostics. The future of underfill is not only about filling the space beneath the die, it is rather about bridging the gap between reliability, functionality, and sustainability in semiconductor integration.

6.4 THERMAL INTERFACE MATERIALS

6.4.1 Heat Dissipation Roles and Types

TIMs play a pivotal role in modern electronics packaging by mitigating the thermal resistance at the interfaces between heat-generating components and heat-dissipating structures. In high-performance computing, power electronics, and advanced communication systems, managing thermal energy effectively is essential to ensure device functionality, longevity, and reliability. At the micro-scale, die surfaces and heat spreaders or lids exhibit surface roughness and non-uniformities that hinder direct contact. TIMs are designed to fill these microscopic air gaps and improve thermal contact, thereby enabling efficient heat flow away from active devices. Without a well-optimized TIM, localized overheating—commonly referred to as hot spots, shown in Figure 6.17—can degrade device reliability, reduce performance, or even lead to catastrophic failures.

TIMs must simultaneously offer high thermal conductivity and sufficient mechanical compliance to maintain intimate contact over the lifecycle of the device. This includes accommodating thermomechanical stresses due to mismatched CTE, enduring thermal cycling, and resisting degradation from oxidation, moisture ingress, and

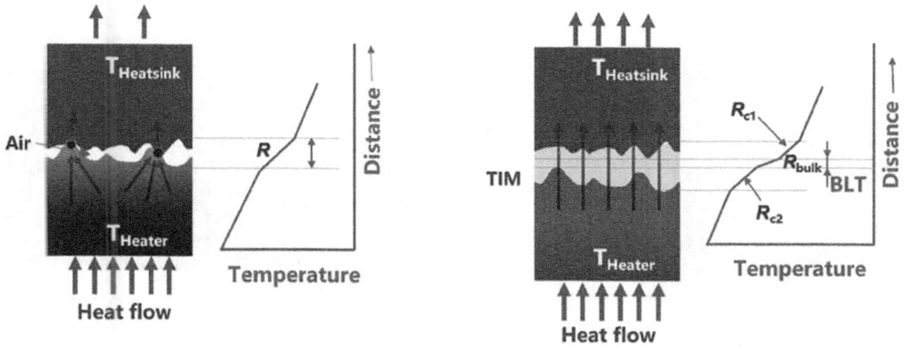

FIGURE 6.17 Heat accumulation differences between an electronic device without TIM (a) and with TIM (b). (Reproduced from Ref. [25] with permission from John Wiley and Sons.)

long-term heat exposure. A well-designed TIM must preserve interfacial conformity under both static mechanical loading and dynamic operational conditions, such as repeated heating-cooling cycles, vibration, and external mechanical shock.

The landscape of TIMs is broad, and materials are generally classified based on their phase (solid, liquid, gel), functionality (permanent or reversible), and application location within the package. Solid-state TIMs, including cured epoxies, pads, or thermally conductive greases, offer stable performance over long durations and are widely used in CPU/GPU lid attach and power electronics modules. These materials often combine a high filler content with minimal fluidity, ensuring mechanical stability and reproducibility. Liquid TIMs, such as silicone oils or low-melting-point alloys like gallium-based liquid metals, provide superior conformability to surface asperities but present challenges such as potential migration, oxidation, and pump-out under shear or vibration.

Phase Change Materials (PCMs) offer an adaptive solution, transitioning from a solid to a viscous or liquid state at the operational temperature. Upon heating, the PCM softens to improve wetting and interface contact; upon cooling, it resolidifies to retain position and interface conformity. This reversible phase transition enables low contact resistance and easy reusability, making PCMs highly attractive in reworkable systems and dynamic thermal conditions. However, the potential for phase separation, outgassing, and pump-out remains a concern.

Gelled TIMs, often composed of soft crosslinked silicone matrices embedded with high-conductivity fillers, offer a balanced solution between solid and liquid materials. They combine mechanical compliance, low modulus, and good reworkability, making them particularly well-suited for fragile or stress-sensitive dies such as those found in thin wafer or stacked die configurations. Gel TIMs also resist migration under pressure or thermal cycling, which improves long-term thermal and mechanical stability.

To standardize terminology and clarify functional positioning, the industry defines TIMs by their placement in the heat dissipation chain (Figure 6.18 and Table 6.9).

This classification helps engineers distinguish between application-specific requirements such as reworkability, thermal impedance, mechanical durability,

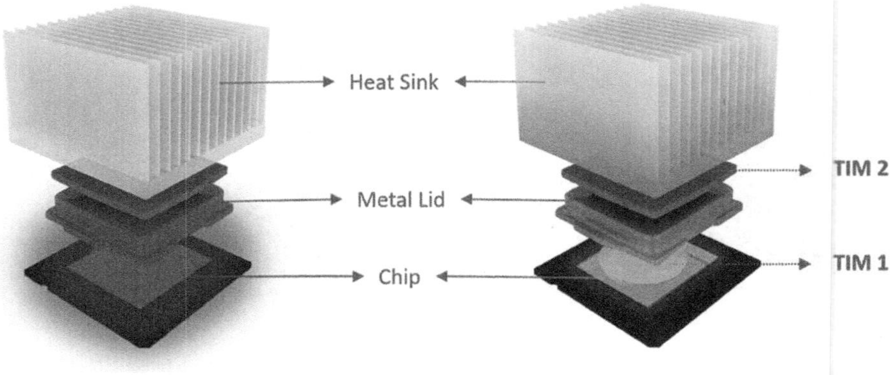

FIGURE 6.18 Schematic diagram of thermal interface materials (right column) that promote heat dissipation in electronic device. (Reproduced from Ref. [25] with permission from John Wiley and Sons.)

TABLE 6.9
Classification of TIMs by Application Location

TIM Type	Interface Location	Typical Non-Metallic Materials
TIM-1	Die to Lid	Grease, Gel, Cured Epoxy
TIM-2	Lid to Heatsink	Pads, Gels, Grease
TIM-3	Heatsink to Chassis/System	Thermal Pads, Phase Change Films, Adhesives

and placement method. For example, TIM-1 typically requires the lowest thermal resistance due to proximity to the heat source, while TIM-2 and TIM-3 can tolerate slightly higher thermal impedance but may demand greater conformability and reworkability. Understanding these distinctions is vital for selecting appropriate TIMs tailored to the specific demands of chip-level, module-level, and system-level thermal management.

6.4.2 Polymer Matrix Systems and Thermal Contact Optimization

The thermal and mechanical performance of TIMs is highly dependent on the choice of polymer matrix, which governs not only the intrinsic material behavior but also its ability to wet, conform, and retain contact across thermal cycles. The polymer matrix acts as the host medium for thermally conductive fillers and plays a critical role in determining viscosity, elasticity, adhesion, shrinkage, and curing behavior. Various classes of polymers are used depending on application demands, with silicone-based materials being among the most prevalent.

Silicones are widely adopted as TIM matrices due to their inherent softness, high thermal stability, low surface energy, and good compression compliance. Their elastic nature allows effective surface conformity with a minimal assembly stress, making them ideal for applications with fragile die, fine-pitch interconnects, or large

CTE mismatch. Moreover, silicone-based TIMs exhibit long-term thermal aging resistance, maintaining stable mechanical and thermal performance under operating temperatures ranging from –50°C to above 150°C. Additionally, silicones are generally chemically inert and possess excellent electrical insulation, adding multifunctionality to TIMs in certain high-voltage or signal-sensitive environments.

Epoxy matrices, on the other hand, provide superior mechanical strength, chemical bonding, and adhesive performance, which are desirable for structural TIMs that double as lid attach adhesives or encapsulants. Epoxy-based TIMs are typically used where permanent attachment, a minimal bondline thickness, and a high thermal conductivity are critical—such as in server processors, automotive ECUs, and embedded lid designs. Their high modulus offers mechanical robustness but can also lead to stress concentration and interfacial fatigue, particularly under aggressive thermal cycling or substrate warpage.

Acrylic systems are often used in UV- or heat-curable applications requiring rapid processing and low-temperature manufacturing. They offer a balance between adhesion, flexibility, and cure speed, and are often found in light-curing TIM pastes or jet-printed adhesives for microelectronic assemblies. Their relatively moderate thermal stability, however, limits their use in high-temperature or high-power applications.

Polyimide (PI) systems are chosen for high-temperature electronics, such as those found in SiC-based power devices, aerospace avionics, or extreme-environment sensors. With thermal stability well above 300°C and excellent dielectric properties, PI-based TIMs enable operations in environments where most organic polymers fail. However, polyimides require precise processing control, often involve solvent-based formulations, and can exhibit a relatively high modulus and lower conformability.

A comparative overview of common polymer matrices used in TIMs is summarized in Table 6.10.

The choice of the matrix also directly affects the material's ability to retain conformal contact across thermal cycles. Parameters such as modulus evolution upon curing, thermal shrinkage, and softening point influence both the thermal resistance and mechanical stress distribution at the interface. A mismatch in shrinkage or modulus between the TIM and the adjacent materials can result in delamination, cracking, or excessive warpage. For instance, B-stageable matrices can be partially cured to achieve a stable shape and then fully cured post-assembly to ensure robust bonding without excessive stress buildup.

TABLE 6.10
Representative Polymer Matrices for TIMs

Polymer Type	T_g (°C)	Typical Modulus (MPa)	Thermal Stability	Conformability	Typical Application
Silicone	–50 to 150	0.1–1.0	Excellent	High	Gel TIMs, Reworkable Pads
Epoxy	80–180	100–3,000	Good	Low–Medium	Lid attach, Structural TIM-1
Acrylic	10–120	10–500	Moderate	Medium	UV-cure TIMs, Jet-dispensed TIMs
Polyimide	>250	500–2,000	Excellent	Low	SiC modules, High-temp Aerospace

Moreover, rheological control—especially in gels or phase change systems—is crucial for maintaining gap filling, wetting, and reflow behavior. By tailoring the viscosity profile, yield stress, and thixotropic recovery, engineers can fine-tune the conformability of TIMs to surfaces with varying roughness or topology. These properties are particularly important in modern 2.5D/3D packaging structures with interposer-based redistribution layers or glass-core substrates, where interface topography becomes increasingly complex. Polymer matrix design is thus not merely about selecting a base resin—it is about engineering the full thermal-mechanical lifecycle of the TIM interface.

6.4.3 Thermally Conductive Filler Systems and Interface Engineering

To significantly enhance the bulk thermal conductivity of TIMs while maintaining mechanical compliance, a variety of thermally conductive fillers are incorporated into the polymer matrix. These fillers serve as the primary channels for heat transport, forming percolated networks that enable rapid phonon or electron transmission across the interface. The type, shape, size, concentration, and surface chemistry of the fillers critically influence the thermal, mechanical, and electrical performance of the final TIM formulation.

Oxide-based fillers such as BN, AlN [26], and SiC [27] are commonly used in TIMs where electrical insulation is paramount. Hexagonal BN, in particular, offers a unique combination of high in-plane thermal conductivity (up to 400 W/m·K), low density, and intrinsic dielectric strength, making it ideal for applications in power electronics and RF systems. AlN provides both high bulk thermal conductivity and close thermal expansion matching with silicon, which reduces the interfacial stress. SiC, though slightly more conductive than AlN, is more abrasive and requires careful control of the surface finish in high-precision assemblies.

Metallic fillers such as silver (Ag) [28], copper (Cu), and nickel (Ni) deliver excellent thermal performance due to their high intrinsic thermal conductivities—Ag reaching nearly 430 W/m·K. However, these materials introduce challenges including potential electrical conduction, corrosion susceptibility, and oxidation under prolonged operation. Surface passivation, core-shell structures, and selective dispersion strategies are often employed to balance thermal benefits with reliability and safety.

Carbon-based materials like carbon nanotubes (CNTs), graphene nanoplatelets, and expanded graphite offer extremely high aspect ratios and intrinsic thermal conductivities that can exceed 1,000 W/m·K in their idealized crystalline forms. While these materials have exceptional potential, their practical performance is often limited by the interfacial resistance with the polymer matrix and by dispersion issues. Nonetheless, hybrid composites that combine carbon fillers with ceramic or metal nanoparticles have been shown to improve both the thermal conductivity and mechanical compliance, leveraging synergistic effects (Figure 6.19).

Achieving a high thermal conductivity in TIMs requires constructing continuous, low-resistance thermal pathways within the matrix. This is typically accomplished through the use of high filler loadings—often exceeding 50 vol%—which approach the percolation threshold. However, an excessive filler content can deteriorate

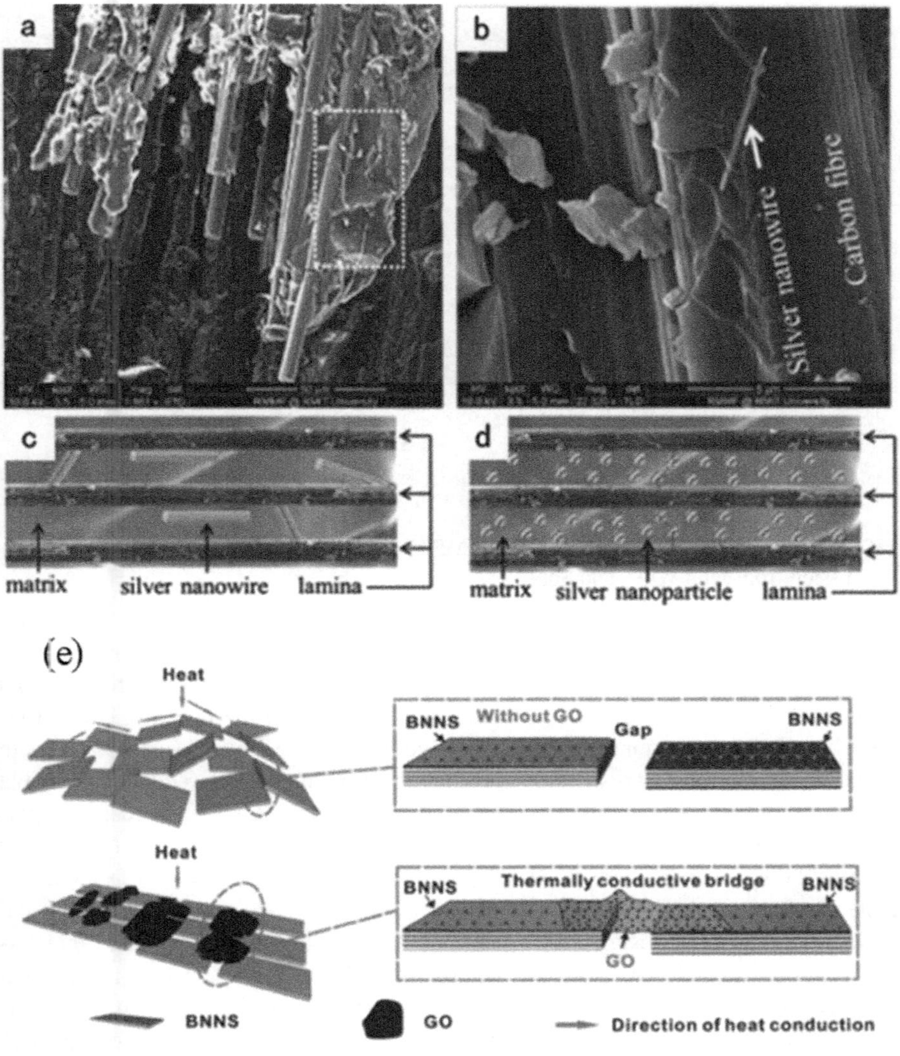

FIGURE 6.19 (a and b) SEM images showing the carbon fiber-reinforced epoxy composite laminates with SNPs and SNWs, respectively. (c and d) SEM images showing that the carbon fiber-reinforced epoxy composite laminates and the fillers (SNPs and SNWs) form 3D networks. (e) Schematic showing that flexible and ultrathin GO sheets are filled in the interspaces between adjacent BNNSs to form the oriented conductive network. (Reproduced from Ref. [29] with permission from John Wiley and Sons.)

mechanical flexibility, increase viscosity, and impair processability. To mitigate this, multiscale filler strategies are employed: mixing micron-sized particles with nanoscale fillers to optimize the packing density, reduce interstitial voids, and facilitate bridge formation across interfaces. Such a hierarchical architecture enhances not only the bulk thermal conductivity but also the stability of the interface during compression or thermal cycling.

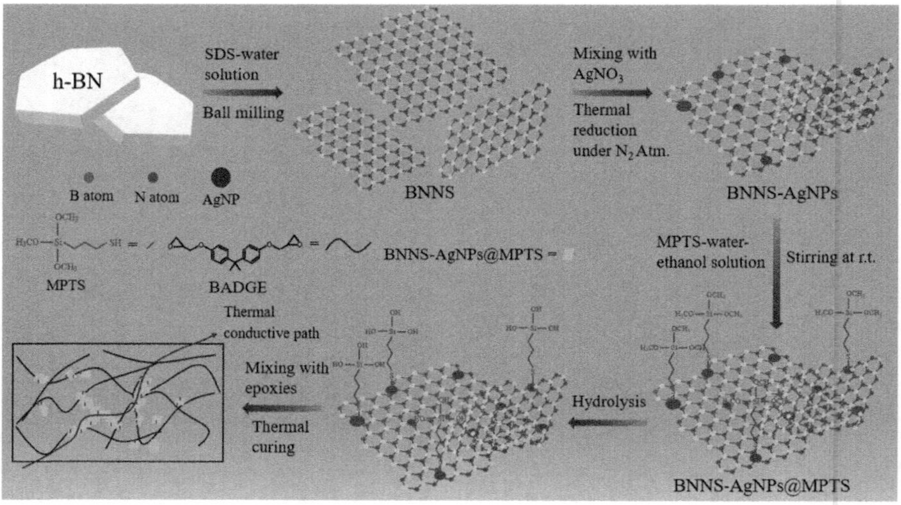

FIGURE 6.20 The fabrication process of MPTS-G-Ag. (Reproduced from Ref. [30] with permission from John Wiley and Sons.)

Surface functionalization of fillers using silane coupling agents, organic surfactants, or polymer grafts is critical to reduce the interfacial thermal resistance (ITR) between the filler and the matrix. Poor thermal coupling at the interface can bottleneck heat flow even when using highly conductive fillers. Modified surfaces promote better wetting, improved dispersion and chemical bonding, which reduce the Kapitza resistance at the polymer-filler junctions. For instance, MPTS (3-mercaptopropyltrimethoxysilane) treated BN or Ag nanoparticles have demonstrated significantly improved dispersion and enhanced composite thermal conductivity in epoxy and silicone matrices, as shown in Figure 6.20.

In advanced TIM designs, the control of the filler orientation plays an increasingly important role. Anisotropic fillers like BN or graphene can be aligned through external fields (e.g., magnetic or electric), shear flow, or printing techniques to maximize in-plane or through-plane conductivity depending on application needs. For example, magnetic alignment of flake-shaped fillers has enabled thermal conductivities exceeding 10 W/m·K in flexible silicone gels while maintaining low modulus.

Another promising strategy involves creating pre-formed 3D thermally conductive networks such as foam scaffolds, aerogels, or interconnected filler frameworks. These structures provide continuous heat pathways independent of the matrix, minimizing reliance on random percolation. Techniques such as freeze-drying, templating, or sintering are used to fabricate such networks before backfilling with compliant polymers. In one case, a vertically aligned graphene foam filled with a soft epoxy exhibited an order-of-magnitude reduction in interfacial thermal resistance compared to conventional composites.

The integration of interfacial bridging agents, such as metallic coatings (e.g., Ag plating on BN) or atomic-layer-deposited ceramic shells, further enhances phonon

TABLE 6.11
Thermally Conductive Fillers for Polymer-Based TIMs

Filler Type	Thermal Conductivity (W/m·K)	Electrical Conductivity	Typical Use Case	Challenges
BN	~200 to 400 (in-plane)	Insulating	Power modules, RF systems	Alignment needed for max performance
AlN	~140 to 180	Insulating	High-temp SiC packaging	Cost, processing sensitivity
Ag	~430	Conductive	High-performance TIMs	Oxidation, cost, migration risk
Cu	~400	Conductive	TIMs for thermally grounded systems	Oxidation, electromigration
Graphene	>1,000 (ideal), ~10 to 50 (bulk)	Semi-conductive	Advanced TIMs, printed electronics	Dispersion, agglomeration
CNTs	~2,000 (along tube)	Conductive	Flexible electronics, ultra-thin TIMs	Orientation control, interfacial bonding
Hybrid (e.g., BN + Ag)	5–20	Tunable	Balanced thermal and electrical specs	Interface complexity

transmission across heterogeneous interfaces. These bridging agents act as phonon-conducting "solders" between filler particles and the surrounding matrix. Additionally, interphase engineering using gradient compositions or compliant interlayers can help tailor stress distributions and prevent delamination at high filler loadings.

A comparative summary of representative filler types is provided in Table 6.11.

The optimization of filler systems is thus a delicate balance between achieving thermal performance, maintaining processability, and ensuring mechanical integrity. As devices continue to scale in power density and shrink in form factor, the design of thermally conductive networks and filler-matrix interfaces will remain a critical frontier in next-generation TIM development.

6.4.4 Thermal Resistance Characterization and Reliability Evaluation

Thermal interface materials are evaluated based on their ability to maintain low thermal resistance, structural integrity, and reliable adhesion under operational stresses. Characterizing these materials requires a combination of thermal, mechanical, and environmental testing techniques tailored to the specific application and geometry.

The most critical parameter for TIMs is the interfacial thermal resistance (R_{th}), typically measured in °C·cm²/W. It quantifies the cumulative resistance to heat transfer across the TIM and the two contacting surfaces. The ASTM D5470 test method (Figure 6.21) is the standard approach for this measurement. In this setup, the TIM is sandwiched between two metal plates with embedded thermocouples, and a known heat flux is applied. By measuring the temperature difference across the interface, R_{th} can be extracted. The method provides consistent and repeatable measurements under controlled pressures and is widely used in both academia and industry.

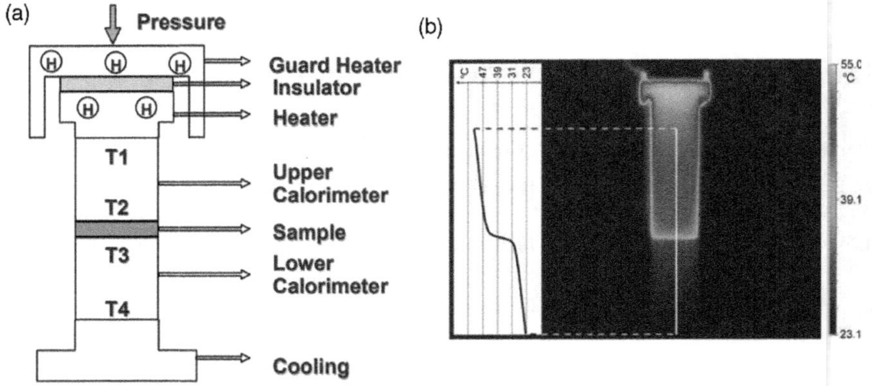

FIGURE 6.21 (a) A diagram of ASTM D5470 guarded heated method tester. (Reproduced from Ref. [31] with permission from Springer Nature.) (b) Exemplary thermal image captured on the surface of HFM blocks. (Adapted from Ref. [32], published in *Journal of Achievements in Materials and Manufacturing Engineering*, under an open-access license.)

Another commonly reported metric is the bulk thermal conductivity (κ) of the TIM, measured using techniques such as the transient plane source (TPS) method (hot disk), laser flash analysis (LFA), or the guarded hot plate method. Hot disk and LFA offer high precision, with the latter suitable for thin cured films and solid composites. However, these techniques typically ignore interfacial effects and serve as a complement to full-stack thermal resistance evaluation.

The TIM thickness after compression is another important parameter. Thinner TIM layers generally yield lower R_{th}, but maintaining uniform thickness becomes more challenging with soft gels and liquids. Profilometry, scanning electron microscopy (SEM) cross-sections, and confocal microscopy are used to assess bond line thickness (BLT). Uneven or excessive BLT can introduce voids or thermal bottlenecks.

Mechanical characterization focuses on assessing the adhesion, elasticity, and structural robustness of TIMs under thermal and mechanical cycling. Shear strength tests, as defined in standards such as ASTM D1002, measure the force required to delaminate the TIM from bonded surfaces. These tests are particularly relevant for cured epoxies and structural adhesives. For compliant TIMs like gels, compression tests are used to determine modulus and compressibility.

Creep and stress relaxation measurements evaluate how the TIM responds to prolonged loading. Compression set tests, where a TIM is compressed under a fixed strain at elevated temperatures for a defined period, are critical for gels and elastomers. A high permanent set may indicate pump-out risk or loss of contact integrity.

Environmental reliability is assessed using thermal cycling (TCT), thermal shock, and accelerated aging tests such as unbiased highly accelerated stress testing (uHAST). These procedures simulate conditions experienced during the real-world operation. For example, TCT involves repeated cycling between –40°C and 125°C for hundreds or thousands of cycles to test the TIM's thermal fatigue resistance. The impact of aging on κ and R_{th} is often tracked in situ using thermal impedance testers or post-mortem D5470 evaluations.

Pump-out, one of the most critical failure modes in compliant or semi-liquid TIMs, as shown in Figure 6.22, is evaluated using high-temperature aging under compressive load and subsequent cross-sectional imaging. Voids, edge migration, or center depletion indicate poor retention and high thermal drift risk. The inclusion of filler settling studies, often tracked through scanning acoustic microscopy (SAM) or X-ray CT, further informs materials optimization.

The following table summarizes key tests for TIM characterization (Table 6.12).

In advanced applications, the interface quality is further analyzed by infrared thermography for heat map uniformity, and time-domain thermoreflectance (TDTR) or frequency-domain thermoreflectance (FDTR) for nanoscale interface characterization. These methods, though sophisticated, offer valuable insights for next-generation nanoscale TIMs, such as those used in high-density 3D-ICs or photonics modules.

While standard characterization and reliability tests discussed above are useful in characterizing the fundamental properties of TIMs, it is highly desirable to replicate the complex thermal environment of real semiconductor devices, and to emulate

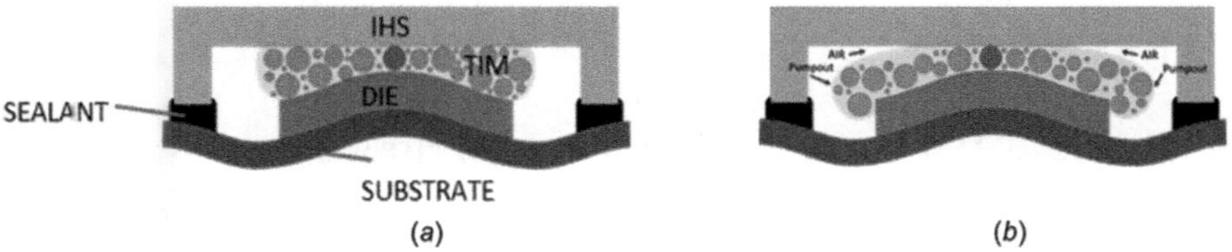

FIGURE 6.22 Schematic of (a) typical IHS package under use condition at the beginning of life depicting good contact of TIM between IHS and die, and (b) schematic of typical end of life IHS package depicting TIM pump out and air gaps that get created. (Adapted from Ref. [33], © ASME. Used with permission.)

TABLE 6.12
Standard Characterization and Reliability Tests for TIMs

Property	Test Method	Instrument Example	Key Insights
Thermal Resistance	ASTM D5470	Thermal Conductance Rig	Includes contact + bulk resistance
Thermal Conductivity (κ)	Hot Disk, LFA	Hot Disk TPS, LFA	Excludes interface effects
Bond Line Thickness	SEM, Profilometer, Confocal		Affects R_{th} uniformity
Shear Strength	ASTM D1002	Mechanical Tester	Delamination resistance of structural TIMs
Compression Set/Creep	ISO 815, Custom Fixture	DMA	Pump-out risk, long-term shape retention
Thermal Cycling (TCT)	JESD22-A104	Thermal Cycler	Fatigue behavior over time
uHAST/Autoclave Aging	JESD22-A118, IEC 60068		Moisture + temperature-induced degradation

dynamic power profiles, non-uniform heating, complex interfacial thermal behaviors, and packaging-induced mechanical constraints that influence TIM performance in actual systems. For heterogeneous integration with multiple chips of various functionalities, sizes, and power distributions all in the same package, characterization of the thermal behavior becomes even more challenging. Silicon-based Thermal Test Chips (TTC) emulate realistic semiconductor chips, including spatially varying power densities (down to 1 mm × 1 mm resolution) and temperature gradients, with embedded in situ sensors enabling high-resolution temperature mapping for accurate localized thermal characterization and improved thermal model validation. By performing thermal measurements and testing (steady state or transient) in a real-life environment, the thermal and reliability performance of TIMs can be accurately assessed, in realistic, multi-chip packaging configurations [34]. More details can be found in Chapter 8 of this book.

Together, a robust characterization and testing protocol ensures that TIMs meet their intended performance across thermal, mechanical, and environmental domains. These evaluations form the foundation of materials qualification for demanding packaging platforms such as high-performance computing (HPC), SiPho modules, and power electronics.

6.4.5 Novel TIM Concepts (PCMs, Printable, Flexible)

As the power density continues to increase in advanced packaging systems, the role of TIMs extends beyond simple gap filling to multifunctional and adaptive materials that respond to thermal, mechanical, and environmental changes. One of the most promising directions is the development of Phase Change Materials (PCMs) for TIM applications [35]. PCMs utilize low-melting organic waxes or eutectic alloys that remain solid at room temperature but transition into a low-viscosity liquid at elevated operating temperatures. This phase transition enables excellent conformability to micro-roughness and enhances interfacial contact. Upon cooling, the PCM solidifies, effectively locking itself in place and maintaining a stable thermal path.

PCMs offer the advantage of low thermal contact resistance and high repeatability across thermal cycles. However, they are susceptible to pump-out due to repeated expansion and contraction, especially when used in vertically oriented modules or under high thermal gradients. Formulation strategies such as encapsulating PCMs within a polymer scaffold or employing self-sealing edge barriers have been explored to mitigate migration. Applications in CPU lid attach and compact mobile systems have demonstrated reduced hotspot formation and improved performance stability over extended cycling.

Another cutting-edge innovation is the development of reworkable TIMs (Figure 6.23). These materials leverage dynamic covalent bonding chemistries, such as Diels–Alder reactions or Schiff base linkages, to allow for reversible crosslinking. This design enables disassembly of components for maintenance or replacement, a critical feature for high-value or mission-critical systems such as satellite payloads and military-grade electronics. Reworkable TIMs also support sustainability goals by extending the service life of components and facilitating materials recovery.

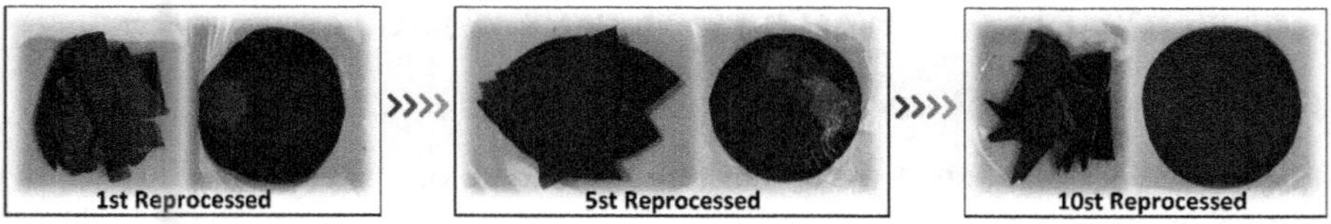

FIGURE 6.23 Photograph of the reworkable (LTGP-15) TIMs after 1, 5, and 10th repeatedly reprocessed at 120°C and 10 MPa for 5 minutes. (Reproduced from Ref. [36] with permission from Elsevier.)

In the realm of flexible electronics and wearable systems, the emergence of printable and conformal TIMs addresses challenges related to non-planar interfaces and dynamic deformation. These TIMs are formulated using UV-curable acrylates or printable silicones, which can be applied via screen printing, inkjet, or 3D printing methods. Their ability to conform to curved geometries while maintaining thermal connectivity opens opportunities in biomedical sensors, soft robotics, and next-generation consumer electronics.

Micro/nanoscale interface engineering is another frontier with profound impact on TIM performance. Advanced techniques such as atomic layer deposition (ALD) and molecular self-assembly are used to create chemically bonded interfaces between filler particles and the matrix, reducing phonon scattering and enhancing thermal continuity. Moreover, interfacial design strategies like thermal expansion gradient layers, functional interphases, and chemical coupling agents allow stress distribution to be tailored in thermally and mechanically mismatched interfaces.

A particularly promising approach is the development of co-curable TIMs, which can be processed simultaneously with adjacent materials such as underfills or EMCs. This synchronized curing process simplifies manufacturing, ensures dimensional compatibility, and enhances adhesion by eliminating interphase boundaries. In 2.5D/3D integration or photonics packaging, such co-curable TIMs offer advantages in warpage control, stress uniformity, and assembly throughput.

In parallel with materials innovations, TIM development is increasingly guided by data-driven design and predictive modeling. Machine learning techniques are being deployed to analyze large datasets from rheology, thermal performance, and reliability tests to predict optimal formulations. AI-assisted design tools help accelerate screening of filler-matrix combinations and optimize processing parameters for custom applications.

The next generation of TIMs will likely integrate multiple functionalities such as self-healing, thermal sensing, and embedded diagnostics. Smart TIMs that can detect delamination or overheating in real-time using embedded piezoresistive or thermoelectric sensors are under active exploration. These developments aim to transform TIMs from passive heat conduits into active elements in the thermal management ecosystem.

These advances mark a transition from TIMs as passive gap fillers to highly engineered, multifunctional interfaces tailored for the demands of heterogeneous integration, high-frequency operation, and extreme reliability in modern electronic packaging.

6.5 POLYMERS FOR WAFER-LEVEL PACKAGING

6.5.1 Functional Roles in RDL and Fan-Out

Polymers are foundational to the design and fabrication of wafer-level packaging technologies, where they serve multifunctional roles spanning mechanical, electrical, chemical, and integration aspects. Their importance extends across different layers and functionalities within advanced packaging architectures. In redistribution layer schemes, polymers act as dielectric insulation between successive layers of copper traces. These polymeric dielectrics not only provide electrical isolation and mechanical support but also enable fine-pitch lithography by offering high planarity and dimensional control. Materials like polyimide and benzocyclobutene are commonly employed here due to their low dielectric constants, high thermal stability, and compatibility with standard semiconductor fabrication steps. Figure 6.24 shows the process of WLP.

In fan-in and fan-out wafer-level packaging (FIWLP/FOWLP), polymers form the structural basis of reconstituted wafers and serve as encapsulants that embed one or more dies. These materials must remain dimensionally stable through various thermal processes including mold cure, lamination, and reflow. Epoxy-based mold compounds with optimized filler content are used to minimize warpage and maintain surface flatness essential for subsequent RDL patterning. Additionally, in Fan-Out Panel-Level Packaging (FOPLP), polymers enable the formation of large-area substrates by filling gaps between dies placed on a carrier and serving as a support matrix during de-paneling and singulation.

Another key function of polymers in WLP is stress buffering. The mismatch in CTE between dissimilar materials like Cu, Si, and mold compound can induce interfacial stresses during temperature excursions. By tailoring the modulus and viscoelastic behavior of polymeric buffer layers, these stresses can be absorbed or

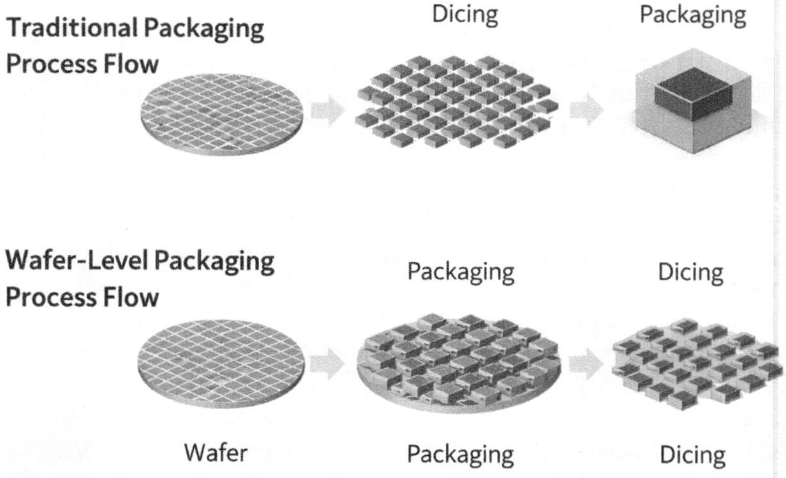

FIGURE 6.24 Wafer-level packaging process flow.

redistributed, thus improving interconnect reliability and package lifetime. Emerging device configurations, such as 2.5D interposers with RDL and embedded components, increasingly require stress-relief capability. In chip-last FOWLP designs such as TSMC's InFO and ASE's FOCoS, the polymer encapsulation around dies and pillars provides dimensional integrity and CTE mitigation, enabling multiple layers of RDL and passive integration.

Polymers also play a role in temporary or sacrificial patterning functions, especially in high-density RDL and through-polymer via (TPV) fabrication. Photosensitive polyimides and epoxy acrylates are used as photoresists or laser-drill masks due to their compatibility with high-resolution lithography and residue-free development. These materials also find application in forming precise dielectric openings for Cu plating or as alignment scaffolds for die placement in hybrid bonding. The use of polymers in such advanced integration tasks reflects their processability and chemical tunability.

In more recent developments, polymer layers also serve as hybrid bonding dielectrics in 3D and heterogeneous integration structures. These layers must provide both ultra-smooth surfaces and low-temperature curability, compatible with Cu-to-Cu bonding and thermocompression processes. Furthermore, polymer performance at the wafer edge (such as crack resistance during dicing) and at the die level (such as laser ablation selectivity for via formation) are also critical design factors. A practical example includes Intel's EMIB technology, where polymer layers ensure both electrical insulation and stress compliance across the embedded bridge.

6.5.2 PI, BCB, PBO and Epoxy Acrylates

The selection of polymers for wafer-level packaging applications is guided by a complex interplay of electrical, thermal, mechanical, and process compatibility criteria. Four major polymer systems are prevalent in current industry use: polyimides (PI), benzocyclobutene (BCB), polybenzoxazoles (PBO), and epoxy acrylates. Each material platform brings unique advantages and trade-offs in terms of the dielectric constant, thermal stability, modulus, moisture resistance, and photopatternability.

PI is perhaps the most widely used polymer dielectric in WLP due to their exceptional thermal resistance, chemical robustness, and mechanical strength. Their high glass transition temperature ($T_g > 300°C$) and excellent adhesion to silicon, copper, and oxide surfaces make them well-suited for high-temperature processing, including multiple reflow steps and plasma etching. Their rigidity ensures structural integrity during RDL formation, though this high modulus can introduce stress concentration at interfaces unless modulated with plasticizers or blends. Additionally, spin-coatable and photo-definable formulations are available, offering fine feature resolution down to 5 μm.

BCB is favored for its ultra-low dielectric constant (~2.65), making it ideal for applications requiring high-frequency signal transmission, such as RF interposers and mmWave modules. Its low moisture uptake (<0.2%) ensures dimensional and dielectric stability even under humid operating environments. BCB also exhibits excellent planarization, simplifying multilayer metallization. However, its mechanical brittleness and sensitivity to oxygen plasma etching necessitate careful stress management and process integration.

PBO offers a compelling combination of flexibility and low water absorption. They form tough, thermally stable films upon thermal curing and are valued in mobile and flexible electronics. PBO layers can accommodate bending and thermal cycling without cracking or delamination, making them suitable for fan-in WLP, stacked memory modules, and advanced flip-chip designs. The relatively low modulus of PBO allows it to absorb thermomechanical stress, thereby protecting fragile interconnects and low-k dielectrics in complex multilayer structures.

Epoxy acrylates, typically UV-curable and photopatternable, have gained traction for applications demanding low-temperature processing and rapid prototyping. Their high resolution and clean development profiles support advanced RDL fabrication with minimal thermal budget. Hybrid systems combining acrylates with epoxy or siloxane backbones can provide tailored curing behavior, adjustable modulus, and improved adhesion. These materials are particularly suitable for chip-first FOWLP processes and temporary bonding/debonding layers in wafer reconstitution.

The following table compares the primary material characteristics of representative polymer systems (Table 6.13).

Each polymer platform must be tailored through formulation or processing adjustments to meet specific packaging demands. For example, PI or PBO can be blended with low-k additives or silane coupling agents to reduce permittivity and enhance adhesion, while BCB may be layered with flexible overcoats to reduce film stress. Epoxy acrylates may incorporate thermal initiators to extend their application to hybrid curing strategies.

Furthermore, copolymer systems and polymer nanocomposites are emerging as customizable solutions. By integrating nanoparticles such as silica, alumina, or fluorinated nanoclays, it is possible to tune dielectric constants, reduce moisture permeability, and reinforce mechanical properties. In particular, core-shell fillers with functional surfaces can enhance polymer-filler compatibility, suppress aggregation, and ensure uniform dispersion during film casting or spin coating.

In summary, polymer materials choice for WLP is not a static decision but one that must adapt to evolving package architectures, interconnect densities, and reliability expectations. The next section will delve into the performance requirements and design metrics that govern this selection process.

TABLE 6.13
Primary Material Characteristics of Representative Polymer Systems

Polymer Type	Dielectric Constant (*k*)	Glass Transition Temp. (°C)	Thermal Stability	Mechanical Modulus	Moisture Absorption	Photopatternability
Polyimide (PI)	~3.1	>300	Excellent	High	Moderate	Medium
BCB	~2.65	~280	Very good	Moderate	Low	High
PBO	~2.8	~300	Good	Moderate	Very low	Medium
Epoxy Acrylate	~3.2	150–200	Fair	Adjustable	Low	Very high

6.5.3 Photo-Patternability and Dielectric Needs

The performance requirements for polymer materials in wafer-level packaging are defined by the demanding environment of advanced microelectronics assembly and operation. These requirements span thermal, mechanical, electrical, and processing domains. To ensure long-term reliability and functional integration in RDL structures and fan-out reconstituted wafers, polymer systems must deliver precise combinations of stability, compatibility, and tunability.

Thermal performance is perhaps the most critical requirement. Polymers in WLP must withstand processing temperatures above 250°C—common during reflow soldering and RDL curing—without degradation, delamination, or loss of mechanical properties. High T_g ensures that polymers remain in a glassy, rigid state during thermal excursions, maintaining dimensional and dielectric stability. Furthermore, polymers must resist oxidation, volatilization, and shrinkage during sequential thermal processes, especially in multi-layer RDL schemes. Figure 6.25 shows the three different types of polymers that undergo a glass transition.

Mechanical properties are equally vital. WLP structures encounter frequent thermal cycling and mechanical loading, including bending, die attach stresses, and assembly warpage. Polymers must balance stiffness (to maintain layer shape and planarity) with toughness and compliance (to buffer stresses between materials with mismatched CTEs). A high modulus may enhance structural integrity, but too rigid a film risks cracking or delamination. Conversely, a low-modulus material may lack shape fidelity or flow excessively during curing. Thus, modulus tailoring through crosslink density control and filler loading is key.

Electrical properties impact signal integrity and power distribution. Polymers should exhibit low dielectric constant ($k < 3.0$) to reduce parasitic capacitance, especially in high-speed digital and RF designs. Low dielectric loss (tan δ) is essential for high-frequency performance, and high dielectric breakdown strength (>200 V/μm) ensures resistance to electrical overstress. Moisture uptake must also be minimized, as absorbed water increases k, degrades insulation, and promotes ion migration. Figures 6.26 and 6.27 show the 2-D and 3-D structure of parasitic capacitance.

Chemical resistance and moisture stability determine packaging robustness. Polymers must survive exposure to process chemicals (such as plasma etchants, solvents, fluxes) and prevent moisture ingress over device life. Moisture not only impacts

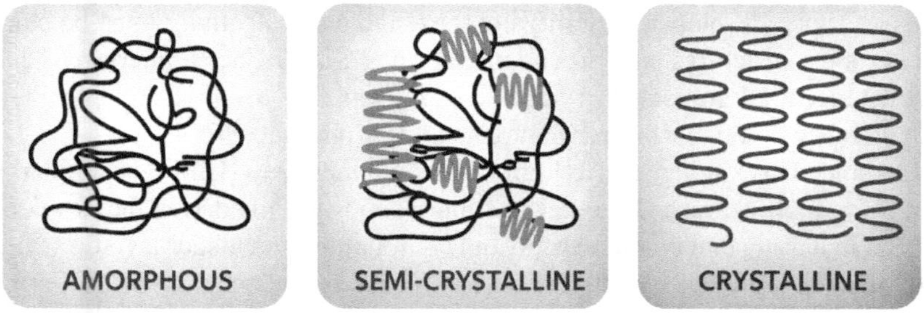

FIGURE 6.25 The three different types of polymers that undergo a glass transition.

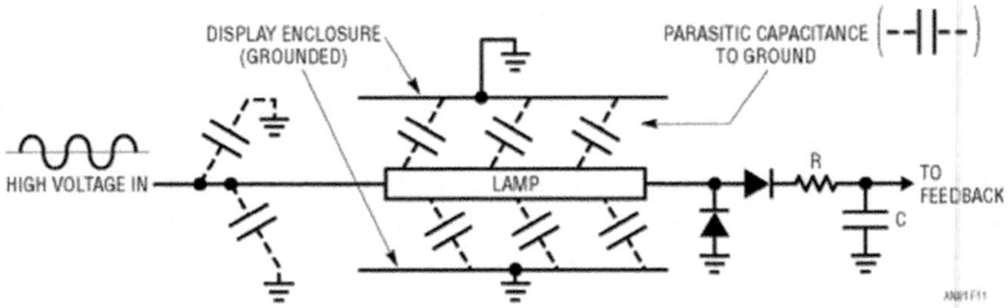

FIGURE 6.26 Parasitic capacitance. (Adapted from Ref. [37] with permission from CRC Press.)

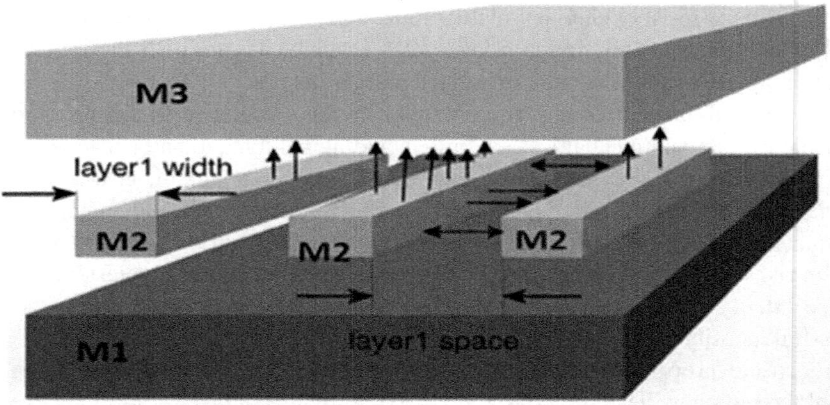

FIGURE 6.27 Parasitic capacitance from sides of metal and from top and bottom of metal.

dielectric behavior but also causes warpage and interfacial debonding during reflow. Low water absorption (<0.5%) and dense crosslinked networks are desirable. The use of hydrophobic side groups (e.g., fluorinated chains) and moisture barriers (e.g., nanoparticle fillers) enhances resistance.

Photopatternability is crucial in RDL lithography and via formation which is showing in Figure 6.28. Modern polymer systems must support fine feature patterning (<10 μm) with steep sidewalls, minimal swelling, and high resolution. Photosensitivity, developer solubility contrast, and etch selectivity must be balanced to yield robust structures post-processing. For multi-layer stacks, interlayer adhesion and resistance to alignment shift become significant.

A summary of key performance targets is shown in Table 6.14.

In advanced packaging such as high-bandwidth memory (HBM) and chiplet-based SoCs, these metrics must be achieved simultaneously. This drives the development of hybrid materials that combine rigid and compliant segments, or layered structures where different polymer types are stacked to serve complementary roles. For example, a high-T_g polyimide base layer may support metallization, while a photosensitive epoxy top layer defines vias and trenches.

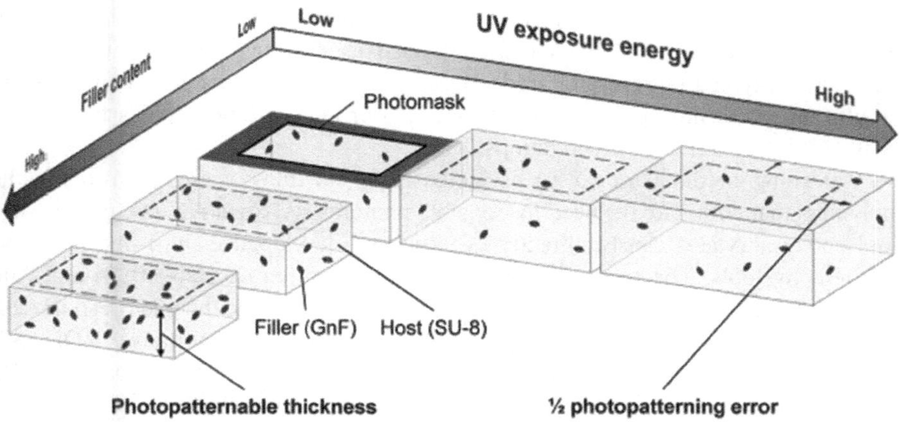

FIGURE 6.28 Prediction on the photopatternable thickness of a photocurable PnC containing 3D randomly distributed carbon-based HAR fillers. (Reproduced from Ref. [38] with permission from Elsevier.)

TABLE 6.14
Typical Target Performance Values for Polymers Used in WLP

Property	Typical Target Value	Evaluation Method
Dielectric constant (k)	≤3.0	LCR Meter, Split-post Resonator
Glass transition temp.	>250°C	DSC, TMA
Modulus	1–10 GPa	DMA, Tensile Testing
Moisture absorption	<0.5%	ISO 62, Gravimetric Analysis
Breakdown strength	>200 V/μm	High Voltage Test
Lithography resolution	≤10 μm	Optical/SEM Profiling

These design strategies will be further discussed in the next section, which explores integration and process compatibility of polymer systems in WLP environments.

6.5.4 Processing and Cu Integration

The successful implementation of polymers in wafer-level packaging hinges on their compatibility with complex, high-throughput fabrication workflows. These include precise deposition, photopatterning, thermal curing, metallization, and multilayer stacking, all of which must be achieved without compromising dimensional control, adhesion, or reliability.

Film deposition is one of the earliest and most critical steps in polymer integration. For round wafers used in Fan-In WLP or Fan-Out on wafer, spin coating remains the standard due to its ability to produce uniform thin films across 200–300 mm substrates. The viscosity of the polymer precursor, typically in the range of 200–2,000 cP, is tuned to control the film thickness and avoid edge bead formation or void entrapment. For large-format reconstituted panels in Fan-Out Panel-Level Packaging, however, scalable techniques such as curtain coating, slit-die coating, and spray coating are preferred. These methods accommodate panel sizes up to 600 mm × 600 mm and allow for coverage over uneven topographies formed by embedded dies and pre-formed bumps.

Patterning of dielectric layers is increasingly shifting toward photo-definable polymers that eliminate the need for separate photoresists. Negative-tone polyimides and epoxy acrylates can be directly exposed and developed using i-line or DUV lithography tools. The resolution of these materials often exceeds 10 μm and can reach sub-5 μm with advanced stepper alignment. Importantly, the photoresponse of these materials must remain stable over multiple exposures in multilayer RDL fabrication, necessitating tight control over exposure dose, post-exposure bake (PEB), and development time. Residual films post-development can degrade Cu plating or increase line-edge roughness, hence developers and rinse chemistries must be optimized for each polymer type. Figure 6.29 illustrates the PEB.

Curing behavior plays a central role in determining final film properties such as modulus, glass transition temperature, and residual stress. For thermally curable systems like PI and BCB, multi-step curing profiles are used: typically beginning with a soft-bake to drive off solvents, followed by imidization or ring-opening polymerization at 200°C–350°C. For UV-curable or hybrid dual-curing materials, exposure to 365–405 nm light initiates radical polymerization, which may be followed by a thermal post-cure to complete cross-linking. These dual-cure systems are advantageous in large-area panel formats where warpage control and stress balancing are required before final heat exposure.

One of the most persistent challenges in WLP integration is adhesion between the polymer dielectric and metal interconnects, particularly copper. Poor interfacial bonding can lead to delamination, cracking, or underplating voids. To address this, polymer surfaces are treated with plasma activation (O_2, Ar, or N_2), chemical silanization using organosilanes like APTES or MPTS, or adhesion promoter coatings (e.g., Ti, Cr, or Ni). The polymer formulation itself may also be modified with

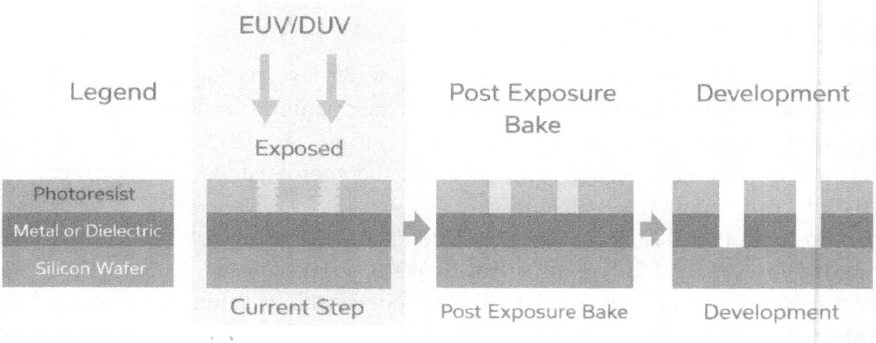

FIGURE 6.29 Post-exposure bake (PEB).

functional groups—such as hydroxyl, amine, or carboxylic moieties—to enhance interfacial bonding through chemical linkage or hydrogen bonding.

In advanced packages such as TSMC's InFO or Intel's EMIB, where multiple layers of RDL and through-polymer vias are formed over organic or molded substrates, maintaining interlayer adhesion and registration is critical. Stress accumulation at each cured layer can lead to via misalignment or delamination during Cu electroplating. Therefore, polymer modulus, shrinkage, and curing sequence must be designed to balance flow, dimensional control, and stability during layer build-up. Lithographic alignment accuracy must be maintained below $\pm 1.5\,\mu m$, particularly for fine-line redistribution (line/space $< 5\,\mu m/5\,\mu m$).

Polymers also play key roles in wafer or panel reconstitution. In chip-first processes, polymer encapsulants must fill die-to-die gaps uniformly and maintain dimensional stability during debonding and RDL formation. Void-free encapsulation is critical, as trapped air pockets may expand during thermal cycling and cause delamination. Degassing procedures, vacuum lamination, and material rheology optimization (viscosity $< 10\,Pa{\cdot}s$ at the molding temperature) are employed to achieve this.

Finally, residue management and etch compatibility are vital for multilayer builds. After development or laser ablation of vias, polymer residues can remain on Cu pads or sidewalls, impeding metallization or increasing contact resistance. Mild plasma ashing, wet cleaning (e.g., dilute HF, TMAH), or proprietary dry etch formulations are used to clean features without damaging underlying structures. In emerging integration schemes involving hybrid bonding, ultra-smooth polymer surfaces ($<0.5\,nm$ RMS) are required for alignment and contact quality.

In summary, the integration of polymers into WLP demands a balance between processability, thermal and dimensional stability, pattern fidelity, and interfacial reliability. Continued innovation in polymer formulation, coating techniques, and surface engineering is essential to meet the escalating complexity and miniaturization of next-generation packaging platforms.

6.5.5 Reliability: Warpage, Delamination, Aging

Reliability remains one of the foremost concerns in wafer-level packaging, where increasingly dense interconnects, thin polymer layers, and diverse material interfaces are all exposed to a wide range of mechanical, thermal, and environmental stressors. Polymers used in these architectures are often the weakest links when it comes to long-term reliability. Their viscoelastic nature, susceptibility to moisture, and interfacial properties all contribute to failure modes that can significantly limit package performance and lifetime.

A primary reliability challenge arises from thermomechanical stresses induced by mismatches in the CTE among different materials. For example, copper traces (CTE $\approx$ 17 ppm/°C), silicon dies ($\approx$ 2.5 ppm/°C), and polymer dielectrics (30–60 ppm/°C) expand and contract at different rates during thermal excursions. This differential movement results in stress concentrations at polymer–metal and polymer–die interfaces, particularly at corners or via transitions. Repeated thermal cycling—during testing, reflow, and operation—can lead to delamination, cracking, and even via extrusion or failure. Interfacial adhesion is often improved using coupling agents or interfacial engineering layers, but intrinsic stress management through modulus tailoring and stress-buffer polymer layers is also essential.

Moisture absorption is another critical issue. Many polymer systems are hydrophilic or semi-hydrophilic, absorbing moisture from the environment over time. Under reflow or high-temperature operating conditions, this moisture can expand rapidly, leading to the so-called "popcorn effect," where trapped moisture causes delamination or even die cracking. Even at lower temperatures, absorbed moisture can diffuse toward metal interfaces, resulting in corrosion, increased leakage current, or electrochemical migration. Moisture-induced degradation is particularly problematic in mobile SoCs, automotive systems, and high-reliability industrial modules, where operation spans a wide temperature and humidity range.

The dielectric properties of polymers can also degrade over time or under specific stress conditions. High-field operation, particularly at GHz frequencies, exposes polymer dielectrics to strong electric fields that can induce dielectric breakdown or long-term polarization effects. Dielectric loss and leakage current can increase with moisture uptake, thermal aging, or UV exposure, all of which are common in standard packaging processes. For instance, UV patterning of photosensitive polymers can result in local chain scission or crosslinking, altering dielectric uniformity. Furthermore, ion migration in the presence of moisture can form conductive paths, leading to premature electrical failure.

Mechanical reliability must also be addressed, especially in thin-film RDL structures. Cracking, crazing, and plastic deformation can occur in polymer layers subjected to repeated bending, vibration, or package warpage. While low-modulus polymers are better at absorbing strains, they may also exhibit cold flow or creep under a constant load, leading to dimensional instability. Conversely, high-modulus materials provide dimensional rigidity but can exacerbate stress transfer to underlying fragile structures, such as thin Cu lines or TSVs. Therefore, the selection of modulus-balanced polymers is critical depending on the specific application environment.

Aging behavior of polymers poses long-term reliability concerns. Over time, crosslinked networks may undergo relaxation, hydrolysis, oxidation, or UV-induced degradation, leading to changes in modulus, T_g, and moisture uptake. These effects accumulate during burn-in, operational use, and prolonged storage, requiring careful materials qualification using accelerated aging protocols. Temperature–humidity bias tests (85°C/85% RH), thermal cycling tests (–40°C to 125°C), and high-temperature storage tests (HTS) are commonly employed to benchmark reliability performance.

To better illustrate the various reliability challenges associated with polymer systems in WLP, Table 6.15 summarizes common failure mechanisms, their root causes, and representative mitigation strategies.

In conclusion, the reliability of polymers in wafer-level packaging must be holistically managed through thoughtful materials selection, processing, and design integration. This includes controlling moisture pathways, minimizing CTE mismatches, tuning the modulus and film thickness, and validating long-term performance under representative service conditions. Future advances in predictive simulation, coupled with accelerated reliability testing and machine-learning-based failure prediction, will help streamline this critical phase of packaging development.

TABLE 6.15
Common Failure Mechanisms and Mitigation Strategy

Failure Mechanism	Root Cause	Mitigation Strategy
Interfacial delamination	CTE mismatch, poor adhesion	Silane coupling, modulus tuning, surface activation
Cracking/Crazing	Residual stress, thermal cycling	Use of stress buffer layers, optimized cure profile
Dielectric breakdown	Moisture ingress, field concentration	Low-D_f polymers, improved encapsulation, moisture barriers
Moisture-induced popcorn	Hygroscopic materials, rapid reflow heating	Low-absorption polymers, pre-bake, void-free encapsulation
Electrical leakage	Ionic contamination, dendrite formation	Clean etch/develop, low-ion polymers, barrier layers
Aging/degradation	UV exposure, oxidation, hydrolysis	Antioxidants, UV stabilizers, low-energy processing

6.5.6 Future Low-k and Eco-Friendly Options

In recent years, polymer materials used in wafer-level packaging (WLP) have undergone rapid innovation to meet the increasing performance demands of heterogeneous integration, finer feature sizes, higher thermal budgets, and emerging panel-level processing techniques. These developments reflect not only materials science progress but also co-evolution with equipment capabilities, integration strategies, and the broader electronics packaging roadmap.

One of the major areas of advancement is the development of ultra-low dielectric constant (ULK) polymers. Traditional polymers such as PI and BCB, while reliable, often present dielectric constants in the 2.6–3.2 range, which limits their suitability for high-speed signal transmission and advanced RF applications. To address this, new polymer chemistries have been engineered to incorporate fluorinated groups, siloxane segments, or micro/nano-scale porosity. These structural modifications reduce the overall polarizability and density, lowering the dielectric constant to values below 2.4. For example, fluorinated polyimides (FPI) and nanofoamed epoxies demonstrate improved high-frequency performance while still maintaining thermal and mechanical stability. However, the introduction of porosity often reduces modulus and adhesion, necessitating careful balancing of electrical and mechanical requirements.

Another frontier is the shift toward multifunctional polymer layers. In advanced WLP configurations, especially in high-density fan-out and chiplet architectures, the same polymer layer may need to fulfill several roles simultaneously—such as serving as a dielectric, a moisture barrier, a stress buffer, and a medium for fine-pitch photo-patterning. Hybrid polymer systems that combine acrylate, epoxy, or benzoxazole building blocks have been designed to integrate these diverse functionalities. For instance, B-staged polymers that exhibit both low-temperature flow and subsequent thermal curing are used in embedded bridge interposers and mold-first fan-out packages to improve warpage control and multilayer integration fidelity.

Patternability and process compatibility have also seen significant improvement. Emerging polymer systems are increasingly designed to be compatible with advanced lithography tools such as stepper and scanner systems. This necessitates narrow processing windows, precise photosensitivity, low post-exposure shrinkage, and excellent etch resistance. New photosensitive polymers, such as siloxane-modified epoxy acrylates or novel negative-tone polyimides, now enable sub-5 μm features and high aspect ratio vias with fewer process steps. Moreover, dual-tone and switchable polarity photoresists have been explored to allow greater design flexibility in multilayer RDL patterning.

Another significant direction is large-area processability, especially relevant in fan-out panel-level packaging (FOPLP). As packaging transitions from wafer-scale to panel-scale substrates, the need for polymers that maintain uniformity across 500 mm × 500 mm or larger areas becomes critical. To address this, spray-coatable or curtain-coated polymer systems have been developed with tuned rheology and leveling behavior. These systems must resist sagging, edge pooling, and thickness nonuniformity, while maintaining adhesion to multiple materials such as mold compound, Cu RDLs, and embedded dies. Solvent-free formulations with rapid UV-curing characteristics are gaining interest for high-throughput manufacturing.

Environmental sustainability and regulatory compliance have also begun to influence polymer development in WLP. Traditional polymers often involve hazardous solvents or generate volatile organic compounds (VOCs) during curing. In response, researchers are investigating green alternatives, such as aqueous-developable polyimides, VOC-free epoxy acrylates, and reworkable mold resins. These new systems aim to support greener packaging lines without sacrificing performance, especially in consumer electronics and automotive sectors where both environmental impact and performance reliability are heavily scrutinized.

Table 6.16 provides a representative timeline of key polymer development trends for WLP, including dielectric constant evolution, integration roles, and patterning capability.

In summary, polymer materials for wafer-level packaging are no longer simple dielectric fillers or stress buffers; they are evolving into highly engineered,

TABLE 6.16
Timeline of Key Polymer Development Trends for WLP

Year	Development Focus	Key Polymer Advancement	Typical Application
2000	Standard WLP/PI use	Spin-coated PI, $T_g > 300°C$	Basic RDL and passivation
2010	High-density fan-in/out	BCB and PBO with improved D_f and adhesion	Mobile SoCs, RF packages
2015	Fine-line patterning + ULK	Fluorinated PI, hybrid acrylates	HPC, SiP, interposers
2020	Panel-level integration	Sprayable UV-curable epoxies	FOPLP, chip-first fan-out
2025	Sustainable, multifunctional materials	VOC-free, self-healing, B-stage polymers	AI/ML, automotive, wearable electronics

multifunctional components that contribute directly to the electrical, mechanical, and environmental performance of the package. Future research will likely focus on adaptive materials that respond to in-field stimuli, ultra-low-K yet mechanically robust systems, and greener chemistries that align with sustainable manufacturing goals.

6.6 POLYMERS FOR 3D/2.5D APPLICATIONS

6.6.1 TSV Liners, Adhesives, Stress Buffers

In the rapidly evolving field of heterogeneous integration, the emergence of 2.5D and 3D packaging paradigms has brought polymer-based encapsulation and support materials to the forefront of advanced materials engineering. In these architectures, polymer systems no longer play a purely protective role, but become essential components that enable structural integration, thermal management, and electrical performance optimization.

One of the most critical functional roles is associated with Through-Silicon Via (TSV) technology, shown in Figure 6.30, where polymers are used as insulating liners and filler materials. TSV liners require conformal deposition into high aspect ratio vias while maintaining excellent dielectric insulation and mechanical flexibility to buffer thermal stresses between Cu and Si. The filler materials must uniformly encapsulate the TSV column, provide mechanical damping, and reduce the formation of voids or cracks that may propagate due to CTE mismatches.

Beyond TSVs, polymers are also used as die attach adhesives or gap-filling materials in chip stacking. These inter-die adhesives must exhibit controlled flow behaviors to minimize voids while achieving a uniform thickness. Their compliance properties are key to accommodating strains between dissimilar materials such as Si, glass, or organic interposers. In organic or glass interposer structures, polymers also act as redistribution dielectric layers and intermediate stress-relieving layers, enabling copper RDL traces to be patterned with precise resolution while maintaining dimensional stability during multiple thermal excursions.

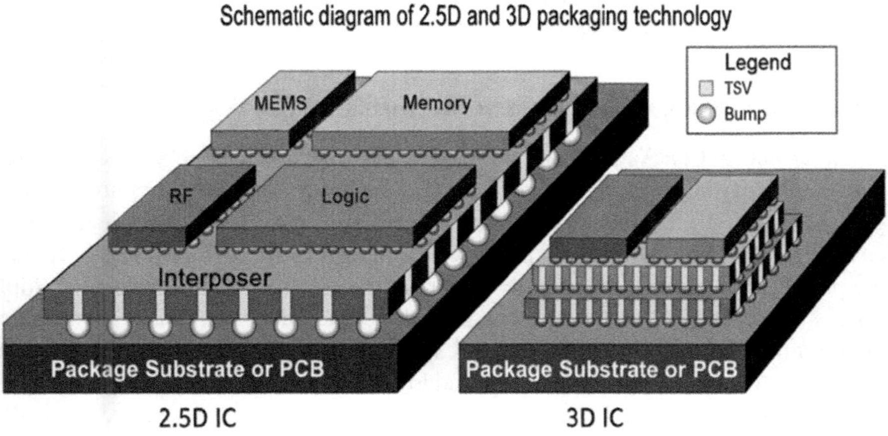

FIGURE 6.30 Through-Silicon Via 3D interconnection.

TABLE 6.17
Representative Polymer Systems for 2.5D/3D Integration

Polymer System	Thermal Stability (°C)	Dielectric Constant (1 MHz)	CTE (ppm/°C)	Typical Use Case
Epoxy (filled)	150–250	3.5–5.0	20–40	Inter-die adhesive, TSV filler
Polyimide	>400	2.9–3.5	30–50	TSV liner, redistribution dielectric
BCB	250–300	2.6–2.9	<20	Low-*k* RDL dielectric, surface passivation
Hybrid epoxy-siloxane	200–300	3.0–4.2	Tunable	Glass interposer bonding, stress relief layers

Moreover, these materials contribute significantly to mechanical decoupling between stacked dies or between dies and interposers. For high-density packages using micro-bumps and hybrid bonding, ultra-thin polymer films—often less than 10 µm thick—serve as bonding layers with well-controlled thicknesses and surface energy, ensuring alignment fidelity and co-cure integration with other encapsulants or thermal interface materials. In many system-in-package (SiP) modules, polymers provide lateral stress absorption, enable electrical shielding compatibility, and act as barrier layers against moisture and ionic contaminants.

To contextualize the functional diversity of these polymers, Table 6.17 summarizes key polymer systems currently used in 2.5D/3D stacking, their materials properties, and application-specific roles.

This diversity of roles and formulations demonstrates that polymers in 2.5D/3D packaging are not monolithic; rather, they serve as carefully engineered materials whose properties must match the mechanical, thermal, and electrical demands of their specific placement within a vertically integrated system. Each polymer system is selected and customized based on its role in TSV encapsulation, bonding, stress buffering, and dielectric routing, and their success relies on balancing functionality with manufacturability across diverse stacking platforms.

6.6.2 Functional and Processing Requirements for Polymers

As 2.5D and 3D packaging technologies scale toward finer pitches, higher interconnect density, and increased die stacking complexity, the design requirements for polymer materials become highly specialized and application-dependent. Unlike conventional encapsulants used in surface-mount packages or wire-bonded devices, the polymers employed in vertically integrated architectures must perform multiple roles simultaneously—structural, thermal, dielectric, and mechanical—while also maintaining compatibility with complex fabrication workflows such as wafer thinning, TSV formation, hybrid bonding, and low-temperature lamination.

A central requirement in inter-die polymer design is the need for mixed-modulus behavior. Within a single vertical stack, the polymer must locally transition between regions of high stiffness (to support structural integrity and maintain coplanarity) and regions of low modulus (to accommodate thermal strains between heterogeneous die materials such as silicon, glass, or GaN). For example, in 3D memory stacks (e.g., HBM or stacked DRAM), thermal cycling from 25°C to 125°C can result in significant stress accumulation due to the mismatch between silicon dies and epoxy-based adhesives. A uniformly rigid polymer may induce delamination or cracking, while an excessively soft system may allow die tilt, bondline collapse, or poor thermal conduction. Therefore, polymers used in these contexts are often formulated with tunable modulus profiles—achieved via filler gradients, dual-network chemistries, or domain-segregated morphologies—to balance rigidity and compliance across functional zones.

Thermal conductivity is another critical driver in materials selection. In vertically stacked dies or Si-interposer modules with dense logic cores, significant heat is generated within a small volume, necessitating that polymers facilitate vertical or lateral heat dissipation. Conventional underfill or gap-filling polymers typically exhibit thermal conductivities of 0.2–0.5 W/m·K. However, advanced formulations may incorporate aligned BN [39], AlN [40], or graphene-based fillers to achieve values exceeding 3 W/m·K, particularly in anisotropic directions. The challenge lies in achieving a high filler loading without compromising rheology, void-free processing, or electrical insulation.

Moreover, the polymers used in 2.5D/3D integration must also meet stringent processing requirements, particularly in bonding schemes that rely on partial curing and controlled flow behavior. In polymer-assisted bonding approaches—such as thermocompression bonding (TCB) with adhesive layers or polymer gap-fill bonding—the polymer is positioned around the Cu-Cu interconnect structures. In these processes, the polymer must enter a B-staged, partially cured state, illustrated in Figure 6.31, prior to bonding so that it retains sufficient flowability for surface conformity and gap filling during alignment and compression, while still maintaining enough mechanical integrity to avoid uncontrolled spreading or resin squeeze-out. The timing of this B-stage window is therefore critical. If the polymer advances too far toward full cure before bonding, its viscosity becomes too high to wet recessed topography or fill nanoscale voids. Conversely, if the polymer remains insufficiently advanced, excessive flow, void formation, or outgassing may occur during the thermal-compression step. Proper control of this curing profile ensures reliable bond formation, void-free gap filling, and compatibility with fine-pitch Cu interconnects during 2.5D/3D assembly.

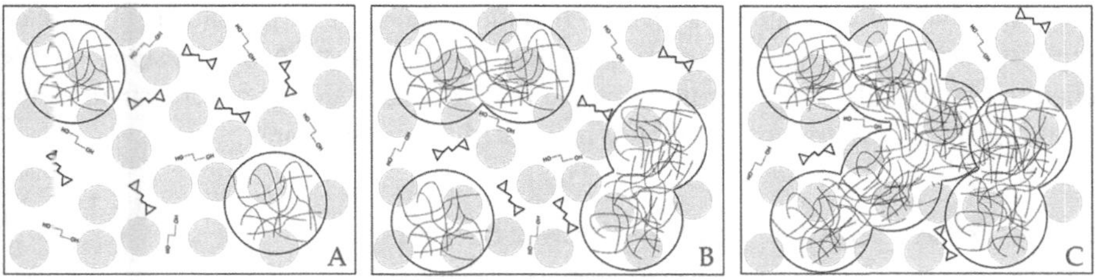

FIGURE 6.31 Different stages of cross-linking formation in EMC. (Adapted from Ref. [41], published by MDPI under the CC BY license.)

Another key requirement is photo-patternability, particularly in redistribution layers or in applications involving wafer-level underfill routing. Polymers such as photosensitive polyimides, BCB, and epoxy-acrylates enable direct photolithographic structuring, offering fine resolution (<10 μm), smooth sidewalls, and compatibility with Cu electroplating processes. These materials are essential in building multilayer interposers or embedded bridge structures, where the routing density and via alignment depend on tight pattern fidelity and a low dielectric loss. Furthermore, materials must maintain shape fidelity during post-exposure bake and subsequent cure, necessitating control of crosslinking kinetics and thermal shrinkage.

Mechanical compliance, quantified by stress-strain behaviors, must be balanced with adhesion to dissimilar materials such as Cu, Si, oxide, and glass. Interfacial reliability often governs the failure threshold of a 3D system under thermal cycling. Silane coupling agents, amine-modified fillers, or in situ interpenetrating networks, are often used to enhance adhesion and minimize interfacial delamination. Simultaneously, polymers must demonstrate dimensional stability during vacuum lamination and compression molding, particularly in FOPLP or panel-level 2.5D integration, where large-area stress uniformity becomes critical.

To summarize, design requirements in 2.5D/3D packaging call for multifunctional polymers that exhibit:

- Controlled modulus gradients, with typical low-modulus zones <50 MPa and reinforcement zones >500 MPa;
- Thermal conductivity ≥1.5 W/m·K, with anisotropic fillers aligned along vertical heat paths;
- B-stage processability, with tack-free storage times >24 hours at room temperature and cure onset at 120°C–160°C;
- Photopatternable resolution <10 μm, with dielectric constants below 3.0 for high-speed RDL applications;
- High adhesion energy (G_c > 200 J/m^2) to metals, oxides, and polymeric dielectrics;
- Dimensional stability, with shrinkage below 1%, and warpage control across die spans >10 mm.

These performance targets are interdependent and often conflicting. For example, increasing the thermal conductivity by adding ceramic fillers may raise viscosity and hinder fine-pitch gap filling. Enhancing the modulus for mechanical strength may introduce interfacial brittleness. Therefore, materials development is increasingly reliant on multi-objective optimization, leveraging molecular design, surface chemistry, and filler engineering to create systems that meet both performance and manufacturing constraints in a highly integrated 3D environment.

6.6.3 Epoxy, PI, BCB with Functional Fillers

The performance of polymer systems in 2.5D/3D stacking applications is intimately tied to the selection of matrix chemistry and the engineering of filler types, distribution, and interfacial behavior. While the base polymer determines processability, adhesion, and intrinsic dielectric properties, the incorporation of fillers offers opportunities to tailor the thermal, mechanical, and even electrical performance across application-specific requirements. Importantly, the synergy between the polymer matrix and the filler system dictates the effectiveness of the material in TSV encapsulation, gap filling, and thermal path optimization.

Among the most widely used polymer matrices are epoxy resins, which offer tunable viscosity, excellent adhesion to silicon and copper, and versatile curing behavior. Epoxy systems can be modified to achieve B-stage behaviors for bonding applications or fully cured thermosets for rigid encapsulation. Additionally, their compatibility with various filler types makes them suitable for thermal conductivity enhancement. Polyimides, particularly spin-on or photo-definable grades, offer exceptional thermal stability ($T_g > 350°C$), chemical resistance, and dielectric strength, making them ideal as TSV liners and thin RDL dielectrics. However, their processing typically requires high-temperature curing and solvent-based handling. BCB is another important matrix for low-*k* applications. Its low dielectric constant (~2.6 to 2.9), low moisture absorption, and excellent planarization make it valuable in high-speed signal routing, though it requires stringent curing conditions and exhibits relatively low fracture toughness.

To improve thermal conductivity and mechanical performance, high-performance fillers are employed. Ceramic-based fillers such as AlN and BN offer high thermal conductivity, low CTE, and electrical insulation. For instance, AlN particles can provide up to 200 W/m·K in bulk, though composite values depend on percolation and dispersion. Boron nitride, particularly in its platelet or nanosheet form, enables thermal conduction along specific axes and contributes to anisotropic thermal transport.

Emerging filler candidates include graphene, reduced graphene oxide, and carbon nanotubes. These materials provide an ultra-high intrinsic thermal and electrical conductivity, though their surface chemistry must be carefully controlled to avoid percolation paths that lead to a leakage current. Surface modifications using silane coupling agents, thiols, or amines are often applied to enhance interfacial bonding with polymer matrices and prevent aggregation. Moreover, hybrid filler systems—combining high-aspect-ratio 2D fillers (e.g., graphene) with spherical particles (e.g., Al_2O_3 or SiO_2)—have been shown to construct multiscale thermal pathways while preserving rheological stability.

Another critical strategy involves filler alignment. Random dispersion leads to isotropic behavior, which is suboptimal for through-thickness heat transfer in stacked die configurations. Techniques such as electric-field alignment (for BNNS or CNTs), shear-induced alignment during spin coating, or magnetic field processing (for functionalized AlN) are employed to induce anisotropic networks. These aligned fillers can yield thermal conductivity enhancements of 2–5× in the preferred direction compared to randomly dispersed counterparts.

To provide a clearer overview, Table 6.18 summarizes the performance attributes of representative filler materials and their suitability in different stacking applications.

In addition to the properties of the fillers themselves, the polymer–filler interface plays a decisive role. Interfacial thermal resistance (Kapitza resistance) can dominate the overall thermal performance if the bonding between the polymer matrix and the filler is weak or incomplete. This is particularly relevant in nanoscale fillers,

TABLE 6.18
Comparison of Key Filler Types for 2.5D/3D Packaging Polymers

Filler Type	Thermal Conductivity (W/m·K)	Electrical Conductivity	Particle Morphology	Typical Applications
AlN	~140–200 (bulk)	Insulating	Spherical or platelet	Thermal paste, underfill
BN	~250 (in-plane)	Insulating	Platelet or nanosheet	Gap filler, anisotropic TIM
Graphene	>2,000	Conductive	Sheet (2D)	Embedded heat spreader (with surface treatment)
CNT	~3,000	Conductive	1D tubular fiber	Hybrid bonding, advanced TIMs
SiO_2	~1.4	Insulating	Spherical	Low CTE gap filler, dielectric control

where the surface-to-volume ratio is large. Tailored interfacial chemistry, including covalent grafting, dendritic functionalization, and hydrogen-bonding networks, are currently active areas of research aimed at minimizing this resistance and enhancing phonon transfer across interfaces.

Overall, matrix materials and filler selection must align with multiple priorities: thermal conductivity, electrical insulation, low CTE, low moisture uptake, fine-pitch flowability, and co-cure compatibility. The integration of multiscale fillers, the control of filler orientation, and the design of advanced polymer–filler interfaces are shaping the future of multifunctional encapsulants in high-performance 3D integration.

6.6.4 Challenges at Ultra-Fine Pitch

As 2.5D and 3D integration technologies move toward finer bump pitches and denser interconnects, the complexity of polymer processing increases dramatically. Achieving void-free encapsulation, uniform filler dispersion, and precise thickness control becomes significantly more challenging under the constraints of sub-100 μm features, multi-material interfaces, and hybrid bonding schemes. Processing challenges in these advanced packaging systems are often rooted in the dynamic interplay between materials rheology, filler behaviors, and interfacial compatibility across temperature and pressure gradients.

A primary concern in polymer processing is achieving effective dispersion of thermally conductive fillers without inducing aggregation or clogging. As thermal conductivity targets continue to rise—often requiring filler loadings >50 wt%—the risk of agglomerate formation increases, especially with high-aspect-ratio materials such as BNNS or CNTs. These agglomerates can lead to viscosity spikes, poor flow under capillary forces, or even nozzle clogging during stencil printing or jet dispensing. Uniform dispersion is essential not only for the thermal performance but also for avoiding dielectric inhomogeneity and localized mechanical stresses. To mitigate this, dispersants, surfactants, or surface-functionalized fillers are employed. High-shear mixing, ball milling, and ultrasonication are often integrated into the process flow to disrupt agglomerates and ensure a homogeneous prepolymer mixture. However, care must be taken to avoid excessive shear, which may degrade the filler morphology or accelerate premature crosslinking.

Another major challenge is managing sedimentation and void formation during polymer curing and thermal cycling. In systems using micron- or nanoscale fillers with density mismatches relative to the matrix, gravitational settling can occur, particularly during prolonged pot life or low-viscosity processing stages. This leads to filler-rich and filler-depleted zones, compromising thermal and mechanical uniformity. In 3D gap-filling applications, this problem is compounded by the vertical orientation of dies, which exacerbates sedimentation along the z-axis. Formulation strategies such as thixotropic agents, viscosity modifiers, and gelation tuning are used to minimize settling. The curing profile is also designed to include rapid gelation phases that immobilize fillers before migration occurs. Vacuum-assisted underfilling and degassing steps are critical to remove entrapped air and prevent microvoids, especially when processing around TSVs or embedded features where capillary flow may be disrupted.

Adhesion and wetting behavior of the polymer on various surfaces—such as Cu, SiO_2, glass, and dielectrics—represents another significant processing variable. Poor wetting leads to interfacial delamination, void entrapment, or incomplete coverage. Surface activation techniques such as plasma treatment, chemical silanization, or ozone exposure are commonly employed to increase surface energy and promote uniform spreading. Achieving consistent adhesion is further complicated by the coexistence of heterogeneous materials and topographies in advanced packaging, where differences in surface chemistry, roughness, and contamination sensitivity demand carefully tailored polymers with controlled rheology and well-defined interfacial interactions.

As shown in Figure 6.32, die-level warpage during polymer curing also presents processing difficulties. Thermoset shrinkage, thermal expansion mismatches, and capillary-induced stresses can distort die positioning and compromise coplanarity, particularly in chip-last fan-out processes. The mismatch between the CTE of the filler-loaded polymer and the adjacent die materials must be accounted for through careful pre-compensation design, thermal ramp optimization, and mechanical clamping strategies. Die tilting and warping (Figure 6.33) not only affect bond integrity but also misalign subsequent photolithography steps, impacting yield in RDL formation.

Finally, integration with wafer-level or panel-level processing introduces unique scale-related challenges. While fine-pitch materials may perform well at small scales, upscaling to large panel formats increases sensitivity to flow uniformity, edge beading, and drying stresses. Advanced dispensing and coating techniques, such as inkjet printing, curtain coating, and slot-die lamination (illustrated in Figures 6.34 and 6.35), are being explored to provide spatial control of polymer deposition while maintaining process scalability. These techniques must be compatible with the existing cleanroom infrastructure, and their success depends heavily on the polymer's viscoelastic properties, curing kinetics, and environmental sensitivity.

Overcoming these processing challenges requires co-optimization of materials formulation and equipment parameters. Rheological models, CFD simulation of the flow behavior, and real-time monitoring of filler dispersion and cure progression are

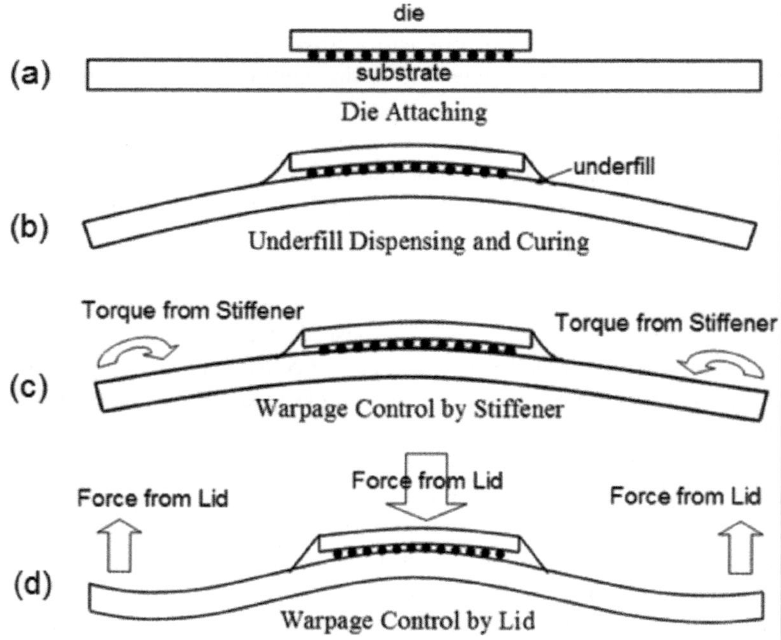

FIGURE 6.32 Assembly processes of flip chip package and mechanism of conventional ways for warpage control: (a) die attach, (b) underfill dispensing and curing, and (c) Stiffener or lid attach. (Adapted from Ref. [42], © IEEE. Used with permission.)

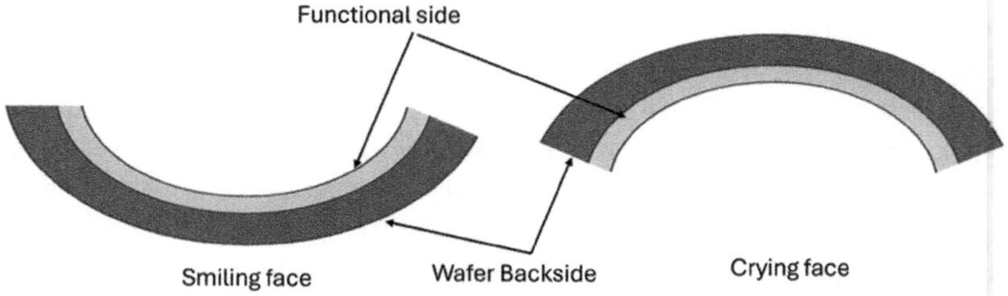

FIGURE 6.33 Definition of warpage orientation. (Adapted from Ref. [43], published in *Frontiers in Electronics*, under the Creative Commons Attribution (CC BY) license.)

increasingly employed to predict and correct processing anomalies. As feature sizes shrink and system integration becomes more complex, the processing of polymer systems in 2.5D/3D packaging must be regarded not merely as a material handling task, but as a core enabling discipline for yield, reliability, and long-term performance.

6.6.5 Reliability of TSV and Interposer Systems

Ensuring long-term reliability is paramount for polymer systems used in 2.5D/3D packaging, given the complex stress states, high thermal loads, and heterogeneous interfaces inherent to these architectures. Unlike traditional planar assemblies,

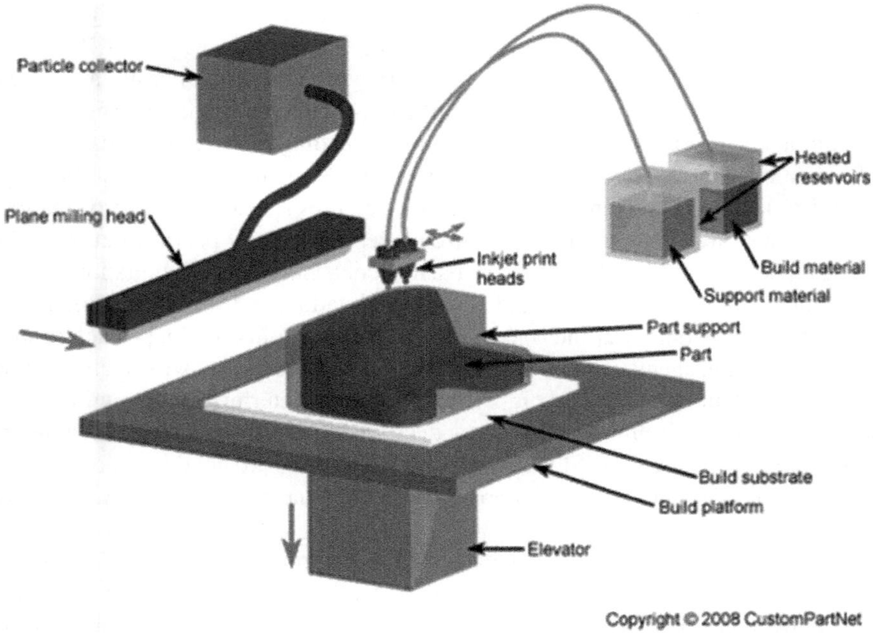

FIGURE 6.34 Schematic diagram of the inkjet printing process. (Adapted from Ref. [44] with permission from Springer Nature.)

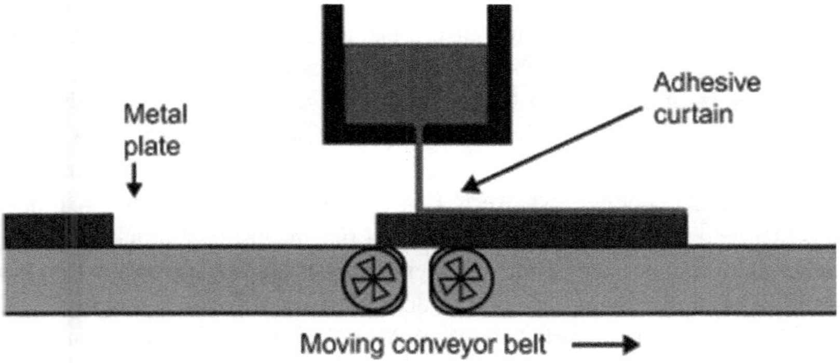

FIGURE 6.35 Schematic diagram of the curtain coating process. (Reproduced from Ref. [45] with permission from Elsevier.)

stacked structures involve multi-directional expansion, highly localized temperature gradients, and sharp CTE mismatches between polymeric encapsulants and rigid materials such as silicon, glass, or copper. These conditions generate unique degradation modes that must be addressed at both the material formulation and integration process levels.

A key reliability concern is the polymer's performance under thermal cycling. With package-level and board-level temperature excursions ranging from –40°C to 125°C or higher, repeated expansion and contraction cycles can induce cyclic shear stresses at the polymer–substrate interface. These stresses may lead to fatigue cracking, particularly at the TSV-polymer, die-to-die interface corners, and RDL step coverage zones. For example, if a polymer with a high crosslink density and corresponding high modulus is used without sufficient compliance zones, thermal fatigue can initiate microcracks that propagate over time, leading to insulation breakdown or delamination. Thus, polymers must demonstrate not only high thermal stability (e.g., $T_g > 150°C$) but also low-temperature flexibility and resistance to crack propagation.

Delamination is another critical failure mode, particularly prevalent at the polymer–Cu and polymer–oxide interfaces. Factors contributing to delamination include poor initial adhesion, filler-induced stress localization, moisture absorption, and chemical incompatibility at interfaces. The introduction of adhesion promoters such as silane coupling agents or thiol-functionalized interfaces has shown effectiveness in mitigating delamination. Still, their performance often deteriorates after aging, especially in humid environments. To assess delamination resistance, JEDEC-standard temperature cycling (e.g., JESD22-A104) and HAST tests are conducted, often complemented by scanning acoustic microscopy (SAM) and interfacial fracture toughness (G_c) measurements.

Another long-term degradation mechanism arises from dielectric breakdown and interfacial aging. As interposer systems operate at higher frequencies and voltages, the polymer dielectric must maintain a low loss and a high breakdown strength (>200 V/μm) over extended time. However, polymer aging—accelerated by heat, UV exposure, and moisture—can result in crosslink network scission, ion migration, and an increased leakage current. These effects are exacerbated in filler-rich composites if conductive fillers (e.g., silver nanoparticles or poorly isolated CNTs) create percolation paths. Thus, formulation must ensure electrical isolation and stable dielectric constant, typically <3.5, over 10^4–10^5 hours of operation.

In the context of TSV encapsulation, polymers must accommodate the thermal expansion of large-diameter Cu vias during power cycling. Cu exhibits a high CTE (~17 ppm/°C) compared to silicon (~3 ppm/°C), resulting in radial expansion of the TSV core and a hoop stress exerted on the surrounding polymer. If this stress exceeds the yield strength or fracture energy of the encapsulant, cracking or interfacial delamination may occur. Finite element analysis studies have shown that polymers with a moderate modulus (~100 to 300 MPa) and a fracture toughness >2 kJ/m^2 provide the best stress-buffering effect. Moreover, co-curing behaviors with Cu plating or oxide deposition processes must be carefully managed to avoid thermal stress accumulation during post-TSV processing.

Aging under moisture, bias, and elevated temperature—commonly referred to as MBB (moisture-biased bias) testing—reveals another dimension of polymer reliability. Hydrophilic components in the polymer matrix can absorb moisture, leading to swelling, hydrolysis, or even pop-corning during reflow soldering. Moisture also accelerates metal ion migration, which can form leakage paths or dendritic growth in high-density interposers. Low moisture uptake (<0.2 wt%), high crosslink density, and chemically inert backbones (e.g., polyimide or fluorinated resins) are design targets in such applications.

Finally, co-cure compatibility and cross-contamination with adjacent materials (e.g., bonding films, underfills, or TIMs) must be considered. In highly integrated

TABLE 6.19
Common Reliability Tests and Performance Targets for 2.5D/3D Polymer Systems

Test Type	Conditions	Typical Performance Target
Thermal cycling	–55°C to 125°C, 1,000–3,000 cycles	No delamination or crack formation
HAST/PCT	130°C, 85% RH, 2 atm, 96–168 hours	<5% change in electrical/ mechanical properties
Die shear/Peel strength	Room temp or 85°C testing	>20 MPa
Breakdown voltage	1 kHz or DC ramp	>2 kV
Leakage current	25°C and 85°C under 100 V bias	<10 nA
Moisture uptake	85°C/85% RH exposure or gravimetric test	<0.2 wt%

packages, simultaneous or sequential cure cycles are used to streamline process flows. However, mismatched cure kinetics can lead to incomplete curing, stress localization, or interface weakness. Differential scanning calorimetry (DSC) and dynamic mechanical analysis (DMA) are employed to match T_g, storage modulus, and degree of cure between co-processed materials, minimizing residual stress accumulation.

To ensure reliability, a comprehensive suite of tests is typically employed, as summarized in Table 6.19.

Overall, reliability is not simply a post-fabrication metric, but an embedded requirement in the design of polymer encapsulants for 2.5D/3D systems. Through coupling of materials chemistry, interface engineering, and predictive modeling, polymers can be tuned to deliver not just temporary functionality, but multi-year operational stability in increasingly complex system architectures.

6.6.6 Outlook: Smart and Hybrid Polymers

As heterogeneous integration evolves toward finer pitches, higher interconnect densities, and multifunctional stacking, the role of polymers will continue to expand beyond conventional mechanical and electrical functions. Future encapsulation materials in 2.5D/3D applications must not only satisfy existing reliability and processing requirements but also adapt to emerging trends in system architecture, sustainability, and intelligent functionality.

One prominent trend is the integration of polymer materials with glass interposers and ultra-thin substrates. These platforms offer reduced warpage and improved dimensional stability compared to organic build-up substrates, but pose significant adhesion and moisture management challenges. Future polymers must exhibit tailored adhesion to smooth, inert glass surfaces while maintaining low moisture uptake and high dielectric performance. Advanced silane or organophosphonate-based adhesion promoters may play a key role, and polymer backbones with inherently low polarity and hydrolytic stability—such as fluorinated epoxy or siloxane hybrids—will gain traction.

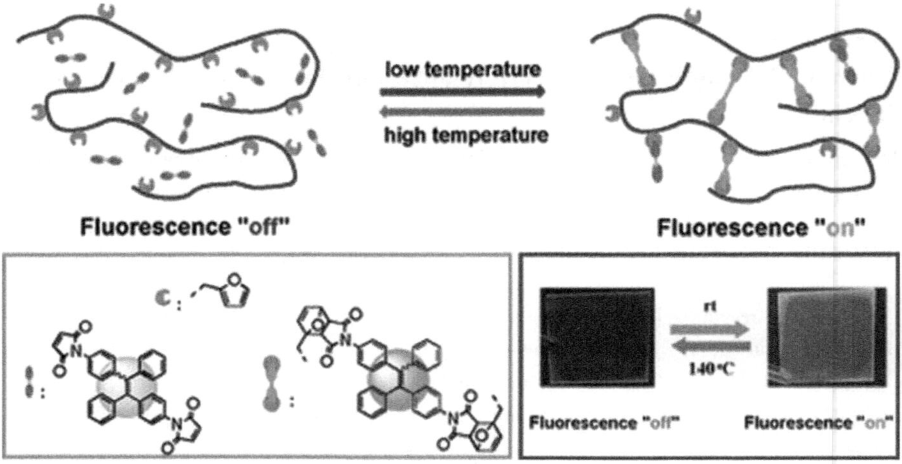

FIGURE 6.36 Reversible Diels–Alder bonds or disulfide crosslinks. (Reproduced from Ref. [46] with permission from John Wiley and Sons.)

Another critical direction is the development of smart polymers with embedded sensing, self-healing, or degradable functionalities. For instance, polymers incorporating reversible Diels–Alder bonds or disulfide crosslinks can autonomously repair microcracks induced by thermal cycling or mechanical stresses, as explained in Figure 6.36. In mission-critical aerospace or medical applications, materials capable of signaling early-stage degradation via colorimetric, fluorescence, or electrical change are under active research. These materials offer the potential for real-time health monitoring of package integrity, enabling predictive maintenance and failure avoidance in the field.

The future also calls for the hybridization of functional layers—combining adhesive, thermal interface, and underfill roles into a single formulation. These "bonding-underfill-TIM" hybrids aim to reduce process complexity and minimize interfacial defects between layers. Designing such multifunctional polymers requires a careful balance between the thermal conductivity, modulus, flow behavior, and curing compatibility. This is especially important in ultra-dense chiplets where space constraints preclude the use of separate encapsulation and thermal layers.

Sustainability and environmental compliance will become increasingly important, particularly as advanced packaging expands into consumer electronics and automotive markets. This includes the use of halogen-free, low-VOC, and potentially recyclable polymer matrices. Green chemistry approaches—such as bio-based resins, solvent-free curing systems, and degradable additives—are actively being explored. As regulations tighten and lifecycle assessments become part of materials qualification, the polymer industry will need to innovate to meet both environmental and performance standards.

Furthermore, simulation-driven design and AI-assisted formulation are poised to revolutionize how polymer systems are developed. Multiphysics modeling tools that couple rheology, thermal conduction, mechanical stresses, and cure kinetics are now enabling materials engineers to simulate void formation, interfacial delamination, and filler percolation behavior prior to fabrication. When combined with high-throughput experimentation and machine learning algorithms, these tools will accelerate the discovery of optimal polymer–filler systems with minimized trial-and-error. This is particularly vital for highly customized packages with unique thermal and mechanical demands.

Finally, as 2.5D/3D packaging is increasingly deployed in emerging application domains such as quantum computing, neuromorphic processors, and optical *I/O*, polymers must adapt to more extreme environments—ranging from cryogenic operations to high-frequency photonic integration. In such contexts, ultra-low dielectric loss, minimal coefficient of thermal expansion, and optical transparency may all be concurrently required. Meeting these multifunctional demands will require the convergence of novel chemistries, precision processing, and adaptive design methodologies.

In summary, future directions for polymer encapsulation materials in 2.5D/3D packaging are shaped by converging needs for higher integration density, multifunctionality, long-term reliability, and sustainability.

6.7 CONFORMAL COATINGS AND EDGE PROTECTION MATERIALS IN ADVANCED PACKAGING

6.7.1 Protective Roles in Harsh Environments

In the domain of advanced microelectronic packaging, conformal coatings and edge protection materials have emerged as indispensable elements that ensure the long-term functionality, mechanical resilience, and environmental integrity of electronic assemblies. These materials, though often applied as ultra-thin layers, serve critical purposes across various types of package configurations. As electronic systems continue to increase in complexity and integration density, and as packaging moves toward compact, multifunctional architectures such as System-in-Package, wafer-level packages, and heterogeneous integrations, the need for robust, high-performance conformal and edge-coating materials becomes more pressing.

Conformal coatings are thin, continuous polymeric layers applied over the surfaces of assembled electronic components. Their primary role is to protect sensitive circuits and interconnects from environmental degradation factors, including moisture ingress, ionic contamination, corrosive gases, dust, UV radiation, etc. These coatings conform closely to the topography of the underlying substrate and components, thereby ensuring comprehensive coverage without leaving vulnerable uncoated zones. This function becomes increasingly vital in surface-mounted assemblies and fine-pitch interconnect layouts, where even minute contamination or leakage paths can result in performance loss or catastrophic failure. Figure 6.37 shows examples of physical defects related to conformal coatings.

Edge protection materials, although similar in composition, are typically formulated with a higher viscosity, enabling them to act as physical barriers along exposed die edges, chip sidewalls, or laminate boundaries. Their localized application serves to prevent mechanical chipping, crack initiation, or propagation from the vulnerable package corners or interfaces, particularly in die-stack configurations, ultrathin

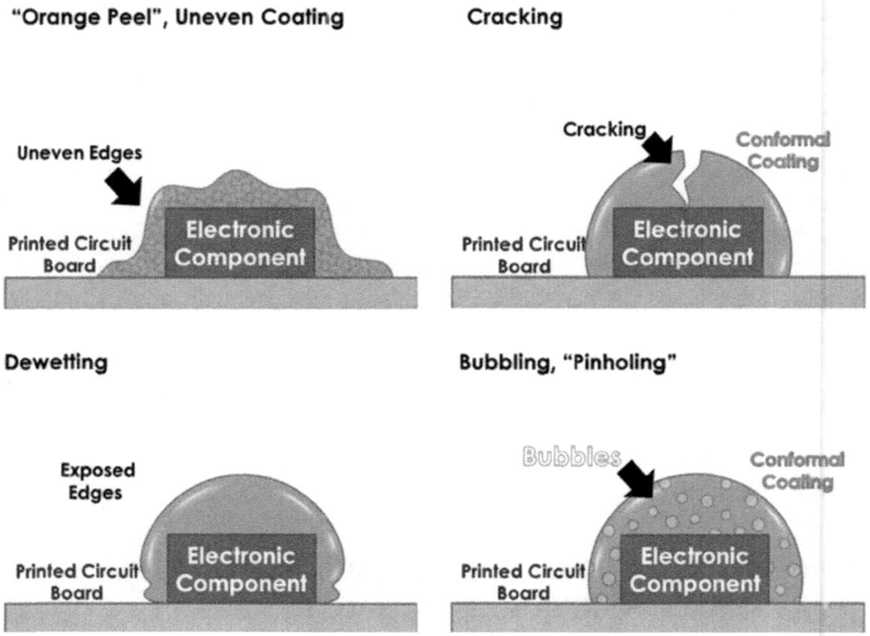

FIGURE 6.37 Key physical metrics related to conformal coating performance.

substrates, or mechanically fragile MEMS systems. In stacked-die 3D packages or high-aspect-ratio interposer structures, edge protection serves to decouple thermo-mechanical stresses between different materials and dimensions.

These coatings find widespread use in SiP architectures, where multiple dies, passives, and sensors are integrated into a compact module with minimal clearance. In such assemblies, conformal coatings prevent moisture and chemical infiltration, especially around die passivation edges, wire bonds, or ball-grid array interfaces. MEMS devices, which incorporate movable structures or exposed cavities, rely on ultra-thin conformal coatings for mechanical damping, hermetic sealing, and long-term calibration stability without impeding functional motion. RF modules, particularly those operating at high frequencies (e.g., 24–77 GHz in automotive radar), require coatings with low dielectric loss and uniform dielectric constants to preserve impedance matching while protecting waveguides and shielding cavities. Furthermore, conformal coatings are heavily used in automotive and aerospace applications, where devices must survive wide temperature cycles (–55°C to +150°C), high humidity, salt spray, corrosive oils, and mechanical vibration over decades of operation.

6.7.2 RTV, UV, Acrylic and Epoxy Coatings

Polymer systems used in conformal coatings and edge protection must be carefully selected to balance flexibility, adhesion, environmental resistance, electrical insulation, and optical transparency. A variety of polymer chemistries are available, each suited to different packaging environments and reliability requirements. Among the most prevalent material classes are RTV (room-temperature vulcanizing) silicones [47], UV-curable acrylates [48], solvent-borne acrylics, thermosetting epoxies, and increasingly, hybrid systems that merge advantageous characteristics from multiple families.

RTV silicones are among the most widely used materials for conformal coatings in high-reliability applications such as automotive, aerospace, and MEMS packaging. They are known for their inherently low modulus (typically <5 MPa), which provides superior flexibility and mechanical damping. This allows them to absorb stresses from CTE mismatches during thermal cycling without delamination or cracking. Their chemical structure imparts excellent resistance to UV degradation and oxidation, while their moisture barrier properties remain effective even after prolonged exposure to 85°C/85% RH conditions. RTV silicones cure at ambient temperature and are ideal for applications where thermal loading must be minimized, such as temperature-sensitive MEMS components.

UV-curable acrylates are often preferred for high-throughput assembly lines, where rapid cycle times and precise process control are crucial. These materials cure within seconds under UV irradiation, making them highly suitable for surface-mount assemblies and wafer-level modules that require localized application. With tunable viscosity and refractive index, UV acrylates can be formulated to achieve high optical transparency (>90% transmittance), which is particularly beneficial in applications such as camera modules, image sensors, and optoelectronic devices. Their relatively higher modulus (20–100 MPa) compared to silicones can pose challenges for stress-sensitive structures but can be compensated by reducing the coating thickness or modifying the formulation.

Acrylic conformal coatings, typically solvent-based, offer excellent coverage and ease of application using dip, spray, or selective dispensing techniques. These materials adhere well to a wide range of substrates, including FR4, polyimide, and metallic leads, and provide strong moisture resistance and moderate mechanical robustness. However, due to their solvent content, they may pose environmental and processing challenges, including the need for fume extraction and post-application drying. Their long-term aging performance and chemical resistance, although generally adequate, may not match that of silicones or epoxies in highly aggressive environments.

Thermosetting epoxy coatings are selected for applications requiring high modulus, chemical resistance, and structural protection. These coatings exhibit excellent adhesion to metals and silicon, making them ideal for die edge reinforcement in 3D stacked packages and interposer-based systems. Epoxies also provide high barrier performance against moisture and ionic contaminants. However, they tend to have higher cure temperatures (typically >120°C) and greater internal shrinkage, which can introduce residual stresses if not properly controlled.

Hybrid coating systems, such as silicone-epoxy blends or dual-cure polysiloxanes, represent an emerging category that aims to bridge the trade-offs between mechanical flexibility, barrier performance, and processing speed. For example, UV-curable siloxane acrylates offer both fast curing and excellent thermal cycling performance, making them ideal candidates for SiP and flexible device encapsulation.

The following table provides a comparative overview of these polymer systems (Table 6.20).

TABLE 6.20
Comparison of Polymer-Based Encapsulation and Coating Material Systems for Electronic Packaging

Material System	Elastic Modulus (MPa)	Curing Method	Optical Clarity	Moisture Resistance	Processing Speed	Typical Applications
RTV silicone	<5	Room-temperature cure	High	Excellent	Slow	MEMS, Automotive Modules, Edge Sealing
UV-curable acrylate	20–100	UV irradiation	High	Moderate	Very fast	Optical Sensors, Camera Modules, WLP
Solvent-borne acrylic	50–200	Solvent evaporation	Medium–High	Good	Moderate	PCB Surface Coating, RF Modules
Thermosetting epoxy	>300	Thermal cure	Low	Excellent	Slow	Die Edge Reinforcement, Harsh Environments
Hybrid systems	Tunable	Dual-cure or co-poly	Medium–High	Excellent	Fast–moderate	SiP, Flexible Devices, Selective Coating Zones

In summary, the selection of polymer coatings and edge protection systems must be tailored to the specific reliability, processing, and design constraints of the packaging platform. Whether prioritizing optical clarity, low modulus for stress relief, rapid throughput, or high barrier performance, the trade-offs between performance attributes require careful evaluation in the context of the end-use application and integration process.

6.7.3 Barrier Properties and Optical Requirements

The suitability of a polymer system for conformal coating or edge protection hinges on meeting a rigorous combination of electrical, mechanical, thermal, optical, and environmental performance targets. These requirements must be tuned to the application's functional context and lifecycle conditions.

Mechanically, the coating should be sufficiently flexible to absorb and redistribute stresses generated during thermal excursions, mechanical shocks, or vibration events. A low elastic modulus (typically <10 MPa) is essential for minimizing the interfacial stress in conformal applications, while a higher modulus may be desirable for edge protection where structural integrity is needed. Elongation at break above 50% is generally considered desirable for coatings subjected to flexural or tensile strains. The viscoelastic nature of silicones and polysiloxanes makes them particularly effective in this role.

Thermally, coatings must endure wide temperature ranges—often from –55°C to +150°C—without embrittlement, delamination, or performance drift. This necessitates

TABLE 6.21
Key Performance Requirements and Target Values for Advanced Electronic Packaging Polymers

Requirement	Typical Target Values	Relevance
Elastic modulus	<10 MPa (conformal); >100 MPa (edge)	Mechanical compliance and structural integrity
Elongation at break	>50%	Flexibility under thermal and mechanical strain
Thermal stability	–55°C to +150°C	Automotive, aerospace, and industrial environments
Moisture absorption	<0.5%	Long-term environmental durability
Dielectric constant (*k*)	≤3.0	Signal integrity and RF performance
Breakdown strength	>100 V/μm	Electrical insulation between fine-pitch interconnects
Optical transmittance	>90%	Imaging, sensing, and optical communication modules
Cure shrinkage	<2%	Dimensional control and stress minimization

high glass transition temperature (T_g), low thermal expansion, and stability against oxidation or outgassing. Materials should exhibit minimal shrinkage during curing to avoid warpage or crack formation, especially in sensitive WLP or SiP configurations.

Moisture and chemical resistances are critical parameters in preventing corrosion, delamination, or electrical leakage. The water vapor transmission rate and ionic permeability determine the longevity of protection, particularly under 85°C/85% RH or salt fog conditions. RTV silicones and epoxy hybrids typically demonstrate low WVTR values, with epoxy-siloxane hybrids often achieving <1 $g/m^2/day$.

From an electrical standpoint, a high surface and volume resistivity, low dielectric constant and low dissipation factor are required to maintain signal fidelity and prevent leakage current. The dielectric breakdown strength should exceed 100 V/μm for most high-density modules. Materials used in RF, radar, or high-speed data transmission must ensure uniformity and minimal loss across GHz frequencies.

Optically, coatings for photonic or sensing applications must maintain >90% transmittance across the visible or infrared spectrum, minimal birefringence, and UV stability to prevent yellowing or haze. Fluorinated or cycloaliphatic acrylates are often used in optical windows and camera modules for these reasons.

The following table summarizes key performance targets (Table 6.21).

Meeting these requirements often involves trade-offs. For example, lowering the modulus may increase the moisture permeability, while enhancing the thermal stability may reduce the optical clarity. Thus, formulation and process optimization are essential for achieving balanced performance suitable for advanced wafer-level and fan-out packaging architectures.

6.7.4 Application Methods and Patterning

Application techniques for conformal coatings in advanced packaging are not merely ancillary steps in the manufacturing process; rather, they are foundational to the successful realization of robust environmental barriers and mechanical protection layers. The choice of the coating method is governed by a complex interplay of factors, including the physical geometry of the device, materials compatibility, production throughput, uniformity requirements, and the thermal or UV sensitivities of the components being protected [49]. As packaging architectures diversify—with fine-pitch interconnects, multi-die stacking, and exposed microstructures—there is a corresponding need for greater precision and adaptability in conformal coating processes.

Spray coating remains a dominant method for applying conformal coatings, particularly in large-volume applications such as printed circuit boards and flat modular assemblies. In this process, the coating material is atomized into fine droplets and applied via programmable robotic nozzles. The robotic systems ensure reproducibility and a consistent film thickness, with adjustable parameters such as nozzle speed, atomization pressure, standoff distance, and spray pattern. However, despite its versatility, spray coating poses challenges in achieving uniform coverage over topographically complex substrates. Overspray, for instance, is a common issue that not only leads to materials waste but also risks contaminating adjacent sensitive regions, necessitating careful masking and sometimes extensive post-process cleaning. To improve spray uniformity, the use of electrostatic spray nozzles and multi-pass deposition strategies are considered.

Dip coating is well-suited for components with intricate geometries or vertically protruding structures, such as wire-bonded dies, leaded packages, and multi-tiered modules. The entire assembly is immersed in a coating solution and then withdrawn at a controlled speed, allowing a continuous film to form via gravitational drainage and solvent evaporation. The film thickness is modulated by the withdrawal rate, solution viscosity, and ambient conditions (e.g., humidity and temperature). This method ensures 360-degree coverage and is valued for its ability to penetrate recessed areas, yet it is prone to pooling in cavities and edges, which can create localized thickness variations and curing defects. Addressing these requires careful formulation of the coating's rheological profile—balancing flowability with leveling and edge retention—as well as pre-conditioning steps such as surface plasma treatment to enhance wettability.

Selective dispensing is indispensable in heterogeneous packaging modules that incorporate sensitive MEMS components, high-frequency RF elements, or optically active regions that must remain uncoated. This technique uses micro-needle dispensing heads or jetting systems to apply coating materials only to designated areas. Coupled with real-time vision systems and high-precision motion stages, selective dispensing allows for micron-scale targeting and tailored coating paths. In fan-out wafer-level packages, for instance, selective coating is applied to edge regions around chiplets or RDLs to mitigate delamination while leaving central areas exposed for further processing. One of the main technical challenges in selective dispensing lies in managing the dot size consistency, maintaining adhesion on sloped or irregular surfaces, and ensuring seamless merging of dispensed lines to form a uniform layer without gaps.

Advanced patterning techniques—such as inkjet printing, screen printing, and stencil deposition—are being increasingly adopted in next-generation packaging formats that require microscale resolution and materials efficiency. Inkjet printing offers a non-contact, digital approach that enables precise deposition of low-viscosity coatings in customized patterns. Its compatibility with flexible substrates, curved surfaces, and multi-material stacks makes it ideal for wearable electronics and photonic devices. However, the formulation must be optimized to avoid nozzle clogging, satellite droplet formation, and insufficient film continuity. Screen printing, in contrast, supports higher-viscosity materials and is more tolerant to particulate loading (e.g., nanofillers), but requires physical masks and is less adaptable to geometry variations. Stencil printing occupies a middle ground, allowing moderate resolution and a consistent film thickness for features such as edge guards and underfill dams in flip-chip packages.

The curing process is a critical integration point that can determine both the functionality and reliability of the coating. Thermal curing, which relies on elevated temperatures to drive crosslinking reactions, is suitable for robust substrates such as ceramic packages or high-T_g laminates. However, it may induce thermal stresses or component warpage in thin wafers or temperature-sensitive assemblies. UV curing, which utilizes photoinitiated polymerization, offers much faster cycle times and is compatible with low-temperature processing. Its effectiveness, however, is contingent on optical access, limiting its use in shadowed areas or thick coatings where light penetration is insufficient. To address this, dual-cure systems have emerged, combining initial UV exposure with a thermal post-cure to ensure full-depth polymerization and improved mechanical integrity. This hybrid approach also supports staged processing, enabling temporary fixation during alignment or inspection prior to final curing.

Table 6.22 provides a comparative overview of the major coating methods used in advanced conformal coating applications.

In conclusion, the selection and optimization of coating and dispensing methods for conformal materials must account for not only materials properties and substrate complexity, but also integration constraints, yield implications, and the evolving requirements of miniaturized electronic systems. Advanced dispensing technologies and hybrid curing protocols represent the direction of future developments, aiming to deliver ultra-precise, environmentally stable, and scalable coating solutions across increasingly diverse packaging formats.

TABLE 6.22
Comparison of Coating Techniques for Electronic Packaging Applications

Coating Method	Coverage Pattern	Suitable Geometries	Process Flexibility	Material Utilization	Common Applications
Spray coating	Area-based	Flat, uniform surfaces	High	Moderate–Low	PCBs, modular electronics
Dip coating	Full immersion	Complex 3D or multi-height	Moderate	High	Wired dies, pin packages
Selective dispensing	Localized	Cavities, edges, openings	High	High	MEMS, chip edges, EMI shielded zones
Inkjet/screen printing	Patterned	Flat/curved, micro-patterned	High	High	WLP, photonics, localized microstructures

6.7.5 Reliability Issues: Cracking, Ionic Migration

In the landscape of advanced microelectronics packaging, the long-term performance of conformal coatings and edge protection materials is inextricably linked to their reliability under diverse operational stresses. As electronic components are deployed in increasingly harsh or dynamic environments—ranging from automotive and aerospace to industrial and consumer applications—the ability of coatings to maintain their protective functions over extended lifespans becomes a fundamental qualification criterion. Reliability assessment thus extends beyond basic adhesion and coverage quality, delving into the multifaceted interactions between material properties, processing conditions, and external stimuli.

A dominant category of reliability failures is mechanical degradation, where cracking and delamination emerge as primary concerns. Cracking typically results from either excessive shrinkage during cure or thermomechanical mismatch between the coating and underlying materials. For example, coatings with high crosslink density or high glass transition temperatures may develop residual stresses upon cooling from cure temperatures, particularly when applied over substrates such as copper, glass-fiber laminates, or silicon dies with disparate CTE. These internal stresses can accumulate over thermal cycles, eventually leading to microcrack formation. Microcracks act as ingress pathways for moisture and ionic species, undermining the integrity of the protective barrier and accelerating further degradation. Delamination, on the other hand, can originate from poor surface preparation, contamination by processing residues (e.g., flux or fingerprint oils), or inadequate wetting and flow of the coating material. This interfacial failure not only compromises the environmental seal but also enables capillary ingress during humidity exposure or pressure cycling, a particularly concerning scenario in high-reliability applications. Figure 6.38 shows the process of delamination.

Moisture and gas permeation constitute another major reliability vector. Despite being marketed as moisture-resistant, many coatings exhibit non-zero permeability to water vapor, oxygen, and other corrosive gases over extended durations. In humid or salt-laden atmospheres, such as coastal or automotive environments, even slow ingress can result in corrosion of aluminum bond pads, copper traces, or solder joints. Chloride and sulfate ions, once introduced, can migrate under bias and initiate electrochemical corrosion. To assess this risk, accelerated environmental tests such as 85°C/85% RH, highly accelerated stress testing (HAST), and salt fog testing are employed. Materials with poor moisture barrier performance show early-onset discoloration, ionic migration, or metallization breakdown under such tests. In advanced assessments, moisture transmission rate and water uptake (by weight or volume) are quantified to predict coating lifespan in target applications.

Ultraviolet (UV) degradation represents a third failure mechanism, particularly in optically exposed modules such as image sensors, optical MEMS, or LiDAR components. UV photons can break down polymer chains, resulting in yellowing, haze, or embrittlement. For instance, coatings based on acrylics or certain urethanes are especially prone to photodegradation unless stabilized with UV absorbers or hindered amine light stabilizers (HALS). In high-irradiance conditions, UV-induced oxidation can reduce the mechanical flexibility of the coating, leading to surface

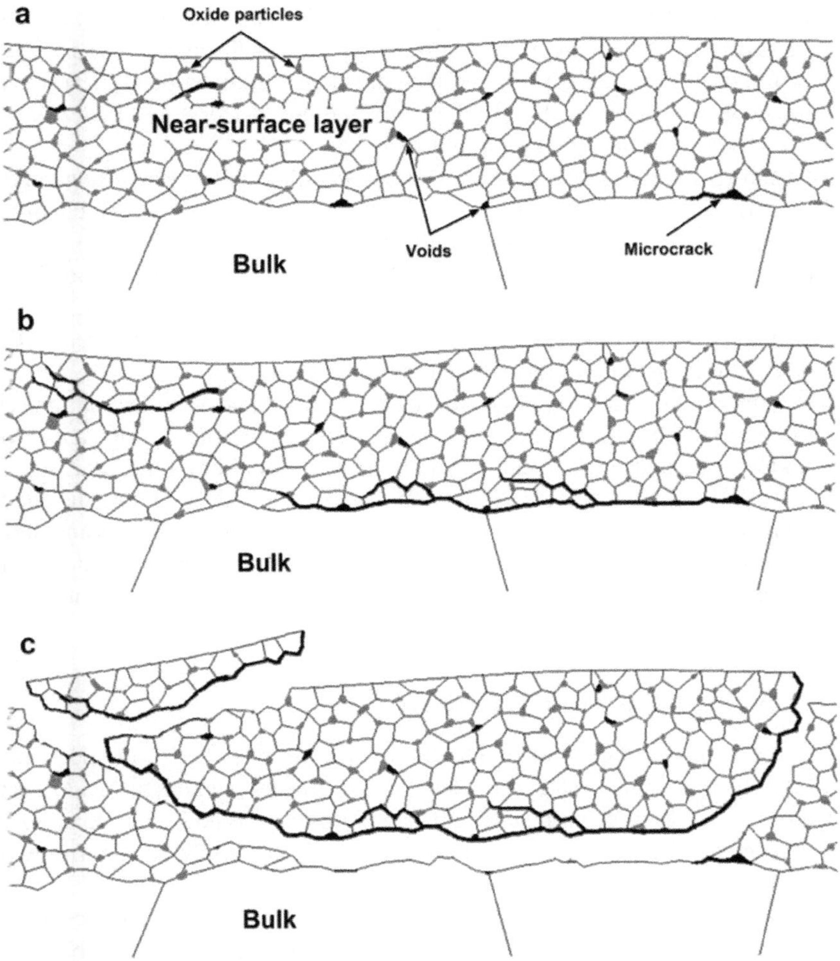

FIGURE 6.38 Schematic diagrams showing the delamination process of the near-surface layer: (a) crack initiation; (b) crack propagation; (c) fracture. (Reproduced from Ref. [50] with permission from Elsevier.)

cracks or edge peeling. The loss of optical clarity further deteriorates signal quality in optical interconnects or camera lenses, thereby compromising the overall system performance.

Electrical failure due to ionic contamination and dielectric breakdown are particularly problematic in high-density RF and power modules. Trace ionic residues in the coating formulation, or introduced during handling, can migrate under an electric field, forming conductive filaments or dendrites. Over time, this leads to leakage currents, arcing, or insulation breakdown, especially under high voltage or high humidity conditions. Partial discharge is a critical concern in power electronics, where local electric fields exceed the dielectric strength of the coating, resulting in plasma formation and irreversible polymer degradation. To mitigate this, coatings must demonstrate a low ionic content (typically <10 ppm for Na+, K+, Cl−, etc.), high volume resistivity (>10^{14} Ω·cm), and strong dielectric withstand properties (>2 kV for a 50 μm film). Additionally, electrical characterization under damp heat bias and pressure cooker testing provides insights into long-term performance under simultaneous electrical, thermal, and humidity loads.

Thermal reliability is assessed through repeated thermal cycling, often between −55°C and +125°C or beyond, to simulate real-world operating conditions. Coatings with poor flexibility or high modulus may undergo fatigue cracking at the coating-substrate interface. Stress accumulation at corners, edges, or bond wires leads to fracture initiation, especially in thin or underfilled regions. The inclusion of thermomechanical modeling during material development helps predict these high-stress zones and guides formulation improvements.

In summary, the reliability of conformal coatings and edge protection materials is not determined solely by their initial properties, but by their performance over time under combined mechanical, thermal, electrical, and environmental stresses. The interplay between materials formulation, application processes, and the end-use environment necessitates holistic testing and design strategy. Addressing failure modes such as cracking, delamination, permeation, UV degradation, and electrical breakdown is essential to ensure that these protective layers fulfill their critical roles in safeguarding next-generation electronics packaging systems.

6.7.6 Trends in Green and Nano-Coatings

The field of conformal coatings is rapidly evolving in response to the demand for multifunctionality, higher integration density, and greener manufacturing. Materials research is focusing on performance-enhanced formulations and smart coatings that move beyond passive protection. Figure 6.39 demonstrates synthesis of partially bio-based polymers and integrated biobased polymer.

Next-generation UV-curable polymers are being tailored for nanosecond-scale curing, a ultra-low modulus, and enhanced elongation, enabling compatibility with flexible electronics, wearable devices, and bio-integrated sensors. These materials offer better compliance without sacrificing processing speed.

Nanocomposite coatings are incorporating functional nanofillers such as SiO_2, Al_2O_3, graphene oxide, or boron nitride to improve mechanical toughness, barrier properties, and even introduce directional thermal conductivity. These systems offer new performance dimensions while maintaining compatibility with traditional coating methods.

Multifunctional coatings that integrate EMI shielding, thermal conduction, and even energy harvesting functions are under development. For example, coatings with embedded conductive networks or phase-change elements can serve dual roles in RF module isolation and thermal regulation.

Sustainability is an increasingly important driver. Solvent-free, low-VOC, and waterborne coatings are being adopted to comply with environmental regulations and reduce processing hazards. Bio-derived and degradable polymers are being explored for disposable or consumer electronics where the end-of-life impact is significant.

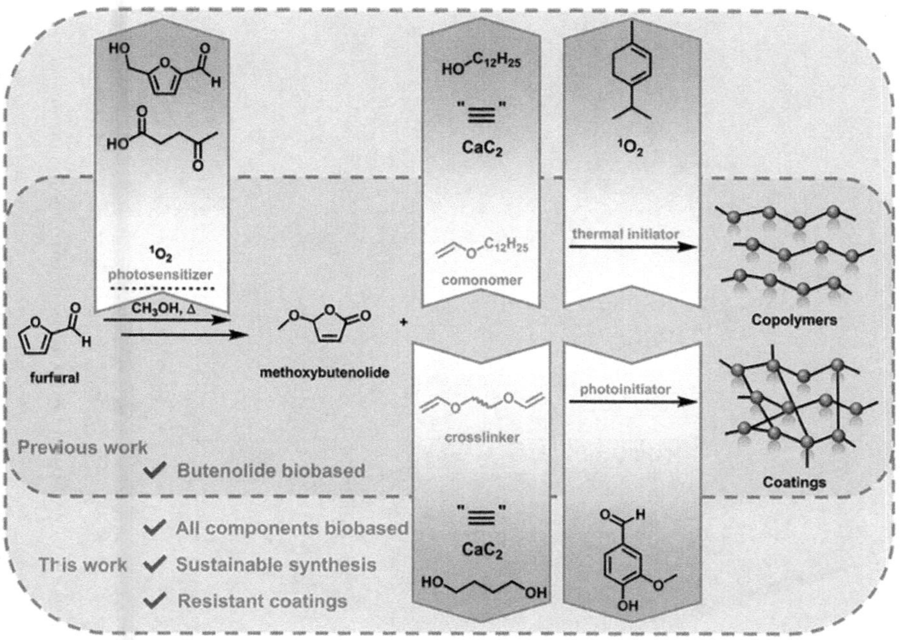

FIGURE 6.39 Synthesis of partially biobased polymers and coatings (blue). Integrated bio-based polymer and coating system with all components derived from biorenewable resources (green). (Adapted from Ref. [51] with permission from the Royal Society of Chemistry.)

For space-constrained systems, ultrathin (<10 µm), optically clear coatings with tailored refractive index, minimal haze, and robust thermal/mechanical properties are enabling protection without compromising form factors. These coatings are essential for next-generation optical and optoelectronics packaging, where performance and transparency must coexist with durability and manufacturability.

In summary, conformal coatings and edge protection materials are no longer ancillary elements but integral parts of the advanced packaging design paradigm. Their selection, integration, and long-term performance have direct implications on system-level reliability, environmental endurance, and functional diversity. Future development will continue to emphasize hybrid performance, processing efficiency, and sustainable innovation to meet the demands of increasingly miniaturized and multifunctional electronic systems.

6.8 MATERIALS FOR HETEROGENEOUS INTEGRATION AND PHOTONICS

6.8.1 Materials for New Integration Scenarios (SiP, Co-Packaged Optics)

Heterogeneous Integration (HI) refers to the co-integration of multiple device technologies—such as logic, RF, photonics, MEMS, and sensing—within a unified packaging platform, -while 2.5D/3D technologies emphasize the physical stacking

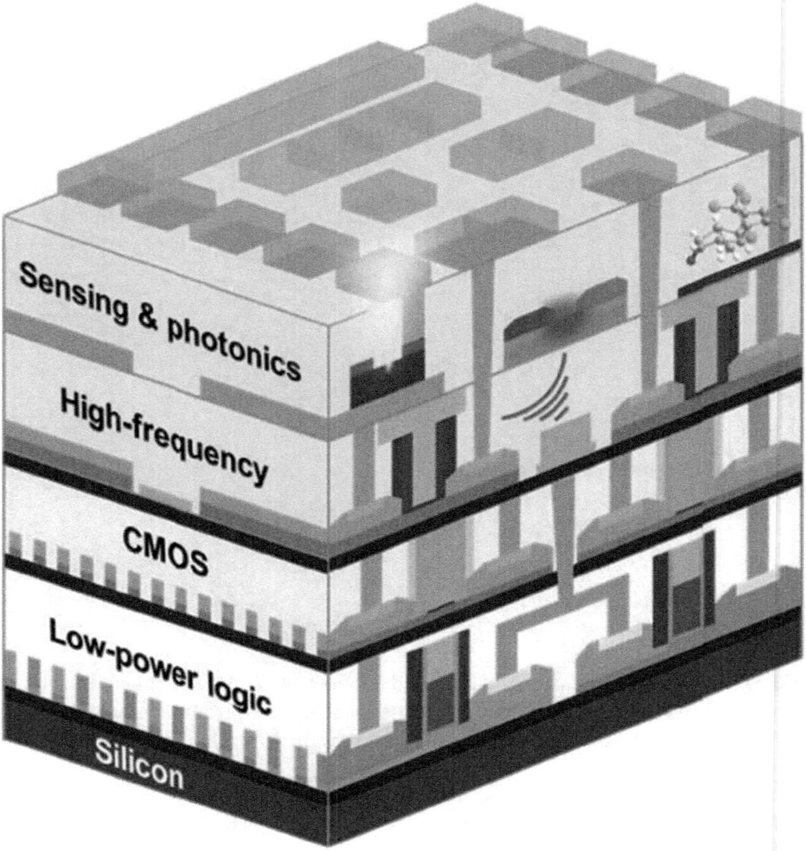

FIGURE 6.40 Illustration of a heterogeneous 3D integrated system combining multiple functional layers, including low-power logic, CMOS logic, high-frequency RF, sensing, and photonic components. Such multi-functional stacking exemplifies system-level heterogeneous integration across diverse device technologies. (Adapted from Ref. [52], © IEEE. Used with permission.)

or interposer-based side-by-side placement of multiple dies to achieve higher interconnect density and bandwidth. Figure 6.40 illustrates one representative HI architecture, where low-power logic, CMOS circuits, high-frequency RF devices, photonic components, and sensing elements are vertically integrated within a single multi-layer system. This type of functional stacking highlights the essence of HI: bringing together heterogeneous device technologies to achieve system-level capabilities that cannot be realized through conventional die stacking alone exemplifies functional diversification and the ability to combine dies fabricated in different process nodes or wafer technologies.

In HI architecture, the encapsulation materials must support chiplet-based designs with extremely tight die-to-die spacing, varying die thicknesses, and multiple vertical integration tiers. Materials solutions must conform to small pitches, ultra-thin layers, and complex layouts, while preserving electrical performance and minimizing thermal expansion mismatches. The deployment of HI in high-performance computing (HPC), AI accelerators, and RF front-ends further compounds the demand for warpage control, co-planarity, and thermomechanical reliability.

Photonics packaging, though often considered a subset of HI, introduces unique requirements due to the co-integration of electronic and optical functions. Unlike purely electrical systems, photonic modules require optical transparency, controlled refractive indices, and extremely low surface roughness at interfaces. The packaging must accommodate components like laser sources, waveguides, photodetectors, and optical fibers, often with sub-micron alignment tolerances. Additionally, photonics packages are more sensitive to moisture ingress and UV degradation, which can impair optical clarity and coupling efficiency. As photonic integration becomes central to next-generation interconnects (e.g., Co-Packaged Optics or CPO), the convergence of optical, electrical, and thermal domains within a single packaging stack places unprecedented demands on materials.

Given these differences, it is essential to approach HI and photonics packaging not as incremental extensions of 2.5D/3D systems, but as fundamentally new packaging paradigms that require integrated material-function co-design.

6.8.2 Optical-Grade Polymers and Low-Loss Resins

Encapsulation and structural materials used in HI and photonics packaging must fulfill highly specialized and often conflicting requirements across multiple domains: optical transparency, electrical insulation, high-frequency signal integrity, thermal conductivity, mechanical reliability, and chemical resistance. This complexity demands a multipronged approach to materials selection and formulation.

In photonic integration scenarios, optical-grade polymers are paramount. These include UV-curable acrylics, silicone elastomers, and epoxy resins engineered for high transmittance in the near-infrared spectrum (850–1,550 nm), where most silicon photonics and VCSELs operate. These polymers must not only be optically clear (transmittance >90%) but also exhibit low haze (<1%), minimal birefringence, and stable refractive index over time and temperature. This ensures minimal signal distortion and consistent optical coupling. Moisture uptake must be tightly controlled (<0.1 wt%) to prevent delamination, haze formation, and optical path degradation.

From the electrical perspective, RF and high-speed digital modules demand low-k, low-loss encapsulants to preserve signal fidelity, and dielectric constant values below 3.0 and loss tangent values under 0.002 are typical targets. These values reduce parasitic capacitance and insertion loss, especially critical for mmWave and 5G applications. Popular materials include BCB, liquid crystal polymers (LCP), and cyclo-olefin polymers (COP), which combine a low dielectric loss with high thermal stability and low moisture uptake.

Thermal management is another core challenge. HI packages often include high-power logic, RF front-ends, or optical sources that generate localized heat. Encapsulants must integrate thermally conductive but electrically insulating fillers such as BN, Al_2O_3, or surface-modified silica. These fillers enable heat dissipation pathways while maintaining electrical isolation. However, in photonic areas, filler selection must avoid refractive index mismatches and scattering, which would

TABLE 6.23
Application-Oriented Materials Requirements for Electronics Packaging

Application Focus	Property Target	Example Material Systems
Optical encapsulation	>90% transmittance (850–1,550 nm), low haze	Silicone, UV acrylics
RF signal integrity	$D_k < 3.0$, $D_f < 0.002$	BCB, LCP, COP
Thermal management	>2 W/m·K, electrically insulating	Epoxy + BN/Al_2O_3
Mechanical compliance	Modulus < 1 GPa, CTE-matched	Flexible silicones, hybrid resins

degrade optical performance. The filler particle morphology, size distribution, and surface treatment become critical to balancing thermal performance with optical clarity.

Mechanically, the wide range of components within HI modules—including MEMS structures, antennas, and thin dies—requires encapsulants with adaptable modulus and CTE. For example, materials must be compliant enough to avoid damaging fragile MEMS actuators (<1 GPa modulus), yet stiff enough to suppress warpage in large-area interposers. Glass transition temperatures (T_g) above 150°C are also needed to ensure dimensional stability through reflow and extended thermal aging.

A comparative table of key application-driven material metrics is shown in Table 6.23.

These diverse demands often preclude the use of a single material for the entire package. Instead, hybrid or layered encapsulant systems are adopted, where optical-grade materials protect photonic components, low-*k* polymers insulate RF modules, and thermally enhanced resins dissipate heat in high-power regions. Such heterogeneous materials architectures, tailored to the functional domains within the package, are becoming the new norm in both HI and photonics packaging.

6.8.3 Advanced Substrates (LCP, COP, Hybrid)

To meet the escalating performance and integration demands of heterogeneous integration (HI) and photonics packaging, materials innovation has increasingly shifted toward multi-functional, application-specific encapsulant systems. Rather than relying on conventional homogeneous polymers, emerging systems are now based on advanced formulations that exhibit spatially tunable properties, engineered interfaces, and synergy between optical, thermal, electrical, and mechanical functions.

One key development in this direction is the use of acrylics and silicones tailored for optical encapsulation. In particular, UV-curable acrylic resins have gained traction due to their rapid curing, controllable viscosity, and high optical clarity. These materials are commonly used to encapsulate vertical-cavity surface-emitting lasers (VCSELs) (Figure 6.41), photodiodes, and optical fibers in co-packaged optics (CPO) or silicon photonics modules. Compared to thermally cured epoxies, UV acrylics offer better process flexibility and lower thermal budgets, which is critical for thermally sensitive photonic components. Silicones, on the other hand, are preferred in high-reliability applications such as automotive LiDAR and aerospace systems,

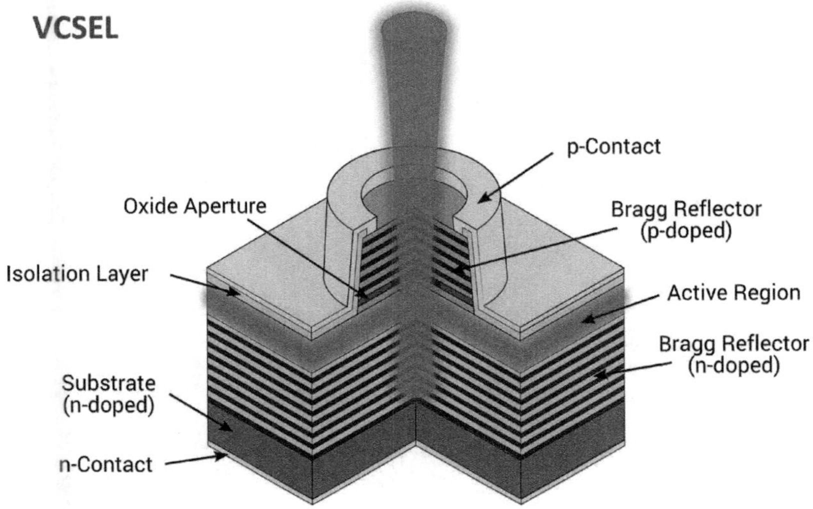

FIGURE 6.41 Schematic diagram of a typical VCSEL device formed by an active layer of semiconductor material between two Bragg reflectors. (Adapted from Ref. [53] with permission from the American Physical Society.)

where long-term UV stability, a wide service temperature range (−55°C to 200°C), and a low modulus are needed to accommodate mechanical vibrations and thermal cycling.

Beyond base polymers, researchers and industry developers are turning to functionally graded polymers (FGPs), which spatially modulate composition and cross-link density within a single encapsulant body. For instance, the area directly above a photonic waveguide may consist of a low-refractive-index and high-transmittance segment, while the surrounding regions incorporate thermally conductive and mechanically reinforced domains. These gradients are often achieved through multi-step processing, localized UV exposure, or in-situ migration of functional nanoparticles. FGPs enable performance zonation without requiring discrete material transitions, thus avoiding interfacial discontinuities and adhesion concerns that commonly plague multilayer systems.

Another major advancement lies in the adoption of novel substrate-compatible polymers, particularly for flexible and high-frequency modules. LCPs and COPs are two standout systems (as shown as Figures 6.42 and 6.43). LCPs, composed of aromatic polyesters with rigid mesogenic units, provide a low dielectric loss (D_f < 0.002) and excellent dimensional stability even under reflow conditions. They are increasingly used as build-up layers in flexible interposers and antenna modules for mmWave applications. COPs, which are non-polar hydrocarbon-based thermoplastics, exhibit high transparency, low water uptake, and good thermal resistance. Their ability to be injection molded or extrusion cast into thin films makes them well-suited for photonic circuit substrates, especially where low birefringence is critical.

An important strategy in thermal management within HI systems is the integration of aligned filler networks. Rather than relying on randomly dispersed particles,

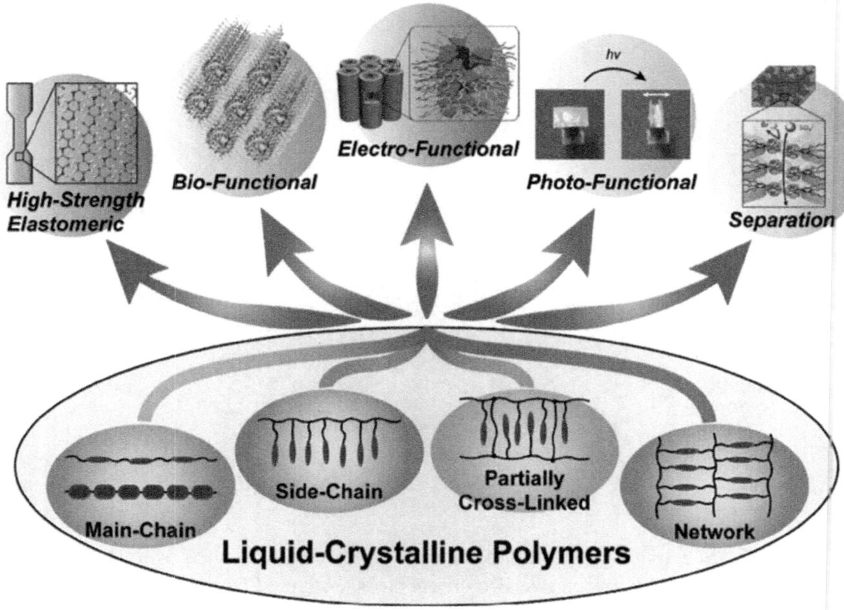

FIGURE 6.42 Design and applications of LCPs. (Reproduced from Ref. [54] with permission from Springer Nature.)

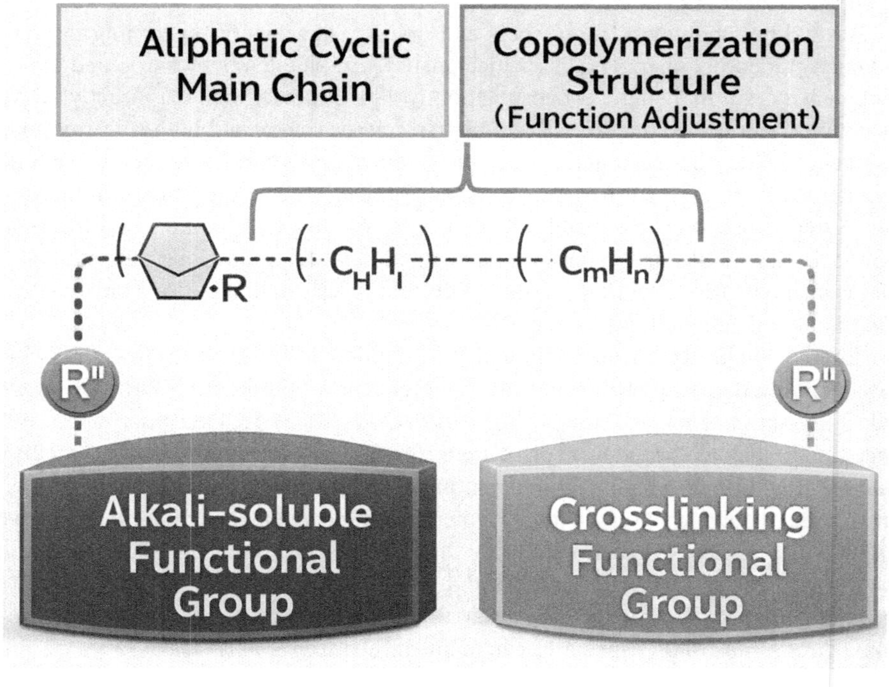

FIGURE 6.43 Structural design of linear-chain COPs.

TABLE 6.24
Functional Material Systems for Emerging Photonics and Packaging Applications

System Type	Functional Benefit	Key Use Cases
UV-curable acrylics	Fast processing, high optical clarity	VCSEL encapsulation, fiber attach
Silicones (optical-grade)	UV durability, mechanical flexibility	LiDAR, MEMS sensors, automotive photonics
Functionally graded polymers	Spatial control of modulus, RI, thermal pathways	Photonic interposers, hybrid optical-electrical
LCPs and COPs	Low-k, low-loss, low moisture	mmWave antennas, photonic substrates
Aligned filler networks	Anisotropic thermal conductivity	Chiplet modules, RF front-end, AI accelerators

thermal fillers such as h-BN, graphite nanoplatelets (GNP), and AlN are now being directionally aligned during processing to form anisotropic heat conduction pathways. This is particularly beneficial in chiplet-based HI modules where localized hotspots—such as AI compute cores—require lateral or vertical heat extraction toward heat spreaders or integrated thermal vias. Techniques such as magnetic field alignment, shear-induced orientation, or 3D printing are employed to realize filler alignment. However, for photonic packaging, where optical path clarity is essential, such fillers must be carefully confined to thermal zones away from waveguide interfaces.

These advanced materials systems increasingly demand co-design with process technologies. For example, in wafer-level integration of photonic interposers, a hybrid stack might involve: (i) a UV-curable acrylic coating for VCSEL lensing; (ii) a BCB layer for dielectric insulation of electrical interconnects; and (iii) a thermally conductive epoxy film for chip backside encapsulation. Each material must be compatible with its neighbors in terms of cure schedule, shrinkage behavior, and chemical stability. This necessitates the use of multi-physics simulation tools and in-line metrology to ensure that materials interact synergistically without introducing voids, warpage, or delamination.

To summarize, the following table outlines several categories of advanced polymer systems now being deployed or explored in cutting-edge HI and photonics packaging (Table 6.24).

In essence, advanced systems for heterogeneous and photonics packaging are no longer defined merely by bulk materials properties, but by how precisely materials can be engineered, patterned, and integrated in space and function. As packaging becomes a key enabler of system-level innovation, the encapsulation materials themselves evolve into critical design levers.

6.8.4 Processing Challenges in Optics and HI

The complexity of heterogeneous integration and photonics packaging extends beyond materials selection into the domain of processing, where even minor deviations can lead to performance degradation or outright failure. Unlike conventional electronics packages, HI and photonics modules demand tight control over the flow behavior, curing kinetics, alignment, and post-cure dimensional stability across multiple materials within a confined and functionally segmented space.

In photonics packaging, the processing of optical encapsulants necessitates bubble-free application and high optical uniformity. Bubbles, voids, or embedded particulates can scatter light or distort beam profiles, leading to losses in coupling efficiency and photonic signal integrity. To mitigate this, low-viscosity, degassed resins are often dispensed under vacuum conditions or within nitrogen-purged environments. In addition, capillary-driven self-filling behavior is leveraged in narrow waveguide trenches to ensure full wetting without mechanical displacement of the photonic elements. Some photonic systems use jet-dispensing or stencil-printing techniques with custom-designed patterns to define resin coverage zones without affecting nearby metal traces or vias.

Light-transmissive polymer selection must be tightly coupled with exposure processes. For example, UV-curable optical resins should have absorption spectra well-matched to the UV source (typically 365–405 nm), ensuring complete and uniform polymerization. This requires consideration of optical penetration depth to avoid under-cured regions that might soften under heat or pressure during subsequent assembly. For materials used in VCSEL lens molding or fiber attach regions, surface tension must also be optimized to maintain controlled dome profiles or lens curvature upon curing. In these scenarios, UV flood exposure or proximity lithography tools are employed alongside thermal post-curing to drive the crosslink density while minimizing shrinkage-induced distortion.

In HI systems, multi-chip encapsulation processes often require materials that are compatible with sequential curing. For example, in chiplet-based integration with redistribution layers, encapsulants may be dispensed and partially cured to provide temporary stability before final alignment and lamination. B-stage materials are particularly important when encapsulation must occur prior to high-temperature reflow steps or fan-out redistribution. These materials should exhibit minimal bleed, controlled rheology, and good tack at intermediate cure stages. Furthermore, their final cure must proceed without excessive volumetric shrinkage to avoid stress buildup at die edges or vertical interconnects.

Thermal and mechanical stresses during cure must be modeled and managed through both resin formulation and process tooling. Materials with steep viscosity drop-offs near the gel point can lead to filler sedimentation or die shift, especially in thin gap encapsulations (e.g., <50 μm). To counteract this, processing strategies may include stepwise ramping of temperature, vibration-assisted leveling, or filler morphology tuning (e.g., using spherical, low-aspect ratio particles). For photonic regions, the use of transparent molds or silicone release liners is common to avoid surface defects during demolding.

Multi-layer integration and co-curing further complicate processing. In advanced modules, photonic, logic, RF, and memory dies may each require distinct materials with dissimilar cure temperatures or thermal expansion profiles. Careful scheduling of sequential lamination, partial cure and final global cure must be devised to prevent warpage, delamination, and reflow-induced misalignment. Crosslinking chemistry

TABLE 6.25
Processing Requirements and Constraints for Advanced Packaging Polymers

Processing Element	Requirement	Notes
Bubble control in optics	Voids < 10 μm tolerated	Use vacuum dispensing or degassed materials
UV-curing compatibility	Full cure at specific wavelengths (365/405 nm)	Consider penetration depth and cure-through thickness
Sequential die encapsulation	B-stage behavior with stable tack	Needed for chiplet stacking or reflow compatibility
Dimensional stability	<50 μm warpage across 20 mm package span	Depends on CTE match, shrinkage, and mold uniformity
Mixed material co-curing	Compatible chemistries, similar T_g or cure kinetics	Epoxy-silicone, acrylate-epoxy combinations must be tuned

compatibility is essential: acrylates, epoxies, and silicones must not inhibit each other's cure or undergo interfacial slippage. In this context, simulation-assisted process planning is increasingly adopted to predict flow fronts, curing gradients, and stress distribution under real mold geometries and cure profiles.

The following table summarizes some key process parameters and requirements in photonics and heterogeneous encapsulation (Table 6.25).

Overall, the processing of encapsulants in HI and photonics is no longer a passive downstream step—it is a key design constraint and enabler that must be co-optimized with materials and package architecture. The interplay of flow control, cure sequencing, and interfacial fidelity governs not only device yield but also long-term optical and electrical reliability.

6.8.5 Reliability in High-Frequency/Optical Modules

Reliability in HI and photonics packaging is governed by a unique combination of mechanical, thermal, optical, and electrical stressors that differ significantly from those seen in conventional packaging. The increasingly multifunctional and compact nature of HI systems, as well as the sensitive optical pathways in photonics modules, pose complex reliability threats that demand holistic materials design and predictive failure modeling.

One of the key reliability risks in these systems stems from UV aging and optical degradation. Materials exposed to high-intensity light, particularly in systems involving lasers or VCSELs, may undergo photochemical changes such as yellowing, crosslink breakdown, or surface crazing. These changes not only affect visual clarity but also reduce optical transmittance and introduce scattering centers, which impair optical alignment and signal integrity. To counteract this, optical encapsulants must incorporate UV stabilizers or be intrinsically UV-resistant, such as silicones or fluorinated resins with robust molecular backbones. Long-term exposure simulations under controlled UV irradiation and elevated temperature/

humidity conditions (e.g., 85°C/85%RH with UV cycling) are often used to screen materials for photonic reliability.

Moisture penetration remains a second major concern, particularly at the interfaces between dissimilar materials or near embedded optical components. Water absorption can trigger refractive index shifts, haze formation, delamination, and even corrosion in metallized waveguides or contacts. For high-reliability photonics modules, water uptake must be minimized (typically below 0.1 wt%), and encapsulants must form tight barrier interfaces with optical substrates such as glass or SiN. Strategies to address this include the use of nano-silica reinforcement to reduce polymer free volume, hydrophobic surface treatments, and multilayer moisture barrier stacks. Moisture reliability is evaluated using Highly Accelerated Stress Tests, Pressure Cooker Tests (PCT), and Fourier-transform infrared spectroscopy (FTIR) to monitor bond hydrolysis.

Delamination and cracking remain prevalent in both HI and photonics packaging due to the use of dissimilar materials with large mismatches in CTE. For example, a photonic chip on a polymer substrate may experience interface shear or debonding during thermal cycling or under board-level reflow. In HI designs with tall chip stacks, through-mold vias (TMVs), or embedded passive components, stress concentrations at corners and interfaces are common failure initiation points. Encapsulants must therefore exhibit high interfacial adhesion, moderate modulus, and sufficient fracture toughness. Materials are often tested via temperature cycling (−40°C to 125°C), drop shock testing, and thermal mechanical analysis (TMA) to simulate realistic usage environments. Additionally, cohesive cracking within the encapsulant itself must be minimized by avoiding filler agglomeration and by incorporating toughening phases or interpenetrating networks.

Electrical degradation—especially in high-frequency signal paths—is another growing concern. Photonics-enabled RF packages operating at millimeter-wave or sub-THz frequencies are particularly sensitive to increases in dielectric loss and changes in impedance caused by material aging or contamination. Even trace ionic impurities introduced via moisture ingress or outgassing can raise the leakage current or cause signal attenuation. Therefore, materials must maintain a low dielectric constant and loss over time, and exhibit high volume resistivity ($>10^{14}$ Ω·cm) even after prolonged thermal aging. Surface cleanliness and ionic contamination testing (e.g., ion chromatography) are important for quality assurance.

Reliability evaluation in this field increasingly combines traditional environmental testing with advanced simulation tools. Finite element analysis (FEA) is used to model warpage, interfacial stresses, and crack propagation, while multiphysics simulations may model optical distortion, temperature gradients, and moisture diffusion across layered materials. These predictive techniques enable virtual qualification before fabrication, accelerating design iteration and reducing risk.

In summary, reliability in HI and photonics packaging requires materials that can withstand multi-domain stresses: UV radiation, moisture ingress, mechanical cycling, and high-frequency electrical fields. The integration of experimental testing, interface engineering, and predictive modeling forms the basis of reliable encapsulation in these emerging advanced packaging technologies.

6.3.6 Opportunities for Multi-Functional Materials

As heterogeneous integration and photonics packaging become increasingly central to the evolution of high-performance electronic systems, the demand for next-generation encapsulation materials continues to grow. These applications require materials not only with superior thermal, electrical, and mechanical properties, but also with new functionalities tailored to the co-packaging of disparate devices, including optical components, MEMS, sensors, and logic chips. In this context, several emerging opportunities are shaping the trajectory of materials development.

One key area of opportunity is the design of triple-functional materials—those that simultaneously offer optical transparency, high thermal conductivity, and electrical insulation. Such materials are critical in optoelectronic systems where thermal dissipation is essential, signal fidelity must be preserved, and optical transmission must remain unaffected. Conventional materials often exhibit trade-offs among these three properties, but research into novel polymer systems and nanofillers—such as surface-modified boron nitride, aluminum oxide nanoparticles, and high-clarity silicones—aims to address this need. For example, incorporating aligned boron nitride nanosheets within optically clear silicone matrices can create a composite that maintains >90% visible light transmission while achieving thermal conductivities above 1.5 W/m·K and volume resistivities over 10^{14} Ω·cm.

Artificial intelligence (AI)-assisted materials design is also opening new frontiers in encapsulant development. Using machine learning algorithms trained on large datasets of polymer structures, filler characteristics, and reliability outcomes, researchers can now predict materials behavior, optimize formulations, and even suggest novel combinations of monomers and additives. These predictive models are accelerating the innovation cycle and reducing reliance on trial-and-error methods. In the context of heterogeneous integration, where materials interactions across multiple domains (thermal, electrical, optical, mechanical, etc.) must be simultaneously balanced, AI provides a powerful tool for co-optimization.

Sustainability is another critical dimension. As regulatory pressures and corporate ESG goals intensify, there is an urgent push to develop environmentally friendly encapsulants. This includes halogen-free materials, low-VOC (volatile organic compound) formulations, and bio-based polymers derived from renewable resources. For photonics packaging, the challenge is even greater, as optical-grade polymers must maintain extreme purity and optical clarity. Research into bio-derived silicones, polyimides, and cycloolefin polymers is beginning to yield promising candidates that match the performance of petroleum-based systems while offering improved life-cycle sustainability.

A further area of innovation lies in the integration of functionality directly into the packaging layers. Rather than treating encapsulants as passive protective shells, future materials may serve active roles such as sensing, self-healing, or thermal regulation. For example, materials with embedded thermochromic or mechanochromic agents could provide visual feedback on local heating or stress. Similarly, microencapsulated healing agents or reversible dynamic bonds could allow materials to repair microcracks autonomously, extending device lifetimes without external intervention. In high-frequency or photonics modules, encapsulants with tunable

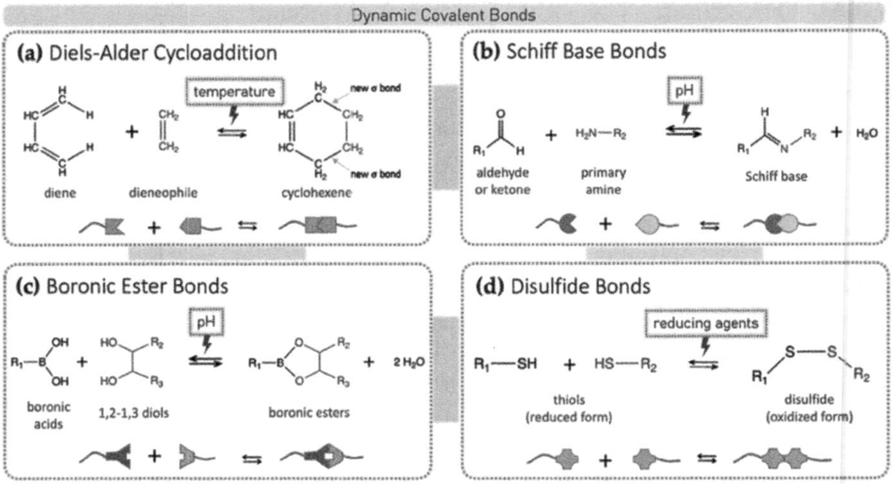

FIGURE 6.44 Reversible reactions of main dynamic covalent bond formation as (a) Diels-Alder cycloaddition, (b) Schiff base bonds, (c) boronic ester bonds, and (d) disulfide bonds. (Reproduced from Ref. [55], published by MDPI under the CC BY license.)

dielectric or refractive indices could enable reconfigurable signal paths or adaptive focusing. Figure 6.44 demonstrates reversible reactions of main dynamic covalent bond formation.

Lastly, co-design of materials and packaging processes will become increasingly important. The move toward chiplet-based systems, large-area fan-out integration, and panel-level packaging demands that encapsulants be designed not just for their final properties but for compatibility with all stages of manufacturing. This includes printability, B-stage processing, reworkability, and alignment precision. Encapsulants that can cure under multiple triggers (e.g., UV + thermal), adapt their rheology during placement, or selectively bond to specific interfaces will offer significant advantages.

In conclusion, the future of encapsulation materials for heterogeneous integration and photonics packaging lies in multi-functionality, sustainability, and intelligent design. The convergence of novel chemistries, AI-enabled formulation, and advanced processing will define the next generation of high-performance packaging platforms, enabling new levels of system integration, reliability, and functionality.

6.9 SUMMARY: CROSS-CUTTING CHALLENGES AND FUTURE DIRECTIONS

6.9.1 Thermal–Mechanical Trade-Offs in Filled Polymers

The interplay between the thermal conductivity and mechanical compliance remains one of the most fundamental and difficult-to-resolve trade-offs in the design of advanced polymer-based encapsulation materials. As packaging architectures become more power-dense and thermally constrained, especially in 3D stacked dies, high-performance processors, and photonics-integrated modules, efficient heat dissipation becomes a priority. This is typically achieved by introducing a high-volume fraction of thermally conductive inorganic fillers, such as alumina, aluminum nitride, boron nitride, or silicon carbide, into the polymer matrix.

However, increasing the filler loading proportionally increases the composite's elastic modulus and decreases its fracture toughness, making the material more brittle and more prone to cracking or delamination. This can be particularly problematic during thermal cycling, where the mismatch in CTE between the encapsulant, die, and substrate induces interfacial shear and tensile stresses. The resulting mechanical strain can compromise solder joint reliability, delaminate interfaces, or induce warpage in thin substrates—ultimately reducing device yield and lifetime.

The severity of this trade-off is not only dependent on the filler loading but also on the particle morphology, size distribution, interfacial adhesion, and dispersion uniformity. Agglomerated fillers or weak fillers–matrix interfaces serve as crack initiation sites under mechanical or thermal stresses. Therefore, materials engineers are increasingly focusing on improving filler surface treatments, such as silane coupling agents or polymer grafting, to maintain robust thermal percolation paths without excessively sacrificing mechanical flexibility.

One promising direction is the development of functionally graded materials (FGMs) and hierarchical architectures. In FGMs, the filler distribution is engineered to vary spatially, such that regions near the die hot spots are highly filled and thermally conductive, while regions near sensitive interconnects or interfaces maintain a lower modulus and greater compliance. This gradient design can be achieved through multi-step casting, selective curing, or even additive manufacturing, enabling localized property optimization. Figure 6.45 demonstrates different types of functionally graded materials.

Another advanced strategy involves using percolated networks of 2D or 1D nanomaterials—such as graphene, CNTs, or BNNS—that can form thermally conductive

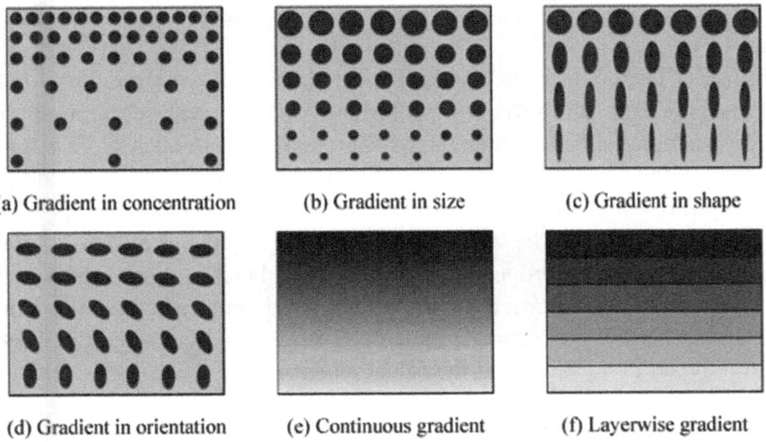

FIGURE 6.45 An overview of FGMs. (a) Gradient in concentration, (b) Gradient in size, (c) Gradient in shape, (d) Gradient in orientation, (e) Continuous gradient, and (f) Layer-wise gradient. (Reproduced from Ref. [56] with permission from Elsevier.)

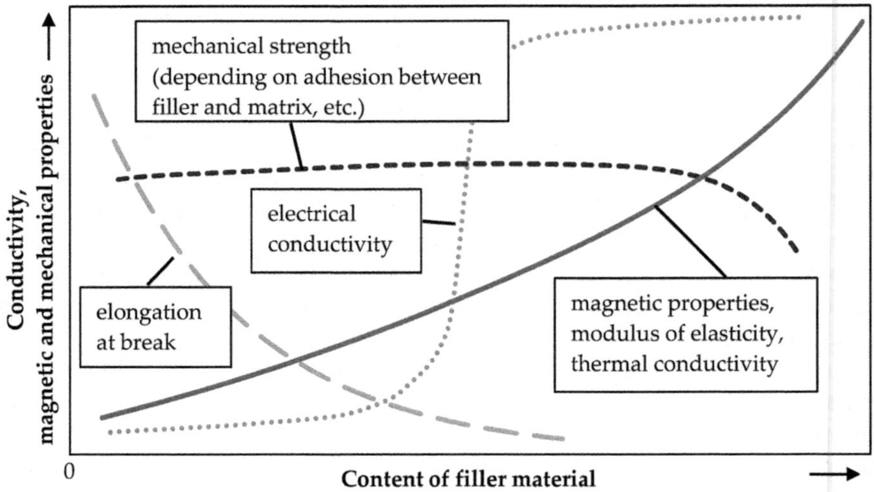

FIGURE 6.46 Relationship between properties of polymer composites and filler content. (Reproduced from Ref. [57] with permission from Elsevier.)

paths with minimal impact on modulus. These materials offer a high intrinsic thermal conductivity and high aspect ratios, allowing lower loading thresholds while preserving mechanical compliance. When aligned or interconnected into ordered structures, they can provide anisotropic thermal conductivities, directing heat away from hotspots without reinforcing the entire volume.

Furthermore, new hybrid fillers that combine a high thermal conductivity with structural softness—such as core-shell particles with soft elastomeric coatings, or encapsulated ceramic particles—are being developed to mediate stress transfer while retaining thermal function.

Polymer matrices themselves are also evolving, with flexible backbones, dynamic crosslinking, and molecular designs that allow better energy dissipation under cyclic loading.

Ultimately, overcoming the thermal–mechanical trade-off requires multi-scale design thinking: from atomistic tuning of filler–polymer interfaces, to mesoscale network construction, to macroscale process engineering. Tools such as FEA for stress–strain prediction, combined with thermal simulation and machine learning–based property optimization, are becoming integral to materials development pipelines.

This challenge reflects the larger theme of packaging materials design in the heterogeneous era: no single property can be maximized in isolation. Instead, trade-offs must be intelligently balanced through integrated materials selection, interface design, and structural innovation. Figure 6.46 shows the relationship between properties of polymer composites and the filler content.

6.9.2 High-Frequency and 6G Materials Needs

As the packaging of integrated circuits and modules extends into the frequency regimes of 5G, 6G, and advanced radar systems, the materials requirements for dielectric performance become increasingly stringent. In these high-frequency domains—spanning from 30 GHz to beyond 300 GHz—the propagation of electromagnetic signals becomes acutely sensitive to even minor variations in dielectric properties. A low dielectric constant reduces capacitive loading, while a low dissipation factor minimizes energy loss due to dielectric polarization and relaxation, both of which are essential for maintaining signal integrity and reducing the insertion loss.

However, most conventional packaging polymers—such as epoxy resins or polyimides—possess dielectric constants in the range of 3–4 and dissipation factors of 0.01 or higher, which are inadequate for sub-THz performance. Furthermore, these values are often not static: they exhibit strong frequency and temperature dependencies, introducing uncertainty in performance over a broad operational envelope.

Materials scientists are therefore engineering new classes of low-*k* polymers, such as fluorinated poly(arylene ether)s, polytetrafluoroethylene (PTFE), and cyclic olefin copolymers (COC/COP), which exhibit dielectric constants below 2.5 and D_f values under 0.002. In parallel, inorganic-organic hybrids and nanoporous dielectrics are being pursued, although the latter often suffer from poor mechanical strength or moisture ingress due to a high free volume.

A key issue in high-frequency regimes is the role of interfacial dielectric discontinuity. Modern packages are inherently heterogeneous, with multilayer interfaces between RDLs, build-up films, passivation layers, and adhesive dielectrics. These interfaces can cause local field enhancements, parasitic reflections, and modal distortion—effects that are negligible at DC or MHz regimes but catastrophic at mmWave. Consequently, high-frequency materials development now requires interface-aware design: interlayer adhesion, surface roughness, filler interphase chemistry, and local polarization all influence the system-level dielectric behavior.

Furthermore, characterization of materials at high frequencies presents nontrivial challenges. Conventional capacitance-based methods lose resolution and accuracy as frequencies increase. Resonant cavity techniques, coplanar waveguide extraction, and terahertz time-domain spectroscopy (THz-TDS) are emerging as essential tools to measure dielectric properties under actual operating frequencies and conditions. The absence of universally adopted test standards has led to inconsistency in materials qualification, prompting industry alliances such as IPC, IEEE EPS, and JEDEC to develop new protocols specifically for high-frequency dielectric testing.

From a design standpoint, accurate modeling of dielectric behaviors across a range of temperatures, humidities, and bias voltages is critical. Advanced finite-element electromagnetic solvers are now being coupled with molecular dynamics (MD) and density functional theory (DFT) simulations to predict materials responses from first principles. Such multiscale approaches allow developers to explore new chemistries and formulations with a deeper understanding of frequency-dependent polarization mechanisms.

In summary, the shift toward mmWave and sub-THz electronics demands not only new materials with intrinsically superior dielectric properties but also new strategies to model, measure, and integrate these materials in increasingly complex 3D package architectures. The co-optimization of dielectric function, mechanical robustness, and process compatibility remains an active frontier for electronics packaging science.

6.9.3 Long-Term Reliability under Coupled Stressors

Ensuring long-term reliability of polymer-based encapsulation materials has become increasingly complex as package architectures shrink, the stacking density increases, and operating conditions become harsher. Unlike traditional discrete components, modern heterogeneous systems operate under continuous stresses from multiple physical domains—thermal, mechanical, electrical, and environmental—simultaneously. As such, reliability challenges are no longer isolated to single failure modes but arise from compounded degradation mechanisms that evolve over extended lifespans, often in unpredictable ways.

One critical source of long-term failures is moisture-driven degradation. Despite advances in moisture barrier formulations, permeation through ultra-thin coatings, microcrack networks, or imperfect interfaces remains a significant concern. Moisture ingress accelerates electrochemical corrosion, dielectric breakdown, and interface delamination, especially in high-humidity, bias-stressed environments. To counteract this, researchers are exploring low-polarity matrix resins, hydrophobic nanoparticle fillers, and fluorinated surface treatments that limit water affinity while preserving adhesion.

Thermal cycling is another major contributor to long-term degradation, especially in packages with materials that exhibit mismatched coefficients of thermal expansion (CTEs). Repeated expansion and contraction induce cyclic shear and peel stresses at critical interfaces, including die–underfill, EMC–substrate, or interposer–mold compound boundaries. Over time, this leads to interface fatigue, void growth, and eventual delamination. Finite element modeling (FEM) is widely used to simulate thermal stress accumulation over thousands of cycles, but must now account for time-dependent viscoelastic properties and evolving interface strength under aging conditions.

Additionally, filler-related phenomena such as sedimentation, phase separation, and microvoid formation emerge during long-term storage or thermal excursions. In underfills and molding compounds with high filler loadings, gravitational settling or poor surface treatment can result in inhomogeneous mechanical behavior, creating localized stress concentrations and cracking. Advanced rheological control, surface functionalization, and hierarchical filler architectures are being introduced to mitigate these effects.

Recent developments in self-healing materials offer new pathways toward enhancing lifetime reliability, and Figure 6.47 shows methods of self-healing. These systems use reversible dynamic bonds (e.g., Diels–Alder, disulfide exchange, or hydrogen bonding) or encapsulated repair agents to autonomously repair microcracks or seal pathways after mechanical or chemical damage. While still under evaluation for high-reliability applications, these approaches hold promise for reducing inspection and rework costs in mission-critical environments such as aerospace, automotive, and medical electronics.

Predictive reliability modeling is also undergoing a paradigm shift. Traditional lifetime estimation relied heavily on empirical acceleration factors (e.g., Arrhenius or Coffin-Manson models). However, such models often fall short in capturing the multiscale, multiphysics nature of modern packaging failures.

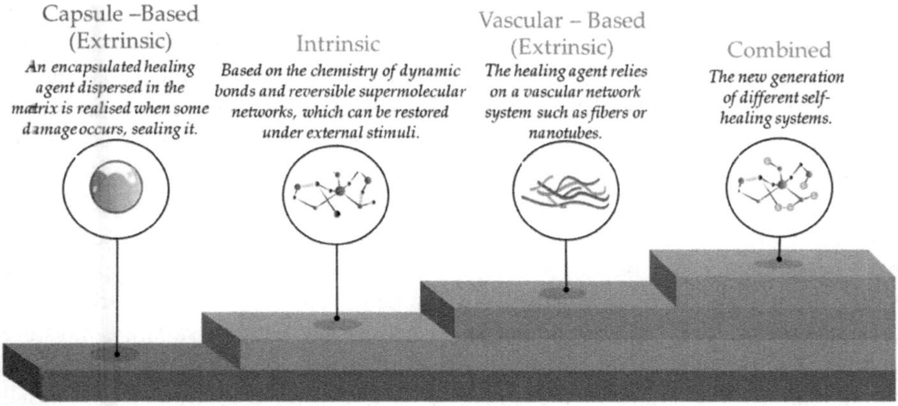

FIGURE 6.47 The self-healing methods. (Reproduced from Ref. [58] with permission from Elsevier.)

As a result, hybrid approaches combining physics-based aging simulations, machine learning models trained on field return data, and in-situ monitoring sensors (e.g., embedded resistance/strain sensors) are gaining traction. These allow for more accurate, dynamic assessment of reliability margins throughout a product's operational life.

Finally, the push toward heterogeneous integration and high-density packaging introduces new interface materials and geometries, which often lack historical reliability data. In such cases, the interfacial adhesion strength, coefficient of moisture diffusion, dielectric integrity, and chemical resistance must all be characterized under accelerated conditions and co-optimized during materials design. Advanced testing protocols—such as interfacial fracture energy measurements, four-point bending under humidity bias, and localized nanoindentation—are essential tools for identifying weak links in novel material stacks.

In summary, long-term reliability is no longer an end-stage concern but a design imperative that must be integrated from the earliest stages of encapsulant and packaging material development. Addressing it requires a confluence of smart materials engineering, rigorous interface science, and predictive, data-informed assessment frameworks.

6.9.4 Toward Sustainable and Green Packaging Polymers

The demand for sustainable materials and processes in electronics packaging is intensifying, driven by global environmental regulations, corporate ESG commitments, and the rapid turnover of consumer electronics. As advanced packaging continues to scale in volume and complexity, the environmental footprint of polymer encapsulants, coatings, and adhesives becomes a significant concern. Traditional epoxy, silicone, and acrylic systems are often derived from petroleum feedstocks, require solvent-based processing, and pose challenges for recycling and safe disposal.

One key direction in sustainable materials development is the replacement of halogenated flame retardants and solvent-based chemistries with halogen-free and solvent-free formulations. These alternatives aim to reduce toxic emissions during manufacturing and end-of-life incineration. For instance, phosphorus- or nitrogen-based flame retardants, dispersed in waterborne matrices, are increasingly adopted in encapsulants used for consumer electronics and power modules.

Biodegradable and bio-based polymers are also gaining traction. Poly(lactic acid) (PLA), polyhydroxyalkanoates (PHA), and modified cellulose or starch derivatives have been adapted for use in low-temperature, low-power packaging applications. While these materials offer improved environmental compatibility and carbon footprint reduction, their intrinsic properties—such as low thermal stability, moisture uptake, and brittleness—limit their use in high-reliability sectors. To overcome these limitations, hybrid strategies are emerging. These involve reinforcing bio-based matrices with inorganic fillers (e.g., clay, silica, or nanocellulose), crosslinking with bio-derived epoxides, or blending with engineered polymers to tune barrier and mechanical properties.

Another important sustainability vector is recyclability and disassembly. Conventional thermoset materials pose a major barrier to electronics recycling due to their irreversible crosslinked networks. Researchers are now investigating dynamic covalent chemistries—such as transesterification or imine exchange—that enable reprocessable or depolymerizable encapsulants. These materials can facilitate component separation or enable encapsulant recovery during recycling, which is particularly valuable in high-cost applications like servers, electric vehicles, and telecom infrastructure.

Sustainable packaging also intersects with manufacturing energy efficiency and emissions. UV-curable, low-temperature, and solvent-free formulations not only reduce processing energy but also minimize volatile organic compound (VOC) emissions in production facilities. Additionally, dry deposition and additive manufacturing techniques such as inkjet coating and vapor phase polymerization are being developed to reduce materials waste and improve process precision.

Beyond materials, life-cycle assessment (LCA) is becoming a standard tool to quantify the environmental impact of packaging materials from cradle to grave. Parameters such as global warming potential (GWP), water usage, toxicity index, and end-of-life fate are now used to benchmark candidate materials. Companies are increasingly integrating LCA metrics into materials selection decisions, balancing performance with sustainability objectives.

In summary, sustainability in semiconductor packaging is no longer an aspirational target but a design requirement. It demands innovations in green chemistry, bio-derived polymer science, reprocessable networks, and systems-level thinking around recyclability and process emissions. The challenge lies in achieving these goals without compromising performance, especially as microelectronics packages become more complex and application-specific. Future success will depend on synergizing materials innovation with regulatory alignment, supply chain readiness, and circular economy principles. Figure 6.48 demonstrates several sustainable and biodegradable polymers.

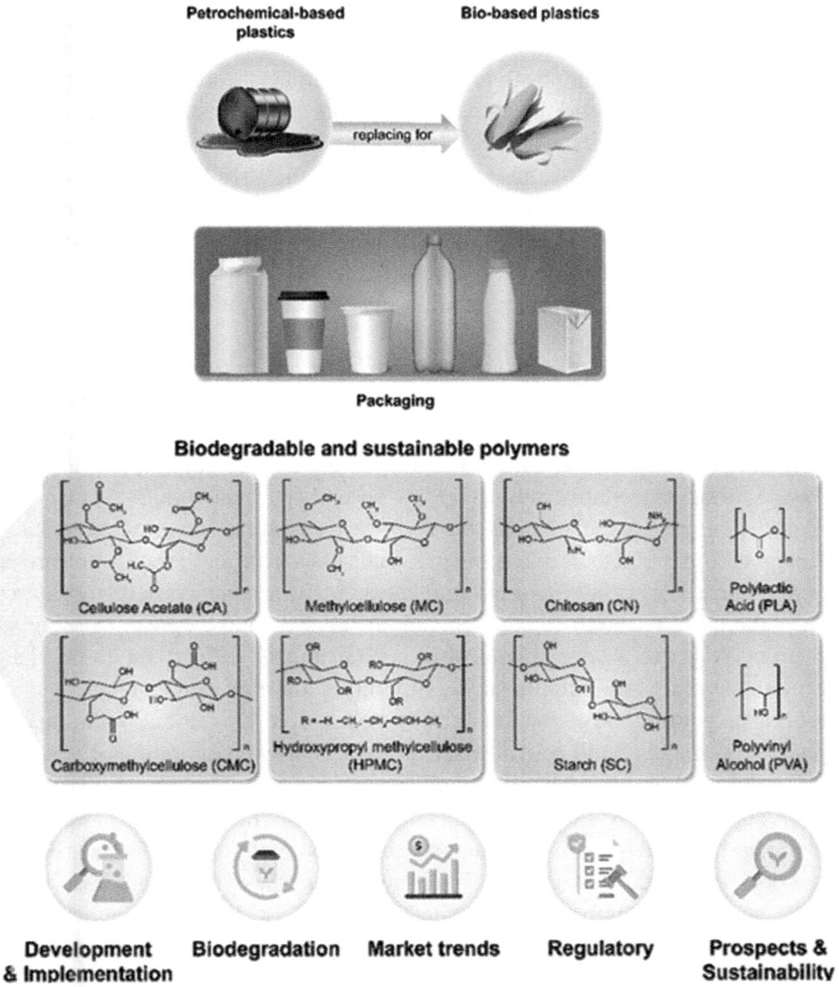

FIGURE 6.48 Sustainable and biodegradable polymers. (Reproduced from Ref. [59] with permission from Elsevier.)

6.9.5 AI and Digital Tools for Materials Development

The increasing complexity of packaging materials and processes necessitates a shift toward data-driven design. One of the most transformative trends is the integration of multiphysics simulation platforms capable of simultaneously modeling thermal, mechanical, and electrical interactions within the package structure. These simulations account for materials anisotropy, interfacial behavior, and curing-induced stress, enabling precise co-optimization of materials, geometries, and process conditions. For example, virtual prototyping tools can predict warpage, delamination, and thermal hotspots before physical fabrication, reducing costly design iterations.

A more advanced form of this modeling involves full coupling of materials, structure, and process parameters into unified simulation workflows. For instance, a thermosetting epoxy composite may undergo volumetric shrinkage during curing, which interacts with the thermal expansion of adjacent materials, generating internal stresses that can influence warpage or even interfacial debonding. By embedding the viscoelastic behavior of the polymer matrix, anisotropic filler properties, and the sequence of thermal profiles into the model, researchers can simulate the full evolution of stress and strain across the package during assembly and operation. This material–structure–process (MSP) coupling captures real-world interdependencies that are often missed in simplified models, making the simulation outputs more predictive and actionable.

These simulations also support sensitivity analysis and uncertainty quantification, which are critical for identifying process windows and evaluating design robustness. In manufacturing environments where materials batch-to-batch variability or tool-induced deviations are common, MSP-coupled simulations provide a digital backbone for quality control and design-for-reliability frameworks.

High-throughput experimental frameworks combined with machine learning (ML) algorithms are also accelerating materials discovery and process optimization. By systematically varying input parameters such as monomer ratios, filler types, or curing schedules, researchers can generate expansive datasets encompassing structure–property–performance relationships. ML models trained on these datasets can rapidly predict key outputs such as modulus, thermal conductivity, and moisture uptake, guiding experimental focus and resource allocation. Techniques like Bayesian optimization and active learning allow efficient exploration of vast formulation spaces with minimal experimental runs.

Such approaches are further enhanced by digital twin ecosystems, where simulated materials behaviors are continuously calibrated with real-time sensor data from production lines or field operations. These systems allow dynamic adaptation to processing variability, support in-line defect detection, and facilitate predictive maintenance through anomaly detection. The fusion of digital simulation, experimental automation, and AI promises a paradigm shift from empirical materials design to agile, closed-loop innovation pipelines.

6.9.6 From Passive Encapsulation to Active Platforms

The future of encapsulation materials lies in their transformation from passive protective layers into system-level functional enablers. Rather than viewing these materials solely through the lens of barrier properties or mechanical compliance, they are increasingly designed as integrated components that influence thermal management, electrical integrity, processing dynamics, and environmental resilience.

This chapter has emphasized that solving performance bottlenecks requires holistic strategies. The conventional material-property trade-offs—such as conductivity versus modulus, or barrier performance versus processability—are no longer sufficient. Instead, packaging innovation demands co-engineered architectures where materials interact dynamically with the structure and process. These interdependencies require deeper mechanistic understanding and predictive modeling capabilities.

Moreover, long-term reliability and sustainability are not afterthoughts but must be encoded into the material's design logic from inception. Emerging failure modes in advanced systems—whether driven by filler segregation, ion migration, or UV aging—demand materials that are responsive, adaptive, or even self-regenerating under operational stress.

Finally, a digital-centric approach is becoming essential. Multiphysics modeling, AI-guided design, and high-throughput experimentation are breaking traditional development cycles, making it possible to navigate increasingly complex formulation spaces with precision and speed. These tools not only enable optimal performance matching but also enhance manufacturability and supply chain resilience.

The next generation of packaging materials will be judged not only by how well they protect, but also by how well they perform in collaboration with the entire electronic system. Their role is evolving—becoming a nexus of performance, reliability, and innovation.

REFERENCES

1. M. Töpper, Wafer level chip scale packaging, in D. Lu and C.- P. Wong, Eds. *Materials for Advanced Packaging*, Cham, Switzerland: Springer, 2017, ch. 15, Available: https://doi.org/10.1007/978-3-319-45098-8_15. Accessed: Nov. 23, 2025.
2. J. Oberhammer, F. Niklaus, G. Stemme, Selective wafer-level adhesive bonding with benzocyclobutene for fabrication of cavities, *Sensors and Actuators A: Physical*, vol. 105, pp. 9772–9780, April 24, 2003, Available: https://doi.org/10.1016/S0924-4247(03)00202-4. Accessed: Nov. 23, 2025.
3. K. Yamamoto and T. Hirano, Application for WLP at positive working photosensitive polybenzoxazole, *Journal of Photopolymer Science and Technology*, vol. 15, no. 2, pp. 173–176, 2002, Available: https://doi.org/10.2494/photopolymer.15.173. Accessed: Nov. 23, 2025.
4. C. L. Gan, M.-H. Chung, L.-F. Lin, C.-Y. Huang, and H. Takiar, Evolution of epoxy molding compounds and future carbon materials for thermal and mechanical stress management in memory device packaging: A critical review, *Journal of Materials Science: Materials in Electronics*, vol. 34, no. 30, p. 2011, Oct. 23, 2023. Available: https://doi.org/10.1007/s10854-023-11388-5. Accessed: June 25, 2025.
5. Z. Plachý, A. Pražanová, K. Dušek, and A. Géczy, Underfill: A review of reliability improvement methods in electronics production, *Polymers*, vol. 17, no. 16, p. 2206, Aug. 13, 2025. Available: https://doi.org/10.3390/polym17162206. Accessed: Nov. 23, 2025.
6. S. L. Liu, G. Chen, and M. S. Yong, EMC characterization and process study for electronics packaging, *Thin Solid Films*, vols. 462–463, pp. 454–458, 2004, Available: https://doi.org/10.1016/j.tsf.2004.05.080. Accessed: Nov. 23, 2025.
7. G.-Y. Heo and S.-J. Park, Effect of coupling agents on thermal, flow, and adhesion properties of epoxy/silica compounds for capillary underfill applications, *Powder Technology*, vol. 230, pp. 145–150, Jul. 2012, Available: https://doi.org/10.1016/j.powtec.2012.07.022. Accessed: Nov. 23, 2025.
8. S. M. Firdaus and M. Mariatti, Fabrication and characterization of nanofiller-filled epoxy composites for underfill application, *Journal of Materials Science: Materials in Electronics*, vol. 23, pp. 1293–1299, 2012, Available: https://doi.org/10.1007/s10854-011-0587-3. Accessed: Nov. 23, 2025.
9. R. D. Guha, E. O. Danilov, K. Berkowitz, O. Oluwajire, and L. R. Grace, Consequences of humidity cycling on the moisture absorption characteristics of epoxy resins with

different network architectures, *ACS Applied Polymer Materials*, vol. 5, no. 1, pp. 400–411, 2023, Available: https://doi.org/10.1021/acsapm.2c01570. Accessed: Nov. 23, 2025.
10. M. L. Auad, S. R. Nutt, P. M. Stefani, and M. I. Aranguren, Rheological study of the curing kinetics of epoxy–phenol novolac resin, *Journal of Applied Polymer Science*, vol. 102, no. 5, pp. 4430–4439, 2006. Available: https://doi.org/10.1002/app.24674. Accessed: June 28, 2025.
11. Z. Lin, Z. Sun, W. Fu, Y.-C. Lin, K.-S. Moon, and C. P. Wong, Thermally conductive and electrically insulative alumina/epoxy composites for advanced electronic packaging applications: A comprehensive review of filler morphologies and surface modifications, *Materials Today*, vol. 86, pp. 393–413, Jul. 2025, Available: https://doi.org/10.1016/j.mattod.2025.03.011. Accessed: June 28, 2025.
12. W.-S. Lei and A. Kumar, Delamination and reliability issues in packaged devices, *Reliability of Electronic Packages*, vol. 7, pp. 267–312, 2014, Available: https://doi.org/10.1002/9781118831373.ch7. Accessed: Nov. 23, 2025.
13. S. P. Phansalkar, C. Kim, and B. Han, Effect of critical properties of epoxy molding compound on warpage prediction: A critical review, *Microelectronics Reliability*, vol. 130, p. 114480, 2022, Available: https://doi.org/10.1016/j.microrel.2022.114480. Accessed: June 28, 2025.
14. T. Aziz, A. Ullah, H. Fan, et al., Recent progress in silane coupling agent with its emerging applications, *Journal of Polymers and the Environment*, vol. 29, pp. 3427–3443, 2021, Available: https://doi.org/10.1007/s10924-021-02142-1. Accessed: June 28, 2025.
15. Y. He, B. E. Moreira, A. Overson, S. H. Nakamura, C. Bider, and J. F. Briscoe, Thermal characterization of an epoxy-based underfill material for flip chip packaging, *Thermochimica Acta*, vol. 357–358, pp. 1–8, Aug. 2000. Available: https://doi.org/10.1016/S0040-6031(00)00357-9. Accessed: June 29, 2025.
16. Z. Zhang, P. Zhu, and C. P. Wong, Flip-chip underfill: materials, process, and reliability, in D. Lu and C. P. Wong, Eds. *Materials for Advanced Packaging*, Cham, Switzerland: Springer, 2017, pp. 331–371. Available: https://doi.org/10.1007/978-3-319-45098-8_8. Accessed: June 29, 2025.
17. S. Katsurayama, H. Suzuki, J. W. Nah, M. Gaynes, and C. Feger, High performance wafer level underfill material with high filler loading, in *Proceeding of the 2011 IEEE 61st Electronic Components and Technology Conference (ECTC)*, Lake Buena Vista, FL. May 31–Jun. 3, 2011, pp. 370–374. Available: https://doi.org/10.1109/ECTC.2011.5898540. Accessed: June 29, 2025.
18. S. S. Zamir and B. Fathi, Organo-titanates and zirconates coupling agents in polymer composites: A review, *Journal of Elastomers and Plastics*, vol. 57, no. 6, pp. 1072–1106, 2025. Available: https://doi.org/10.1177/00952443251345123. Accessed: July 2, 2025.
19. R. S. Hebbar, A. M. Isloor, and A. F. Ismail, Chapter 12 – Contact angle measurements, in N. Hilal, A. F. Ismail, T. Matsuura, and D. Oatley-Radcliffe, Eds. *Membrane Characterization*, Amsterdam, Netherlands: Elsevier, 2017, pp. 219–255. Available: https://doi.org/10.1016/b978-0-444-63776-5.00012-7. Accessed: July 2, 2025.
20. M. G. Veena, N. M. Renukappa, K. N. Shivakumar, and S. Seetharamu, Dielectric properties of nanosilica filled epoxy nanocomposites, *Sādhanā*, vol. 41, no. 4, pp. 407–414, Apr. 2016. Available: https://doi.org/10.1007/s12046-016-0473-z. Accessed: July 2, 2025.
21. J. Taweeplengsangsuke and R. A. Pearson, Development of processing diagrams for underfill resins, in *Proceedings of the 4th International Conference Joining and Coating Technology in Electronics Manufacturing*, Como, Italy, Jun. 18–21, 2000, pp. 174–181. Available: https://doi.org/10.1109/ADHES.2000.860594. Accessed: July 2, 2025.
22. Y. Deng, Q. Wang, J. Ma, J. T. Oh, and Z. Chen, The combined impact of voids and thermal aging on the mechanical reliability of epoxy resin evaluated by statistical analysis, *Polymer Degradation and Stability*, vol. 215, p. 110455, Sep. 2023. Available: https://doi.org/10.1016/j.polymdegradstab.2023.110455. Accessed: July 2, 2025.

23. Y. Wen, C. Chen, Y. Ye, Z. Xue, H. Liu, X. Zhou, Y. Zhang, D. Li, X. Xie, and Y.-W. Mai, Advances on thermally conductive epoxy-based composites as electronic packaging underfill materials—A review, *Advanced Materials*, vol. 34, no. 52, p. 2201023, 2022, Available: https://doi.org/10.1002/adma.202201023. Accessed: July 11, 2025.
24. Y.-C. Cheng, Y.-H. Chen, H.-H. Hung, et al., Accurate numerical simulations of capillary underfill process for flip-chip packages, *Engineering with Computers*, vol. 41, pp. 4043–4060, 2025, Available: https://doi.org/10.1007/s00366-024-02022-x. Accessed: July 11, 2025.
25. B. Wei, W. Luo, J. Du, et al., Thermal interface materials: From fundamental research to applications, *SusMat*, vol. 4, no. 6, p. e239, 2024. Available: https://doi.org/10.1002/sus2.239. Accessed: July 11, 2025.
26. H. Yu, L. Li, T. Kido, et al., Thermal and insulating properties of epoxy/aluminum nitride composites used for thermal interface material, *Journal of Applied Polymer Science*, vol. 124, pp. 669–677, 05 October 2011. Available: https://doi.org/10.1002/app.35016. Accessed: Nov. 23, 2025.
27. W. Dai, L. Lv, J. Lu, et al., A Paper-like inorganic thermal interface material composed of hierarchically structured graphene/silicon carbide nanorods, *ACS Nano*, vol. 13, pp. 1547–1554, 2019. Available: https://doi.org/10.1021/acsnano.8b07337. Accessed: Nov. 23, 2025.
28. J. Xu, A. Munari, E. Dalton, A. Mathewson, and K. M. Razeeb, Silver nanowire array–polymer composite as thermal interface material, *Journal of Applied Physics*, vol. 106, no. 12, p. 124310, Dec. 15, 2009, Available: http://doi.org/10.1063/1.3271149. Accessed: Nov. 23, 2025.
29. P. Zhang, J. Zeng, S. Zhai, Y. Xian, D. Yang, and Q. Li, Thermal properties of graphene filled polymer composite thermal interface materials, *Macromolecular Materials and Engineering*, vol. 302, no. 9, p. 1700068, 2017. Available: https://doi.org/10.1002/mame.201700068. Accessed: July 12, 2025.
30. Z. Lin, A. King, X. Hu, et al., In situ generated silver nanoparticles on boron nitride nanosheets followed by silane modification in high-performance epoxy nanocomposites for advanced packaging applications, *Advanced Electronic Materials*, vol. 12, p. e00245, 2025. Available: https://doi.org/10.1002/aelm.202500245. Accessed: July 15, 2025.
31. X. Pan, X. Cui, S. Liu, et al., Research progress of thermal contact resistance, *Journal of Low Temperature Physics*, vol. 201, pp. 213–253, 2020, Available: https://doi.org/10.1007/s10909-020-02497-0. Accessed: July 15, 2025.
32. Z. Buliński, S. Pawlak, T. Krysiński, W. Adamczyk, and R. Białecki, Application of the ASTM D5470 standard test method for thermal conductivity measurements of high thermal conductive materials, *Journal of Achievements in Materials and Manufacturing Engineering*, vol. 95, no. 2, pp. 57–63, 2019. Available: https://www.journalamme.org. Accessed: July 15, 2025.
33. V. Subramanian, J. Sanchez, J. Bautista, et al., Mechanical characterization of thermal interface materials and its challenges, *Journal of Electronic Packaging*, vol. 141, no. 1, p. 010804, 2019. Available: https://doi.org/10.1115/1.4042805. Accessed: July 27, 2025.
34. R. J. Warzoha, A. A. Wilson, B. F. Donovan, et al., Measurements of thermal resistance across buried interfaces with frequency-domain thermoreflectance and microscale confinement, *ACS Applied Materials and Interfaces*, vol. 16, no. 31, pp. 41633–41641, 2024, Available: https://doi.org/10.1021/acsami.4c05258. Accessed: July 2, 2025.
35 R. Kandasamy, X.-Q. Wang, and A. S. Mujumdar, Application of phase change materials in thermal management of electronics, *Applied Thermal Engineering*, vol. 27, no. 17–18, pp. 2822–2832, 2007, Available: https://doi.org/10.1016/j.applthermaleng.2006.12.013.
36 Y. Liu, L. Hu, M. Lyu, et al., Reprocessable thermal interface materials with optimized phonon transport based on Diels-Alder click chemistry, *Composites Part A: Applied*

Science and Manufacturing, vol. 171, p. 107563, Aug. 2023. Available: https://doi.org/10.1016/j.compositesa.2023.107563. Accessed: Aug. 10, 2025.

37. W. Li, Wireless considerations in ocular implants based on microsystems, in D. Uttamchandani, Ed. *Handbook of MEMS for Wireless and Mobile Applications*, New York: Woodhead Publishing, 2013, pp. 424–462. Available: https://doi.org/10.1533/9780857098610.2.424. Accessed: Aug. 13, 2025.
38. M. Bae, S. Woo, J. M. Lee, W. Lee, and S.-H. Yoon, A prediction model for photopatternable thickness of photocurable polymer nanocomposites containing carbon-based high-aspect-ratio fillers, *Composites Science and Technology*, vol. 218, p. 109207, Feb. 2022. Available: https://doi.org/10.1016/j.compscitech.2021.109207. Accessed: Aug. 15, 2025.
39. H. He, W. Peng, J. Liu, et al., Microstructured BN composites with internally designed high thermal conductivity paths for 3D electronic packaging, *Advanced Materials*, vol. 34, p. 2205120, Aug., 2022. Available: https://doi.org/10.1002/adma.202205120. Accessed: Nov. 23, 2025.
40. S. Kume, I. Yamada, K. Watari, et al., High-thermal-conductivity AlN filler for polymer/ceramics composites, *Journal of the American Ceramic Society*, vol. 92, no. s1, pp. S153–S156, 2009, Available: https://doi.org/10.1111/j.1551-2916.2008.02650.x. Accessed: Nov. 23, 2025.
41. J. Vogelwaid, M. Bayer, M. Walz, F. Hampel, L. Kutuzova, G. Lorenz, A. Kandelbauer, and T. Jacob, Optimizing epoxy molding compound processing: A multi-sensor approach to enhance material characterization and process reliability, *Polymers*, vol. 16, p. 1540, 2024, Available: https://doi.org/10.3390/polym16111540. Accessed: Sept. 7, 2025.
42. Y. Shen, L. Zhang, and X. Fan, Achieving warpage-free packaging: A capped-die flip chip package design, in *Proc. 2015 IEEE 65th Electronic Components and Technology Conference (ECTC)*, San Diego, CA, May 26–29, 2015, pp. 1546–1552. Available: https://doi.org/10.1109/ECTC.2015.7159803. Accessed: Sept. 8, 2025.
43. P. Praful and C. Bailey, Warpage in wafer-level packaging: A review of causes, modelling, and mitigation strategies, *Frontiers in Electronics*, vol. 5, p. 3389, Feb. 2025. Available: https://doi.org/10.3389/felec.2024.1515860. Accessed: Sept. 11, 2025.
44. R. Dancel, Case study paper on additive manufacturing (3D printing technology), *Philosophy of Technology*, Apr. 2019. Available: https://www.researchgate.net/publication/335733535. Accessed: Sept. 11, 2025.
45. S. Ebnesajjad, Adhesive coating and lamination of polyvinyl fluoride films, in S. Ebnesajjad, Ed. *Polyvinyl Fluoride*, New York: William Andrew Publishing, 2013, pp. 213–249. Available: https://doi.org/10.1016/b978-1-4557-7885-0.00009-0. Accessed: Sept. 11, 2025.
46. Y. Jiang and N. Hadjichristidis, Diels–alder polymer networks with temperature-reversible cross-linking-induced emission, *Angewandte Chemie International Edition*, vol. 60, no. 1, pp. 331–337, 2021. Available: https://doi.org/10.1002/anie.202013183. Accessed: Sept. 11, 2025.
47. C. Lamont, T. Grego, K. Nanbakhsh, et al., Silicone encapsulation of thin-film SiO_x, SiO_xN_y and SiC for modern electronic medical implants: A comparative long-term ageing study, *Journal of Neural Engineering*, vol. 18, no. 5, pp. 055003, April 2021. Available: https://doi.org/10.1088/1741-2552/abf0d6. Accessed: Nov. 23, 2025.
48. M. Azani and A. Hassanpour, UV-curable polymer nanocomposites: Material selection, formulations, and recent advances, *Journal of Composites Science*, vol. 8, no. 11, p. 441, 2024. Available: https://doi.org/10.3390/jcs8110441. Accessed: Nov. 23, 2025.
49. B. Fotovvati, N. Namdari, and A. Dehghanghadikolaei, On coating techniques for surface protection: A review, *Journal of Manufacturing and Materials Processing*, vol. 3, p. 28, 2019. Available: https://doi.org/10.3390/jmmp3010028. Accessed: Nov. 23, 2025.

50. J. Wang, X. Zhou, G. E. Thompson, J. A. Hunter, and Y. Yuan, Delamination of near-surface layer on cold rolled AlFeSi alloy during sheet forming, *Materials Characterization*, vol. 99, pp. 109–117, Jan. 2015. Available: https://doi.org/10.1016/j.matchar.2014.11.011. Accessed: Sept. 17, 2025.
51. J. G. H. Hermens, T. Freese, G. Alachouzos, et al., A sustainable polymer and coating system based on renewable raw materials, *Green Chemistry*, vol. 24, no. 24, pp. 9772–9780, 2022. Available: https://doi.org/10.1039/D2GC03657F. Accessed: Sept. 17, 2025.
52. D. Caimi, P. Tiwari, M. Sousa, K. E. Moselund, and C. B. Zota, Heterogeneous integration of III–V materials by direct wafer bonding for high-performance electronics and optoelectronics, *IEEE Transactions On Electron Devices*, vol. 68, no. 7, Jul. 2021. Available: https://doi.org/10.1109/TED.2021.3067273. Accessed: Nov. 21, 2025.
53. P. V. Paulau, C. McIntyre, Y. Noblet, et al., Adler synchronization of spatial laser solitons pinned by defects, *Physical Review Letters*, vol. 108, p. 213904, 2012. Available: https://doi.org/10.1103/PhysRevLett.108.213904. Accessed: Sept. 20, 2025.
54. T. Kato, J. Uchida, T. Ichikawa, and B. Soberats, Functional liquid-crystalline polymers and supramolecular liquid crystals, *Polymer Journal*, vol. 50, no. 1, pp. 149–166, Jan. 2018. Available: https://doi.org/10.1038/pj.2017.55. Accessed: Sept. 20, 2025.
55. I. Condò, S. M. Giannitelli, D. Lo Presti, B. Cortese, and O. Ursini, Overview of dynamic bond based hydrogels for reversible adhesion processes, *Gels*, vol. 10, no. 7, p. 442, 2024. Available: https://www.mdpi.com/2310-2861/10/7/442. Accessed: Oct. 2, 2025.
56. M. Sam, R. Jojith, and N. Radhika, Progression in manufacturing of functionally graded materials and impact of thermal treatment—A critical review, *Journal of Manufacturing Processes*, vol. 68, pp. 1339–1377, Aug. 2021. Available: https://doi.org/10.1016/j.jmapro.2021.06.062. Accessed: Oct. 2, 2025.
57. J.-H. Imholze and H. Glade, Corrosion-resistant polymer composite tubes with enhanced thermal conductivity for heat exchangers, *Inventions*, vol. 9, no. 5, p. 111, 2024. Available: https://www.mdpi.com/2411-5134/9/5/111. Accessed: Oct. 2, 2025.
58. A. Kontiza and I. A. Kartsonakis, Smart composite materials with self-healing properties: A review on design and applications, *Polymers*, vol. 16, no. 15, p. 2115, 2024. Available: https://www.mdpi.com/2073-4360/16/15/2115. Accessed: Oct. 2, 2025.
59. S. C. Teixeira, T. V. de Oliveira, N. de Fátima Ferreira Soares, and P. A. Raymundo-Pereira, Sustainable and biodegradable polymer packaging: Perspectives, challenges, and opportunities, *Food Chemistry*, vol. 470, p. 142652, Apr. 2025. Available: https://doi.org/10.1016/j.foodchem.2024.142652. Accessed: Oct. 2, 2025.

7 Interconnect Reliability for High-Performance Semiconductor Packages and Devices

Tae-Kyu Lee

7.1 INTRODUCTION AND THE LANDSCAPE OF HIGH-PERFORMANCE INTERCONNECT RELIABILITY

7.1.1 Quality and Reliability

An electronic device or system deployed into the field is expected to perform its intended function reliably. However, once deployed, the performance of the device or system, under a constant load, regardless of whether it is mechanical, thermal, or electrical, will degrade over time and ultimately will reach a point where the performance is not acceptable and the device or system needs repair, upgrade, or replacement. This performance cycle depends on the loading conditions and how robust the device and system are designed to be. An additional factor is the quality of the device and system, which defines the starting point of the device and system performance.

It is important to point out that we often face confusion between quality and reliability. Quality and reliability are different terms with a common ground. If a device's functionality and performance are not at the full expectation at the start of using it, we often call it a quality issue, and when the expected performance is impacted and resulted in a shorter lifetime than expected, we call it a reliability issue. Quality and reliability are often seen as complementary yet distinct pillars of electronic system performance. Quality is a measure of how well a product meets its specifications, requirements, and user expectations at the time of manufacturing and delivery. Reliability, in contrast, refers to the ability of a product or system to perform its required function over time under stated conditions, with emphasis on long-term operation and sustainability [1]. In other words, quality reflects manufacturing excellence at time-zero, while reliability reflects the robustness of design and materials choices to ensure long-lasting functionality under real-world stresses.

As shown in Figure 7.1, quality and reliability have close interactions and contribute to the product lifetime expectation. Quality is affected by the incoming components and materials. It is also influenced by the manufacturing process. If the process is optimized, the outcome quality will be as expected. If not, the function rate will be lower at the product life start, resulting in a shorter lifetime. On the other

DOI: 10.1201/9781003587637-7

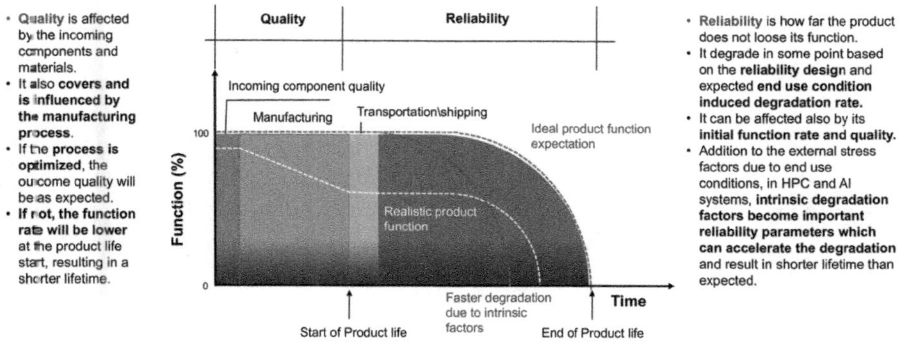

FIGURE 7.1 Quality and reliability, and their impact on product lifetime and expected functionality.

hand, reliability is how long the product continues to perform its intended function. It degrades at some point based on the reliability design and at the expected end-use condition induced degradation rate. But it can be affected also by its initial function and quality. Adding to the known external stress factors, which are the known end-use condition induced stress factors, unexpected events, which are not considered in normal end-use conditions but can occur during the product lifetime, can affect the reliability. These factors should also be considered in a proactive and preventive reliability design, which needs to be implemented in the early product design stage seeking reliability robustness. A robust design increases the lifetime against known and unknown external factors, but a weak quality and reliability, and their impact on product lifetime and expected functionality design, can cause a shorter than expected lifetime, which is indicated as 'Realistic product function' in Figure 7.1. There are also intrinsic factors in addition to the external stress factors. Even though the product is designed based on all possible known external factors, the degradation rate can occur faster than expected if the intrinsic factor induced degradation rate is not correctly estimated, which is indicated as 'faster degradation due to intrinsic factors' comparable to the 'Ideal product function expectation'. In HPC and AI network systems, intrinsic degradation factors become more important reliability parameters, which can accelerate the degradation and result in a shorter lifetime than expected.

Regardless of whether the stress factors arise from external sources or intrinsic driving forces, it is essential to identify which factors accelerate or slow the degradation mechanisms, to accurately estimate and predict the product lifetime and prevent unexpected failures. In earlier generations of electronics, systems could still achieve relatively long service lives even when the delivery quality was imperfect, because conservative design margins compensated for variations in end-use conditions, and intrinsic factors remained relatively controllable. For example, the power requirements per component and system generally stayed within a manageable thermal range, and interconnects could accommodate the power demand. The mechanical stress levels in components of a relatively simple structure with a smaller substrate and body size compared to the current HPC and AI network system components with a multi-die complex structure and larger body size, were in a manageable range

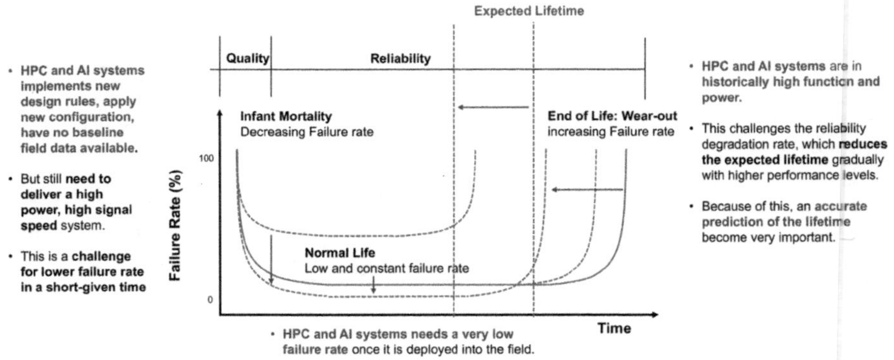

FIGURE 7.2 Bathtub curve for HPC and AI network system quality and reliability, which shows the three different regions related to infant mortality, normal life, and wear-out.

through additional structural reinforcement or design modifications. However, with the transition to HPC and AI network systems, where both the functional density and the power demand have reached unprecedented levels, newly implemented designs and configurations have introduced stress factors that challenge previously manageable parameters in thermal, mechanical and electrical performances. Furthermore, HPC and AI systems must maintain extremely low failure rates once deployed, as their high-speed functions and mission-critical tasks demand continuous, uninterrupted operation. All of this is occurring under an accelerated development pace, leaving limited time to fully understand emerging failure and degradation mechanisms or to implement proactive, preventive design strategies.

Figure 7.2 is a bathtub curve, which explains the failure rate over time correlation and identifies the three regions: infant mortality, normal life, and wear-out. In general, the infant mortality rate decreases as the product development stages progress, once the new design concepts are implemented with revisions, continued with pilot-builds to eliminate design flaws, and by applying corrective actions to improve and assure reliability of the new product. However, in the domain of HPC and AI network system development, achieving a zero-failure rate is challenging. This difficulty arises not only from the introduction of new design architectures but also from functional demands and increasingly compressed development cycles. As a result, it becomes difficult to eliminate all potential failures before field deployment, especially given the need for extensive testing under both known and newly emerging stress factors. In this context, predictability becomes a critical dimension of reliability assessment. Even when a product initially achieves a zero-failure rate, determining how long that state can be maintained before various degradation mechanisms begin to activate and impact performance, becomes essential for accurate reliability and quality assurance.

Traditionally, product lifetime qualification has relied on estimated operational life under defined end-use conditions, which are evaluated primarily through thermal and mechanical stress testing. Such a qualification ensured that a product met the minimum lifetime requirement, regardless of whether it lasted significantly longer or only just reached the target lifespan. In contrast, HPC and AI network systems require a clearly defined full-function target lifetime, where degradation rates must be accurately assessed to predict when performance will begin to decline. This enables proactive replacement or upgrade before the system enters the wear-out region, thereby ensuring continuous full-function operation throughout its intended service life.

7.1.2 High-Performance Computing (HPC) and AI Network Systems

Modern electronics reliability must be understood within the convergence of two major technology trends: high-performance computing (HPC) and artificial intelligence (AI) network systems. Each introduces distinct reliability challenges, yet together they define the trajectory of next-generation computing architectures. Although no single numerical threshold formally distinguishes these systems from conventional computing platforms, both share key characteristics that set them apart.

HPC systems are designed for sustained, efficient performance across large-scale parallel workloads. They are validated through standard benchmarks and enabled by high-bandwidth memory, low-latency interconnects, and large component and system footprints. Typical HPC devices include CPUs, GPUs, compute accelerators, domain-specific accelerators, and ASICs. These high-performance components feature high core counts, large cache capacities, and substantial memory bandwidth, often employing high-bandwidth memory (HBM), along with high power consumption and advanced thermal and power management schemes. They are primarily deployed in supercomputers and hyperscale data centers. By pushing the limits of transistor scaling and memory bandwidth, high-performance chips deliver maximum computational capability. However, these same attributes introduce reliability challenges arising from high power densities, thermal cycling, and interconnect stresses. To maximize the functionality, heterogeneous integration extends performance at the system level by combining diverse dies within a single package. Although this approach enables functionalities beyond monolithic scaling, it also introduces new reliability risks such as interconnect degradation, warpage-induced stresses across multiple layers, thermal coupling, and yield variability among heterogeneous materials and chip types.

AI network systems, similarly, lack a strict quantitative definition compared with conventional network systems but are characterized by ultra-high link speeds, massive bandwidth requirements, and sensitivity to signal loss and tail latency. HPC and AI networks demand both flawless quality and exceptional reliability. Even minor defects or slight performance deviations can propagate into large-scale failures, compromising data integrity, system availability, and mission-critical operations, which are issues that now permeate nearly every sector of our modern life. AI network chips represent the intersection of compute and connectivity, accelerating AI workloads while providing high-speed, low-latency networking that links thousands of processors across hyperscale data centers.

Both HPC and AI network systems operate under extreme conditions with unprecedented power densities, ultra-high frequencies, densely packed interconnects, and large component scales. Under these stresses, even latent defects originating during manufacturing can trigger accelerated failure mechanisms during operation, consequently impacting the entire interconnect hierarchy from solder joints and redistribution layers to through-silicon vias (TSVs) and high-speed optical or chiplet-links and inducing degradation and performance loss.

7.1.3 New Challenges in Reliability and Quality per High-Performance Computing (HPC) and AI Network Systems

Figure 7.3 presents the intrinsic characteristics of HPC and AI network system structures along with the reliability consideration factors in a nutshell. Globally, HPC and AI network systems are built on thick, highly complex printed circuit boards (PCBs) designed to deliver substantial power to advanced processing chips. The silicon die area, whether in single or multi-die configurations, continues to increase each year, with the substrate size expanding accordingly. This growth inherently leads to greater coplanarity variations and warpage behavior, stemming from the large die-to-substrate area ratio. Modern PCBs, composed of multiple copper layers and diverse via structures, generate uneven z-axis stresses across localized regions. This stress, combined with compressive loads from heat sink attachments and thermal management assemblies, impacts interconnect joint stability.

The periphery of HPC and AI network system chips typically incorporates multiple optical engines (OEs) for co-packaged optics (CPO) or direct Cu cable connectors for co-packaged copper (CPC) configurations.

Future heterogeneous integration trends point toward packaging architectures that include organic interposers with silicon bridges and redistribution layers (RDLs), or silicon interposers featuring through-silicon vias (TSVs) and potentially glass-core substrates on the horizon. With forced air/free air cooling, liquid cooling or immersion cooling, the HPC and AI network systems need testing in corrosive environments for reliability along with a high current-induced Joule heating and extreme temperature conditions, which can accelerate materials degradation, solder joint and bump degradation and trace degradation where the current flows through. Mechanically, we are dealing with multi-axis loading conditions not only at room temperature but also at elevated temperatures. Thermal interface materials (TIM) need a combined cross-related evaluation and assessment methodology under mechanical loading and thermal loading, where the thermal assessment needs a correlation to electrical performance.

The demand for the required reliability in nowadays HPC and AI network systems is challenged by multiple interrelated degradation mechanisms and driving forces. These elements must align correctly and precisely to guarantee consistent operation over years of deployment. Any form of degradation, whether it is due to current stressing, thermal exposure, thermomechanical fatigue, or materials instability, cannot be tolerated in HPC and AI network system environments where system uptime and data correctness are crucial and intercorrelated.

Quality and reliability requirements in HPC and AI systems represent a paradigm shift with convergence, correlation, and expanded reliability scope. Traditional reliability considerations, which once centered on mechanical, thermal, and chemical/environmental domains, are no longer sufficient. Instead,

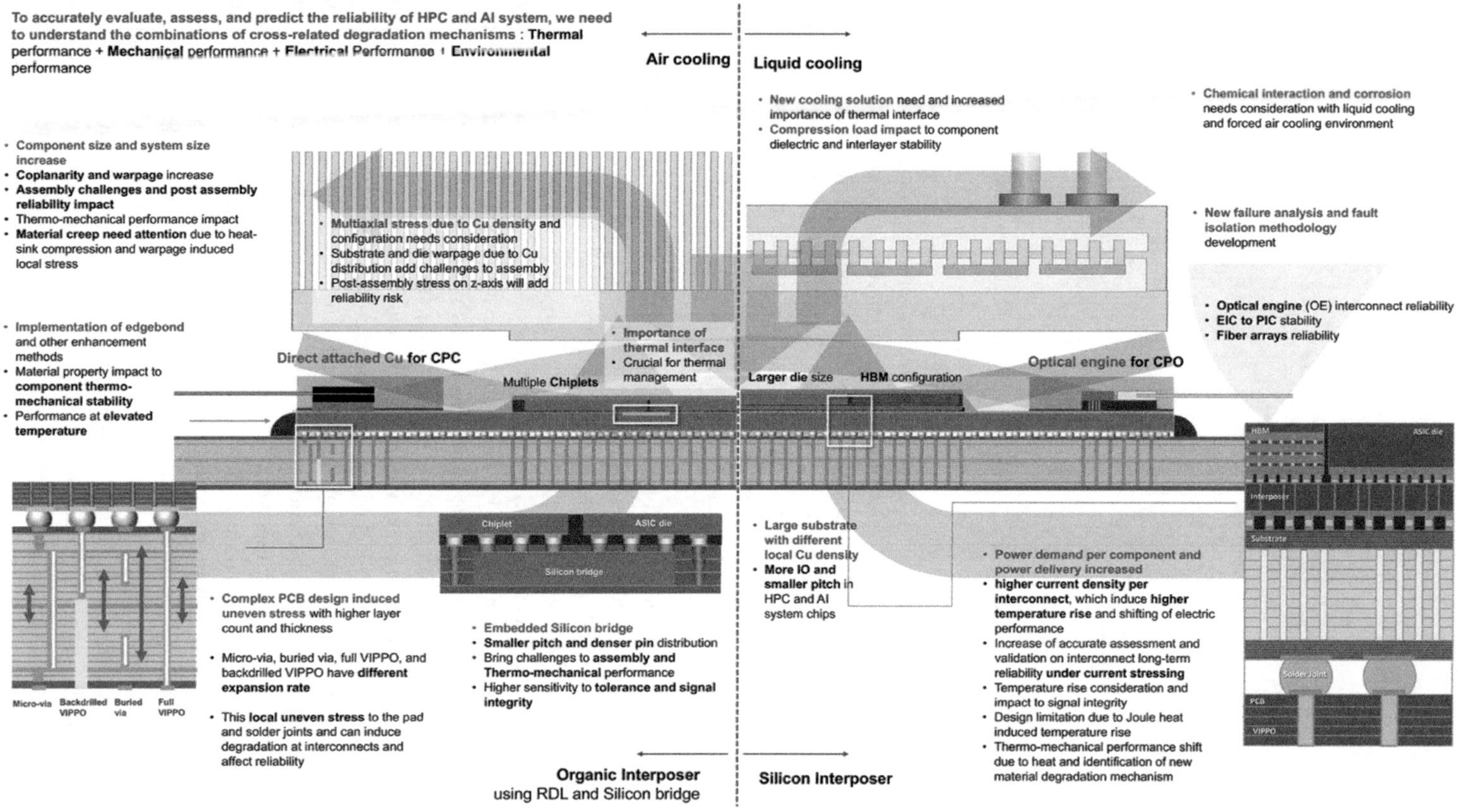

FIGURE 7.3 Intrinsic characteristics of HPC and AI network system structures along with the reliability consideration factors in a nutshell.

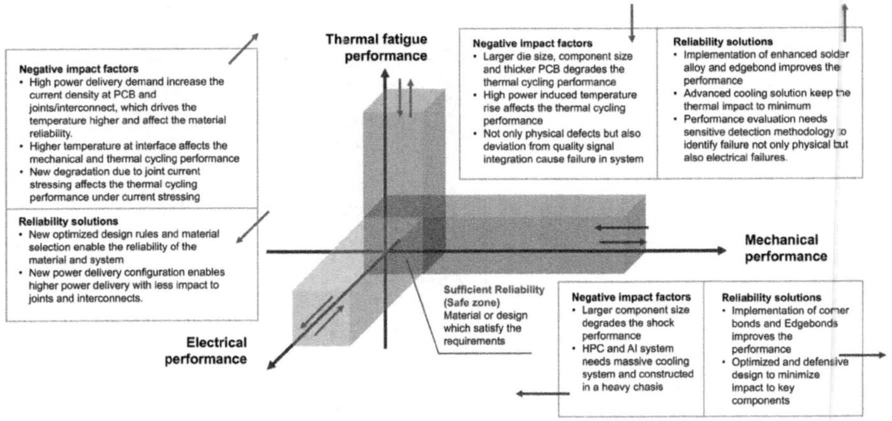

FIGURE 7.4 Intertwined reliability correlation between thermal, mechanical and electrical performances.

the modern reliability landscape has expanded to encompass complex electrical, interconnect, and system-level phenomena that were previously overlooked. Historically, the accelerated stress testing was designed to simulate worst-case operating conditions, but today's extreme boundary conditions in high-performance computing chips often align with actual use environments. This shift not only redefines the scope of reliability assessment but also introduces new defect modes and degradation mechanisms that were not critical in earlier generations of devices.

In general, electronic device and system reliability has been categorized into thermal, mechanical, and electrical domains. However, with the advent of new technologies, including high-performance chips, heterogeneous integration, and AI network chips, the framework has grown increasingly interconnected and complex. Figure 7.4 shows the intertwined reliability correlations between thermal, mechanical, and electrical performances. Thermal cycling performance is closely correlated to the component size and die size ratio, the PCB thickness, and the solder alloy configuration. High power Joule heating induced temperature rises affect the thermal cycling performance, which not only initiates physical defects but also affects signal quality. Against these negative factors, implementation of enhanced solder alloys and edgebond/cornerbonds on component corners can improve the thermal fatigue performance along with the mechanical performance, which is also intertwined with electrical performance. Having several end-use conditions and related accelerated test conditions reveals the nature and behavior of the component and provides the expected lifetime derived from its acceleration factors. The correlations to mechanical performance are not unidirectional and can be affected by multiple factors. For example, if the solder alloy is modified to improve the mechanical shock performance by reducing the Ag content, the thermal cycling is expected to be lower in performance. If the environmental temperature is higher, the mechanical shock and bend performance is expected to be better, while the thermal cycling performance is expected to be poorer.

Once the electrical performance is included in the correlation, the interactions become more multidirectional. High electrical performance per power delivery leads to a higher current density per interconnect, resulting in a higher temperature rise at the PCB, interconnect and system levels. This can either negatively or positively impact the component and system thermal cycling and mechanical performances. Higher and lower density of Cu traces and vias in some localized areas of the system to enable higher level of power and signal integrity per electrical performance can induce uneven stress distribution in the system which affects the thermal and mechanical performance.

The target is to widen the common reliability ground to cover all directions of the performance levels the HPC and AI network system requires, for reliability and performance. This not only includes the package-level reliability but also the system-level reliability, where compute and data movement must operate seamlessly across distributed infrastructures. This convergence, correlation, and expanded reliability scope creates a reliability environment where failure modes once considered secondary, such as signal degradation, warpage-induced interconnect fatigue under a high current, have become primary design considerations. Compared to the classical definition of failures, the failure criteria for HPC and AI network systems are more signal integrity driven. Before reaching a physical defect, which can be visually observed, the failure is more like a deviation from the function signal and is measured by signal deviation and resistance increase rather than physical disconnection.

7.2 DEGRADATION AND FAILURE MECHANISMS IN HPC AND AI NETWORK SYSTEMS

For conventional electronic devices and systems, thermal fatigue-induced driving forces, which degrade the stability of solder joints and solder bumps, and mechanical fatigue and shock-induced strains, were factors affecting the reliability testing parameters to define and validate the lifetime of the device and system in the field, based on the known end-use conditions. Industry standards are available and implemented. For thermal cycling performance, JESD22-A104F 'Thermal cycling' is in use, which defines conditions and procedures for devices, components, and their interconnections. For mechanical performance, JESD22-B111 'Board Level Drop Test Method of Components for Hand-held Electronic Products' is applicable, which defines the standardized methods for drop/shock testing of surface-mount electronic components on printed circuit boards (PCBs) based on handheld electronic devices. For larger devices and components, IPC/JEDEC-9703 'Mechanical Shock Test Method for Solder Joint Reliability of Surface-Mount Components' is the right one to refer to; this specification includes the mechanical bending test methodologies for large components like FCBGAs.

However, considering the larger sizes we see in recent HPC and AI systems, for chips and components, and the associated system with heavy heatsinks and structure complexity, the limitation of these specifications are driving the industry to private methodology and assessment criteria by each individual company. A broad specification on large system level assemblies and systems is listed in the MIL-STD-810

'Environmental Engineering Considerations and Laboratory Tests', which provides a comprehensive set of test methodologies associated with shock and vibration. But a detailed specification on the recent large body size components with the given structure complexity and the level of warpage-induced stress factors on sensitive interconnections, is not yet available as a public and common document but exists only in a form of individual industry internal specifications and guidelines. Even though the reliability testing and test methodology are critical for recent HPC and AI systems, the existing standards are reaching their limitations per new failure modes, large mass loading conditions like heatsinks and complex and sensitive packaging configurations. The variety of geometry and design factors that each company is embracing makes it also difficult to find a common and unified standard, which can apply to the wide range of available packaging configurations in HPC and AI systems. Currently, the effective way is to generate the standards and guidelines in-house, per individual company, with an expectation that the standardization of these reliability test methodology and assessment criteria will become available following the vast demand and broad industry consensus.

7.2.1 Thermal Cycling Performance in HPC and AI System Interconnections

Thermal cycling reliability has long been a cornerstone of electronics qualification, traditionally evaluated through coefficient of thermal expansion (CTE) mismatches between a single die, its substrate, and the supporting PCB. The JESD22-A104F standard 'Thermal Cycling' defines multiple thermal cycling test conditions, which are implemented to get the thermal cycling performance converted into expected lifetime per component and configuration.

However, in high-performance computing (HPC) and AI network systems, the situation is far more complex. The interaction of large body-size components, heterogeneous integration, power delivery demands, and mechanical reinforcements introduces multiple stress-driving forces that redefine how thermal cycling must be assessed. Figure 7.5 presents an overall landscape of thermal cycling performance per component configuration. All test data are based on SAC305 BGAs and paste tested with 0°C–100°C thermal cycling profile. It is hard to present a characteristic lifecycle number per component body size since there are various factors including the die size, die size to substrate ratio, substrate material and thickness, PCB thickness and material, component placement, and of course solder alloys. The general trend we can see in Figure 7.5 is that the components on thicker PCBs are degrading faster and resulting in a lower characteristic lifecycle number to failure. Components with no substrates like Wafer-level-chip-scale-package (WLCSP) or with thinner substrates like Chip-array-ball-grid-array (CABGA) have lower lifetime cycle numbers compared to thicker substrate components like FCBGAs. The thermal cycling performance constantly improves with larger flip-chip BGA (FCBGA) components but that also changes to a downward trend once the die size increases and the component size increases. The coefficient of thermal expansion (CTE) mismatch between the PCB and the substrate and between the substrate and the die are considered the main driver, which affects the interconnect degradation.

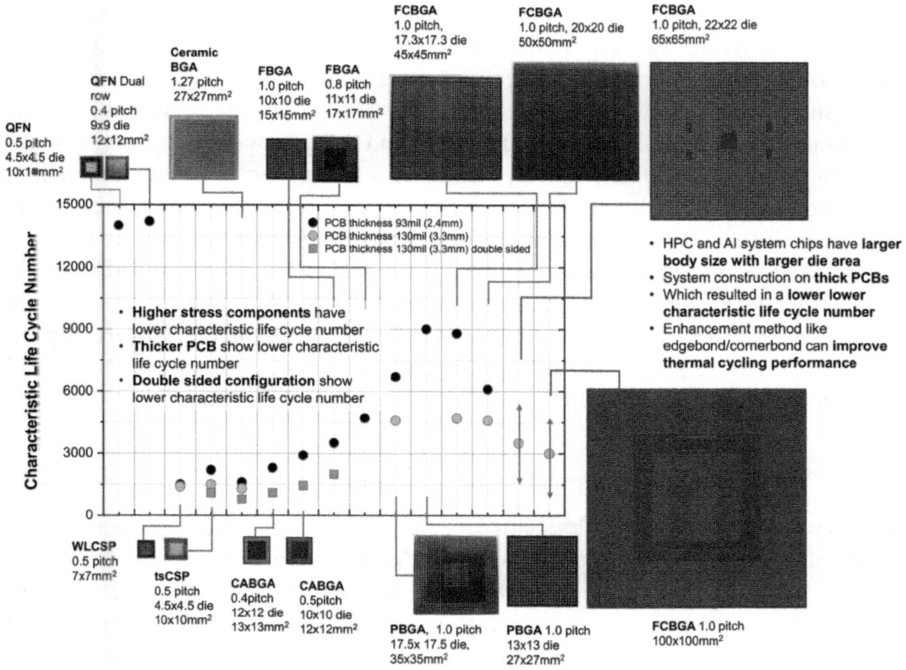

FIGURE 7.5 Landscape of thermal cycling characteristic life cycle numbers for selected components with various PCB thicknesses. All test data based on SAC305 BGAs and paste tested with 0–100°C thermal cycling profile.

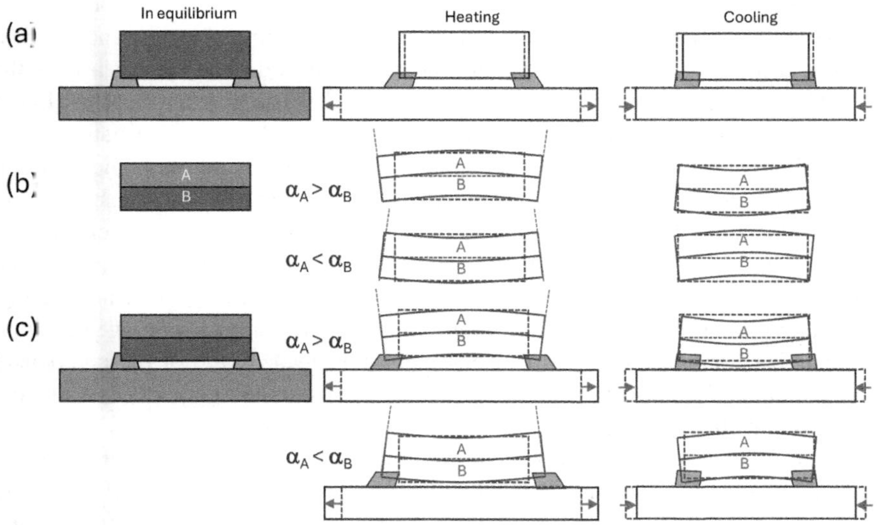

FIGURE 7.6 Effects of a temperature change on the stress and strain in solder joints. (a) single component attached on ends, (b) warpage in a two-layer assembly, (c) effect of bilayer warpage on the mechanical stress in attachments on the end [2].

Figure 7.6 presents the effects of a temperature change on the stress and strain in solder joints. While we have a single homogeneous component like WLCSP, the expansion is unidirectional, however with materials stack-up of different CTEs, the mismatch induces warpage which not only affects the joint reliability but also the component structural stability. Since HPC and AI system chips and components are in large sizes with an enormously large silicon die area, often with multiple dies including chiplets, the thermal cycling performance has a strong downward trend. Of course, the edgebond application and various design features optimizing the CTE mismatch can enhance the stability of the component/board interaction and bring the thermal cycling performance to a required level. But with a large size component and thick PCB combination and intrinsic PCB complexity, additional factors are impacting the thermal cycling performance, including the warpage of the component, and the PCB via multi-axis stress impact during assembly and function.

7.2.2 Warpage and Coplanarity Induced Localized Stress Concentration

Warpage and coplanarity are critical mechanical factors that directly impact the reliability of advanced semiconductor packages. As device architectures evolve toward larger die sizes, multi-die configurations, and heterogeneous integration, the mechanical interactions among components create complex deformation patterns that can no longer be overlooked in reliability assessment. Unlike general single-die components, high-performance computing (HPC) devices often consist of multiple silicon dies within a single package. Each package exhibits a unique warpage behavior due to its unique substrate thickness, materials properties, and die area. When assembled, these differences create non-uniform stress distributions at the bump and solder joint levels. The result is an elevated risk of solder fatigue, dielectric layer cracking, or delamination during both assembly and field operation. Unlike earlier systems, modern HPC and AI system components enter operation with pre-existing warpage and residual stresses. During assembly, the substrates and dies deform to a scale that affects the assembly yield. Also, during the assembly process, the component behaves like a stand-alone component, but now with the solder and paste phase transformations from solid to liquid and back to solid, with limited opportunity to return to their natural coplanarity state, with the solder alloy remaining in a non-equilibrium transition state. Once in this state, the external static and cyclic loading conditions in tension and compression imposes persistent stresses on the solder joints, bumps, laminates, and dielectric layers. Such warpage behaviors become superimposed on thermal cycling, accelerating fatigue mechanisms. Adding to this already complex warpage behavior, the heat sink configuration that provides a constant load on top of the component to maintain the thermal conductivity the system needs, induces a multi-axis loading condition that changes over performance cycles and over time.

Figure 7.7 presents the warpage behavior of a relatively small DDR memory component, and a larger FCBGA component test vehicle with a $100 \times 100\,mm^2$ body size. The height distribution for the DDR component is actual measurement data using a profilometer, while the FCBGA warpage map is a schematic construction of general warpage behaviors from similar size test vehicles. Both components show the BGA side coplanarity z-axis height distribution at the room temperature, an intermediate

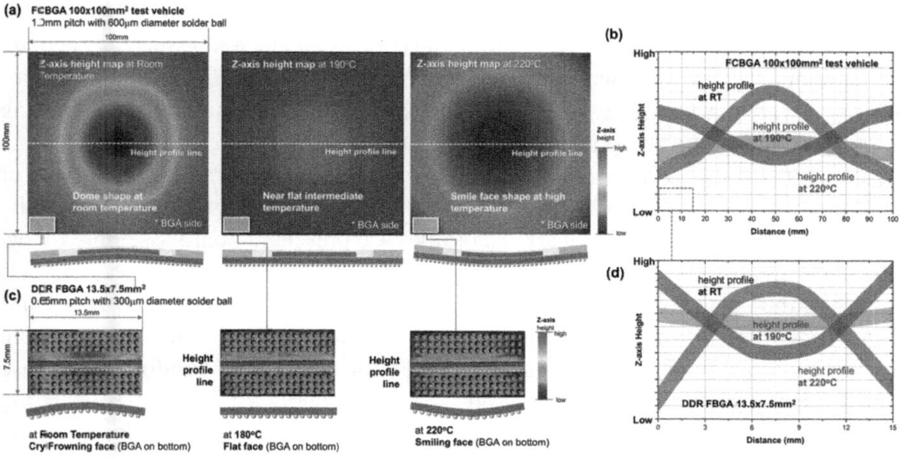

FIGURE 7.7 Warpage behavior of a DDR memory component, and a larger $100 \times 100\,mm^2$ body size FCBGA component test vehicle.

temperature, and near the solder ball melting temperature. The large component at the room temperature has the center region lower in the *z*-axis height compared to the four corners. The BGA side facing the PCB shows a dome shape from the side view. The warpage level decreases while heating the component, which flattens out at reaching around 190°C in this case. Given that this FCBGA test vehicle has its own die size and Cu distribution and substrate configuration, other FCBGA components are expected to show their own coplanarity and warpage behavior. For this component though, the warpage at above 220°C developed towards the reverse direction compared to the room temperature coplanarity, as the center region began to increase in height in the *z*-axis direction compared to the corner regions. Considering the BGA side facing the bottom, we are referring this shape as a smiling face, while the dome shape at room temperature is often referred to as a crying or frowning face. Figure 7.7b shows the line profile per component at each temperature condition. The *z*-axis height difference in the DDR memory components, while smaller in size, show the same trend as the large component with a crying/frowning face at room temperature, flattening out during heating, then transforming to a smiling face near the peak reflow temperature, as shown in Figure 7.7c. Even though the trend is the same between the large and small component size configurations, the larger components have a higher *z*-axis height difference scale compared to the smaller size components. The height difference in the DDR components is usually less than a few tens of micrometers, while the large components have several hundreds of micrometers height difference during reflow temperature cycles.

Current HPC and AI network system chips are designed into large size structures, constructed with multiple component elements including a stiffener ring, multi-die configuration, chiplets and embedded silicon bridges, with different local Cu density distribution, which all contribute to the component complex warpage characteristics. The stiffener ring is stable and flat, holding the perimeter of the large component, while the center of the component has a large die area with an initial warpage at the room temperature on top of a substrate which has its own coplanarity. This complex warpage condition induces stress points at die corners and locations such as the main die to chiplet interfaces, and at localized re-distribution layers. Once the component is elevated in temperature, the warpage characteristics change the pre-existing warpage and lead the component body to a flat surface, but with further increases in temperature, the once flat component begins to show warpage from a cry to smile or smile to cry face configuration depending on the component structure characteristics. These thermally induced warpage profiles are present at board assembly reflow and under thermal cycling conditions.

While the impact of warpage and coplanarity has been discussed above per component and PCB, the assembled components are fixed or frozen in each warpage state during board assembly reflow at the time when the solder joint solidification phase transformation occurs from liquid to solid. As shown in Figure 7.8, the free-standing component and PCB has its own coplanarity. Once the component is placed on the solder paste screen printed PCB, the intrinsic coplanarity of the component (crying face) is revealed as expected. During heating, the warpage of the component changes and flattens out when the temperature is elevated. Assuming the solder paste and solder balls are Sn-3.0Ag-0.5Cu (wt.%) (SAC305) in this case, the substrate is flattened out near the solder paste and ball melting temperature. Once the solder pastes and balls are melted and transformed to liquid solder, the shape of the solder joint follows the z-axis height changes along the component warpage towards the smiling face configuration at the assembly peak temperature. Since the solder joints are in liquid phase, no stress is built up at the joints yet. During cooling, the component warpage follows its own warpage characteristic towards flattening out, then to a crying face, but during this cooling, the solder phase transformation from liquid to solid occurs at the temperature lower than the solid to liquid phase transformation temperature during heating. This is due to an additional energy requirement for nucleation during the liquid to solid phase transformation, which results in a specific amount of

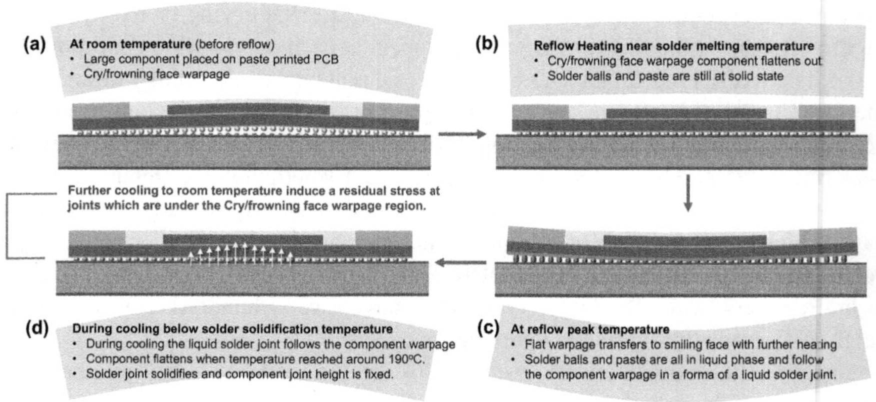

FIGURE 7.8 Component warpage and board interactions during the reflow cycle. (a) at room temperature before reflow, (b) during reflow heating near solder melting temperature, (c) at reflow peak temperature, and (d) during cooling below solder solidification temperature to room temperature.

undercooling. The solidification happens near the temperature where the warpage is flattened out and moving toward a crying face. Once the solder joints are fixed and frozen in a specific height, a localized stress is beginning to build up due to a continuous movement of the component toward a crying face while the solder joints' heights are fixed. So even though the component is fixed in a certain shape, the continuous nature of the component induces residual stresses to the solder joints, solder bumps and dielectric regions, which potentially tend to be released over time. These are localized intrinsic stresses in the components can be released when the component is freely movable and follow the intrinsic warpage character. In other words, if the component can return to its original warpage character at room temperature, the intrinsic local stress should be released. But since the solder solidification occurs before the component original and intrinsic warpage character is reached, the localized stresses are fixed and frozen limited by the moment of the solder solidification timing during reflow. Combining these localized residual stresses which are in a form of tension and the external compression force which are induced later by the heat sink on top of the component, the compression force at certain locations is increased and needs to be considered for long-term reliability assessments. Once assembled on a board, the component with these localized stresses during function in the field can migrate and will eventually accumulate at weak interface regions, inducing additional stress and affecting the structure stability.

It is not only the component side substrate which has its own warpage characteristics; the PCB also shows similar behaviors at elevated temperatures. The stack-up of multiple Cu layers in a thick PCB configuration with via structures for signal and power delivery induces similar multi-axis stress distribution, resulting in a complex warpage behavior. Each structural element introduces its own deformation profile. This warpage behavior not only challenges the assembly process but also affects the long-term thermal cycling performance. The combination of these CTE mismatches intensifies stress transfer across interfaces and accelerates thermal cycling degradation mechanisms. Beyond global warpage, localized stress concentrations are important reliability factors. High Bandwidth Memory (HBM), main dies, and chiplets often exhibit concentrated strain regions at specific structural features, such as corners or die edges. Due to the corner stress concentration in an HBM stack, localized deformation can occur as the initiation point for cracks or delamination. As package body sizes continue to increase, these localized stress concentrations remain persistent, scaling proportionally with package dimensions. The larger the package footprint, the more pronounced the stress distribution gradients, thereby increasing the probability of mechanical and interconnect failures. The challenge of warpage and coplanarity is expected to grow with the adoption of new component configurations such as Co-Packaged Optics (CPO) and Co-Packaged Computing (CPC). These architectures increase package complexity by introducing additional interfaces and larger surface areas, amplifying the likelihood of thermo-mechanical mismatch. Furthermore, the introduction of next-generation connectors and high-speed interfaces imposes tighter coplanarity requirements. Even slight deviations in flatness in substrates and PCBs can degrade the electrical performance, reduce signal integrity, or cause premature thermo-mechanical failures during thermal cycling and mechanical loading.

7.2.3 PCB Complexity and Uneven Stress Distribution

In addition to large body-size components, printed circuit board (PCB) complexity has become a critical factor influencing system reliability. As semiconductor packages grow and functionality expands, the supporting PCBs must incorporate increasing layer counts, greater thickness, and finer interconnect structures. While these advances enable higher performance and a greater integration density, they also introduce new mechanical and thermal reliability concerns. PCBs for HPC and AI systems with higher layer counts and greater thicknesses add another dimension to thermal cycling reliability. Complex PCB designs with increased warpage from thicker, multilayered structures undergo multiaxial expansions under temperature changes, and not only induce via reliability concerns, including barrel cracking and pad lift, but also amplify stresses on local interconnects and induce solder joint fatigue acceleration due to compounded horizontal and vertical stress factors. *z*-Axis expansion introduces vertical strains that interact with horizontal shear forces, due to thermo-mechanical interactions that vary with the PCB stack-up design, resin content, and copper distribution resulting in non-uniform fatigue profiles across the package.

Figure 7.9 shows an example of a representative via structure configuration which can be found in complex PCB configurations. This structural variation per localized regions introduces risks associated with the local *z*-axis coefficient of thermal expansion (CTE) mismatch. Not only during board assembly reflow but also during thermal cycling, the difference in expansion rates between the PCB materials and the solder joints leads to *z*-axis height expansion, which can trigger solder separation defects, and stress vias, solder joints, and interconnects. Historically, reliability models emphasized the horizontal shear stress as the dominant failure driver in PCB/component interactions. However, the vertical (*z*-axis) expansion is equally significant. Thermal cycling reliability, once evaluated primarily by lateral stress distributions, must now account for multi-dimensional deformation. As shown in Figure 7.9a, a PCB test board with full via-in-pad-plated-over (VIPPO) and mixed VIPPO configurations were put at elevated temperatures for *z*-axis pad height measurement. The full VIPPO configuration is a PCB with VIPPO at each BGA pad location, while the mixed VIPPO configuration is a mix of full VIPPO pads and backdrilled VIPPO pads in one component footprint. The height measurement via a surface profilometer is shown in Figure 7.9c. The pads at the backdrilled VIPPO location revealed a higher *z*-axis height increase at higher temperatures compared to the full VIPPO configuration sample PCB. The associated solder separation failure mechanism which induces solder failure at the second reflow is shown schematically in Figure 7.9d. The backdrilled VIPPO pads at elevated temperatures expand in the *z*-axis inducing a tension force at the full VIPPO location joint; once the component reaches the solder melting temperature, solder separation occurs at the full VIPPO location joint. Although assembly process optimization and defensive design is implemented, which prevents this solder separation defect occurrence, the driving force remains and can affect the thermal cycling performance.

Overall, thicker PCBs for HPC and AI systems with higher layer counts experience more pronounced *z*-axis strains, potentially accelerating solder joint degradation

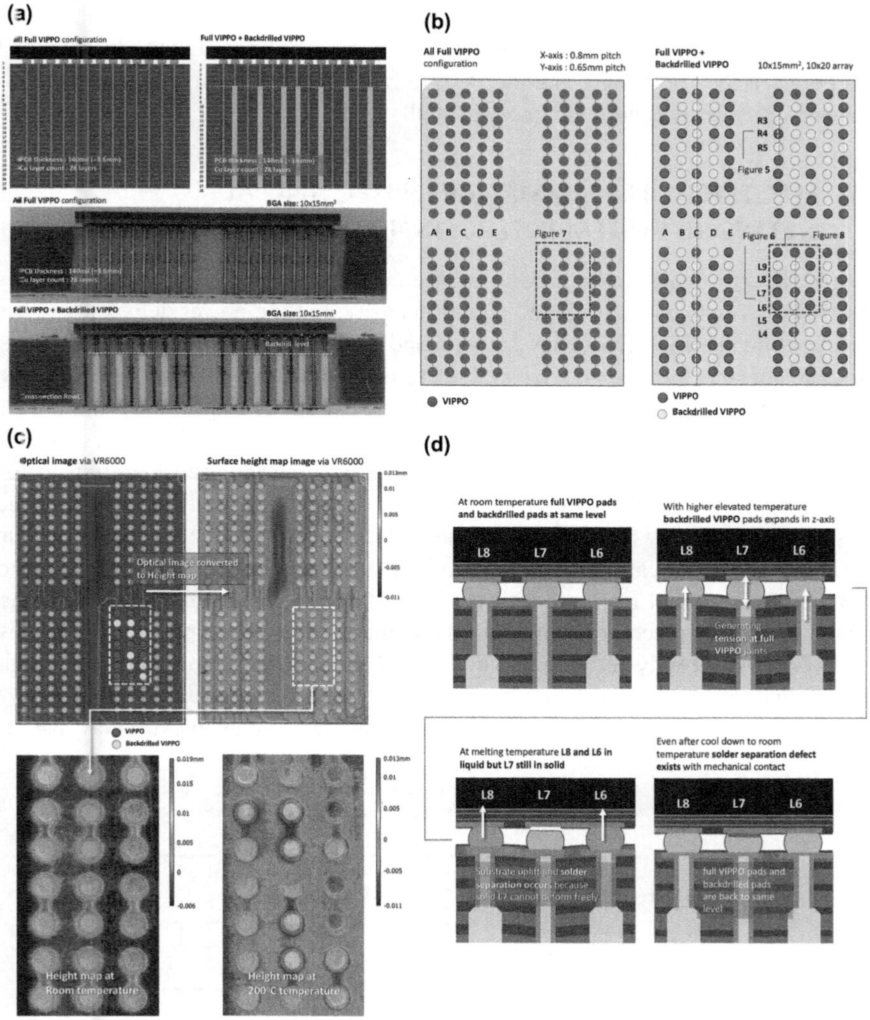

FIGURE 7.9 (a) Full VIPPO and Full VIPPO+backdrilled VIPPO configuration, (b) PCB pad distribution map, which shows the Full VIPPO and backdrilled VIPPO location per sample configuration. (c) z-axis height measurement map using profilometer on Full VIPPO+Backdrilled VIPPO configuration at room temperature and at 200°C. (d) Schematic diagram explaining the full VIPPO+backdrilled VIPPO height difference induced joint stressing mechanism [3].

during power and environmental cycling. An earlier study indicates that the z-axis expansion plays a critical role in reliability degradation, particularly under high-frequency signal transmission and a higher temperature environment, which are typical for HPC and AI network system workloads. To address these challenges, reliability assessment methodologies need to expand from the packaging level to the system level to capture the full spectrum of the component and PCB deformation and stress. Advanced finite element modeling, enhanced materials characterization including the z-axis expansion coefficients, updated qualification standards that reflect vertical stress contributions in addition to shear stresses, and co-design strategies between the component and the PCB to minimize the mismatch and distribute strains more evenly, need to be implemented.

7.3 ELECTRICAL PERFORMANCE IN HPC AND AI NETWORK SYSTEM INTERCONNECTIONS

Electrical performance in high-performance computing (HPC) and AI network system chips is fundamentally tied to signal integrity and the power delivery that sustains continuous, high-frequency signals, and high-current operation across dense interconnect structures. As device architectures evolve toward higher levels of integration and extreme power densities, the electrical stress imposed by the current flow through PCB layers, solder joints, micro-bumps, redistribution layers (RDLs), and substrates has become one of the crucial and important consideration factors for HPC and AI system reliability. As shown in Figure 7.10, the trend towards larger die sizes per HPC and AI system chip and the demand for higher power per square millimeter is dramatically increasing simultaneously, and an efficient power delivery architecture is crucial for these systems [4]. Advanced silicon process nodes, once at 7 nm, is rapidly reaching 2 nm. With the nominal core operating voltages at these process nodes currently between 0.75 and 0.85 V, to meet the performance workloads, the nominal current consumption has increased, with currents exceeding 600 A at steady-state and 1,000 A at peak with a continuing increase at an extremely fast rate

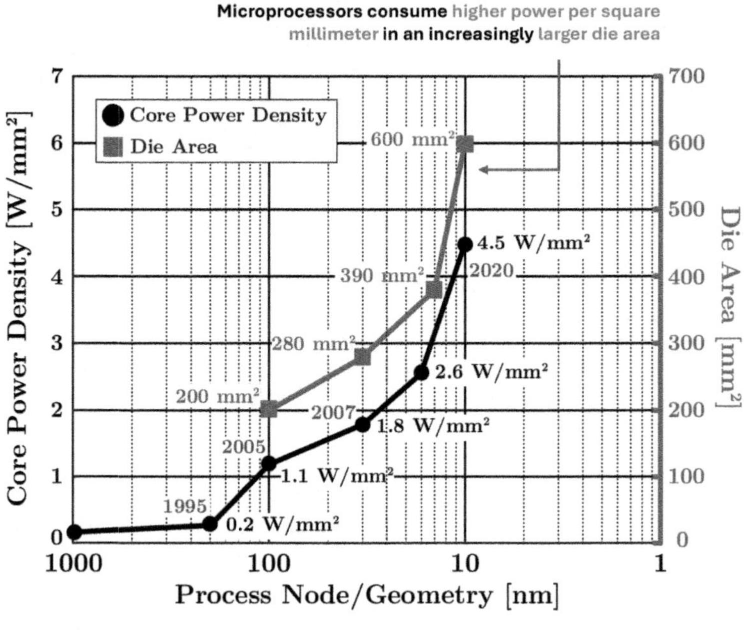

FIGURE 7.10 Correlation between silicon process node and geometry versus core power density and die area [4].

year over year [5]. The current density per interconnect has increased dramatically due to this rising power demand and shrinking bump and via dimensions, which intensifies Joule heating induced temperature rise, electromigration, and localized thermal gradients, each of which accelerates degradation mechanisms at the solder and interconnect interfaces.

7.3.1 Current Density, Joule Heating and Temperature Rise

Increased power delivery translates directly into a higher current per component and proportionally higher Joule heating throughout the system. At the interconnect level, this leads to a measurable temperature rise that propagates through solder joints, redistribution layers (RDLs), and PCB dielectric layers. Elevated temperatures amplify known reliability risks such as solder fatigue and delamination, while also triggering emerging failure mechanisms not previously seen at lower current densities.

Figure 7.11 presents the current density per interconnection configuration. With a given applied current, the current density per interconnect structure varies based on

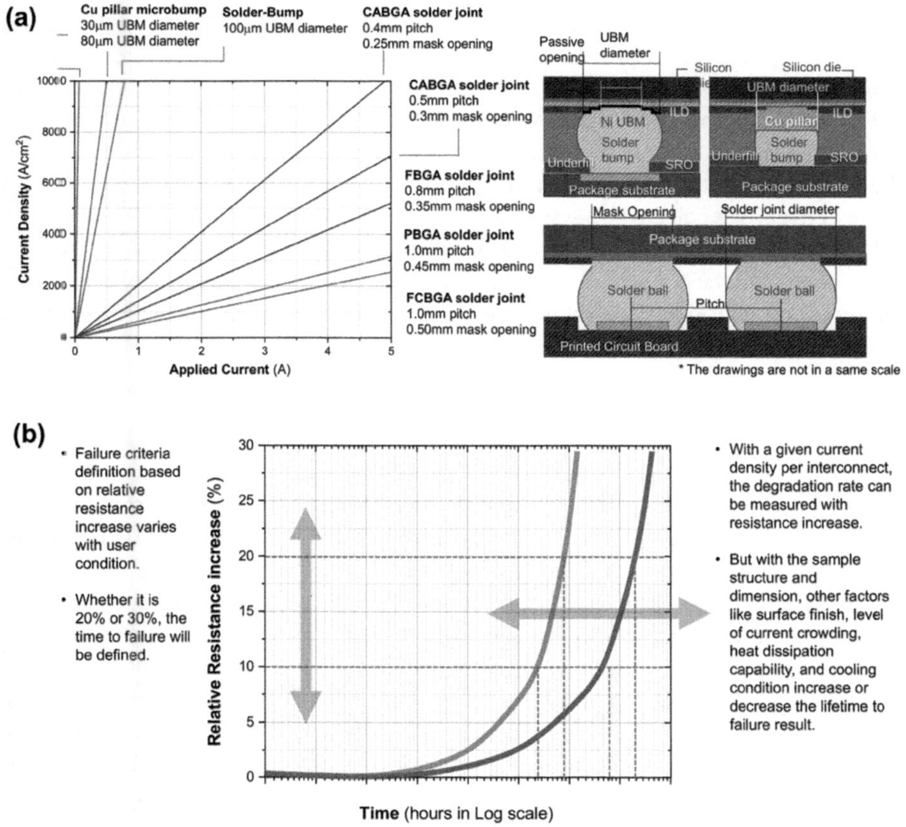

FIGURE 7.11 (a) Current density versus applied current correlation per interconnect (solder joint) configuration. (b) Current stressing induced time-to-failure curve for a given current density. Failure definition based on relative resistance increase percentage.

the dimension of the current flow area. For smaller micro-bumps with a 30 μm diameter, 1A applied current can bring the micro-bump to 1.4×10^4 A/cm^2, while solder joints with a 500 μm pad opening only experience 500 A/cm^2. For micro-bumps and solder-bumps, electromigration testing is typically at a higher level of current density for reliability testing and evaluation, while for solder joints, electromigration testing is typically at a lower level of current density in a range of a few hundred to a few thousand A/cm^2. But this does not mean that the solder joints are robust against current stressing in HPC and AI systems; rather, it means that there are different factors which need to be considered for interconnect reliability. Even though the maximum current density is in the lower 2,000 A/cm^2 with 4 A applied current, this current stressing can affect the long-term reliability of the solder joint if the system defense design is not accurately established and implemented. Figure 7.11b is an example of a current stressing induced time-to-failure curve for a given current density, which is based on the relative resistance increase percentage failure definition. With a given current density per interconnect, the degradation rate can be measured with the resistance increase, and whether 20% or 30% relative increase is a failure or not depends on the defined failure criteria, per interconnect functionality. If the specific interconnect is sensitive to signal deviation, a lower failure criterion will be defined, while a higher resistance is applied as a failure criterion for less sensitive interconnects. With a given joint structure, material and dimension, the current density is a constant value, but it is impossible to derive the time to failure only based on the maximum current density, as the curve is also affected by other factors like the surface finish, the level of current crowding, heat dissipation capability, and cooling conditions. So far in conventional electronic systems, the applied current and current density at the solder joint level is not within the range of reliability concern, but in HPC and AI systems along with the converged, correlated and expanded reliability scope, it becomes an important consideration factor for long-term reliability. The correlation between the current density (J) and Joule heating (Q) is shown as follows:

$$Q = J^2 \rho$$

Where Q is the Joule heating power density (W/cm^3), J is the current density (A/cm^2), and the ρ is the electrical resistivity (Ω·m). It shows that Joule heating increases quadratically with J, while J increases if the joint contact area shrinks and the trace width reduces. A given current density induced Joule heating can elevate the temperature at specific local hot spots where the current density is high. But without considering the heat dissipation factor, assessing the right temperature increase is difficult. For example, if you have a high current density solder joint on a thick PCB which has limited Cu layers, the heat dissipation will be limited, and the generated Joule heat will result in an increased joint temperature. But with the same given solder joint current density, if the thick PCB has multiple Cu layers and via configuration which provide heat dissipation paths, the joint temperature increase will be limited resulting in a lower joint temperature. Figure 7.12 shows an example of the solder joint temperature rise per current density and per PCB via configuration [6]. Three selected PCB via configurations, with all via connected pads, without

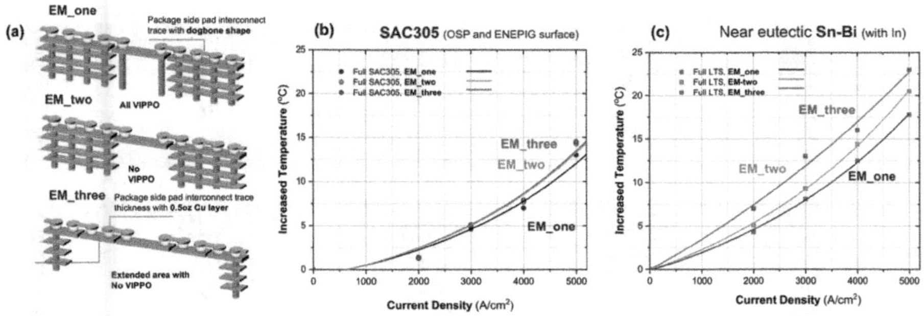

FIGURE 7.12 Solder joint temperature rise per current density and per PCB via configuration. (a) Three selected PCB via configuration, EM-one with all via connected pads, EM-two without via under the center pads, and EM-three without via under multiple pads, (b) SAC305 temperature rise per current density and via configuration, (b) Near eutectic Sn-Bi temperature rise per current density and via configuration [6].

via under the center pads, and without via under multiple pads, show different joint temperature rises for SAC305 and near eutectic Sn-Bi solder joints, even though the component and solder joint sizes are identical. The temperature measurement was performed using temperature coefficient to resistance (TCR) measurement method, which enables the direct temperature measurement of the target joint by measuring the resistance change of the target joint while under current stressing. Shown in Figure 7.12b for a given current density, solder joints located on the via connected pad are lower in the joint temperature compared to joints on top of the no-via location. The higher joint temperature compared to SAC305 with near eutectic Sn-Bi in Figure 7.12c indicates that a material with a higher resistance has a higher temperature rise due to higher Joule heating.

7.3.2 Localized Joule Heating and Thermal Cycling Performance

Thermal cycling is no longer driven solely by global temperature changes. Power delivery and current stressing introduce localized Joule heating, accelerate elemental diffusion, and create steep thermal gradients within the package. Localized hotspots in interconnect layers and joints raise temperatures unevenly. The elevated temperature due to current stressing not only affects the joint stability itself but also accelerates the diffusion of any elements in the interconnect parallel to the electron flow, resulting in current-induced elemental migration. For example, Figure 7.13 shows a current stressed solder joint which presents elemental diffusion aligned with the electron flow [7]. It indicates the phenomenon of elemental migration within solder joints, which is driven by directional electron flow, which can alter the thermal cycling and mechanical stability. With a steady state constant current applied, the 2,000 A/cm^2 current density at the package side interface region generated a Cu diffusion path parallel to the electron flow. Although the steady state solder joint under current stressing did not reveal a void or crack formation, it showed that the Cu diffusion occurred. The same elemental diffusion also occurred under in-situ current stressing and thermal cycling conditions, which induced a thermal cycling

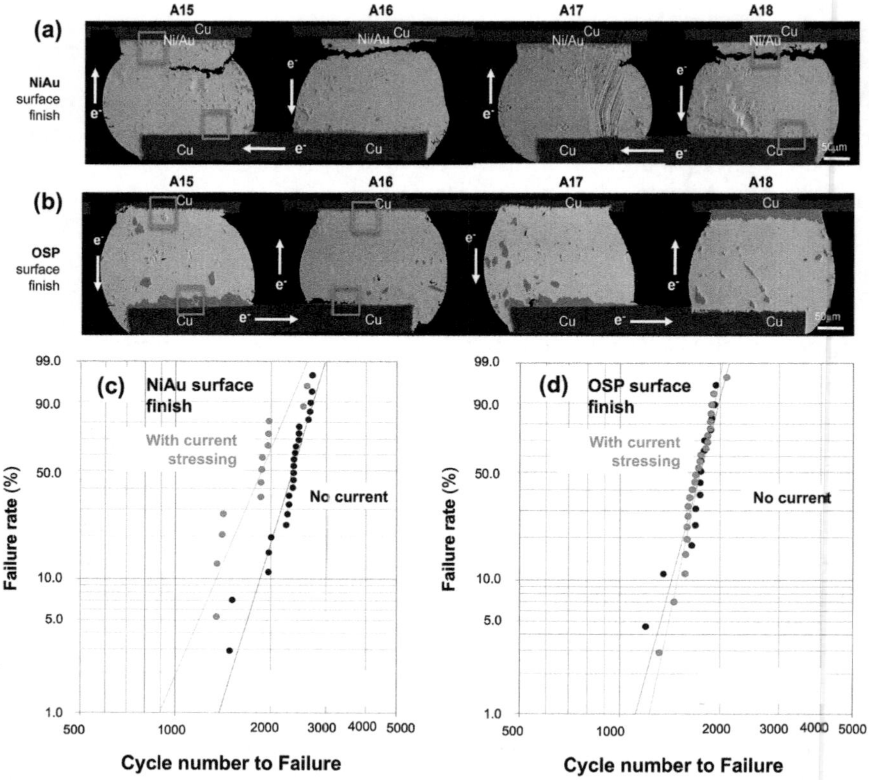

FIGURE 7.13 SAC305 solder joints with Cu-OSP and NiAu surface finish thermal cycled with in-situ current stressing. (a) NiAu surface finish sample and (b) OSP surface finish sample with 0°C–100°C thermal cycling and 2,000 A/cm^2 current stressing. Weibull plot for (c) NiAu surface finish (d) OSP surface finish samples, with and without current stressing [7].

performance shift in NiAu surface finished BGA samples, compared to a minimum deviation for OSP surface finished samples. As shown in Figure 7.13a, the NiAu surface finished sample had an anisotropic character and showed a degradation when the electron flow was from the package side to the boards side, and an enhancement interface region when the electron flow was from the board side to the package side. The Cu depletion zone near the package side interface resulted in a weak interface region which allowed the sub-grain boundary development and crack initiation faster than the no current stressed samples. On the other hand, if the surface finish is Cu, even with an intermetallic phase growth at the interface, sufficient Cu can diffuse into the solder joint and maintained a Cu rich joint, thereby showing no difference between current stressed and no current stressed sample configurations.

This elemental migration induced by electron flow is one of the new degradation mechanisms the HPC and AI system chips will experience due to the higher power demand and higher current density at interconnects. Figure 7.14 presents a set of data on SAC305 solder joints with Cu OSP surface finish and ENEPIG surface finish BGAs [8]. By using a test methodology of removing the PCB side of current stressed

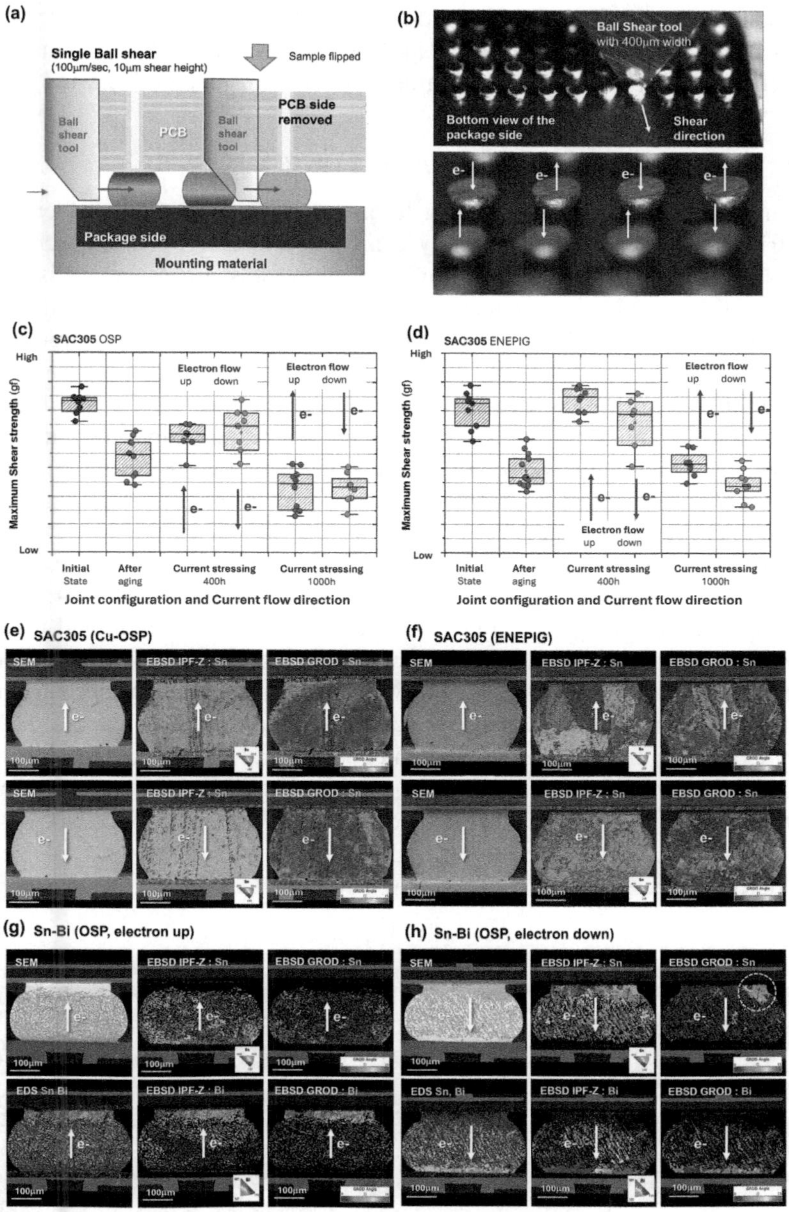

FIGURE 7.14 SAC305 solder joints with Cu-OSP and ENEPIG surface finish shear test on the package side interface after 2,000 A/cm^2 current stressing at 110°C joint temperature. (a) shear test sample configuration and test method, (b) Optical image of the shear test sample after removing the PCB side, (c) Shear test results on SAC305 OSP surface finish, which do not show identifiable deviation, (d) SAC305 ENEPIG shear test results which show indication of current flow dependency, (e) and (f) SEM image, IPF image and GROD map per SAC305 OSP and ENEPIG sample configuration. (g) and (h) SEM image, IPF image and GROD map per Sn-Bi OSP electron up and down configuration [8].

migration of vacancy concentration and coalescence of voids, intermetallic growth, and micro-crack initiation within the solder joint or along the under-bump metallization (UBM) interface, an important degradation mechanism to evaluate and validate.

7.3.3 Redistribution Layers (RDL) and Copper Structures

The impact of current stressing is not limited to solder joints and solder bumps. Redistribution layers (RDLs), typically constructed from copper, are increasingly vulnerable as the pitch scales downward and line widths shrink. For redistribution layers and fine-pitch Cu interconnects, the situation is further compounded by grain boundary electromigration. As pitch dimensions approach the mean free path of electrons, electron wind forces dominate atomic transport, weakening the structural integrity of the Cu-RDL network. The ratio of the Cu grain size to the line width becomes a determining factor, as smaller grain structures exhibit faster migration, leading to early resistance drift and electrical performance instability. Figure 7.15 illustrates how the Cu grain size and structure affect reliability under high current densities. As shown in Figure 7.15a and b, the Cu grain size per Cu stacked via is small with a random orientation. The interface between vias need a closer analysis for reliability assessments. With finer-pitch RDLs, the line width-to-grain size ratio decreases, meaning that grain boundaries dominate electrical and mechanical behaviors. This scaling effect not only increases susceptibility to electromigration but also introduces size-dependent variations in degradation mechanisms. In Figure 7.15c, a Cu-RDL is current stressed for a given time duration. As shown in the Cu-EBSD-IPF image, most of the grain boundaries are vertically arranged. The EBSD-GROD, which reveals the residual stress per Cu grain, indicates a region with a high residual stress along with regions of plastic deformation as the EBSD strain contour map indicates. As technology advances toward sub-micron RDL geometries, defect migration and boundary-driven failures become key reliability risks.

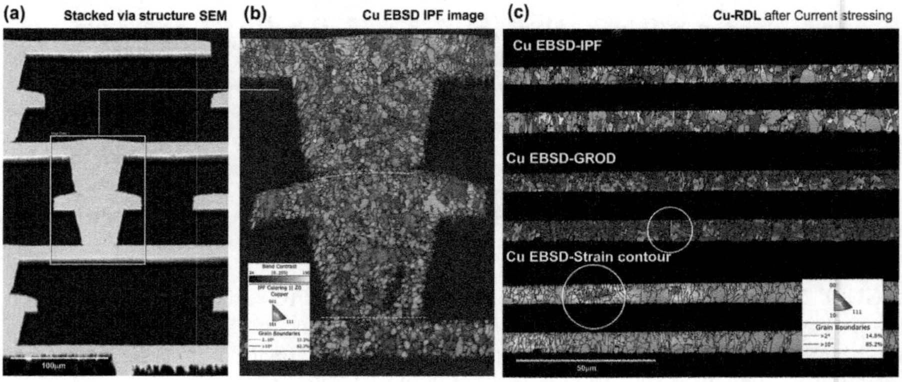

FIGURE 7.15 EBSD analysis on a via stacked structure and Cu-RDL after current stressing. (a) SEM of a as-received two stacked-via, (b) the EBSD-IPF image of the region indicated by the white box in (a), and (c) Cu-EBSD-IPF, GROD, and Strain contour maps of the current stressed RDL.

samples, it allows us to shear test the BGA at the package side interface after selected current stressing conditions. As shown in Figure 7.14c, the OSP surface finish sample does not show a deviation of the maximum shear force per electron flow direction. Whether the electron flow is from the board side to the package or reversed, the maximum shear strength shows a similar value, while an isothermal aging effect lowers the overall shear strength. But ENEPIG surface finished samples in Figure 7.14d show a deviation per electron flow direction. As observed with the in-situ thermal cycling under current stressing data in Figure 7.13, the Cu depletion at the ENEPIG surface finished packaging interface side indicates a lower shear strength compared to a Cu sufficient interface where the electron flow is from the board to the package side and delivers abundant Cu to the interface. Although the deviation amount is not dramatic under the 2,000 A/cm^2 current density, an accelerated shift of mechanical property is expected once the current density is getting higher per BGA ball. Figure 7.13e and f shows the SEM and EBSD images for ENEPIG and OSP surface finish solder joints, which experienced the current stressing. No clear elemental build-up or depletion are observed. But with the same current stressing condition at 110°C joint temperature and 2,000 A/cm^2 current density per solder joint for 400 hours, if the solder alloy is eutectic Sn-Bi, the elemental migration is easy to observe. Eutectic Sn-Bi is not only higher in resistance, which generates more Joule heat, it also has a mobile Sn and Bi which are affected by the electron flow direction. As shown in Figure 7.14g and h, the Bi migrates along with the electron flow and resulted in a Bi depletion region at the package side interface when the electron is from the package to the board, which shows an enhanced shear performance compared to the Bi accumulated region which shows a weak shear stability, as a Bi accumulated interface is built when the electron flow is from the board to the package side with a dramatic reduction in shear strength. Like these examples, current stressing can migrate the elements with the electron flow stress and temperature rise, based on the diffusion rate in these solder joints, resulting thermal cycling performance shifting.

Having the elemental migration in mind, Joule heating induced temperature rises, interface stability shifting, and solder bulk materials property changes, make the once simple thermal cycling reliability test complicated. The industry has implemented in-situ current stressing and thermal cycling for a while, which is power cycling, where interconnects experience surges of Joule heating during functional operation, which differs from overall package thermal cycling and accelerates degradation at specific joints. With the HPC and AI system demanding higher power delivery and higher temperature conditions, the importance of power cycling in an expanded application scope is increasing. The combination of global temperature swings and local Joule heating results in multi-scale thermal cycling stresses that act differently on each interconnect structure. In addition to that, non-uniformities in current distribution can cause localized warpage and coplanarity variations due to the localized heat generation, which can affect the solder joint and interconnect stability, leading to accelerated fatigue, resulting in a structure instability. Solder and solder-bump interconnections are particularly sensitive to combined electrical and thermal stresses. Under repeated power cycling, local temperature oscillations caused by dynamic workload transitions produce thermo-electrical fatigue, where current stressing amplifies thermal expansion mismatches. Over time, this results in

These effects are minimal at larger feature sizes but now represent primary considerations in design and qualification.

7.3.4 Ensuring Long-Term Electrical and Power Delivery Reliability in HPC and AI Systems

Ensuring reliable power delivery in next-generation HPC and AI network systems requires a multi-pronged strategy, which includes design for Reliability, which can be achieved by incorporating current density limits into design rules at the package and interconnect levels, with advanced materials engineering, which, for example, can be driven by optimizing the Cu RDL grain structure. Thermal management integration is an equally important segment by leveraging new materials, heat spreaders, and liquid cooling solutions to limit Joule heating-induced temperature rise. All needs to be done with updated qualification standards, by expanding the scope and looking into the convergence and correlation beyond classical electromigration models to capture grain-size scaling effects and 3D current crowding phenomena. From a system perspective, the power delivery reliability is no longer defined purely by voltage regulation at the package level. As shown in the examples above, the reliability concern extends to current uniformity and heat dissipation across every interconnect path. The interaction between the electrical stress and local temperature gradients forms a feedback loop, with a higher current density leading to local temperature rises, which in turn accelerates elemental diffusion, further degrading electrical conductivity. This self-reinforcing degradation cycle can ultimately lead to system failures or intermittent resistance changes, which is catastrophic in high-speed data environments. Ensuring long-term electrical and power delivery reliability in HPC and AI systems therefore requires accurate current distribution modeling across solder joints, RDLs, and vias using electro-thermal co-simulation, materials innovations, such as reinforced solder alloys and smart structures that minimize current crowding and optimize power delivery.

In essence, for electrical performance in next-generation HPC and AI systems, solder and interconnect structures must now endure unprecedented electrical and thermal stresses, requiring a unified understanding of power delivery, current stressing, and interfacial degradation mechanisms to ensure sustained functionality and long-term system stability.

7.4 MECHANICAL PERFORMANCE IN HPC AND AI NETWORK SYSTEM INTERCONNECTIONS

While electrical and thermal reliability challenges have intensified with the scaling of high-performance chips, heterogeneous integration, and AI network devices, mechanical reliability remains a foundational factor to assure the reliability of these chips and systems. As packages and PCBs grow larger and more complex, mechanical stresses induced during assembly, testing, and operation can become critical failure drivers. Mechanical reliability issues typically arise from mismatches in materials properties and dimensional scaling effects. Key contributors include Coefficient of

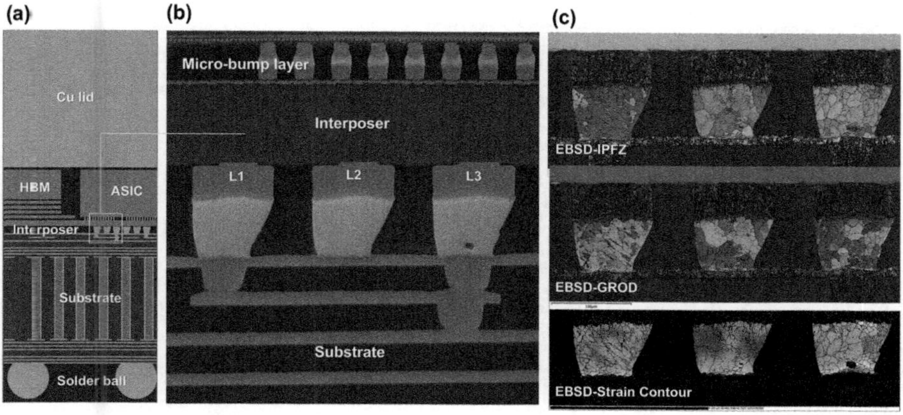

FIGURE 7.16 Solder-bump between the interposer and substrate near the edge of the ASIC and HBM after heat-sink attachment process. (a) The location of the three solder-bumps, (b) SEM of the solder-bumps in interest, (c) EBSD-IPF of the solder bumps Sn grain orientation, EBSD-GROD showing the residual stress per bump, and EBSD-strain contour, which reveals the plastic deformation region per solder-bump [9].

Thermal Expansion (CTE) mismatches between silicon, underfill, solder joints, and PCB laminates, which induce fatigue during thermal cycling. Warpage and coplanarity deviations create uneven stress distributions at interconnects and solder joints as discussed in Sections 7.2.2 and 7.2.3. Assembly-induced stresses are caused by board assembly reflow, edgebond application, and heat sink attachment. For the system level, the operational stresses, including vibration, shock, and board flexing in large-form-factor systems like servers or hyperscale racks movements, warrant consideration. Figure 7.16 presents an example of stress build-up at a large silicon interposer chip internal solder-bumps near the edge of the main die and HBM after heat sink attachment [9]. Based on location and baseline structures, the solder-bumps show different stress states, which are revealed via EBSD in Figure 7.16c. Those under-the-via stacked structures have both strain and stress increases compared to the "no via" bump location. These solder bumps are also located at the edge of the large die, which is a location where a larger bending can exist. Like this example, the compression load on an already warpage-induced structure can further induce localized stress variations, which later can be either released or developed to a defect accumulation site.

7.4.1 Mechanical Performance under Elevated Temperature

With the trend toward larger body sizes and higher power demand, elevated operating temperatures play an increasingly important role in mechanical performance, as temperature affects both materials properties and stress distributions, altering reliability outcomes. More critically, an elevated temperature impacts solder joint reliability, particularly under thermal cycling as well as in mechanical performance. Figure 7.17 demonstrates that solder joints exhibit altered mechanical shock performance under elevated temperature conditions. Figure 7.17a and b presents board

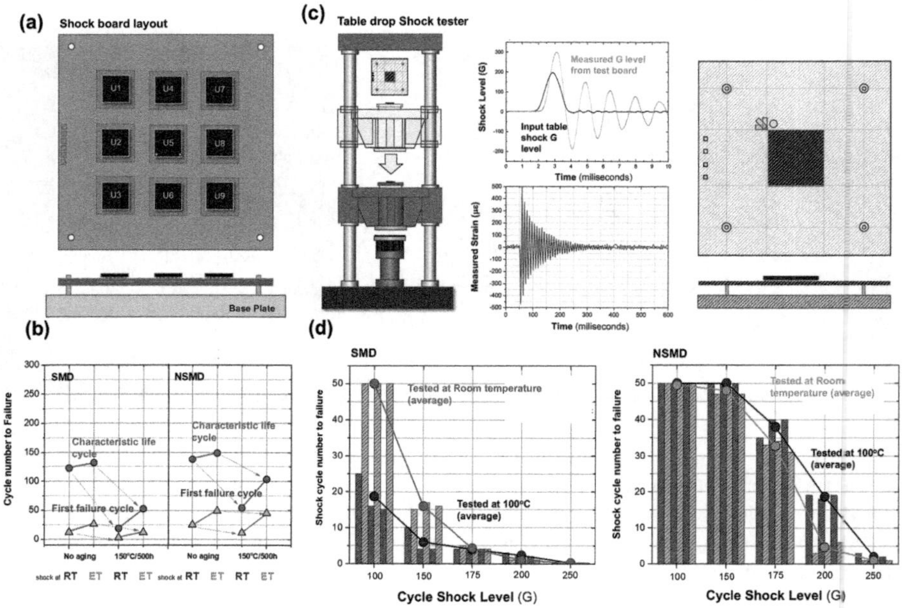

FIGURE 7.17 1500G High-G shock test on 12×12 mm² CABGA at room temperature and 100°C. (a) Test board and (b) Cycle number to failure per test condition. (c) 52.5×52.5 mm² FCBGA shock test sample and set-up. Impact G-level from 100 to 250G. (d) Shock test results per SMD and NSMD pad configuration tested at room temperature and 100°C [10,11].

level shock test data at the room temperature and at 100°C [10]. The solder alloy is SAC305. For both SMD and NSMD solder mask configurations, the shock performance increased with temperature, due to the increased ductility and toughness of the SAC305 joint, enabling the shock absorption and thereby increasing the cycle number to failure at 100°C. Compared to the smaller size component, the shock test at 100°C with a 52.5×52.5 mm² large FCBGA in Figure 7.17c and d shows a degradation in shock performance if the pad configuration is NSMD, while SMD still shows higher performance at elevated temperatures [11]. It is important to recognize that the phenomenon is reversed compared to smaller components, because of different stress concentration points and structure configurations. It is often the case that a larger component size shows a vastly different degradation mechanism.

The widespread use of edgebond materials, which is intended to enhance package stability, changes the strain behavior in critical regions, as the strain rate increases near the edges where the edgebond restrains the corners, creating localized stress concentration zones. While in some cases the temperature effects can be beneficial by enhancing ductility as discussed above, the overall stability of edgebonded regions requires close monitoring. The use of edgebond materials alters the stress distribution during thermal cycling and mechanical shock. While edgebond improves global stability, it also restrains corners, leading to localized stress concentration and altered strain dynamics. This dual effect must be carefully considered. Along with the mechanical properties and characteristics at the room temperature,

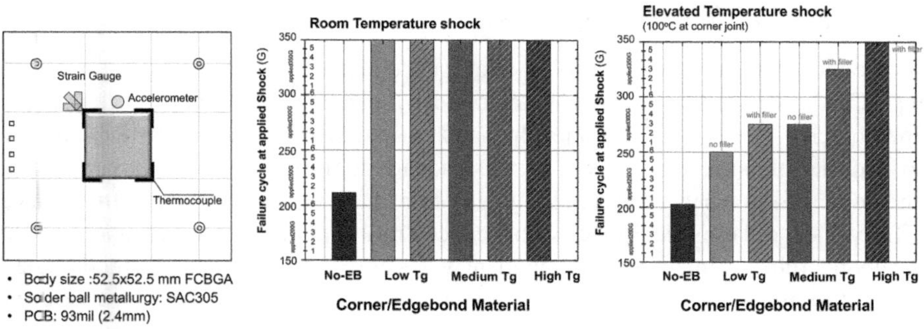

FIGURE 7.18 Mechanical shock test of FCBGA at room temperature and 100°C, with various T_g and CTE edgebond materials applied.

the edgebond material stability at elevated temperatures needs a closer look. For example, Figure 7.18 shows that the shock performance of FCBGAs with an edgebond material applied is stable at the room temperature, but at elevated temperatures, the edgebond material itself shows weakness and deviates from the expected performance. Materials with a lower glass transition temperature (T_g) tend to have lower strength at the elevated temperature of 100°C, while the filler content defines the CTE of the edgebond which will affect the thermal cycling performance. Based on these observations, a material with the combination of a higher T_g and a lower CTE can be a good candidate for structure enhancement applications. In addition to that, the interface between the edgebond and the PCB surface needs a close evaluation to assure the mechanical performance. This very interaction introduces new qualification requirements and testing methodologies that differ from classical reliability assessments.

Another consideration factor with the expanded scope of interfaces which need mechanical performance assessment and evaluation is the component to heat-sink interface. In HPC and AI network systems, thermal interfaces must remain fully operational throughout the system runtime to ensure mechanical and interconnect stability. If thermal interface materials (TIMs), heat spreaders, or cooling attachments degrade, even temporarily, the resulting rise in the junction temperature triggers a cascade of reliability problems. The initial defect generation at the TIM layer not only induces a thermal run-away failure but also increases the overall temperature which affects compression load of the component against the heat sink and generates additional compression load to the multi-layers inside the component and transfers uneven stresses to the solder joints thus accelerating fatigue. It increases the probability of delamination at the boundary between the die, the substrate, and the underfill, along with degradation of signal integrity due to mechanical shifts in the interconnect geometry and residual stress. Like this chain of events, mechanical reliability in advanced systems cannot be considered independently of thermal interface reliability. Stable and efficient thermal interfaces are a prerequisite for maintaining mechanical integrity, ensuring that interconnects operate within their designed stress limits under prolonged workloads.

7.4.2 Future Mechanical Challenges: Combined Driving Forces

As technologies progress toward co-packaged optics (CPO), co-packaged computing (CPC), and ultra-fine-pitch interconnects, mechanical reliability challenges are expected to intensify. Thinner dies and finer redistribution layers (RDLs) will reduce the tolerance for stresses. Larger package footprints will exacerbate global and localized warpage. 3D stacking will introduce new interfaces with distinct CTE values, creating additional stress concentration points. Increased adoption of edgebond reinforcement will demand new specifications and test methodologies tailored to multi-interface stabilization. Continuous operation of thermal interfaces will become a design-critical requirement, as any instability directly translates into accelerated thermo-mechanical failures. Mechanical reliability will continue to be a multidisciplinary challenge, requiring co-optimization across design, materials, and processes. Moving forward, solutions will include adoption of low-CTE laminates and advanced underfills to reduce CTE mismatch-induced stresses, refined edgebond design and monitoring to ensure that edge reinforcement does not introduce new stress concentration sites, and use of warpage-mitigation strategies, including metal stiffener rings, dummy bumps, and balanced stack-ups. Integration of thermal interface stability as a primary reliability metric alongside classical solder fatigue and warpage, with an effort of embedding mechanical performance criteria into Design for Reliability (DfR) and Design for Failure Analysis (DfFA) frameworks, is needed to ensure proactive mitigation during early design. A combination of mechanical and thermal factors is crucial for design considerations in the HPC and AI system environment. The segmented stand-alone property and performance is no longer valid for an accurate reliability assessment. It is a complex reliability bundle that needs to be considered.

7.5 ENVIRONMENTAL RELIABILITY

7.5.1 Environmental Reliability in HPC and AI Systems

As high-performance computing (HPC) and AI network systems advance, environmental reliability has become an increasingly critical item for evaluation and assessment. Unlike traditional data centers, which operate in controlled conditions, next-generation deployments of network systems also extend to edge computing, autonomous systems, and harsh-field environments. These new contexts expose advanced packages and interconnects to uncontrolled thermal, chemical, and mechanical conditions, which can introduce new degradation pathways. Data centers typically maintain a regulated temperature, humidity, and air quality. Reliability risks in these settings are largely tied to power density, cooling, and mechanical stresses within the system. But edge computing requires high-performance systems located near the event site to enable real-time decision making. Unlike data centers, edge environments expose HPC devices to fluctuating temperatures, humidity, contaminants, and vibration. These harsh and less predictable conditions drive the need for new defect mitigation and reliability planning. Chemical reliability is an emerging challenge in both data centers and edge environments. Forced air cooling introduces particulate and chemical contaminants into systems. Humidity ingress accelerates corrosion, oxidation, and electrochemical migration. In extreme cases, moisture triggers failure reactions that directly impact component functionality. These effects are significantly more severe in edge deployments, where exposure to outdoor or industrial atmospheres is unavoidable. This makes humidity and chemical mitigation essential for long-term reliability.

Another environmental dimension is the potential exposure of HPC and AI network systems to sub-zero or cryogenic operating conditions. While not yet common, certain specialized applications are beginning to demand extreme low-temperature performance. At cryogenic temperatures, materials exhibit fundamentally different mechanical and thermal responses. For example, Copper (Cu), with its face-centered cubic (FCC) lattice, shows relatively stable behavior at low temperatures. Tin (Sn), with its body-centered tetragonal (BCT) lattice, undergoes a ductile-to-brittle transition, dramatically shifting its mechanical performance and raising risks for solder reliability. Although the materials used in systems under cryogenic and sub-zero temperature environment will not be the exact same as used in the room temperature setting, the different mechanical behavior based on the materials lattice structure provides us a glimpse of the different mechanical characteristics at cryogenic and sub-zero temperature environments. Figure 7.19 presents a simple shear test on SAC105 solder ball at the room temperature and cryogenic temperature (–196°C) [12]. The results in Figure 7.19b show that the fracture surface of cryogenic temperature shear tested solder balls have brittle trans-granular and inter-granular fracture modes, as compared with the ductile surface with room temperature shear, which

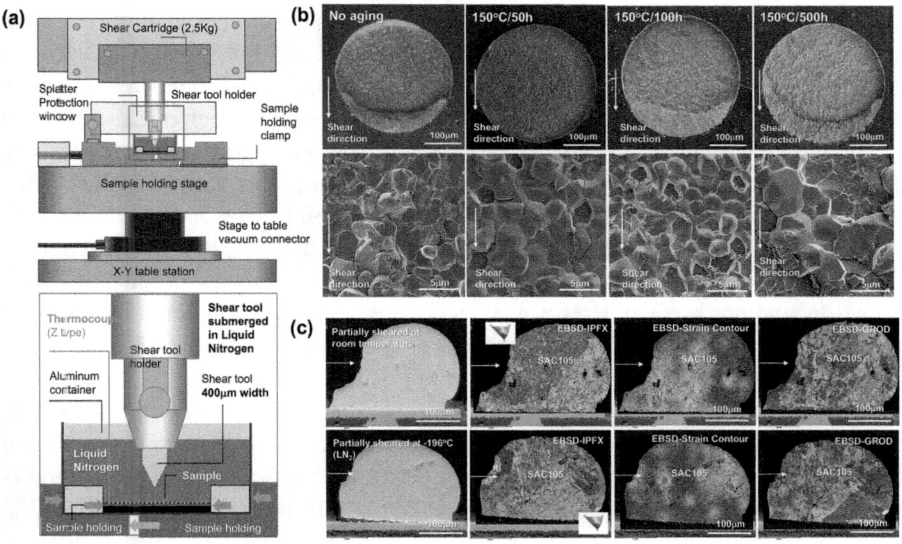

FIGURE 7.19 (a) Cryogenic shear test set-up. (b) Solder joint fracture for sample with Cu-OSP surface finish observed with surface scanning electron microscopy after the shear test in a –196°C temperature environment per isothermal aging condition. (c) Cross section view of partially sheared balls at room temperature and at cryogenic temperatures. EBSD IPF, Strain contour and GROD map per test joint configuration [12].

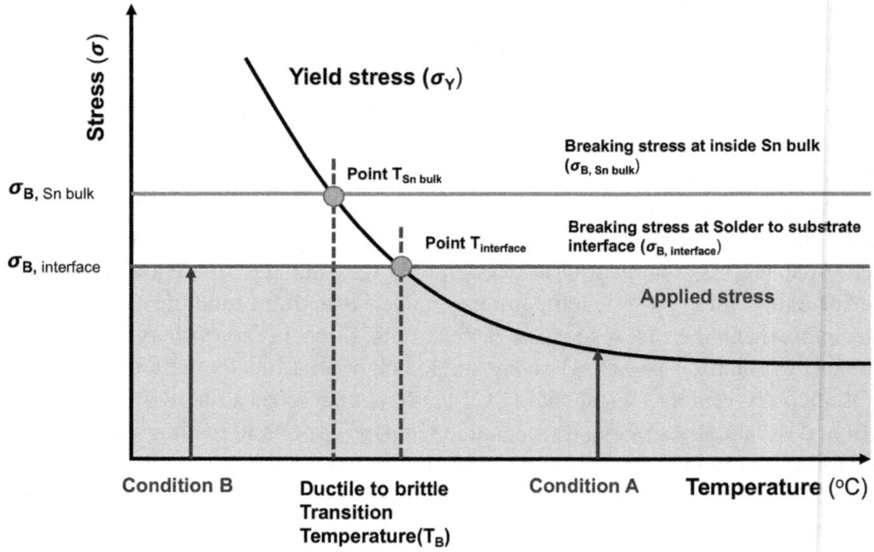

FIGURE 7.20 A schematically described temperature-to-stress correlation Yoffee diagram, which defines the ductile-to-brittle transition temperature [12,13].

can be seen in the cross-section images in Figure 7.19c. The same solder alloy balls had little plastic deformation at cryogenic temperatures due to their temperature dependent yield strength based on the crystal structure. This phenomenon can be explained by a Yoffee diagram shown in Figure 7.20 [13]. From the room temperature toward a low-temperature environment, the SAC105 bulk material shows an increase in yield strength (σ_y), where ductile fracture and plastic deformation occur. Competing with the yield strength is the breaking stress (σ_B), which is the stress level where the Sn bulk material experiences crack initiation and propagation toward full fracture, which is not affected by the temperature and more affected by the allowable crack propagation path. At a relatively high temperature, as indicated by condition A, if the sample experiences an increased applied stress, the yield strength is reached first, and plastic deformation occurs, resulting in crack propagation with a ductile fracture surface. However, if the temperature is below the cross point of the yield strength and the breaking stress, as indicated as condition B, the breaking stress ($\sigma_{B,\ \mathrm{interface}}$) is reached before the applied stress reaches the yield strength ($\sigma_{y,\ \mathrm{Sn\ bulk}}$) with an increased applied stress at that temperature, resulting in a brittle fracture mode because the crack tip region cannot plastically deform and maintain a sharp crack tip radius. Thus, the cross point of the yield strength and the breaking stress, at Point T, defines the ductile to brittle transition temperature (DBTT). The phenomenon of the fracture location shift in the solder joints for shearing at room-temperature and cryogenic temperatures can thus be explained. What this phenomenon implies is that at sub-zero and cryogenic temperatures, the material set behaves differently. Thus, systems exposed to extreme low temperatures require new test methodologies and specifications to capture performance under these unconventional yet emerging scenarios.

7.5.2 Intersection with Environmental Reliability: Edge Computing

Much like quantum computing architectures, which require multiple interconnect layers to support different functional blocks at different temperatures, HPC and AI systems are also trending toward denser interconnect stacks. With increasing reliance on chiplets, 3D stacking, and heterogeneous integration, the number of interconnect interfaces exposed to environmental stressors continues to rise. This makes environmental reliability not only a packaging-level issue but also a system-wide interconnect challenge. The expansion of edge computing reinforces the importance of environmental reliability in HPC and AI systems.

Edge computing refers to the deployment of computational resources closer to the location where data is generated, rather than relying solely on centralized cloud or data center infrastructure. The "edge" can be a factory floor, a vehicle, a telecommunications base station, or even wearable medical devices. By placing processing power near the source of data, edge computing reduces latency, enables real-time decision making, and decreases the amount of raw data that must be transmitted over long distances. In contrast to hyperscale data centers, which are highly controlled environments, edge computing environments are often less predictable, harsher, and more varied. This makes them an increasingly important focus for reliability engineering in HPC and AI systems. Modern AI and HPC workloads, such as autonomous driving, industrial automation, and real-time analytics, require low latency and immediate response. These applications cannot always tolerate the delay of sending data back to centralized cloud systems. Thus, advanced processors, accelerators, and networking devices must be deployed at the edge. Deploying HPC and AI hardware at the edge introduces new reliability challenges compared to traditional data center environments. Unlike controlled server rooms, edge deployments may face high humidity, wide temperature swings, vibration, dust, or chemical contaminants. Cooling is less predictable at the edge, requiring robust thermal interface materials (TIMs), fans, or passive heat spreaders. Systems deployed in vehicles, telecom towers, or industrial machinery, are subject to vibration and shock loading. A system level reliability test approach is needed to assure the performance of these edge computing units and systems. To cover these new reliability challenges, test methodologies and specifications need to be updated as new degradation mechanisms are revealed and explored, adding to the existing industry test standards. Selected current environmental test specifications are listed in Figure 7.21.

7.6 SUMMARY

As highlighted in this chapter, interconnect reliability is central to the performance and long-term functionality of high-performance computing (HPC) and AI network systems. As devices grow larger, more powerful, and increasingly interconnected, reliability can no longer be evaluated through traditional, narrow categories. Instead, it must be approached as a multidisciplinary challenge, integrating mechanical, thermal, electrical, and environmental considerations. A key message of this chapter is that reliability assessment must return to the basics and fundamentals. Interconnect reliability is not simply about catastrophic failure modes such as cracks

Application target	Target performance	Standard specification	Test conditions	Qualification criteria
Package/Die/Interposer (HBM, 2.5D)	Moisture-bias corrosion and Ionic migration	JEDEC JESD22-A101, JESD22-A110/A118	A101: 85 °C/85%RH, biased, 1000 h; A110: 130 °C/85%RH, biased, 96 h; A118 (uHAST) 110–130 °C, 85%RH, 96–264 h	No corrosion induced failures
	Preconditioning (reflow + soak) prior to moisture/corrosion tests	JESD22-A113 + J-STD-020	MSL-qualified soak, 3× reflow to peak per package rating	No delamination and no opens
	High-Temp Storage	JESD22-A103	150–175 °C, 1000 h	No material degardation
	Temperature Cycling	JESD22-A104	–40↔125 °C, 15 min dwells, 500–1000 cycles (table reference)	No solder/interconnect crack propagation to failure
	Power Temperature Cycling	JESD22-A105	Device power pulsing, ΔTj 60–100 °C, 10–30k cycles	No open/void growth
	High-Temp Operating Life	JESD22-A108	125–150 °C, biased, 1000 h	No corrosion induced failure
Substrate and Board level assembly	Electrochemical migration (ECM) / SIR screening	IPC-TM-650 2.6.3.3 / 2.6.3.7	Test patterns, contaminant loading per method; 85 °C/85%RH, DC bias, 168–500 h	No dendrites/whisker growth
	Ionic cleanliness	IPC-TM-650 2.3.25	ROSE ≤ 1.56 μg NaCl-eq/cm²	Meets qualification limit
	Mixed Flowing Gas (MFG) corrosion	IEC 60068-2-60	Class III severity: 40 °C/75%RH with SO_2/NO_2/H_2S/Cl_2 for 20–30 days; copper/silver coupons co-exposed	Meets qualification limit, no functional failure
	Single-gas corrosion as needed	IEC 60068-2-43	Gas and level tailored (e.g., H_2S 10–20 ppb for DC floor)	No open
	Temp-humidity bias at assembly level	JESD22-A101	85/85 biased, 500–1000 h	No leakage
	Thermal shock / rapid temp	JESD22-A106	–55↔125 °C, ≤10 s transfer, 300–500 cycles	No solder/interconnect crack propagation to failure
	Vibration & shock (board level)	MIL-STD-810H Method 514.8 (random vib), 516.8 (shock)	Data-center profile (rack-level PSD) + functional; shock 20–40 g, 6 ms as applicable	No solder/interconnect crack propagation to failure
System/Rack/Chassis	Corrosive atmosphere / datacenter air quality	Telcordia GR-63-CORE contaminants, ISA-71.04 classification	Deploy coupons (Cu/Ag) in representative airflows for 30–90 days or accelerate via chambers	Meets G1/G2 targets; no residue-induced faults
	Salt fog (if coastal / marine logistics)	MIL-STD-810H Method 509.7	35 °C, 5% NaCl fog, 48–96 h + recovery	No corrosion causing malfunction; cosmetic per spec
	Combined environment synergy	MIL-STD-810H Method 520.5 (Temp-Humidity-Vibration-Altitude)	Tailored mission profile	Functional & parametric pass; no latent damage
	Operational robustness	MIL-STD-810H Methods 501/502/507/510 (hot/cold, humidity, dust) as needed	Tailored to site	Functional pass; filtration/derating validated
Connector, Sockets, Cable	Gas corrosion of contacts	IEC 60068-2-60 / -2-43, ISA-71.04	MFG 20–30 days, single-gas where relevant	Low contact resistance drift; no fretting/corrosion opens
	Mixed humidity + bias	IPC-TM-650 2.6.3.x	85/85, biased mated pairs	SIR/contact stability

FIGURE 7.21 Environmental test items/parameters and related industry specifications, test conditions, and qualification criteria.

or open circuits. In modern systems, signal degradation and deviations from acceptable performance windows have become equally critical failure modes. Thus, signal integrity, cross-functional optimization, and system-level interconnect stability must all be viewed as reliability metrics.

We discussed a few major reliability drivers in this chapter. (i) Warpage and coplanarity, once considered secondary, are now primary concerns in heterogeneous systems. Both global deformation and localized stress concentrations dictate solder fatigue and delamination risks. (ii) PCB complexity, increased thicknesses and higher layer counts drive z-axis expansion alongside traditional shear stresses, reshaping thermal cycling and fatigue behaviors. PCB reliability must be re-evaluated with this vertical expansion in mind. (iii) Current stress induced reliability risks must be considered, as rising power densities elevate current stressing and Joule heating, and accelerate elemental diffusion, defect migration, and Cu grain-boundary degradation. Reliability frameworks must evolve to account for these realities.

As summarized in Figure 7.22, the degradation mechanism in HPC and AI systems are complex and combined. Thermal cycling performance is no longer defined solely by CTE mismatches, and modern thermal cycling needs to reflect the combined effects of global thermal factors, localized Joule heating, multiaxial PCB expansion, and mechanical constraints. This demands new multi-physics models and test methods. The mechanical performance and environmental performance are converged, correlated and expanded in a larger scale of the reliability scope.

Ultimately, the evolution of HPC and AI network systems makes interconnect reliability not just a packaging issue, but a system-wide performance assurance strategy.

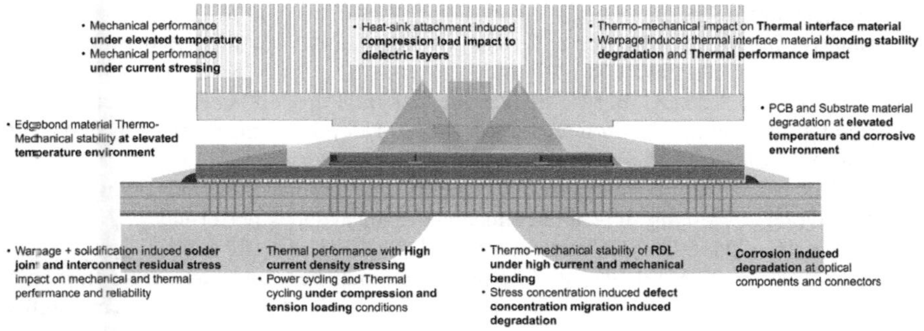

FIGURE 7.22 Complex and combined degradation mechanisms in HPC and AI systems.

By combining fundamentals with forward-looking design and testing methodologies, identifying new degradation mechanisms, developing new ways to proactively detect and predict the failure mechanism in an extremely high power and functional environment, the industry can ensure reliable operation in the face of unprecedented power density, integration complexity, and environmental diversity.

REFERENCES

1. T. -K. Lee, T. Bieler, and C. -U. Kim, *Fundamentals of Lead-Free Solder Interconnect Technology* (Chapter 1), Springer, Berlin, 2014.
2. T. -K. Lee, T. Bieler, and C. -U. Kim, *Fundamentals of Lead-Free Solder Interconnect Technology* (Chapter 5, Figure 5.1), Springer, Berlin, 2014.
3. T. -K. Lee, Y. Park, G. Ramakrishna, J. Nam, D. Yoon, and H. Roh, "Complex board via structure induced uneven stress/strain impact on interconnect stability: SAC305, full and hybrid LTS comparison," *2024 IEEE 74th Electronic Components and Technology Conference (ECTC)*, Denver, CO, USA, 2024, pp. 1124–1130.
4. J. Baek, Y. Elasser, K. Radhakrishnan, H. Gan, J. P. Douglas, H. K. Krishnamurthy, X. Li, S. Jiang, C. R. Sullivan, and M. Chen, Vertical stacked LEGO-PoL CPU voltage regulator. *IEEE Trans Power Electron.* 37 (6), 6305–6322 (2022).
5. P. Yeaman, Powering High Performance Computing, EEPower, November 16, (2021) https://eepower.com/technical-articles/powering-high-performance-computing/
6. T. -K. Lee, Y. Park, G. Ramakrishna, P. Gothe, Y. Lee, H. Kim, S. Lee, and C. Kim, "Impact of current induced Joule heating variation on long-term low melting temperature solder joint stability," *2024 IEEE 74th Electronic Components and Technology Conference (ECTC)*, Denver, CO, USA, 2024, pp. 1439–1444, https://doi.org/10.1109/ECTC51529.2024.00234.
7. T. -K. Lee, Impact of electrical current on the long-term reliability of fine-pitch ball grid array packages with Sn-Ag-Cu solder interconnects. *J. Electron. Mater.* 42, 599–606 (2013). https://doi.org/10.1007/s11664-012-2292-2
8. T. -K. Lee, Y. Park, G. Ramakrishna, Y. Lee, H. Kim, S. Lee, and C. Kim, "Impact of current stressed SAC305 and Sn-Bi interconnect anode and cathode side thermomechanical stability," *2025 IEEE 75th Electronic Components and Technology Conference (ECTC)*, Dallas, TX, USA, 2025, pp. 1369–1374,
9. A. Ahari, A. Hsiao, G. Baty, P. Su, and T. -K. Lee, "Microstructure signature evolution in solder joints, solder bumps, and micro-bumps interconnection in a large 2.5D FCBGA

package during thermo-mechanical cycling," *2019 IEEE 69th Electronic Components and Technology Conference (ECTC)*, Las Vegas, NV, USA, 2019, pp. 1099–1105, https://doi.org/10.1109/ECTC.2019.00171
10. T. -K. Lee, Z. Chen, G. Baty, T. Bieler, and C. Kim, Impact of an elevated temperature environment on Sn-Ag-Cu interconnect board level high-G mechanical shock performance. *J. Electron. Mater.* 45 (12), 6177–6183 (2016).
11. T. -K. Lee, Z. Chen, C. Guirguis, et al., Impact of isothermal aging and testing temperature on large flip-chip BGA interconnect mechanical shock performance. *J. Electron. Mater.* 46, 6224–6233 (2017). https://doi.org/10.1007/s11664-017-5650-2
12. A. Kitamura, T. Matthews, R. Contreras, D. Routledge, and T.-K. Lee, Impact of cryogenic temperature environment on single solder joint mechanical shear stability. *J. Electron. Mater.* 50, 723–734 (2021). https://doi.org/10.1007/s11664-020-08456-5
13. J.W. Morris Jr., Materials science. Stronger, tougher steels, *Science* 320 (5879), 1022 (2008).

8 Characterization, Testing, Simulation, and Failure Analysis of Materials in Advanced Semiconductor Packaging

Yan Li

8.1 INTRODUCTION

Advanced semiconductor packaging enables heterogenous integration, thus improving device performance with accelerated product development cycles, reduced costs, and enhanced yields. The development of packaging technologies greatly depends on materials such as substrates, underfill, molding compound, flux, and interconnects (including solder joints, microbumps, Through-Silicon Vias (TSVs), and Cu-Cu bonded pads).

At the early design stage of product development, simulations including thermal, mechanical, signal, and power aspects need to be conducted to ensure that the package meets the required product performance expectations. Building an accurate simulation model depends on the use of precise materials properties, including modulus, coefficient of thermal expansion (CTE), dielectric constant, and glass transition temperature (T_g), which can vary non-linearly with temperature, moisture, and other environmental or process conditions. Therefore, extensive materials property characterization is critical to provide accurate inputs for simulation models. However, in some cases, it may not be practical to measure materials properties directly at certain process or environmental conditions. In such situations, material simulation methodologies, such as molecular modeling and molecular dynamics, can offer insights into materials behaviors after chemical reactions or physical transformations at the atomic and molecular levels.

During the assembly process and reliability testing, materials degradation and interactions may cause interface delamination or cracks in traces and vias, resulting in potential yield loss and reliability concerns for advanced semiconductor packages. Fault isolation and failure analysis of defective units are essential for identifying defects. Comprehensive materials analysis of these defects is crucial for thorough root cause investigations and solution development to resolving the issues.

DOI: 10.1201/9781003587637-8

8.1.1 Overview of Materials Property Test Methodologies

The bulk properties of packaging materials, such as T_g, CTE, density, mechanical properties, and electrical properties, need to be measured both at room temperature and under high-temperature packaging process conditions to ensure their accuracy. Such validation is critical during the product design phase, as accurate materials data are essential for product performance simulations related to thermal management, thermomechanical reliability, signal integrity, power integrity, yield, and overall product performance. Detailed methodologies for bulk material property measurements are provided in Section 8.2.1.

Warpage of the die, wafer, package, and PCB can significantly impact assembly yield and product reliability. Therefore, accurate warpage measurements at both room temperature and under process conditions are essential. These dynamic warpage behaviors can be simulated. However, experimental warpage measurements need to be performed to validate the simulation models. Methodologies for package dynamic warpage measurement are described in Section 8.2.2.

Interfacial materials properties, particularly adhesion strength between layers, are very important to ensure package structural integrity during the assembly process and subsequent reliability testing. These properties can vary significantly depending on the specific materials and processes used by different products and processes. The methodologies for interfacial property testing are outlined in Section 8.2.3.

Interconnects, including microbumps and Cu-Cu bonded pads between stacked dies, First Level Interconnect (FLI) solder joints between the die and the organic substrate, Second Level Interconnect (SLI) solder joints between the package and the Printed Circuit Board (PCB), are important components in advanced semiconductor packages. To ensure that each interconnect type, with its particular Under Bump Metallurgy (UBM) design, can meet reliability requirements, bump strength testing (such as bump shear and bump pull tests) should be conducted prior to assembly. Detailed metrologies for interconnect testing are discussed in Section 10.2.4.

Materials surface properties (including surface energy, roughness, wettability, and chemical compatibility) significantly influence the interfacial behavior at various interfaces within semiconductor packages. These properties are crucial for yield enhancement in the assembly process including interconnect bonding, underfill dispensing, and lamination. They are also essential to achieving optimal product performance and long-term reliability. Detailed metrologies for surface property measurements are presented in Section 8.2.5.

Advanced packaging technologies, such as multi-die stacking and heterogeneous integration, enable higher integration densities with greater computational capabilities, resulting in increased power densities and higher levels of heat generation. Furthermore, the exponentially growing AI market is driving demand for microelectronic devices with Thermal Design Power (TDP) exceeding 1,000 W. Consequently, effective thermal management is becoming increasingly critical in semiconductor packaging. Section 8.2.6 provides an in-depth discussion on Thermal Interface Material (TIM) characterization methodologies, utilizing Thermal Test Chips (TTC) and Thermal Test Vehicles (TTV).

8.1.2 Overview of Materials Simulation and Characterization

Materials properties may exhibit non-linear variations with changes in temperature, moisture, and other environmental or process conditions. Therefore, materials simulation techniques are essential, particularly under conditions where experimental measurement is challenging. Section 8.3 discusses the applications of Molecular Dynamics (MD) and Artificial Intelligence (AI) in simulating the properties of packaging materials.

A wide range of materials characterization techniques, including Energy-Dispersive X-Ray Spectroscopy (EDX), Wavelength Dispersive X-Ray Spectroscopy (WDX), X-Ray Fluorescence (XRF), Fourier Transform Infrared Spectroscopy (FTIR), Atomic Force Microscopy-based Infrared Spectroscopy (AFM-IR), Optical Photothermal Infrared Spectroscopy (O-PTIR), X-Ray Photoelectron Spectroscopy (XPS), Time-of-Flight Secondary Ion Mass Spectrometry (TOF-SIMS), and Electron Backscatter Diffraction (EBSD), are frequently employed for process monitoring and failure root cause analysis. Section 8.4 presents detailed case studies demonstrating the application of these techniques in advanced semiconductor packaging development.

8.1.3 Overview of Assembly, Testing, and Reliability

Advanced semiconductor packaging involves significantly more complicated assembly processes compared to conventional packaging methods. These processes include Through Silicon Via (TSV) fabrication, Die-on-Wafer (DoW) and Wafer-on-Wafer (WoW) bonding, Hybrid Cu Bonding, Thermal Compression Bonding (TCB), and multiple solder reflow steps. To address the substantial challenges in ensuring product quality and reliability, due to the increased complexity of advanced semiconductor packages, extensive testing is conducted both at the End of Line (EOL) and after the reliability stress tests to verify product integrity. Section 8.5 provides a detailed discussion of reliability testing methodologies, including interconnect electromigration (EM) testing and package level reliability tests.

8.1.4 Overview of Fault Isolation and Failure Analysis

Fault Isolation (FI) and Failure Analysis (FA) techniques are critical for identifying defects that lead to electrical failures, whether observed at the EOL or after reliability testing. Once defects are identified, FA methodologies are applied to investigate root causes, hypothesize underlying failure mechanisms, and provide actionable solutions to resolve issues. FI and FA are key components in the development of semiconductor packaging technologies. Section 8.6 offers a thorough review of FI and FA techniques in advanced semiconductor packages. A generic FI-FA workflow is proposed to enable the detection of subtle defects in complex packages within a relatively short Through Put Time (TPT). Detailed case studies are discussed to illustrate the applications of those FI and FA techniques in revealing subtle defects inside the complex semiconductor packages.

8.2 TEST METROLOGIES OF MATERIALS PROPERTIES

8.2.1 Bulk Materials Property Measurement Metrologies

8.2.1.1 Differential Scanning Calorimetry (DSC)

The glass transition temperature (T_g), one of the most critical properties of organic materials, is the temperature at which a polymer material transitions from a rigid, "glassy" state to a flexible, rubbery form. Critical materials properties, such as volume, enthalpy, strength, and modulus, change significantly at T_g, resulting in substantial stress buildup within organic packaging materials and potential package failures [1]. Consequently, T_g is a key consideration when selecting materials for specific semiconductor packaging applications, as it typically defines the maximum operating temperature of organic materials.

Differential Scanning Calorimetry (DSC) is widely employed for T_g measurement due to its simplicity and short TPT. There are two main types of DSC, i.e., Heat-Flux DSC and Power Differential DSC. Heat-Flux DSC measures the difference in heat flux between a sample and a reference, while Power Differential DSC measures the difference in power input required to maintain identical heating rates for the sample and the reference. The schematic of a Heat-Flux DSC setup is shown in Figure 8.1. In this system, separate heaters independently control the sample pan and the reference pan, having the polymer material sample placed in the sample pan [2]. A control system with a computer monitors the temperature and regulates the heat flow to ensure that both pans are heated at the same rate (e.g., 10°C per minute). Since the polymer sample typically needs additional heat flow to have the same temperature with that of the reference, a graph of heat flow difference vs temperature can be generated. Figure 8.2 presents an example heat flow curve obtained by DSC [3].

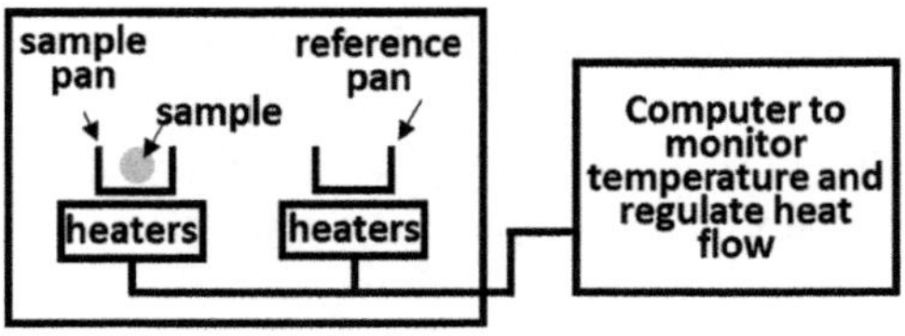

FIGURE 8.1 Schematic of a heat-flux DSC setup. (Adapted from Ref. [2].)

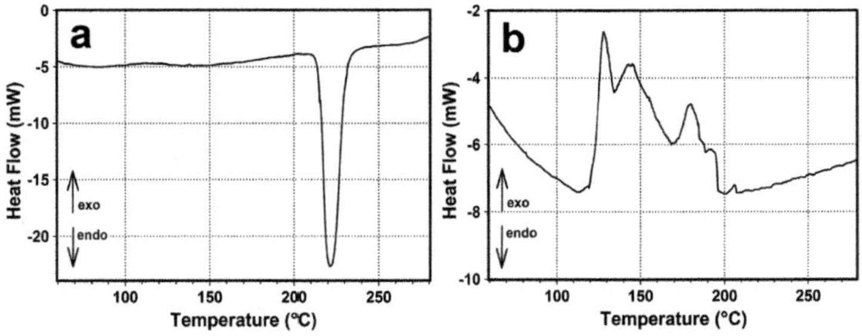

FIGURE 8.2 An example heat flow curve obtained by DSC, illustrating the detected glass transition event of a polymer material. (Adapted from Ref. [3].)

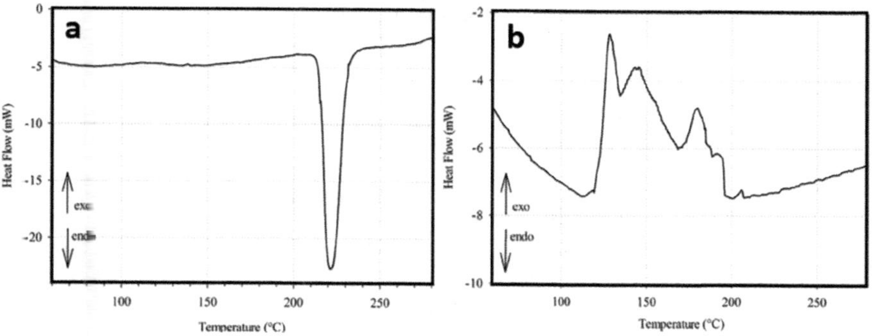

FIGURE 8.3 (a) A DSC curve collected from SnAg nanoparticles, indicating a melting point near 214°C. (b) The corresponding cooling curve, where a series of broad exothermic peaks were observed between approximately 200°C and 120°C, with a distinct peak near 140°C, potentially corresponding to solidification. (Adapted from Ref. [4].)

The glass transition is determined by the freezing of molecular motions, and the dynamics of these motions are time-dependent. Therefore, the T_g value can vary with the experimental time scale. Slower heating or cooling rates generally yield lower measured T_g values, because molecular rearrangements occur more readily. As a result, T_g is not a single point but a range influenced by the involved experimental time scale [1–3]. As illustrated in Figure 8.2, T_b indicates the initial detection of a change in heat capacity, T_1 reflects the onset of the transition, T_g represents the midpoint (the most commonly reported value), T_2 corresponds to the end set temperature, and T_e shows the very end of the detected glass transition event [3].

In addition to T_g measurements, DSC is often employed to define the melting points and determine the various phases of solder alloys used in microbumps, FLIs, and SLIs. By analyzing the heat flow changes during the transition from the solid state to the liquid state of the sample under controlled heating, the melting point or phase change temperature can be precisely measured. Similarly, the solidification behavior of the sample can be characterized during controlled cooling [4]. Figure 8.3a presents a DSC plot collected from SnAg nanoparticles, showing a melting point near 214°C. Figure 8.3b demonstrates the corresponding cooling curve, where a couple of broad exothermic peaks are detected between ~200°C and ~120°C, with a distinct peak around 140°C, potentially corresponding to solidification. Those broadened peaks could be attributed to the presence of the SnO_x surface layer on the SnAg nanoparticles, which can act as a nucleation barrier, preventing complete nanocrystal coalescence [4].

8.2.1.2 Thermomechanical Analysis (TMA)

Thermomechanical Analysis (TMA) is typically employed to identify the CTE of materials. TMA can also be used to determine the T_g of materials, because the molecular free volume change induced thermal expansion increases during the material glass transition [5]. Figure 8.4 shows the schematic of a typical TMA setup. The sample is put inside a furnace and then contacted by a probe, which is connected to

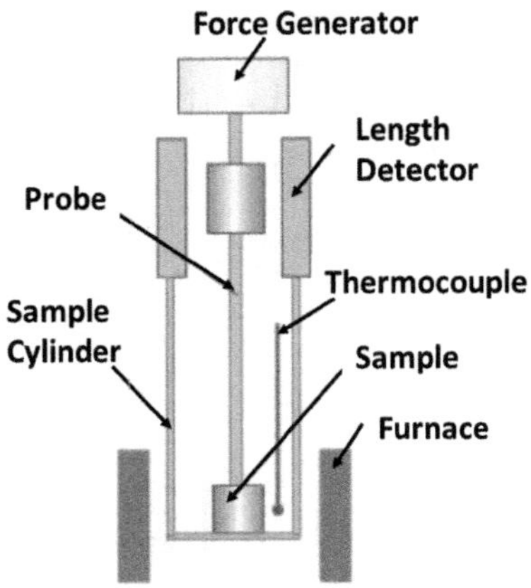

FIGURE 8.4 The schematic of a typical TMA setup. (Adapted from Ref. [6].)

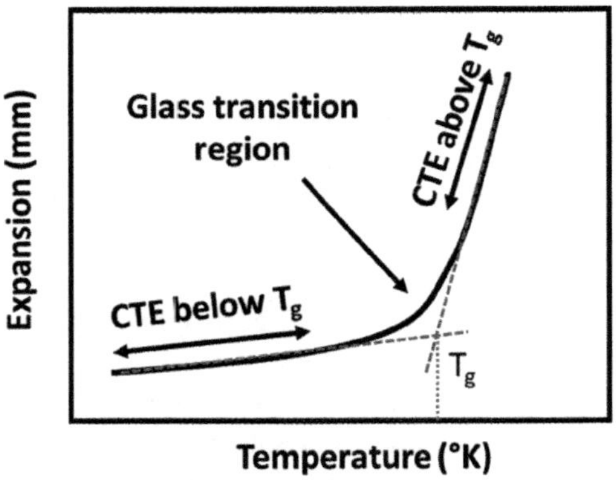

FIGURE 8.5 A representative TMA curve depicting sample expansion as a function of temperature showing how the CTE and T_g are measured. (Adapted from Ref. [5].)

a displacement detector and a force generator. A thermocouple positioned near the sample monitors the temperature throughout the test. The sample is heated at a controlled rate, and the dimensional changes of the sample vs temperature variations are recorded by the probe.

Figure 8.5 demonstrates a representative TMA curve showing the sample expansion as a function of temperature. As illustrated in Figure 8.5, the slope of the tangent before the glass transition of the polymer sample indicates the CTE below T_g, while the slope of the tangent after the glass transition represents the CTE above T_g. The glass transition takes place over a range of temperatures, depicted by the deviation points where the tangent lines separate from the curve. Conventionally, the T_g is identified as the intersection point of the two tangents, as illustrated in Figure 8.5. Additionally, the width of the glass transition region is as important as the T_g value itself [5].

8.2.1.3 Dynamic Mechanical Analysis (DMA)

Figure 8.6 shows the schematic of a Dynamic Mechanical Analysis (DMA) setup, which is a technique used to measure how materials deform under an applied force in a controlled environment [7]. During a DMA test, the sample is put into the measurement head of the instrument, where a sinusoidal force is applied through a probe. The resulting deformation of the sample is detected and the correspondence between the applied force and the resulting strain is recorded. From these measurements, properties such as elasticity and viscosity are derived and plotted as functions of temperature or time. DMA enables the testing of various polymer materials using different deformation modes shown in Figure 8.7, including tension,

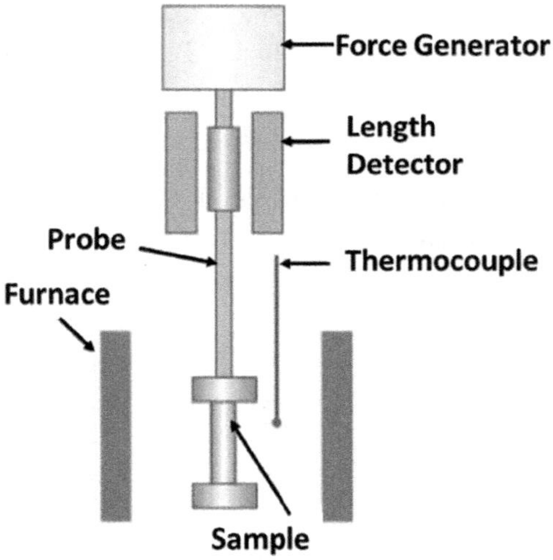

FIGURE 8.6 The schematic of a typical DMA (tensile mode) setup. (Adapted from Ref. [7].)

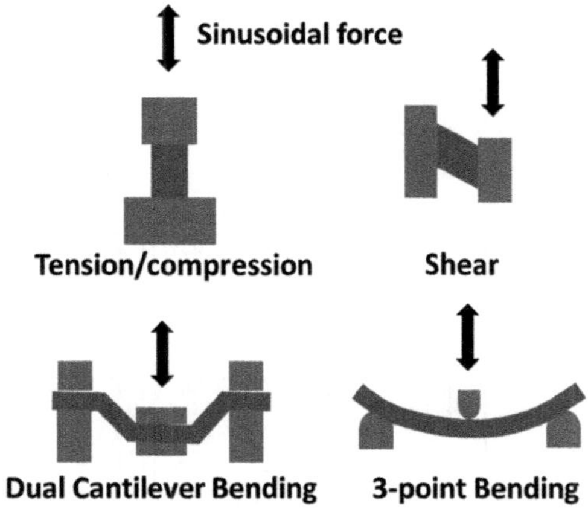

FIGURE 8.7 The schematic of various types of DMA deformation modes. (Adapted from Ref. [7].)

compression, dual cantilever bending, three-point bending, and shear. The specific testing mode is selected based on the sample's shape, modulus, and the objective of the measurement [7].

Table 8.1 summarizes the key terms and variables related to DMA measurements. Viscoelastic properties, including storage modulus (E′, G′) representing the elastic component, Loss Modulus (E″, G″) representing the viscous component, and the Loss Tangent ($\tan \delta = E''/E'$), can be assessed. Their dependence on temperature and frequency provides insights into the material behavior [8].

As illustrated in Figure 8.8, when a material is subjected to a sinusoidal stress or strain (either axial or torsional), its response is measured. An ideal elastic material reacts immediately without phase shift ($\delta = 0°$), while an ideal viscous material exhibits a 90° phase shift. Viscoelastic materials fall between these two extremes, with a phase angle between 0° and 90°. The mathematical definitions of the terms used in DMA testing are detailed in Table 8.2. Besides mechanical inputs such as

TABLE 8.1
Important Terms and Variables Related to DMA Measurements

Term	Elongation	Rotation/Shear
Modulus	E	G
Complex modulus	E*	G*
Storage modulus	E'	G'
Loss modulus	E''	G''
Loss factor	$\tan \delta$	$\tan \delta$
Strain	E	Γ
Stress	Σ	T
Strain amplitude	ε_A	γ_A
Stress amplitude	σ_A	τ_A
Angular frequency	Ω	Ω
Time	T	t

Source: Adapted from Ref. [8].
* Complex modulus.

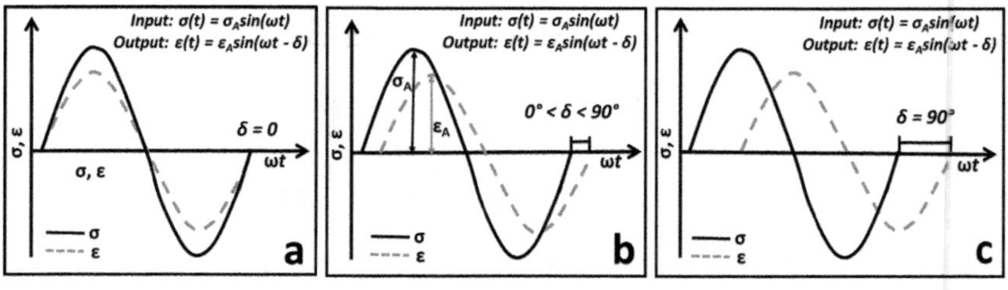

FIGURE 8.8 Stress, strain, and phase shift behaviors of ideally elastic, ideally viscous, and viscoelastic materials in DMA. (Adapted from Ref. [8].)

TABLE 8.2
The Mathematical Definition of Some Important Terms for DMA Measurements

Stress	$\sigma(t) = \sigma_A \sin(\omega t)$
Strain	$\varepsilon(t) = \varepsilon_A \sin(\omega t - \delta)$
Storage modulus	$E' = \dfrac{\sigma_A}{\varepsilon_A} \cos\delta$
Loss modulus	$E' = \dfrac{\sigma_A}{\varepsilon_A} \sin\delta$
Complex modulus	$E^* = E' + iE''$
	$\lvert E^* \rvert = \sqrt{E'^2 + E''^2} = \dfrac{\sigma_A}{\varepsilon_A}$
Loss factor	$\tan\delta = \dfrac{E''}{E'}$

Source: Adapted from Ref. [8].

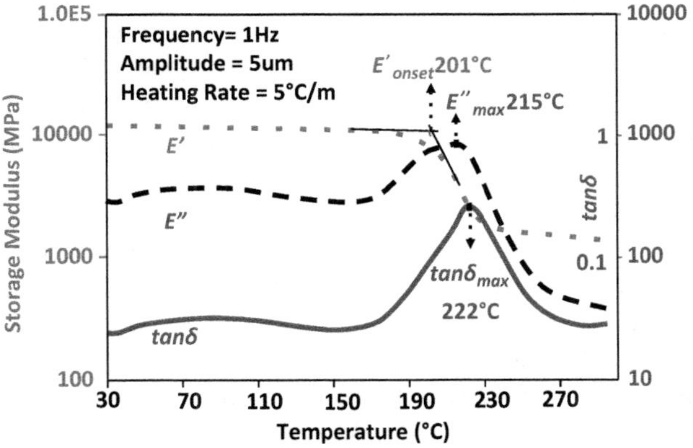

FIGURE 8.9 Dynamic moduli as a function of temperature for a thermoset polymer at 1 Hz measured by DMA, demonstrating a common method for determining T_g by DMA. (Adapted from Ref. [9].)

stress amplitude and frequency, which is provided by the linear or rotational drive of the DMA setup, environmental conditions such as temperature and relative humidity can also be adjusted with an environmental chamber [8].

As demonstrated in Figure 8.9, a commonly used method to determine T_g by the DMA method involves increasing the specimen temperature at a constant rate while keeping a constant frequency (e.g., 1 Hz) to measure the dynamic moduli E′, E″, and tan δ [9]. The dynamic moduli as a function of temperature for a thermoset polymer at 1 Hz is displayed in Figure 8.9 [9]. It is important to report both the measurement method and the convention used, when determining T_g by the DMA method. As illustrated in Figure 8.9, T_g can be identified using three different conventions, i.e., the onset of E' decline (E′ onset), the maximum of E″ (E″ max), and the peak of tan δ.

E′ onset is used for highlighting when mechanical properties start to change, while the tan δ maximum is typically employed in technical datasheets as it yields the highest T_g value. As indicated in Figure 8.9, the glass transition region of a material typically spans about 30°C [9].

8.2.1.4 Comparison of T_g Measurement Metrologies: DSC, TMA, and DMA

As discussed in Sections 8.2.1.1–8.2.1.3 and summarized in Table 8.3, there are three primary techniques often used to measure T_g [5]. DSC measures heat flow changes to determine the heat absorbed or released during the glass transition. It is the simplest and fastest method for identifying T_g. The DSC method does not have a strict sample size requirement. Small and irregularly shaped samples which can be directly collected from semiconductor packages can be used for DSC measurements. However, DSC is not universally applicable to all types of materials. For example, materials with a large filler content, high crosslink densities, or other complex thermomolecular behaviors, can exhibit only minor heat flow changes associated with T_g, making the glass transition difficult or even impossible to detect [5].

TMA defines T_g through changes in the thermal expansion coefficient, by monitoring dimensional changes, detecting shifts in the sample length or thickness as the material is heated. TMA is especially helpful for filled or highly crystalline materials, where dimensional variations around T_g are more obvious than heat flow changes. However, TMA is less sensitive to subtle T_g transitions compared to DMA and could not accurately detect small shifts. Additionally, cure-stress relaxation, which is the gradual release of internal stresses built up during polymer curing, can affect TMA results, shifting the measured T_g to lower temperatures. This relaxation effect depends on factors, such as curing conditions, sample geometry, and cooling rates after curing. Furthermore, TMA measurements could show variabilities within the same sample depending on the testing location (e.g. center vs edge), which could lead to ambiguity in T_g measurements [5].

DMA is generally considered as the most sensitive method for T_g measurement. It detects mechanical property changes, such as storage and loss modulus, across the glass transition. DMA enables precise T_g detection for various materials, including

TABLE 8.3
Comparison of T_g Measurement Metrologies

	Typical Run Time (*m*)	Sample Prep	Repeatability	Dependability	Comments
DSC	20	Easy	Good	Marginal	Easiest method, but many materials do not exhibit clear transitions
TMA	40	Medium	Fair	Good	Results is dependent on sample preparation
DMA	120	Difficult	Excellent	Excellent	Most sensitive method; T_g can be defined in several different ways

Source: Adapted from Ref. [5].

TABLE 8.4
Measured T_g Values of the Same Epoxy Sample Using Three Metrologies

Method	T_g (°C)
DSC	142
TMA	130
DMA	137 (loss modulus)
	146 (tan δ)

Source: Adapted from Ref. [5].

materials with complex structures (such as highly crosslinked polymers), and materials with very low T_g. However, the DMA method is more complex and expensive than DSC, as it typically requires more cautious sample preparation, larger and accurately machined specimens, and more extensive setup [5].

The appropriate T_g measurement technique should be selected based on materials characteristics and specific testing needs. DMA is ideal for highly accurate T_g determination in complex materials. DSC is preferred for general thermal transition screening with simplicity. If dimensional change measurement is needed, especially for filled or crystalline materials, TMA is desired. Table 8.4 demonstrates a case where all three methods are used to test the same sample. The sample is a developmental epoxy encapsulant cured at 145°C for 2 hours after gelling at 100°C. As illustrated in Table 8.4, different T_g values of the same material are obtained by different techniques, with discrepancies as large as 16°C [5]. The T_g value by the TMA method is 130°C, while the DMA method shows 146°C (tan δ) as the T_g value. Hence, when reporting T_g values, it is critical to specify the measurement method and the convention used in order to avoid misunderstanding.

8.2.1.5 Thermogravimetric Analysis (TGA)

Thermogravimetric analysis (TGA) measures the amount and rate of weight variations in a material during either heating at a constant rate or holding at a constant temperature. A thermal analysis technique monitoring mass vs temperature changes over time in samples, TGA is primarily employed to define material composition and forecast its thermal stability. Additionally, TGA can be utilized to analyze materials that experience weight variations because of process conditions, such as decomposition, oxidation, hydration, or dehydration. In those applications, TGA can help expose the underlying physical and chemical process details taking place within the material [10–12].

Advanced semiconductor packages tend to have larger die sizes with smaller and denser interconnects, which brings significant assembly process challenges, such as warpage induced open interconnects and flux residue removal issues. For example, the thin Si die, thin organic substrate, along with asymmetrical bump designs within a package, can significantly reduce the assembly yield if using the conventional solder

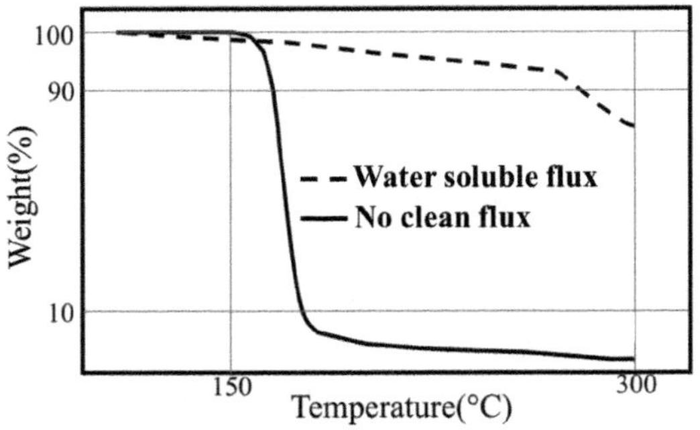

FIGURE 8.10 TGA measurements of a water soluble flux material and a no-clean flux material. (Adapted from Ref. [11].)

reflow process. Novel techniques, including TCB, laser assisted bonding (LAB), and formic acid (or fluxless) reflow, have been developed to address these challenges. During the TCB or LAB process, a wafer or organic substrate with applied flux is often kept at elevated temperatures (100°C–150°C) for extended time, resulting in hardened flux residue that are very difficult to clean after the assembly process [11]. New flux materials and adhesive agents are developed to address this issue. For example, semiconductor-grade ultra-low residue (ULR) no-clean fluxes leave minimal residue and do not require de-flux or flux cleaning process after the assembly, because the minimal flux residue is compatible with the underfill or molding materials used in the succeeding assembly process. In the fluxless (formic acid) reflow process, novel tacky agents are employed to hold the die in place and leave minimal residue compatible with subsequent processing steps, not requiring cleaning post the assembly process [10]. TGA has played a critical role during the development of those novel flux and adhesive materials [11].

Figure 8.10 presents the TGA curves obtained from a traditional wafer soluble flux and a no-clean flux, respectively. Compared with the water soluble flux, the no clean flux has a significant weight loss at temperatures above 150°C. The weight loss is designed by maximizing the solvent content and minimizing acidic activator levels. Once the solvent evaporates at the desired temperature, a thin acid layer is left on the substrate to promote solder wetting and bonding [11,12].

TGA is also utilized as the primary method to determine the decomposition temperature (T_d) of a material. T_d is an inherent material property identifying the temperature where an organic material begins to break down, losing its strength, toughness, and adhesion. For organic substrates used in advanced semiconductor packages, T_d typically is required to be above 350°C [11], because the lead-free solder reflow peak temperature can go up to 260°C.

Hygroscopy represents the ability of a material to absorb or adsorb moisture from its surroundings, and can considerably impact its mechanical properties, particularly for composite materials. Coefficient of Hygroscopic Expansion (CHE) (β), also referred to as Coefficient of Hygroscopic Swelling (CHS) or Coefficient of Moisture Expansion (CME), is used to quantify the effect of moisture sorption on the mechanical characteristics of a material. It can be defined as the following [11,13]:

$$\varepsilon = \beta C \tag{8.1}$$

where ε is the swelling strain (%), C is the moisture concentration and could be either an absolute value in g/cm^3 or a moisture concentration percentage (%). CHE is analogous to the CTE, but is associated with changes in humidity rather than temperature. In advanced semiconductor packages, the CTE mismatch between the die and the organic substrate can lead to thermal mechanical stress induced deformation. Likewise, using materials with significantly different CHE in the same package can result in deformation under moisture exposure, due to expansion differences between materials [11,12].

By combing TGA and TMA techniques, the hygroscopic swelling properties of semiconductor packaging materials can be characterized. As illustrated in Figure 8.11, two identical samples are exposed with the same temperature and humidity conditions. After the pre-conditioning, the same temperature isothermal moisture desorption is performed on two samples using TMA and TGA, respectively. TMA is used to measure the swelling strain (%) vs desorption duration time during moisture desorption, while TGA is employed to measure the corresponding moisture concentration (g/cm^3) vs the same desorption duration time. As demonstrated in Figure 8.11, since the two measurements are synchronized in time, by combining the TMA data with the TGA data, the swelling stain vs moisture content can be obtained as a linear correlation. The slope of the yielded linear line is the measured CHE, as displayed in Figure 8.11 [13].

Employing the metrology depicted in Figure 8.11, the CHE of a couple of underfill materials used in advanced semiconductor packaging is measured [13]. Figure 8.12 demonstrates the details of the CHE characterization process of the underfill material,

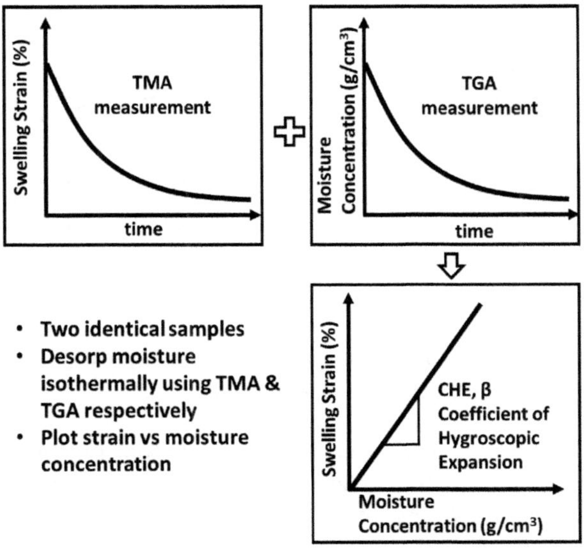

FIGURE 8.11 Setup and technique for characterization of hygroscopic swelling property of materials. (Adapted from Ref. [13].)

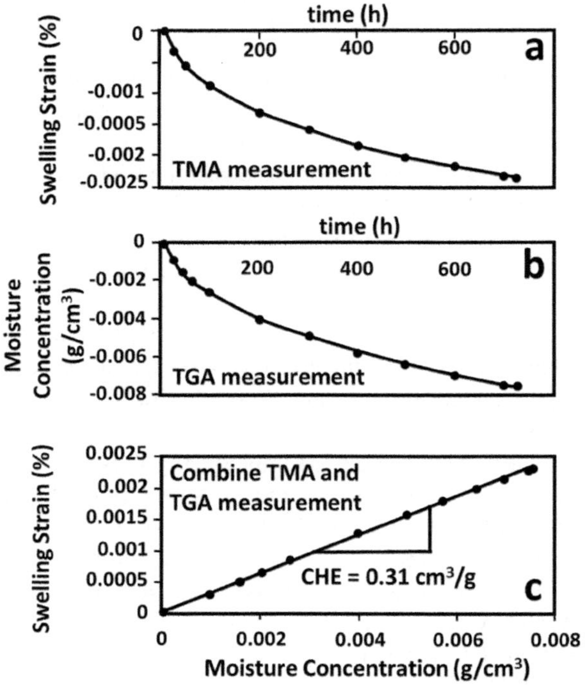

FIGURE 8.12 (a and b) the TGA and TMA measurement of underfill material UF-A at 85°C, respectively. (c) The combination of the two sets of measurement and the computation of CHE. (Adapted from Ref. [13].)

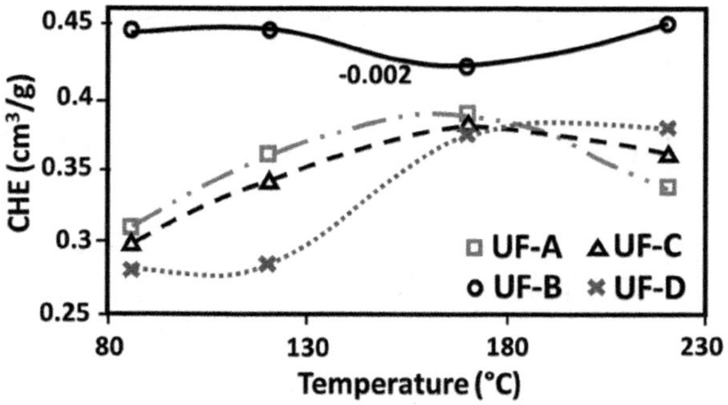

FIGURE 8.13 CHEs of four underfill materials, UF-A, UF-B, UF-C, and UF-D, over temperature range 85°C–220°C. (Adapted from Ref. [13].)

UF-A, at 85°C. The same methodology can be followed for three other underfill materials, including UF-B, UF-C, and UF-D, and at three other temperatures, such as 120°C, 170°C, and 220°C. Figure 8.13 shows the CHEs of the four underfill materials (UF-A, UF-B, UF-C, and UF-D), over temperature ranges from 85°C to 220°C. The result suggests that different underfill materials can have significantly different CHE values, as well as unique non-linear temperature dependence of CHE [13].

Direct and in-situ hygroscopic swelling characterization can be conducted using a DMA with a moisture chamber (DMA-RH system) [12,14]. To ensure the real-time observation of swelling or shrinking during moisture absorption or desorption, the DMA-RH setup enables accurate control of temperature (5°C–120°C) and relative humidity (5%–95%). The system has a strain resolution of 1 nm, and can provide timely hygroscopic swelling characteristic of materials with a very high sensitivity, especially for thin film samples [14].

8.2.2 Warpage Measurement

Advanced semiconductor packages typically have complicated architectures, such as 3D die stacking and heterogeneous integration, where various materials and multiple layers are involved. The CTE mismatch between materials can cause the package to warp as temperatures fluctuate during the assembly process or reliability testing. This "dynamic warpage" needs to be accurately measured, because it can significantly impact the assembly yield, package reliability, and product performance. For example, gross warpage can result in failures, such as solder joint crack, interface delamination, and electrical malfunctions. In multiple die stacking assemblies, even minor warpage could lead to extensive electrical connection interruptions within the 3D package. Therefore, dynamic warpage measurements, particularly under assembly process and reliability testing conditions, can help identify any potential early failures, thus enabling the fine-tuning of package designs and manufacture process parameters, in order to meet the product performance requirements and quality standards.

Dynamic warpage measurement metrologies, which can track the warpage change as a function of temperature, have become more and more important for advanced semiconductor packaging. This section will review four major metrologies employed for dynamic warpage measurement, including shadow moiré, digital fringe projection, confocal microscopy, and digital image correlation (DIC). Each technique has its unique strengths and limitations, depending on the specific applications and measurement requirements [15].

8.2.2.1 Shadow Moiré (SM)

The Shadow Moiré technique is commonly used as a real-time, high-resolution, non-contact, and low-cost method for dynamic warpage measurements [15,16]. As illustrated in Figure 8.14, in the Shadow Moiré setup, the sample is put under the reference grating and a white light beam is projected at a 45° angle through the reference grating, which is a Ronchi ruled glass with line pitches usually ranging from 50 to 500 microns. The interference between the reference grating lines and their shadows generates a contour map, or moiré fringes, which magnifies small surface variations on the sample and provides precise measurement of the surface topography. The out-of-plane displacement of the sample follows the following equation:

$$W = \frac{p}{\tan\alpha + \tan\beta} \tag{8.2}$$

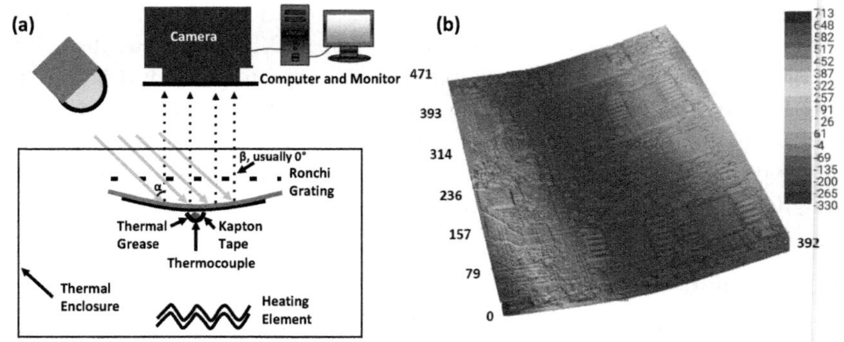

FIGURE 8.14 (a) Schematic of the shadow moiré metrology to measure dynamic warpage. (b) Example of the warpage data obtained by shadow moiré. (Adapted from Ref. [15].)

Where W is the out-of-plane displacement of the sample surface between adjacent fringes, p is the grating pitch, α is the incident light angle, and β is the observation angle [16]. Figure 8.14b demonstrates a typical warpage data obtained by Shadow Moiré [15].

Shadow Moiré can achieve submicron warpage measurement resolution, with its resolution remaining unchanged regardless of the field of view. Additionally, the system is very robust with minimal moving parts, and can provide full-field warpage data in less than two seconds. As presented in Figure 8.14a, to get the dynamic warpage characterization, the sample and reference grating can be placed within a thermal chamber having a glass top. This setup enables the in-situ warpage measurement during the simulated solder reflow process, having the same solder reflow profile as the actual process. The dynamic warpage characterization of semiconductor packages during this simulated solder reflow process can provide valuable insights for the root cause investigation of warpage-induced yield loss and reliability failures, such as various types of solder joint failure mechanisms [17].

8.2.2.2 Digital Fringe Projection (DFP)

As illustrated in Figure 8.15, the Digital Fringe Projection (DFP) technique measures warpage of samples using a similar principle to that of the SM method. Instead of forming an interference pattern in the SM method, DFP projects a series of dark and light lines, or "fringes", onto the specimen by a projector located at a particular position and angle relative to the camera. The fringe density is adjustable (limited to the resolution of the projector). Reference or calibration data of a flat surface collected at multiple distances from the projector is needed in order to measure the warpage of a sample by DFP. A phase-stepping approach is often employed during the warpage measurement [15,18].

For dynamic warpage measurements using the SM technique, as shown in Figure 8.14, a physical reference grating is required to be placed near the sample surface. The reference grating can affect uniform heating of the sample, and lead to warpage data constraints because the testing sample cannot be zoomed in to the point where the grating lines themselves are resolved. Furthermore, surfaces with

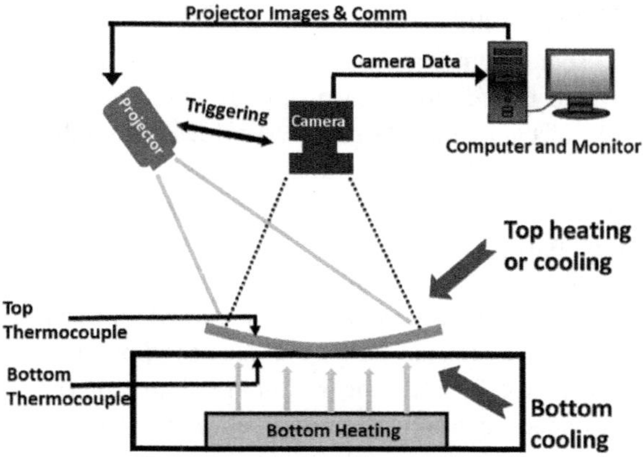

FIGURE 8.15 Schematic of digital fringe projection to measure dynamic warpage. (Adapted from Refs. [15,18]).

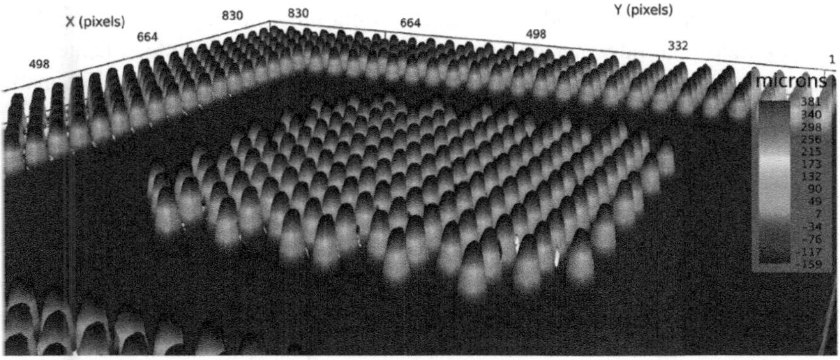

FIGURE 8.16 DFP data of balled BGA surface. (Adapted from Refs. [15,18].)

abrupt height changes, such as solder balls, connectors, or sockets with a complex topography, could not be measured by SM, because it can cause the loss of fringe tracking in the SM warpage measurement methodology.

DFP offers step height measurement capabilities (up to 50 mm) with high data point density, as it can project fringe patterns with varying periods, without the fringe counting and physical grating pitch limitations. Therefore, DFP is often used to inspect connectors, sockets, assembled modules, and localized PCB areas. Unlike SM, the warpage resolution in DFP depends on the field of view. ~2.5 micron resolution could be achieved by DFP with an adjustable field of view from $240 \times 192\,mm^2$ down to $45 \times 36\,mm^2$. Figure 8.16 shows the surface profile of a semiconductor package backside with solder balls, provided by DFP. SM would not be suitable for this type of samples without removing the solder balls first. Additionally, without the need of a physical grating, DFP can help reduce issues that can occur during SM measurements, such as outgassing and uneven heating [15,18].

8.2.2.3 Confocal Technique

Traditional confocal scanning microscopy relies on 2D scanning, small z-axis incremental movements, and the requirement of processing a large number of confocal images. It is capable of capturing 3D surface profiles with sub-micron resolution, including the measurements of surface roughness, bump height, and package warpage. However, it is not good for in-line metrology applications, which typically needs a much higher throughput.

The throughput limitation of the confocal technique has been addressed by the recently developed high-speed confocal technology. As illustrated in Figure 8.17, the novel approach does not use the traditional scanning process. All sensor pixels are simultaneously exposed by utilizing a pinhole array combined with a microlens array. This high-speed confocal technique employs larger step intervals combined with interpolation to maintain high resolution while significantly accelerating 3D image reconstruction, instead of collecting and processing numerous confocal images through the z-axis stepping. Furthermore, z-axis scanning is performed optically rather than mechanically, by inserting a plane-parallel plate into the optical path between the objective lens and the focused plane to adjust the optical path length between the objective lens and the sample without physical movement [19,20].

During the dynamic warpage measurements, the testing sample is heated within a convection oven, with better control of the heating rate [19,20], as demonstrated in Figure 8.17. Typical warpage measurement data by the confocal technique is presented in Figure 8.18, showing the capability to measure warpage from the backside of a semiconductor package without removing the BGA balls [15].

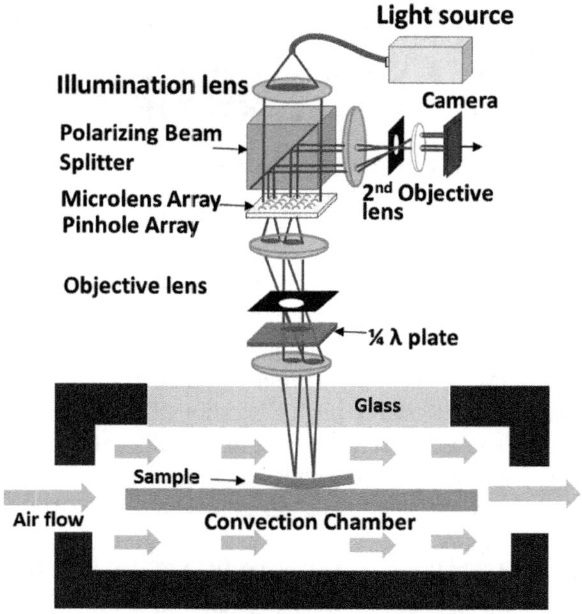

FIGURE 8.17 Schematic of the high speed confocal technique for dynamic warpage measurement. (Adapted from Refs. [19,20].)

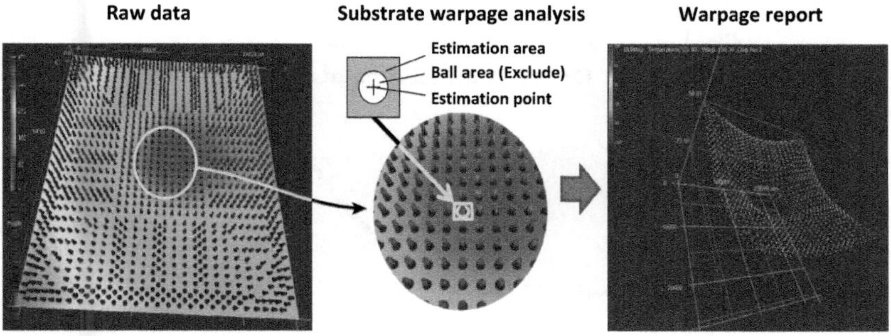

FIGURE 8.18 Example of confocal warpage measurement data. (Adapted from Ref. [15].)

8.2.2.4 Digital Image Correlation (DIC)

As illustrated in Figure 8.19a, Digital Image Correlation (DIC) can measure both in-plane and out-of-plane displacement on a surface by a non-contact, full-field optical method. A high-contrast, random speckle pattern needs to be applied to the sample surface. Two cameras are set up above the oven, imaging the sample from different angles. Two simultaneous images taken from both cameras are digitized and analyzed. The software can identify the same point on the surface from both perspectives, using pattern recognition within a small pixel window. The 3D spatial position of each pixel window relative to the camera can be measured by applying the principle of stereo triangulation. As demonstrated in Figure 8.19b, the mapping of the entire surface displacement in *x*, *y*, and *z* dimensions can be generated by scanning the pixel window across the sample [20].

DIC can be used for the dynamic warpage measurements of semiconductor packages. As illustrated in Figure 8.19a, samples are put inside a convection oven with an optical window. However, taking optical images through the oven's glass window sometimes can introduce light diffraction, which could potentially cause measurement errors. For accurate dynamic warpage data accusation, special sample preparation and prior system calibration are recommended [15].

As DIC can provide in-plane strain measurement with ~ 1.0 μm resolution, it can also be used to determine the average surface CTE of the sample from the stain and

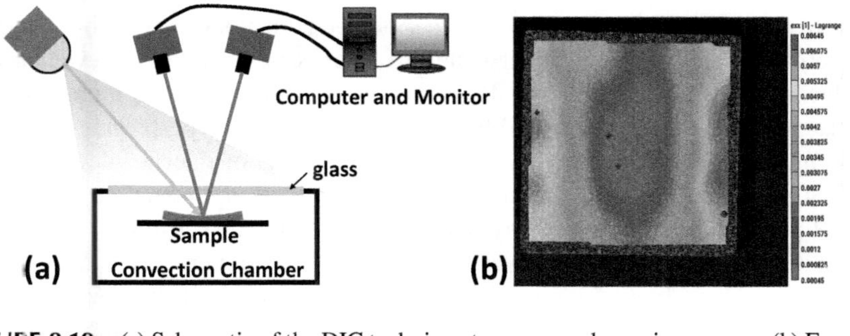

FIGURE 8.19 (a) Schematic of the DIC technique to measure dynamic warpage. (b) Example of the warpage data obtained by DIC. (Adapted from Ref. [20].)

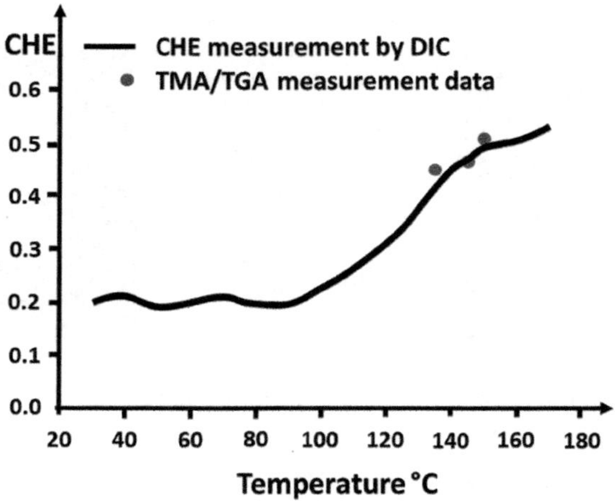

FIGURE 8.20 CHE vs temperature of a molding compound, measured by both DIC and TMA/TGA techniques. (Adapted from Ref. [21].)

temperature data. Additionally, DIC can be employed to measure CHE. For this application, the testing sample is placed in a controlled environment chamber with adjustable temperature and relative humidity, while performing DIC scans through the chamber window [21]. Figure 8.20 shows the measurement results of CHE vs temperature of a molding compound, employing both DIC and traditional TMA/TGA techniques (discussed in Section 8.2.1.5), showing good agreement between the two methods. As presented in Figure 8.20, the CHE of the tested molding compound remains steady below 100°C, but increases abruptly once the temperature reaches above 100°C [21].

8.2.3 Interfacial Material Property Measurement

As novel materials and fabrication processes are introduced to meet the escalating demand for product complexity, integration density, and functional requirements, ensuring the thermomechanical integrity of interfaces becomes increasingly challenging in advanced semiconductor packaging. The CTE mismatch among dissimilar materials in the multilayer architectures and interfacial structures commonly existing in advanced packages, combined with thermal processes during manufacturing and reliability testes, can cause significant thermomechanical stresses within packages that could lead to delamination at materials interfaces. Quantitative evaluation of interfacial fracture toughness and delamination resistance becomes more and more important, as the interface reliability directly affects device performance, product yield, and long-term reliability. Accurate measurements of interfacial strength can help with efficient materials down selection and process optimization, without the need of producing package samples and subsequent reliability testing. This can help accelerate product development cycles and reduce development costs. Additionally, in-depth understanding of interfacial failure mechanisms and the development of predictive simulation methodologies are critical for guiding the design of new materials and enhanced manufactory processes [22].

8.2.3.1 Shear Test

In semiconductor packaging, interfacial strength, such as the adhesion between polyimide (PI) films and die passivation layers, can be evaluated by standard shear test [23]. Figure 8.21 demonstrated the sample preparation and the subsequent shear testing to characterize the adhesion strength between PI and the silicon nitride (SiN) passivation layer. A Si wafer is sequentially coated with a SiN layer and a PI film. The PI film is then patterned by etching to produce the test structures. A horizontal load is applied at a constant rate to the PI structure, and the interfacial adhesion strength is defined by the critical shear load (σ) needed to introduce an interfacial failure.

The critical energy release rate (G_c) is a fundamental material property in fracture mechanics. It signifies the energy per unit area required to initiate and propagate a crack in a material, and is defined by the following equation:

$$G_c = \alpha \frac{\sigma^2 h_f}{E_f} \tag{8.3}$$

Where h_f is the film thickness, E_f is the elastic modulus of the film, and σ is the measured maximum interfacial stress from the shear test. α is a parameter related to the crack configuration, and is typically 0.5 for a crack at the film to substrate interface under steady-state propagation [23]. As an example, the shear test can be conducted on 15 samples, with the average maximum shear stress at failure measured to be 135 MPa [23]. Using Equation (8.3), the critical energy release rate G_c for the PI to SiN interface shown in Figure 8.21 is calculated to be 26.77 J/m^2 [23].

Furthermore, the finite element analysis (FEA) methodology can be used to model the strain energy release rate (G), associated with interfacial cracks in semiconductor packages. The simulated G value can be compared with the experimentally determined G_c. The interfacial crack will remain stable without propagation if $G \leq G_c$. However, if $G > G_c$, the crack is predicted to be "unstable", and will develop into a full delamination [23].

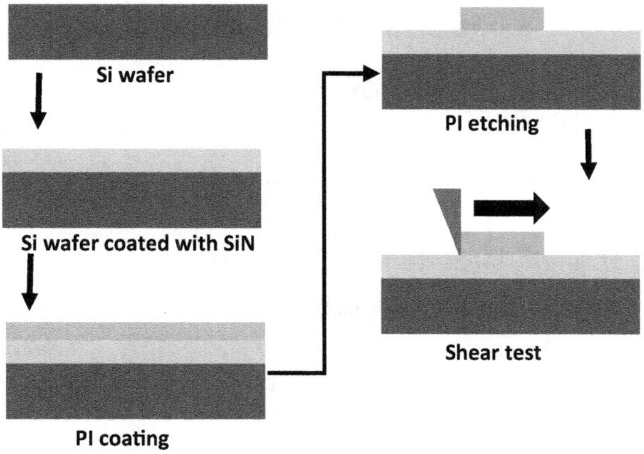

FIGURE 8.21 Sample preparation process and shear test method to evaluate the adhesion strength between PI and die passivation layer. (Adapted from Ref. [23].)

8.2.3.2 Pull Test

Interfacial adhesion in semiconductor packages can also be evaluated by tensile or pull testing. Tensile stresses at materials interfaces of semiconductor packages are commonly developed during the assembly process and reliability testing. Therefore, pull testing can offer a proper and quantitative approach to assess interfacial strength.

The mechanical robustness of Cu to polymer interfaces is becoming more and more critical to ensure the long-term reliability of advanced packages, with the ever-increasing density of copper (Cu) redistribution layers (RDL) within fan-out wafer-level or panel-level packages. Figure 8.22 demonstrates the schematic of the tensile test configuration for assessing the adhesion strength between Cu and a polymer material (Polymer type I). A Si wafer is coated with a layered sandwich structure, Polymer-Cu-Polymer. Epoxy adhesive is used to attach the layered sample surface to a test dolly. A tensile load is then applied perpendicular to the center of the dolly with a controlled rate until any interfacial failure occurs.

As an example, two polymer materials, named "Polymer A" and "Polymer B", are tested. Furthermore, to simulate environmental aging, certain samples are treated with Unbiased Highly Accelerated Stress Test (uHAST) standard conditions before the tensile testing. As summarized in Table 8.5, the post-test analyses record both the

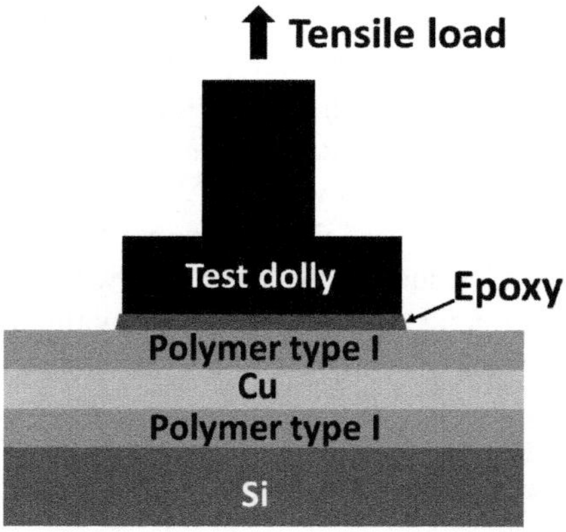

FIGURE 8.22 Schematic of the tensile or pull test of polymer-Cu adhesion strength. (Adapted from Ref. [24].)

TABLE 8.5
Interfacial Strength Measured by the Pull Test

Material	Preconditioning	Interfacial Strength (Kg/cm^2)	Failure Mode
Polymer A	No	690.5	Polymer/Cu interface
	uHAST	611.7	Polymer/Cu interface
Polymer B	No	765	Epoxy
	uHAST	753.9	Epoxy

Source: Adapted from Ref. [24].

adhesion strength and the failure modes. The interfacial adhesion strength is defined by dividing the failure load by the delaminated area. Results shown in Table 8.5 suggest that the tensile test can quantify the interfacial strength at Polymer A and Cu interface, with a clear reduction in adhesion in samples with uHAST preconditioning. However, for Polymer B, the epoxy adhesion is weaker than the polymer to Cu interfacial strength, resulting in adhesive failure at the dolly-epoxy interface, preventing direct measurement of the interfacial strength [24].

8.2.3.3 Double Cantilever Beam (DCB) Test

The double cantilever beam (DCB) test can provide accurate and quantitative measurements of adhesion energy, and is typically used to characterize fracture properties under pure tensile (Mode I) loading conditions. The interfacial fracture energy measured using the DCB method can closely approximate the intrinsic adhesion energy resulting from the chemical bonding across the interface, because of the minimal plastic dissipation and limited asperity effects associated with pure tensile loading. It offers the evaluation of interfacial strength independent of residual stress effects, and enables good correlation between measured fracture energies with fundamental adhesion mechanisms [22,25,26].

Figures 8.23 and 8.24 show the schematic of a typical symmetric DCB specimen and a representative load-displacement curve from a DCB test, respectively [26]. During the DCB testing, the specimen is loaded with a constant displacement rate, and the load-displacement curve is recorded continuously as a function of cantilever separation. The specimen exhibits elastic behavior initially, but eventually has a crack growing within the thin-film multilayer, once the strain energy release rate exceeds the fracture resistance of the interface. As demonstrated in Figure 8.24, at the critical load P_c, the load-displacement curve becomes nonlinear, representing changes in specimen compliance as the crack extends. Afterwards, the DCB specimen experiences partial unloading followed by multiple loading-unloading cycles, shown in Figure 8.24. This cyclic process generates multiple compliance measurements and can provide an accurate measurement of the crack length through a compliance-based relationship.

In DCB loading, it is important to note that the strain energy release rate depends on the crack length a. The following equation can define the critical strain energy release rate G_c:

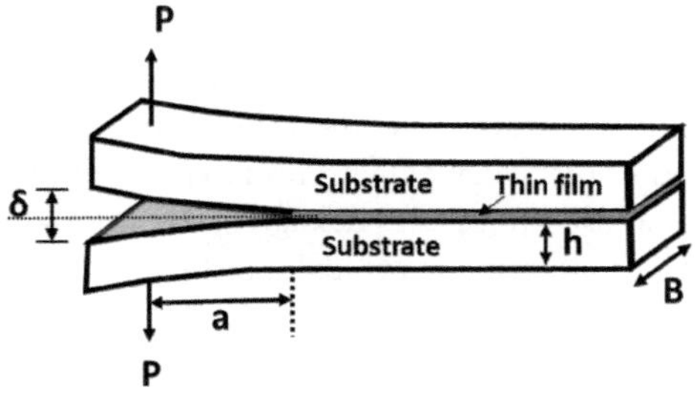

FIGURE 8.23 Schematic illustration of a DCB specimen. (Adapted from Ref. [26].)

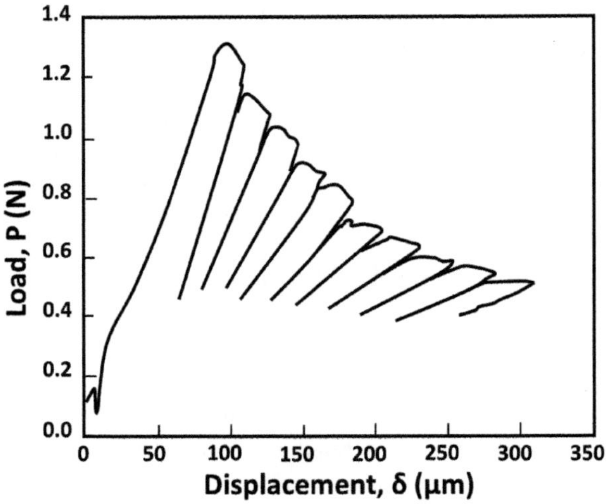

FIGURE 8.24 Applied load versus displacement from a representative DCB critical test, displaying multiple loading and unloading cycles (Adapted from Ref. [26].)

TABLE 8.6
DCB Test Results Showing the Interfacial Adhesion Characterization for the Underfill and Passivated Si Interface

Underfill	Passivation/Si	G_c (J/m²)	Standard Deviation
A	Si_3N_4/Si	15.5	3.4
C	Si_3N_4/Si	14	2.7
E	Si_3N_4/Si	47.1	9.3
B	Si_3N_4/Si	23.4	3.6
F	Si_3N_4/Si	68.5	12.7

Source: Adapted from Ref. [22].

$$G_c = \frac{12P_c^2a^2}{E'B^2h^3}\left(1+0.64\frac{h}{a}\right)^2 \tag{8.4}$$

where B is the specimen width, h is the thickness of a single beam, E' is the plane-strain modulus of the beam, P_c is the critical load, and a is the crack length determined by the load vs displacement curve shown in Figure 8.24 [22,25,26].

The DCB test can be used to select package materials and surface pretreatment methods that can enhance adhesion at critical interfaces within semiconductor packages, because the critical strain energy release rate (G_c) at an interface represents the interfacial adhesion strength [22,25]. As an example, Table 8.6 summarizes the DCB testing results employed to evaluate the adhesion between various underfill materials and a Si_3N_4 passivation layer on silicon substrates. There are five different underfill formulations (A-F) under testing, and the measured G_c values, shown in Table 8.6, suggest that "Underfill F" exhibits the best interfacial adhesion strength [22].

8.2.3.4 Four-Point Bend (FPB) Test

Figure 8.25 illustrates the schematic of a typical four-point bending (FPB) test specimen. The symmetric configuration, where the upper and the lower substrates have identical thicknesses and elastic properties, is often used in the FPB test. Roughly the same magnitudes of normal and shear stresses can develop at the crack tip, causing a loading phase angle of approximately 41°–43°, which is very close to the actual multilayer interface loads within semiconductor packages, including those in the thin film debonding [24,26].

In the FPB test, the specimen is loaded with the four-point test flexure at a constant crosshead displacement rate, and the applied load is continuously recorded as a function of displacement (or time). Figure 8.26 shows a representative load-displacement curve with an FPB test, as an example [26]. In the beginning, the specimen shows an

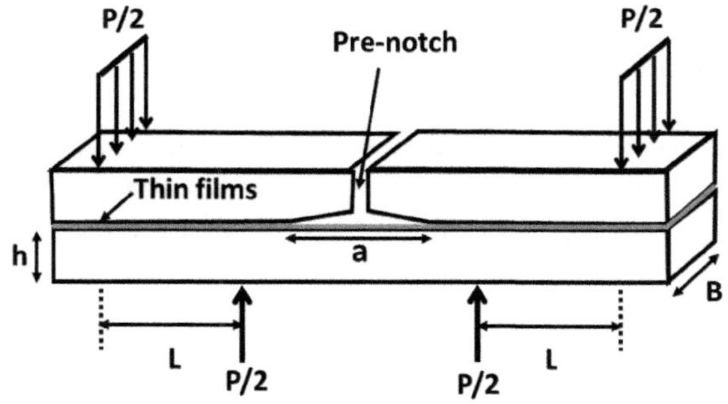

FIGURE 8.25 Schematic illustration of an FPB specimen. (Adapted from Ref. [26].)

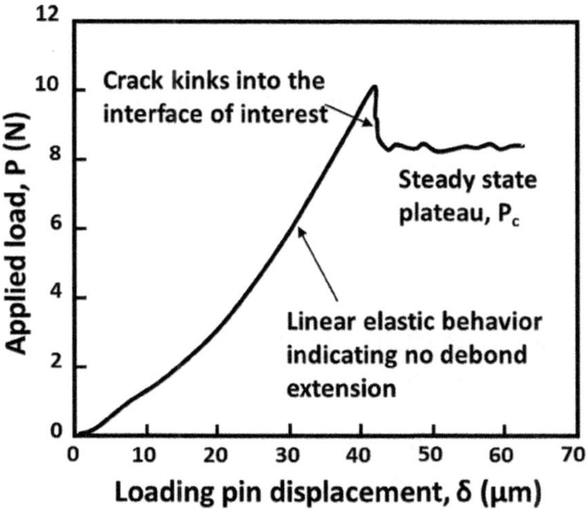

FIGURE 8.26 Load versus displacement from a representative FPB test. (Adapted from Ref. [26].)

elastic behavior, storing strain energy within the substrates. A crack starts from the "pre-notch" in the upper substrate and developed into the thin-film multilayer stack through the upper substrate, as the applied load increases [26]. The crack front typically propagates along a weak interfacial layer of interest until it reaches the inner loading pins where the crack propagation stops. The applied load typically reaches a plateau at the critical value, P_c, throughout this crack propagation process. The critical strain energy release rate G_c of a symmetric four-point bending specimen is given by the following equation [26], based on the classical beam theory:

$$G_c = \frac{21P_c^2L^2}{16E'B^2h^3} \tag{8.5}$$

where L is the applied moment arm length, i.e., the distance from the support point to the loading point, as shown in Figure 8.25. E' is the plane-strain modulus of the substrate, B is the specimen width, and h is the half-thickness of the specimen. It is important to note that in the FPB testing, G_c is independent of the crack length, a. Therefore, it is unnecessary to measure crack length during the experiment, simplifying the practical implementation of the four-point bending method. The FPB test methodology is a powerful tool to evaluate near-interfacial fracture properties of thin-film systems. Because of the mixed-mode loading conditions, the crack typically develops along the interface of interest within the multilayer sample [26].

In summary, the loading mode in the FPB test is a mixture of Mode I (tensile) and Mode II (shear) with a phase angle of about 41°–43°, very close to the realistic interfacial stress conditions encountered in advanced semiconductor packages. The FPB testing has been widely used to select different polymer materials and optimize assembly processes, in order to improve the adhesion between polymers and Cu RDL within advanced semiconductor packages, such as Fan-Out Wafer Level Packaging (FOWLP) and Fan-Out Panel Level Packaging (FOPLP) [24].

8.2.4 Interconnect Bonding Test

Advanced semiconductor packages usually have tens of thousands of interconnects, including Through-Silicon Vias (TSVs), Cu pillar microbumps, FLI, and SLI, with dimensions ranging from a few microns to several millimeters. Quick evaluation of the bonding strength of various types of interconnects can be conducted by the interconnect bonding test metrologies discussed in this section.

8.2.4.1 Bump Shear Test

With smaller sizes and finer pitch as compared with controlled collapse chip connection (C4) solder joints, Cu pillar microbumps have been extensively adopted in advanced semiconductor packages, offering benefits, including high thermal conductivity, increased input-output (IO) density, and better resistance to electromigration. Figure 8.27 illustrates the schematic of a typical Cu pillar, having a lead-free solder material, for example SnAg, electroplated on the top of the Cu pillar (solder cap). As the solder volume is small, intermetallic compounds (IMC) constitute a large portion of the solder cap, especially after undergoing reliability tests, such as thermal

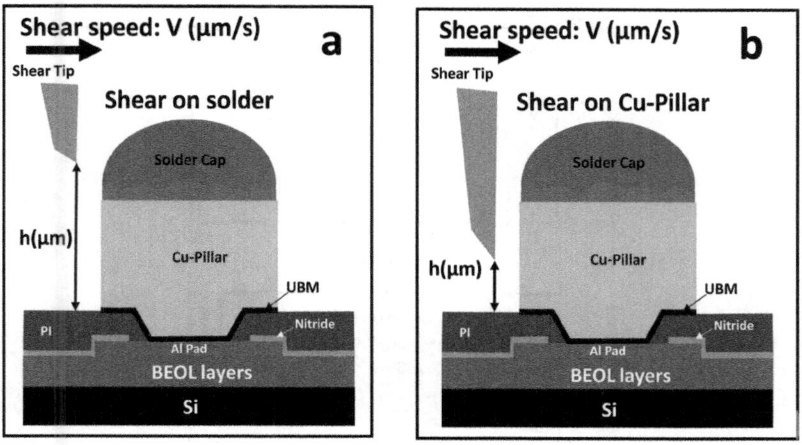

FIGURE 8.27 Schematics of bump shear test for Cu pillar micro bumps. The bump shear test can be performed on both solder cap (a) and Cu pillar (b). (Adapted from Ref. [27].)

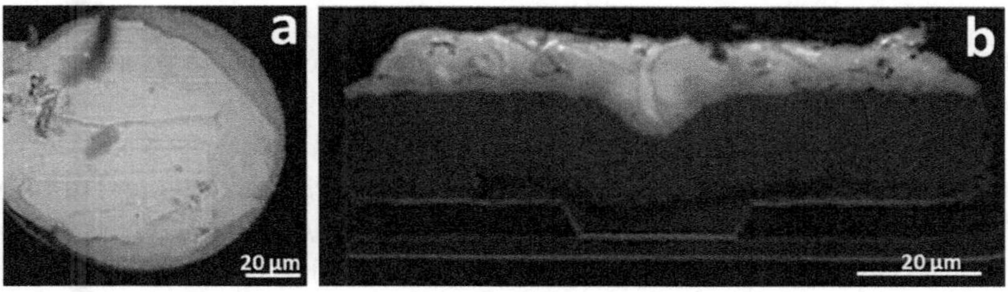

FIGURE 8.28 (a) Top view after the bump shear failure for sample with pad diameter 100 µm, (b) cross-sectional view after failure for the same sample showing that the failing interface is within the bulk solder. (Adapted from Ref. [27].)

aging. Bump shear test is usually used to assess the bonding strength of Cu pillar bumps with different flavors, including various solder cap materials, bump diameters, and aging conditions [27,28].

As displayed in Figure 8.27, a bond tester with a micron size shear tip is typically used to perform the bump shear test. A constant mechanical load, up to 400 g, at a controlled speed, such as 700 µm/s, is applied to shear the Cu pillar. The shear tip position can be adjusted to target either the solder cap (shown in Figure 8.27a), or the Cu Pillar (shown in Figure 8.27b). Bump shear tests are usually conducted at room temperature and need to be repeated on multiple nominally identical samples [27]. The effects of parameters, including solder composition, bump size, and aging time, can be systematically investigated with the shear strength values of the tested Cu pillar bumps [27,28].

As illustrated in Figure 8.28, the post-bump shear test analysis involves determining the failure interface through both top-down and cross-sectional imaging. Figure 8.28a presents the top view of a 100 µm diameter Cu pillar after 1,000 hours of aging at 125°C, showing a ductile failure mode within the bulk solder. After sample

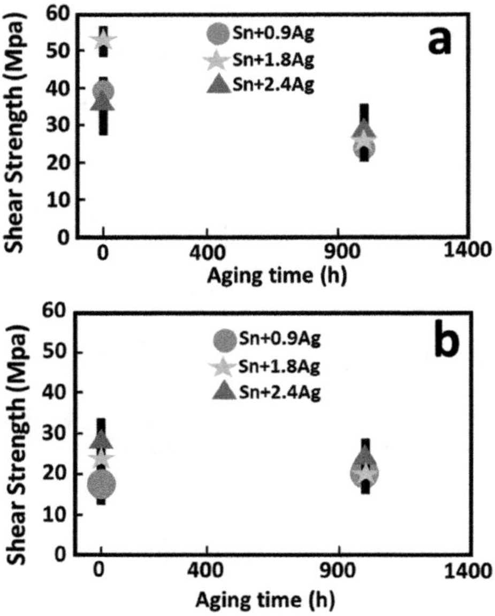

FIGURE 8.29 Shear strength of the solder cap depending on solder composition, pad diameter, and aging time for (a) pad diameter 30 μm, and (b) pad diameter 100 μm. (Adapted from Ref. [27].)

preparation, the cross-sectional image collected from the same bump is illustrated in Figure 8.28b, confirming the bulk solder fracture failure mode [27]. Brittle failure modes within the IMC layer located between the Cu pillar and the solder cap could also occur, depending on the solder cap composition and Cu pillar size [28]. Failures can also initiate at the UBM or the Al pad layer located under the Cu pillar, if interfacial contamination is present. Furthermore, if the shear strength of the BEOL layers is lower than that of the Cu pillar, failures may occur within the BEOL layers [29]. Comprehensive failure analysis, including both top-down and cross-sectional investigations, is highly recommended.

As an example, Figure 8.29a and b demonstrate the dependence of solder cap shear strength on SnAg composition (0.9–2.4 wt% Ag), and aging time (no aging vs over 1,000 hours aging at 125°C) for Cu pillar bumps with 30 μm diameter and 100 μm diameter, respectively, [27]. As illustrated in Figure 8.29a, aging significantly reduced the bump shear strength for Cu pillar microbumps with 30 μm diameter, especially with the "Sn+0.9 Ag" solder cap, while the impact of aging on the bump shear strength of Cu pillar bumps with 100 μm diameter is not as significant [27]. Detailed follow-up microstructural evolution studies of Cu pillar microbumps at 125°C reveal that the size, distribution, and quantity of Ag_3Sn IMC particles within the solder cap could significantly impact the bump shear strength [27].

8.2.4.2 Bump Pull Test

Unlike the bump shear test described in the above section, the bump pull test applies a tensile load directly to each interconnect under test, which can closely simulate the actual stress conditions experienced by interconnects in semiconductor packaging.

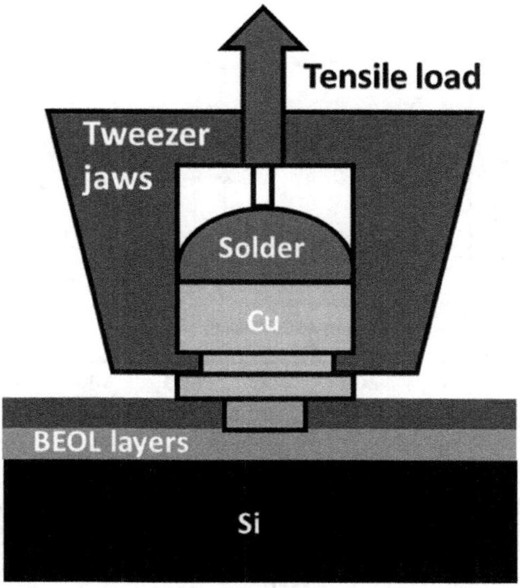

FIGURE 8.30 Schematics of bump pull test setup for Cu pillar micro bumps. (Adapted from Ref. [31].)

The bump pull test is typically selected over the bump shear test if weak interfaces need to be identified, because of the following two key reasons. First of all, interfaces typically exhibit a lower strength under a tensile stress compared to a shear stress. Secondly, a tensile failure is less impacted by external factors, including passivation layer interference and large-scale deformation of the interconnects, etc. [30]. Appropriate tweezer jaws need to be selected in order to firmly grip the "bulk material" of the interconnect, and maintain the quality and accuracy of bump pull test data. Figure 8.30 demonstrates the schematic of a bump pull test setup for Cu pillar microbumps. The tweezer jaws securely hold most of the bump, having an applied constant tensile load with a controlled rate (measured in μm/s) [30,31].

The bump pull test has been extensively used to assess the weakest interfacial layers of interconnects. It is especially effective in identifying interfacial weaknesses, such as contamination-induced delamination at the UBM layer or pad layer, as well as voids at the IMC to bulk solder interface led by IMC growth during the assembly process or aging. Additionally, the bump pull test has been successfully employed to study brittle fracture risks in BGA solder balls related to IMC growth [30]. A strong correlation between the bump pull results and the conventional board-level drop tests is demonstrated in the study, with much lower cost and time spent during the bump pull tests [30].

As an example, Figure 8.31 illustrates the bump pull test results for the BGA solder balls generated with various solder alloys, pad finishes, and aging time at 150°C, at a pulling speed of 5,000 μm/s. A clear degradation of solder joint strength resulting from thermal aging is observed. For instance, for the Sn3.5Ag solder with a Cu Organic Solderability Preservative (OSP) finish, the bump strength decreases by about 40% after 1,000 hours of aging at 150°C. Interestingly, the solder joints with

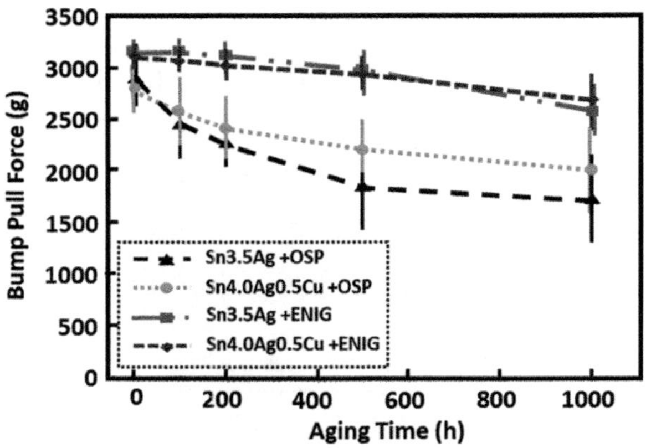

FIGURE 8.31 Effect of thermal aging on the bump pull strength with pull speed 5,000 μm/s for the BGA balls with different types of solder alloy and pad surface finishes. (Adapted from Ref. [30].)

Electroless Nickel Immersion Gold (ENIG) surface finishes constantly show higher bump strength, compared with those with OSP finishes, independent of the solder alloy used. In-depth studies of IMC growth and morphology during aging reveal a strong correlation between the thickness of the IMC layer and the bump strength. Solder joints with thicker IMC layers typically have smaller bump strength. Solder joints with ENIG surface finish demonstrate significantly lower IMC growth rates during aging, compared to those with the OSP surface finish, thus tending to have higher bump strength during aging over time [30]. Furthermore, researchers show that conducting the bump pull test at higher speeds, around 400 mm/s, enables the measurement of the solder ball fracture stress, which correlates well with traditional board-level drop test outcomes [32].

8.2.5 Surface Property Measurements

Comprehensive characterizations of surface properties, such as surface energy, roughness, wettability, and chemical compatibility, are vital for materials selection and process optimization during the technology development of advanced semiconductor packaging, because these material surface properties play key roles in the assurance of product performance and long-term reliability, as well as yield improvement of assembly processes, including interconnect bonding, underfill application, and lamination.

8.2.5.1 Surface Roughness Measurements

Surface roughness measures the microscopic deviations of a surface from its ideal form. A larger deviation means a rougher surface, while a smaller deviation signifies a smoother surface. Two parameters, R_a and R_q, are commonly used to describe surface roughness. R_a is the arithmetic mean of the absolute surface height deviations. R_q is the root mean square average of these deviations, or the RMS roughness. Higher R_a and R_q values mean rougher surfaces. Surface roughness can directly impact hybrid bonding and fusion bonding yield in advanced semiconductor packaging, as it is a key parameter influencing the interface adhesion strength. During organic substrate fabrication, issues, including short circuits and electrical breakdown, can occur due to an excessively rough surface. In high-frequency microelectronic devices, increased surface roughness of traces can create variations in signal travel paths and capacitance, and lead to signal integrity problems [11].

Surface roughness is usually measured by either profilometers or Atomic Force Microscopy (AFM). Contact mode profilometers, or stylus profilometers, use a physical stylus, typically diamond-tipped, to trace and record surface topography. Optical profilometers can measure surface roughness without physical contact, by light-based methods, providing fast and non-destructive analysis. AFM scans surfaces with a sharp tip attached to a cantilever, measuring interaction forces and offering high-resolution topographic maps down to the angstrom level. Either stylus profilometers, optical profilometers, or AFM, can be employed to measure roughness, depending on the measurement resolution requirement and the surface features being examined. In some cases, a combination of all three methods is used. For example, profilometers can be used to survey an entire sample, and measure large surface features at the micron or submicron scale first, while AFMs can be employed for localized nanoscale characterizations at specific areas of interest (AOI) identified by the profilometer scans [33].

The WoW fusion (or direct) bonding process is heavily dependent on the surface roughness of the wafers and the contact pressure applied during bonding. In the glass-to-silicon WoW bonding process by the transmission laser bonding (TLB) technique, the impact of the Si wafer roughness on the bonding strength has been studied extensively [33]. Wafers with different surface roughness are bonded under various contact pressures. The wafer pull test with a micro tensile tester is used to evaluate the bond strength. AFM is used to quantify the surface roughness. Figure 8.32b demonstrates a typical AFM image, showing a R_q value of 25.8 nm for a Si wafer. Figure 8.32a illustrates the significant impact of surface roughness on the WoW bonding strength with a 0.5 MPa contact pressure. According to Figure 8.32a, increasing the surface roughness R_q from 0.58 nm (the first point on Figure 8.32a) to 25.8 nm (the second point on Figure 8.32b) can dramatically decrease the bond strength from about 10 MPa to nearly 5 MPa [33].

8.2.5.2 Surface Energy or Surface Tension Measurements

Surface energy or surface tension (for liquid) characterizes the energy required to break intermolecular and interatomic bonds to create a new surface. Atoms or molecules located at the surface of a material are usually at a higher energy state and are less stable compared to those located within the bulk material. The surface energy determines the liquid wettability to that surface, which eventually impacts the adhesion strength at the interface in multilayer structures. Solid surfaces with a higher surface energy are generally preferred, as they can promote better wetting during the assembly processes, such as spin coating, flux spray, underfill, and interconnect bonding [34,35].

When exposed to the atmosphere, surfaces tend to absorb ambient molecules, leading to a significant reduction in surface energy. Various surface cleaning methods,

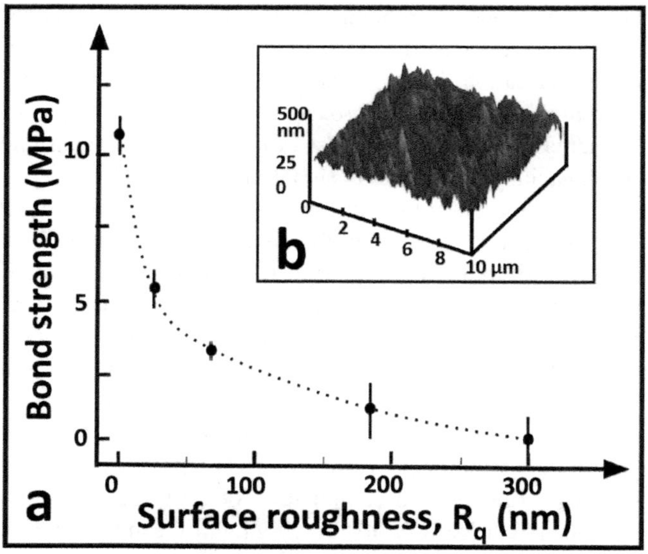

FIGURE 8.32 (a) Surface roughness effects on WoW bonding strength with a 0.5 MPa contact pressure. (b) A typical AFM image showing Si surface roughness measurement with an R_q of 25.8 nm. (Adapted from Ref. [33].)

including solvent cleaning, plasma treatment, chemical etching, and abrasive blasting, can be employed to restore surface energy and improve adhesion [11]. Surface activation, especially through plasma treatment, is critical before bonding processes, such as Die on Wafer (DoW), Chip on Wafer (CoW), and Wafer on Wafer (WoW) hybrid bonding [34,35].

The surface energy is usually defined by measuring the contact angles of pure liquids (probe liquids) placed on the sample surface, as illustrated in Figure 8.33. A larger contact angle means a lower surface energy. Water and diiodomethane are typically used as probe liquids because of their well-defined properties, such as surface tension, polar, and dispersive characteristics. With the measured contact angles and liquid properties, surface energy can be calculated through the developed mathematical models, including the Owens-Wendt-Rabel-Kaelble (OWRK) method.

As an example, Figure 8.34 shows the contact angle measurements during the technology development of the hybrid bonding process, in order to optimize the surface activation. A smaller contact angle reflects a larger surface energy, leading to better interface adhesion after bonding. As illustrated by the time zero sample in Figure 8.34, after extended storage time, organic contaminants tend to accumulate on chip surfaces, resulting in hydrophobic surfaces for materials, such as Cu and SiO_2, with contact angles larger than 70°. An Ar and H_2 mixture plasma activation process is employed to reduce the contact angles of Cu and SiO_2. As demonstrated in Figure 8.34, after about 120 seconds of plasma treatment, the contact angles stabilize at about 14° for Cu and 17° for SiO_2, suggesting a successful surface activation process [35].

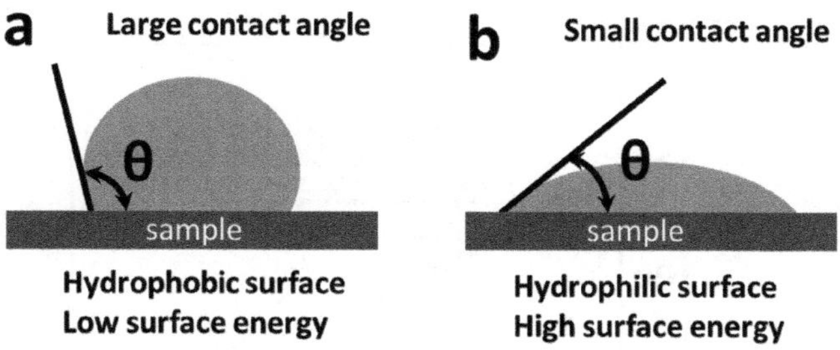

FIGURE 8.33 Schematic illustration of contact angle measurement showing (a) hydrophobic surface and (b) hydrophilic surface. (Adapted from Ref. [34].)

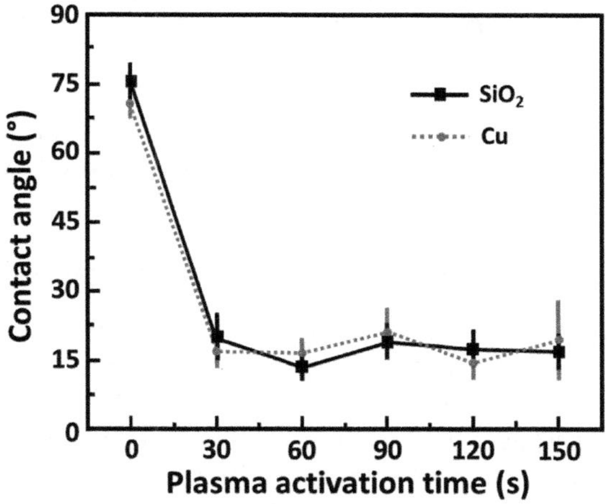

FIGURE 8.34 The water contact angle of Cu and SiO_2 substrates vs the Ar/H_2 plasma surface activation time. (Adapted from Ref. [35].)

8.2.5.3 Surface Wettability

Surface wettability is crucial for the integrity of thin films, dielectric layers, and interconnects in advanced packaging technologies. It is determined by multiple factors, including surface chemistry, composition, roughness, and other process-related effects [11]. The contact angle measurement technique is widely used to evaluate surface wettability, where a smaller contact angle signifies better wettability. To help with solder and flux materials down selection as well as assembly process optimization, the solder wetting behavior can be evaluated by measuring the contact angle between the molten solder and various types of surface finishes, including Cu OSP and ENIG [36]. As illustrated in Figure 8.35b, optical microscopy can be used to record the entire solder wetting process and obtain the surface contact angle vs wetting time plot. As an example, Figure 8.35a shows a graph of contact angle versus wetting time for the 95.5Sn-4.0Ag-0.5Cu solder on a Cu OSP substrate at temperatures of 255°C and 285°C. Compared to that at 255°C, the molten solder contact angle decreases more rapidly at 285°C, indicating better wetting at elevated temperature [36].

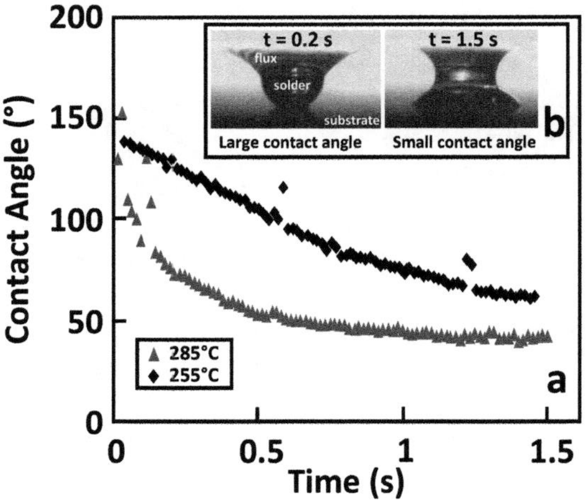

FIGURE 8.35 (a) Contact angle versus wetting time for 95.5Sn-4.0Ag0.5Cu solder on Cu/OSP substrates at 255°C and 285°C. (b) Typical optical images showing solder wetting. (Adapted from Ref. [36].)

8.2.6 Thermal Interface Material (TIM) characterization

Advanced packaging techniques, including 3D die stacking and heterogeneous integration, lead to higher levels of integration and larger computational loads. However, the increased power densities result in greater heat generation within a semiconductor package. The quickly growing AI market further drives the demand of higher thermal design power (TDP) in microelectronic devices, which has exceeded 1,000 W. Without proper thermal management, over 90% of the total heat will be conducted from the front side of the die, through the organic substrate, and into the PCB via solder interconnects, potentially leading to product failures [2]. Because the organic substrate and PCB are poor thermal conductors, this heat buildup can compromise both the performance and reliability of the product. Therefore, efficient and effective thermal management solutions, including the use of heat spreaders, heat sinks, and other advanced cooling techniques, are vital. These thermal managements can help eliminate localized hot spots and facilitate efficient heat dissipation through the die backside, ensuring optimal performance and long-term reliability.

8.2.6.1 Thermal Interface Materials (TIM)

Thermal Interface materials perform as a bridge between solid surfaces, for example, a die and a Cu lid, to enable efficient heat transfer from the heat-generating component (such as a die) to the Cu lid (or a heat sink). As illustrated in Figure 8.36a, in bare die or lidless packages, the TIM between the die and the heat sink is named as TIM 1.5. For lidded packages shown in Figure 8.36b, the TIM between the die and the copper lid is called TIM 1, while the TIM between the lid and the heat sink is called TIM 2. TIMs can have various forms, including polymer-based materials, solder-based TIMs, phase change materials, and conductive adhesives [37].

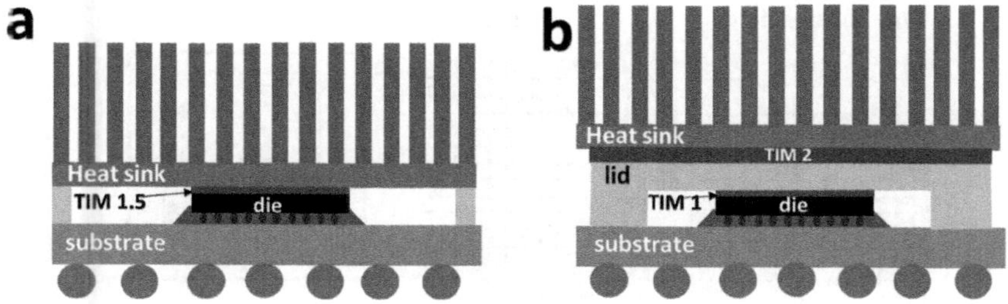

FIGURE 8.36 (a) Schematic of a bare die package and TIM 1.5 between the die and the heat sink. (b) Schematic of a lidded package with TIM 1 between the die and the lid, and TIM 2 between the lid and the heat sink. (Adapted from Ref. [37].)

The overall thermal resistance of a TIM layer depends on multiple factors, such as the surface roughness of the interfaces, the Bond Line Thickness (BLT) of the TIM, the thermal conductivity of the TIM (K_{TIM}), and the contact resistance (R_c) at the interface, which can be described by Equation (8.6):

$$R_{\text{TIM}} = \frac{\text{BLT}}{k_{\text{TIM}}} + R_C \tag{8.6}$$

Where K_{TIM} is the thermal conductivity of TIM, R_c is the contact thermal resistance of the TIM. Reducing BLT is a very effective way to enhance the heat dissipation performance of the TIM. For TIM 1.5 and TIM 2, the BLT is strongly influenced by the loading pressure applied during assembly. However, for TIM1, the BLT depends on the lid attachment process conditions, as well as the physical form and properties of the TIM 1 material. R_c is another important parameter that impacts TIM thermal performance. R_c is determined by the TIM's wettability on the contacting surfaces, which is influenced by surface roughness and the pressure applied during lid attachment or heat sink loading.

8.2.6.2 Thermal Test Chips (TTC)

Thermal measurement and testing are essential for assuring the performance and reliability of advanced semiconductor packages. However, because the final microelectronic products usually need years to develop, thermal assessments need to be conducted during earlier stages of the design process. To enable the thermal evaluations without the actual product chips, TTCs and Thermal Test Vehicles (TTV) are employed. These methodologies physically emulate the power input and thermal behavior of real devices, allowing engineers to assess real-time temperature distribution and overall thermal performance. With them, chip and package design can be optimized, and various thermal management solutions, such as different TIMs, lid designs, heat sinks, or cold plates, can be tested, optimized, and refined, in parallel with product development [38–40], thereby accelerating time to market of new semiconductor devices.

Figure 8.37 illustrates the schematic of a TTC. In this example, the Si-based test chip has a thickness of 700 μm and measures approximately 32 mm × 27 mm.

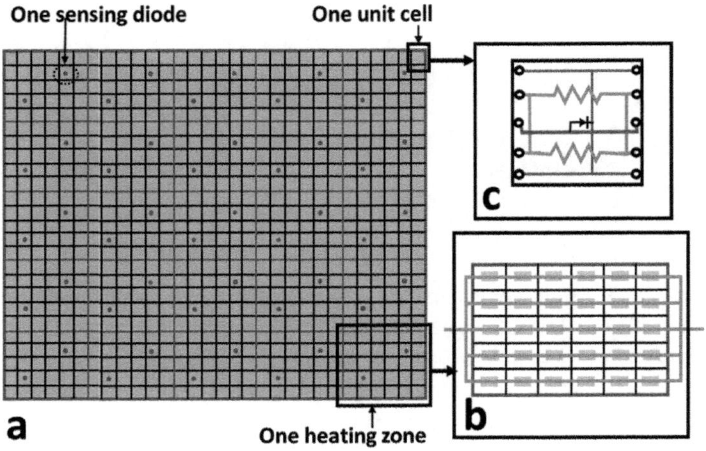

FIGURE 8.37 (a) Schematic of a thermal test chip (TTC) with 25 heating zones and 50 sensing diodes evenly distributed across the TTC. (b) Schematic of a heating zone with 30 unit cells. (c) Schematic of a unit cell containing a resistor and a temperature sensing diode. (Adapted from Ref. [38].)

It has 25 programmable heating zones, each containing 30 unit cells and two sensing diodes. Each unit cell, measuring 1 mm × 1 mm, includes a 10.5 ohm resistor and a temperature-sensing diode accessible via bump pads. The TTC configuration meets Joint Electron Device Engineering Council (JEDEC) JESD51–4a standard. The ability to individually program each unit cell and zone enables precise control of power distribution and in situ real-time temperature monitoring. The methodology facilitates accurate simulation of non-uniform power loads and hot spot conditions in real semiconductor chips using embedded on-chip sensors [38].

8.2.6.3 Thermal Test Vehicles (TTVs)

Consisting of packaged TTCs, Thermal Test Vehicles are configured to represent the intended packaging structure, using the actual TIMs, heat sinks or liquid cooling cold plates, and other components that can impact the heat dissipation and thermal performance of the entire system. They can be built in various package configurations, including BGA, Land Grid Array (LAG), Chip on Board (COB), as well as multi-chips per package to emulate System In Package (SIP) that is usually used in heterogeneous integration. As an example, the TTC schematically illustrated in Figure 8.37 is shown packaged in a BGA format TTV in Figure 8.38. Various types of TIMs, such as liquid metal or a phase change material (PCM), can be applied as TIM 1.5 between the TTC's backside and the liquid cooling plate [38].

TTCs and TTVs are indispensable tools in the thermal characterization of semiconductor devices and advanced packages. They allow flexible control over heat generation and spatially resolved temperature sensing, making it possible to emulate the real-time thermal behavior of actual packages and measure in situ temperatures within the chip. By performing thermal tests using TTCs and TTVs under realistic

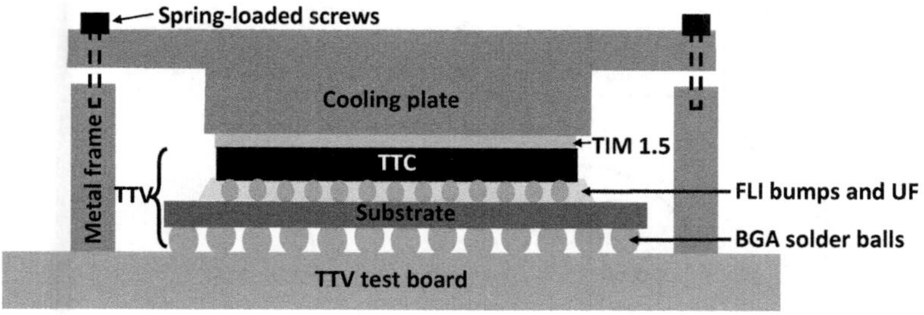

FIGURE 8.38 Schematic of a thermal test vehicles (TTV) with TIM1.5, liquid cooling plate, and TTV test board. (Adapted from Ref. [38].)

conditions, including power inputs, ambient environment, airflow, etc., engineers can map temperature profiles across the chip, including hot spots, evaluate the package's thermal behavior, validate thermal simulation models, and optimize designs, materials, and cooling solutions accordingly.

According to the JEDEC JESD51-1 standard [41], the junction to case thermal resistance (θ_{JC}) is determined as the thermal resistance from the active area of a semiconductor device to the exterior surface of its package or case. There are multiple methods to measure θ_{JC}. For example, the Transient Dual Interface (TDI) method, described in JESD51-14, summarizes the steps for measuring θ_{JC} using thermal impedance analysis. The thermal impedance function, presented as Zth-function $Z_{\theta JC}(t)$, describes how the junction temperature changes over time under a constant power input PH, starting from time zero ($t=0$), when the case surface is properly cooled. It is defined by the following equation:

$$Z_{\theta JC}(t) = \frac{T_J(t) - T_J(t=0)}{\text{PH}} \tag{8.7}$$

It means that the thermal impedance at any moment is the change in junction temperature over time divided by the applied power. It is important to note that variations in the cooling conditions at the case surface should not affect the impedance curve until heat reaches the case interface. However, changes in contact resistance at the heat sink interface will alter the steady state thermal resistance, leading to a divergence between thermal impedance curves. This divergence starts when the external contact resistance begins to affect heat flow, thereby identifying the package case interface.

As an example shown in Figure 8.39, to identify the case surface in transient measurements, two measurements are conducted on the same TTV, one with and the other without TIM 1.5 applied between the die backside (case) and the cooling plate. The observed Zth curves, presented on a logarithmic scale in Figure 8.39, show a clear separation [38]. The curve without TIM (shown in black) has a higher thermal resistance due to microscopic surface roughness between the die and the cold plate, exhibiting the onset of interface heat transfer at time t_s. The thermal impedance value at the separation point t_s, expressed as $Z_{\theta JC}(t_s)$, is referred to as the "transient θ_{JC}", as the separation of the two curves marks the transition of the heat flow from the die back side (case) into the thermal interface layer. In this example,

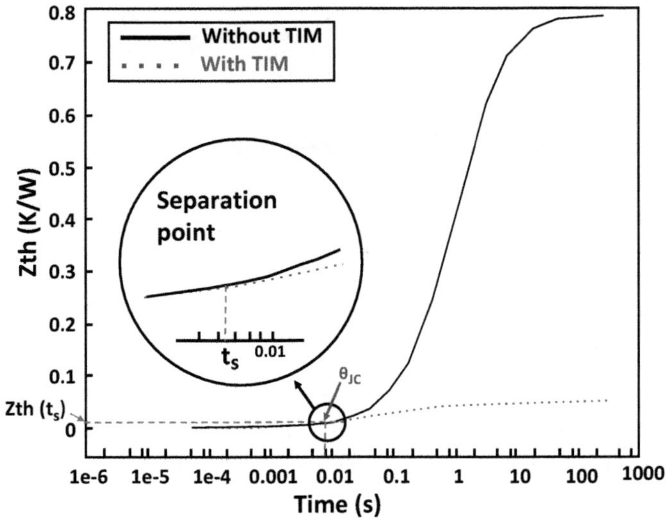

FIGURE 8.39 Zth curves for the TTV package shown in Figure 8.38 measured with and without TIM1.5 at the interface between the die backside and the cooling plate. (Adapted from Ref. [38].)

θ_{JC} is measured as 4.4°C/kW, as shown in Figure 8.39, assuming unidirectional heat flow. Additionally, the junction to ambient thermal resistance (θ_{JA}), as defined in JESD51-1, is the thermal resistance from the device junction to the ambient environment under natural convection. For the TTV shown in Figure 8.38, the θ_{JA} with TIM 1.5 is measured as 49.4°C/kW, as indicated in Figure 8.39 [38], under the specific cooling condition and TIM used in this test. Non-unidirectional heat flow can cause certain uncertainties in the measurement results, which can be properly accounted for through detailed analysis.

TTCs and TTVs play critical roles in evaluating the thermal behavior of semiconductor packages. As prototypical samples, they offer both programmable heating and embedded temperature sensing, and facilitate real time, spatially resolved thermal profiling across the chip. This makes it possible to emulate actual package performance and measure the chip internal temperatures directly. As part of the TIM materials evaluation process, TTVs are also used to assess the reliability of the TIM materials. By conducting thermal tests using TTVs under realistic operating conditions, such as power inputs, ambient environment, airflow, and other variables, engineers can precisely map the temperature distributions (including localized hot spots), assess overall package thermal performance and reliability, evaluate different TIMs and cooling components, and validate thermal simulation models.

8.2.7 Summary

As detailed in Sections 8.2.1–8.2.6, various types of materials are used in advanced packaging technologies, many of which undergo several high-temperature curing or reflow processes. Failures in advanced semiconductor packages, such as interfacial

TABLE 8.7
The Major Failure Modes in Advanced Packaging, Materials Properties Associated with those Failure Modes, and the Corresponding Test Metrologies

Failure Modes	Material Properties	Test Metrologies
Delamination, cracks	CTE	TMA
Delamination, cracks, bad thermal stability	T_g	DSC, TMA, DMA
Delamination, bad thermal stability	T_d	TGA
Delamination, cracks	Moisture content, CHE	TGA, TMA, DIC
Delamination, cracks	Modulus	DMA
Delamination, cracks, assembly yield loss	Dynamic, warpage	Shadow moiré, DFP, confocal techniques, DIC
Delamination, cracks	Interfacial adhesion strength, G_c	Shear test, pull test, DCB, FPB
Interconnect open	Interconnect bonding strength	Bump shear, bump pull
Delamination, bonding failures,	Surface roughness	AFM, stylus profilometers, optical profilometers
Bump open, delamination	Surface energy, wettability,	Contact angle
Device performance drop and failure	TIM characterization, θ_{JC}, θ_{JA}	Thermal tests on TTC and TTV

delamination, cracking in dielectric layers, interconnect degradation, and overall performance loss, can happen during assembly processes and reliability tests. Accurately characterizing the material properties linked to these failure modes is vital during technology and product development. Furthermore, the metrologies can also be used for fast materials down-selection and shortening the technology and product development cycle. Table 8.7 summarizes the key failure modes in advanced semiconductor packaging, the associated material properties, and the corresponding measurement metrologies discussed throughout Sections 8.2.1–8.2.6.

For example, T_g and T_d are characteristic properties of organic materials. Selecting materials with inadequate T_g or T_d can lead to poor thermal stability, delamination, or cracking during package assembly. Techniques such as DSC, TMA, DMA, and TGA, as discussed in Section 8.2.1, are usually used to measure T_g and T_d. CTE and CHE mismatches among materials within a package can also lead to delamination and cracking. TMA is often used to evaluate CTE (refer to Section 8.2.1), while CHE can be assessed using TGA, TMA, or DIC, as detailed in Sections 8.2.1–8.2.2. Mechanical properties such as modulus (including both storage and loss modulus) determine a material's ability to withstand mechanical stresses without causing structural failures. Improper materials selection may result in cracking or delamination, and DMA is the primary metrology for modulus measurement, as detailed in Section 8.2.1. Dynamic warpage of advanced packages under various conditions (room temperature, reflow, or reliability test temperature) can impact assembly yield and reliability. Metrologies for dynamic warpage measurements include shadow moiré, DFP, confocal techniques, and DIC, as explained in Section 8.2.2.

Interfacial materials properties, including adhesion strength and critical energy release rate (G_c), are critical for mitigating delamination and cracks at materials interfaces. Testing methods, such as shear test, pull test, DCB, and FPB are detailed in Section 8.2.3 the integrity of the numerous interconnects in advanced packaging is critical for effective power delivery and signal transmission. Techniques, such as bump shear and bump pull tests, detailed in Section 8.2.4, are commonly adopted to evaluate interconnect strength and verify the adhesion of UBM layers. Surface properties of materials, including roughness, surface energy, and wettability, can dramatically influence interfacial adhesion, interconnect bonding, underfill application, and lamination quality. Non-optimized surface characteristics can cause delamination or interconnect failures. Metrologies, such as AFM, profilometry, and contact angle measurements, are detailed in Section 8.2.5.

Finally, thermal management determines the product performance in advanced packaging. As described in Table 8.7, TIM characterization, including measurements of R_{TIM}, θ_{JC}, and θ_{JA}, is essential to assure thermal reliability. TTCs and TTVs, which replicate the power and thermal behavior of real products, enable early-stage evaluation of device thermal profiles and package-level thermal performance, supporting the development of effective thermal solutions.

8.3 MATERIALS PROPERTY SIMULATION METHODOLOGIES

Advanced packaging enables chiplet integration and improves product performance by stacking or embedding multiple chiplets into one package, supporting faster product development, reducing costs, and enhancing yield. However, as package sizes increase and devices grow more complex, traditional product development methods using samples for physical Design of Experiment (DOE) are becoming less efficient, prompting the adoption of manufacturing process simulation, AI-based process optimization, and predictive simulations for yield and reliability, in order to accelerate yield ramping and assure long-term product reliability [42]. The input of accurate materials properties, such as modulus, CTE, dielectric constant, and T_g, is the key to achieve precise process simulations. Those materials properties often change non-linearly with temperature, moisture, and other environmental or process conditions. Getting materials property measurements under various processing and reliability conditions could be time-consuming, expensive, and in some cases, unrealistic. To address these challenges, materials simulation techniques based on Molecular Dynamics (MD), Machine Learning (ML), and Quantum Mechanics (QM), are being applied to predict key properties of organic packaging materials, such as T_g, CTE, modulus, dielectric behavior, refractive index, and post curing volume shrinkage [43].

8.3.1 Molecular Dynamics (MD)

Molecular dynamics (MD) is a powerful simulation technique that models the time evolution of atomic systems by solving Newton's equations of motion, offering detailed insights into the behavior and energy of complex materials at the molecular level. The modeling enables the calculation of trajectories in a multi-dimensional phase space to define energy and behavior of systems with a large number of atoms. Force fields, which are mathematical models describing the interactions between atoms based on interatomic potentials, is the key for successful MD simulations. They are indispensable for precisely predicting physical properties and behaviors of molecules, polymers, and crystalline materials.

Modern MD software, such as Schrödinger's Materials Science Suite (MSS) and the open source LAMMPS platform, empowers researchers to visualize and simulate molecular behaviors over time. Due to the recent significant progresses in GPU hardware and algorithmic efficiency, it is now possible to screen large libraries of materials, especially polymers, with a relatively short throughput time. This capability is particularly important for predicting properties that require running and analyzing long MD trajectories, such as T_g.

As an example, a GPU accelerated MD workflow built into MSS, uses an Optimized Potentials for Liquid Simulation (OPLS) force field to predict T_g. In this workflow, polymer chains are packed into a simulation box at a low density and equilibrated at 1,000 K and 1 atm. They are then cooled incrementally from 1,300 to 3,00 K in 20 K steps, with each step lasting 5 nanoseconds under a constant pressure. The resulting density data is plotted as a function of temperature, and is analyzed using automated fitting techniques to define T_g. This steamlined process, using MSS tools, including Polymer Builder, Multistage MD, and Thermophysical Properties workflows, can dramatically accelerate materials engineering and reduce reliance on costly or impractical experiments [44,45]. By combining accurate physics-based simulations with scalable computing, this simulation approach allows for rapid and reliable predictions of thermophysical properties, providing faster development cycles and more informed materials selection in advanced packaging and other semiconductor applications.

8.3.2 T_G Simulation

Figure 8.40 summarizes the methodology used to simulate the T_g of Kapton, a PI material. As demonstrated in Figure 8.40a, the Kapton monomer structure is the basis for generating initial polymer configurations using the Disordered System Builder (DSB) tool in the MSS platform. The Tangled Chain algorithm is utilized to construct oligomer chains within the simulation cell by rotating backbone dihedrals through a self-avoiding random walk approach, ensuring realistic chain entanglement and structural diversity. To improve statistical reliability, the replicate simulation cells with unique initial spatial configurations are created for each polymer system. The calculated properties from production runs are averaged over the set of ten replicates. Each polymer chain contains about 2,000 atoms, assuring molecular weight consistency, which is very important for reliable property prediction [46–49]. Each simulation cell has six polymer chains, totaling around 12,000 atoms, having an initial system density set to be 0.5 g/cm^3.

As an example, Figure 8.40b shows an MD snapshot of the system during equilibration, where chains are color coded within a periodic simulation box. The system is cooled from 1,300 to 300 K in 20 K increments, running 5 ns of MD simulation in a constant number of

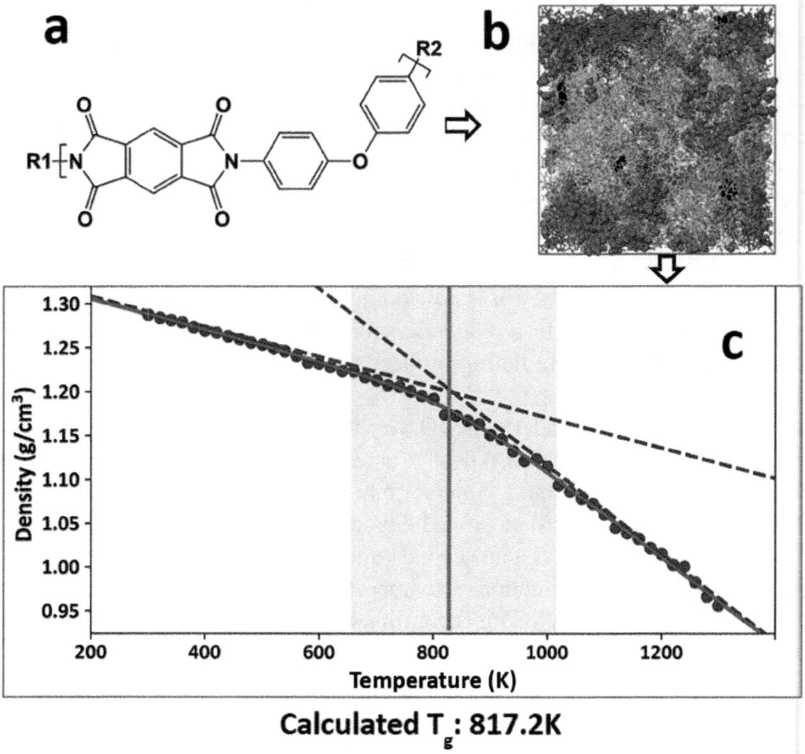

FIGURE 8.40 (a) The Kapton monomer structure. (b) The MD snapshot illustrating multiple polymer chains in distinct colors within a periodic simulation box. (c) T_g was initially estimated at 817.2 K based on automated hyperbolic fitting of the density-temperature plot. (Adapted from Ref. [43].)

atoms, pressure, and temperature at each step while monitoring density. If density fluctuations exceed 5% at any temperature point, additional 5 ns simulations are conducted to assure convergence. The time step is adjusted based on temperature, 1 fs for simulations above 700 K, and 2 fs below that threshold. Temperature control is maintained using the Nosé–Hoover thermostat [50], and pressure is regulated with the Martyna–Tobias–Klein barostat [51]. T_g is estimated using a hyperbolic fit to the density vs temperature data, which offers higher precision and removes the subjectivity of manually identifying transition regions, as compared with the traditional bilinear fitting [52]. As illustrated in Figure 8.40c, the T_g corresponds to the intersection point of the asymptotes in the hyperbolic fit. For Kapton, the initial T_g estimate from the simulation is 817.2 K. Given the rapid cooling rates inherent in the MD simulation methodology, T_g values are usually overestimated. A standard linear calibration for homopolymers [44] is applied to adjust this value to be 654 K, which is within 3% of the experimental T_g (672 K) reported [53]. DuPont datasheet reports T_g values for Kapton ranging from 633 to 683 K, further validating the simulation results [55].

8.3.3 CTE Simulation

The T_g simulation involves calculating the density of the system across a temperature range. This temperature dependent density data is further utilized to evaluate the CTE in both the glassy region (300 K) and rubbery region (800 K). For example, the glassy CTE at 300 K for polymer systems is extracted by performing a linear regression on specific volume versus temperature data corresponding to the ten lowest temperature points. For Kapton, the slope of this linear fit generates a CTE value of 148.2×10^{-6} K^{-1} [43].

8.3.4 Modulus Simulation

Modulus calculations are performed using the Elastic Constants workflow within the MSS platform [45]. This stress-based approach applies successive small strain increments (0.2%) to each replicate's periodic structure, followed by 1 ns of MD simulation at a constant temperature. Strain is applied in uniaxial directions and six biaxial combinations. The resulting stress vs strain data is used to derive the stiffness tensor via linear regression. Assuming isotropic materials behavior, the Young's modulus (E), shear modulus (G), and bulk modulus (κ) are computed from the Lamé parameters (λ, μ) [54].

As an example, at 300 K, the MD-derived Young's modulus of Kapton is 2.4 GPa, within ~13% of the experimentally reported value of 2.76 GPa at 296 K [55]. The bulk modulus is calculated to be 5.04 GPa, and the shear modulus is 0.85 GPa. Additional properties include a universal elastic anisotropy of 0.096, an isotropic Poisson's ratio of 0.42, Lamé's first parameter (λ) of 4.51 GPa, and Lamé's second parameter (μ) of 0.86 GPa [43].

To investigate thermal effects, simulations are also performed at 330 and 400 K, and a clear trend of thermal softening is observed. The Young's modulus decreased from 2.4 GPa at 300 K to 2.27 GPa at 330 K, and further to 2.01 GPa at 400 K, indicating reduced rigidity due to increased molecular mobility at elevated temperatures. In contrast, the bulk modulus remained relatively stable, with values of 5.04 GPa (300 K), 5.11 GPa (330 K), and 5.14 GPa (400 K), suggesting that Kapton maintains volumetric incompressibility across the studied temperature range [43]. The shear modulus shows a more pronounced temperature dependence, decreasing from 0.85 GPa at 300 K, to 0.8 GPa at 330 K and 0.7 GPa at 400 K, consistent with reduced resistance to shape deformation under thermal agitation [43]. The isotropic Poisson's ratio slightly increased with temperature, from 0.42 at 300 K, to 0.426 at 330 K and 0.435 at 400 K, reflecting an enhanced lateral response to axial loading as the material softens. In summary, the simulations capture key trends in Kapton's mechanical behavior, revealing that while compressive stiffness remains relatively unaffected, tensile and shear properties degrade with increasing temperatures. These results are in good alignment with experimental data, demonstrating the reliability of the MD-based approach for predicting polymer mechanics [43].

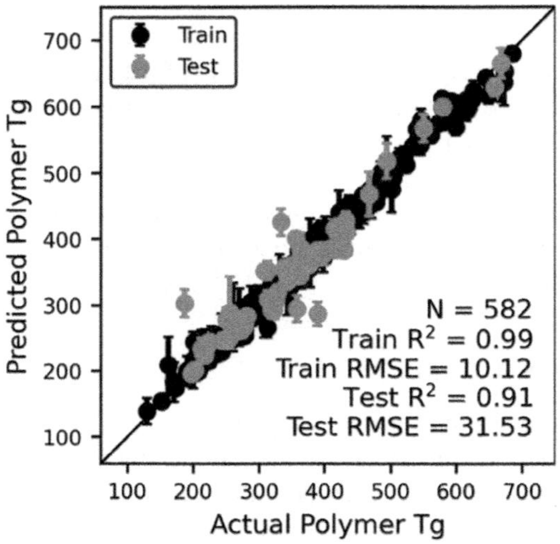

FIGURE 8.41 Performance of the Polymer T_g ML model. (Adapted from Ref. [44].)

8.3.5 Machine Learning (ML)

In addition to the MD simulation techniques discussed earlier, materials properties, such as T_g, can also be estimated using pre-trained ML models in the Machine Learning Property Prediction tool within MSS [45]. These models are built using DeepAutoOSAR, Schrödinger's automated supervised learning framework tailored for molecules and polymers [56,57]. Training data for the models is from existing experimental datasets and simulation studies [44,53]. As illustrated in Figure 8.41, the models achieve high predictive accuracy, with test set R^2 values exceeding 0.9 [44].

As an example, for Kapton, the ML model predicts a T_g of 649.3 K, which aligns closely with the MD results and falls within 4% of the experimental value. In contrast to the MD-based simulation approach requiring several hours of GPU computation, the ML prediction is nearly instantaneous. However, the ML method is primarily applicable to pure polymers resembling those in the training dataset. It is not well suited for complex systems, such as copolymers, branched or crosslinked polymers, or formulated materials. In such cases, the MD-based approach provides critical insights for novel or structurally complex polymer systems where ML models may lack generalizability [42].

8.4 MATERIALS ANALYSIS AND CHARACTERIZATION TECHNIQUES

It is essential to have comprehensive materials analysis and characterization around the defect sites during the in-depth root cause analysis of failures in advanced semiconductor packages. Key information of the defects, such as elemental composition, crystallographic orientation, chemical states, and spectral fingerprints, can be acquired through various advanced analytical techniques. These techniques include energy-dispersive X-ray spectroscopy (EDX), wavelength-dispersive X-ray spectroscopy (WDX), electron backscatter diffraction (EBSD), X-ray photoelectron spectroscopy (XPS), time-of-flight secondary ion mass spectrometry (TOF–SIMS), Fourier-transform infrared spectroscopy (FTIR), optical photothermal infrared spectroscopy (O-PTIR), and atomic force microscopy-based infrared spectroscopy (AFM-IR) [11].

8.4.1 Energy-Dispersive X-Ray Spectroscopy (EDX) and Wavelength-Dispersive X-Ray Spectroscopy (WDX)

Energy-dispersive X-ray spectroscopy (EDX or EDS) is an extensively used analytical technique, usually coupled with electron microscopy, including Scanning Electron Microscopy (SEM) and Transmission Electron Microscopy (TEM), to generate X-ray spectra as a function of energy. When a sample is under an electron beam, the atoms become ionized and emit X-rays with energies characteristic of their elemental identity. Using SEM-EDX, elemental composition information of the sample surface up to a depth of ~ 1–3 μm can be obtained, depending on the beam energy and interaction volume [58]. TEM-EDX, although requiring time-intensive sample preparation, can provide significantly higher spatial resolution than SEM-EDX by using thinner specimens and a smaller interaction volume. While SEM-EDX supports fast, and non-destructive analysis of surface areas, TEM-EDX is better for high-precision microstructural characterization within a small area.

As an example, Figure 8.42 illustrates the use of SEM-EDX in revealing the root cause of a die crack. A foreign material (FM) is found at the impact location on the die back side (Figure 8.42a), which aligns well with the cracking *x*–*y* location observed on the die front side. SEM-EDX analysis of the FM (Figure 8.42b), when compared with the reference spectrum taken from a nearby area without FM, indicates a high aluminum content, matching the composition of the TIM applied between the die and the heat sink. A follow-up SEM-EDX analysis on the heat sink confirms the presence of aluminum in the TIM. The root cause of the failure is considered to be the uneven mechanical loading on the die backside, likely during heat sink installation. Optimizing the loading procedure is recommended to mitigate this failure mode [11].

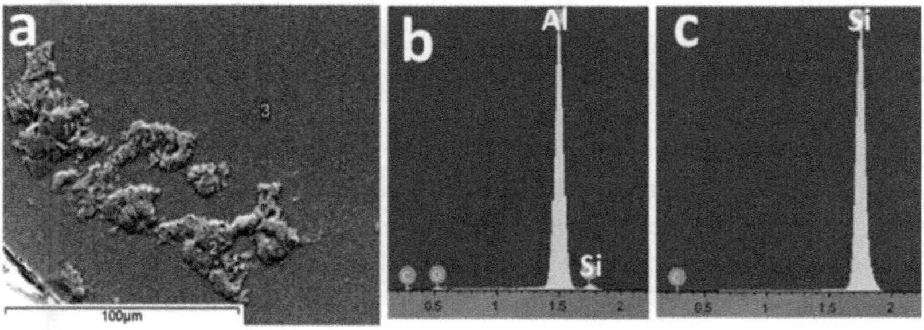

FIGURE 8.42 (a) SEM image of the foreign material (FM) observed around the die back side impact point which leads to a front side die crack. (b) EDX spectra taken from the FM. (c) EDX spectra taken from the area without FM. (Adapted from Ref. [11].)

Wavelength-dispersive X-ray spectroscopy (WDX or WDS) is a complementary technique, offering enhanced resolution and accuracy compared with EDX. It measures X-rays of one wavelength (or element) at a time using crystal diffraction. Although this approach slows down the analysis, it can provide dramatically higher spectral resolution and superior signal-to-noise ratios, which is especially effective for detecting trace elements or resolving overlapping spectral lines. The detection of trace elements down to concentrations of 100 ppm or lower is feasible with WDX. WDX is also very sensitive to low-energy X-rays and is usually employed for light element analysis, including boon (B), carbon (C), sulfur (S), and fluorine (F), across various materials types. In practice, EDX and WDS are usually used in tandem, EDX for rapid, broad spectrum screening, and WDX for high accuracy, element-specific analysis [59].

8.4.2 X-Ray Fluorescence (XRF)

X-ray Fluorescence (XRF) is a non-destructive analytical technique for the elemental composition analysis of materials by detecting the fluorescent (secondary) Z-rays emitted when a sample is irradiated with primary X-rays. XRF can analyze solids, liquids, and powders, and is effective for both qualitative and quantitative elemental analysis. There are two main types of XRF techniques, Energy Dispersive XRF (EDXRF) and Wavelength Dispersive XRF (WDXRF). Each method offers unique advantages depending on the specific application. EDXRF typically detects elements ranging from sodium (Na) to uranium (U), while WDXRF expands this capability to lighter elements, such as beryllium (Be). Elemental concentration characterization can range from 100% down to parts per million (ppm), or even sub-ppm levels in some cases. Detection limits depend on the element and the sample matrix, with heavier elements generally having better detection sensitivity [60].

Besides compositional analysis, XRF is extensively used as a metrology to measure the thin film thickness in advanced packaging. In this application, a fluorescent X-ray thickness analyzer exposes multilayer samples to X-rays, and the resulting fluorescent X-rays are analyzed to determine the thickness of each individual layer. However, this methodology is less accurate when the coating and the base material have similar atomic numbers. Additionally, it reflects the thickness of a pure metal film. In other words, the film density or purity influences the measurement accuracy. By using standard samples to generate calibration curves, XRF enables rapid, non-destructive, and non-contact thickness measurements. In manufacturing environments, especially for alloy plating processes, XRF can measure the film thickness and alloy composition, conduct basic elemental analysis, and support automated measurement with integrated statistical analysis [60].

As an example, Figure 8.43 illustrates copper (Cu) pads in an advanced semiconductor package with electroless Ni/Pd/Au surface finishing. In a high volume manufacturing setting, XRF is routinely used to monitor the thickness of these metal finishing layers. A standard calibration sample, consisting of a Cu substrate with known Ni/Pd/Au layer thickness information, was employed to generate the calibration curves. The XRF measurements are performed under the following conditions: 30 kV primary X-ray excitation voltage, 0.1 mm beam diameter, 60 seconds

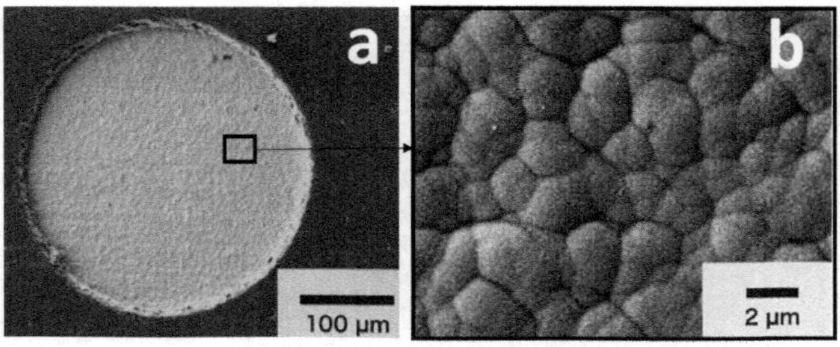

FIGURE 8.43 (a) SEM image of a Cu pad with electroless Ni/Pd/Au surface finishing. (b) Zoom-in SEM image of the pad surface, showing the topography details of the pad. (Adapted from Ref. [60].)

TABLE 8.8
Ni/Pd/Au Thickness Measurement Results by XRF

Measured by XRF	Au (μm)	Pd (μm)	Ni (μm)	Cu (μm)
Mean value	0.049	0.082	3.08	6.31
Standard deviation	0.001	0.002	0.012	0.033

Source: Adapted from Ref. [60].

measurement time, and ambient environment [60]. Table 8.8 summarizes the average values and standard deviations from ten repeated measurements for each layer. The results indicate a ~ 0.05 μm thick Au layer, ~ 0.08 μm thick Pd layer, ~ 3.1 μm thick Ni layer, and ~ 6.3 μm thick Cu pad. The relatively higher standard deviation observed for the Cu layer could be attributed to surface roughness at the interface between the Cu pad and the buildup film, as illustrated in Figure 8.44b [60].

To confirm the XRF measurements, a cross-sectional analysis of the Cu pad with Ni/Pd/Au finishing (Figure 8.43) is conducted using the focused ion beam (FIB) technique. The resulting cross-sectional SEM images are presented in Figure 8.44, and the corresponding layer thicknesses data is reported in Table 8.9. The thickness data by cross-sectional imaging aligns well with the non-destructive XRF measurements, suggesting that XRF can provide fast, accurate, non-destructive and non-contact thickness analysis for thin metal films [60].

8.4.3 Fourier Transform Infrared Spectroscopy (FTIR), Optical Photothermal IR (O-PTIR), and Atomic Force Microscopy-Based Infrared Spectroscopy (AFM-IR)

Fourier Transform Infrared (FTIR) Spectroscopy is extensively utilized in the advanced packaging technique development for chemical characterization. It measures the absorption of infrared light across various IR frequencies by the sample under test, generating a distinct spectral fingerprint [61]. FTIR is especially helpful for identifying and analyzing organic materials. However, due to the long wavelengths

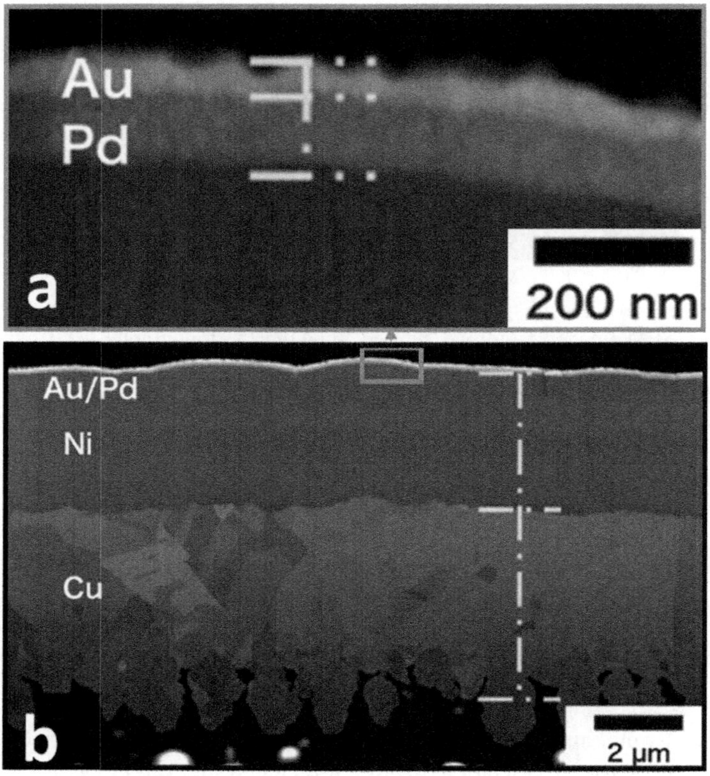

FIGURE 8.44 FIB-SEM cross sectional image of the Cu pad with Ni/Pd/Au surface finishing. (a) Zoom-in image showing details of the Au and Pd layer. (b) Zoom-out image showing the thickness of Ni and Cu pad. Significant roughness was found at the interface between the Cu pad and the buildup film. (Adapted from Ref. [60].)

TABLE 8.9
Comparison of Ni/Pd/Au Thickness Measurement Results by XRF vs FIB-SEM Cross Section

Thickness (µm)	Au	Pd	Ni	Cu
Measured by XRF	0.049	0.082	3.08	6.31
By cross section	0.047	0.094	3.48	6.31

Source: Adapted from Ref. [60].

of IR light (2.5–20 µm), its spatial resolution is limited to a few microns, making it unsuitable for sub-micron scale analysis [61].

As an example, Figure 8.45 demonstrates the application of FTIR in the failure analysis (FA) of advanced semiconductor packages. After undergoing an unbiased Highly Accelerated Stress Test (uHAST) at 130°C and 85% relative humidity,

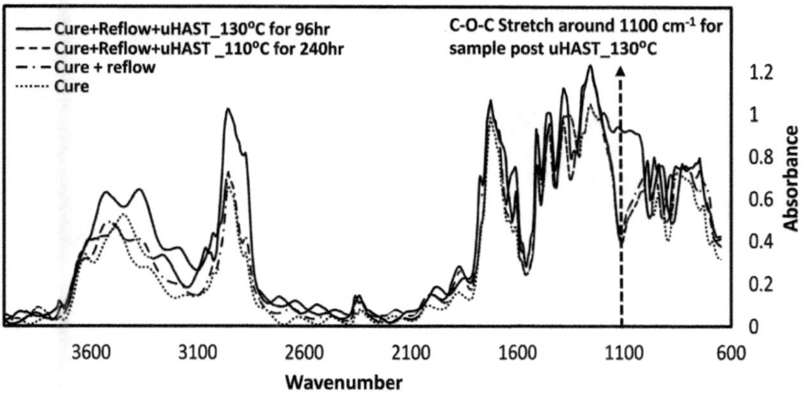

FIGURE 8.45 FTIR spectra taken from the NCF material between HBM stacks (post cure, post cure plus reflow, post cure plus reflow plus uHAST test at 110°C for 240 hours, and post cure plus reflow plus uHAST test at 130°C for 96 hours). The analysis indicates the formation of C–O–C bond only after the 130°C uHAST test. (Adapted from Ref. [62].)

a delamination is observed in the Non-Conductive Film (NCF) material, which is a pre-applied flux underfill used between memory dies in High Bandwidth Memory (HBM) stacks. Interestingly, this failure mode is not found in the samples after uHAST at 110°C under the same humidity conditions [62]. FTIR is employed to analyze the NCF material after various assembly and stress conditions. The results reveal the formation of C–O–C bond only after the 130°C test, indicating chemical changes in the NCF due to aggressive environmental exposure, which could lead to delamination [62].

Optical Photothermal Infrared (O-PTIR) Spectroscopy addresses the resolution limits of traditional FTIR. It leverages the photothermal effect generated when IR-active regions are excited by a tunable IR laser and detected using a visible laser probe. The resulting spectra are similar with that of a transmission mode FTIR, and can be directly compared with known reference [63]. O-PTIR requires minimal sample preparation and enables non-destructive, high-resolution analysis with spatial resolution down to approximately 450 nm, significantly better than the traditional FTIR.

O-PTIR has been applied in advanced packaging FA to investigate FM-induced defects [64,65]. As illustrated in Figure 8.46a, FM located between the solder and the Cu pad caused a "non-wet" microbump open failure. O-PTIR enabled chemical analysis of the FM directly from the cross-sectional sample without additional sample preparation. In Figure 8.46b, spectra collected from both the FM and nearby reference locations facilitated the FM characterization [65], providing critical insight for corrective action to eliminate such failure modes.

Atomic Force Microscopy–Infrared Spectroscopy (AFM-IR) combines the nanoscale spatial resolution of AFM with the molecular specificity of IR spectroscopy, making it ideal for analyzing organic contaminants at sub-micron and nanometer scales [66]. A tunable IR laser targets a localized region near the AFM tip, and the subsequent IR absorption leads to localized thermal expansion.

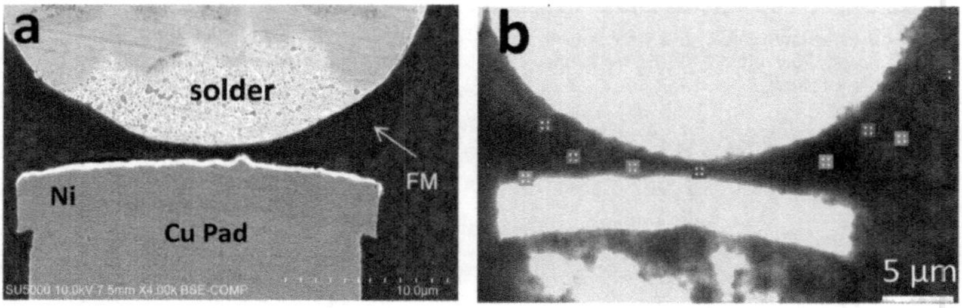

FIGURE 8.46 (a) Microbump non-wet failure caused by an FM between the solder and the Cu pad. (b) O-PTIR analysis can be performed on the cross-sectioned sample at both the FM area and the reference area nearby. (Adapted from Ref. [65].)

The AFM tip detects this expansion with nanometer precision and generates chemical mapping of the material in the nanometer domain, beyond the spatial resolution limit of FTIR and O-PTIR.

8.4.4 X-Ray Photoelectron Spectroscopy (XPS) and Time-of-Flight Secondary Ion Mass Spectrometry (TOF-SIMS)

X-ray Photoelectron Spectroscopy (XPS), which is also known as Electron Spectroscopy for Chemical Analysis (ESCA), is a surface-sensitive analytical technique that determines elemental composition and chemical states by measuring the kinetic energy of photoelectrons emitted from a material's surface when irradiated with monochromatic Al Kα X-rays. XPS typically probes to a depth of roughly 5 nm, providing both quantitative and chemical state information [67]. Time of Flight Secondary Ion Mass Spectrometry (TOF-SIMS) is another high-resolution surface analysis metrology often used to detect elemental, chemical, and molecular information within the top ~1 nm of a sample [68]. The testing sample is bombarded with a finely focused ion beam, which emits secondary ions and clusters that can be analyzed using a time-of-flight mass spectrometer. TOF-SIMS also enables depth profiling of thin films, when combined with ion sputtering.

Unlike SEM-EDX, which provides materials analysis to a depth of 1–3 μm, both XPS and TOF-SIMS are usually employed to analyze thin layers and organic materials in the 2–5 nm surface layer, often not feasible if using SEM-EDX. XPS and TOF-SIMS are critical tools for detecting surface and interfacial impurities and contamination, which can lead to failures in advanced semiconductor packages [69–71]. For example, TOF-SIMS has been employed to investigate the root cause of separation failures between the Cu layer and the Cu_3Sn IMC layer found in lead-free solder joints after bake reliability tests [71]. As presented in Figure 8.47, the TOF-SIMS analysis revealed that the Cu layers electroplated with a current density of 10 mA/cm^2 contained chlorine (Cl) impurities at levels nearly 200 times higher than those plated at 0.5 mA/cm^2 (Figure 8.47a and b). After a 168 hours bake at 175°C, cross-sectional SEM images aken from the samples show significant Cu to IMC separation for the Cu layers with high impurity, while no separation occurred in the low impurity counterparts. The results highlight the critical role of impurity

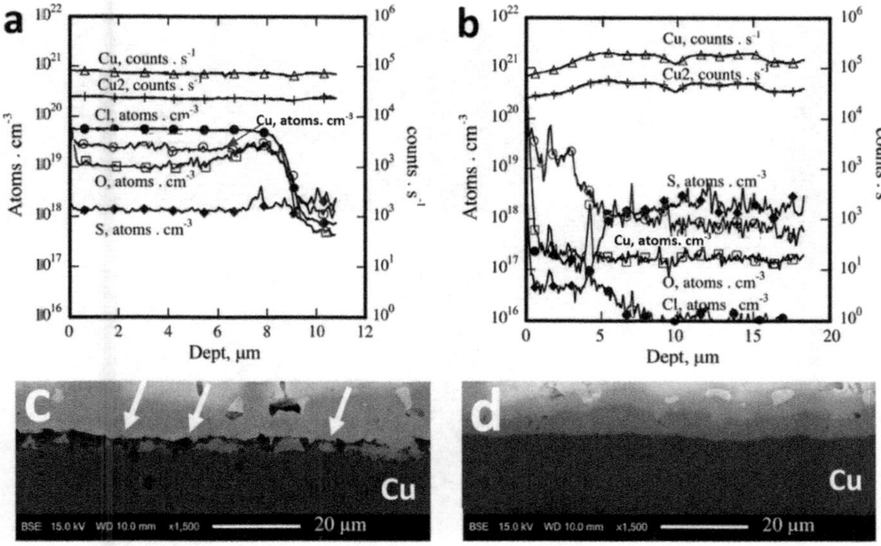

FIGURE 8.47 TOF-SIMS analysis of electroplated Cu layers with an electroplating current density of 10 mA cm^{-2} (a), and 0.5 mA cm^{-2} (b). SEM images of solder joints assembled on the electroplated Cu layers with an electroplating current density of 10 mA cm^{-2} (c), and 0.5 mA cm^{-2} (d) post 168 hours of bake at 175°C. (Adapted from Ref. [71].)

levels in electroplated Cu in modulating the mechanical integrity of solder joints. More importantly, they underscore the importance of surface-sensitive characterization techniques, such as TOF-SIMS and XPS in root cause failure analysis of advanced packaging [71].

8.4.5 Electron Backscatter Diffraction (EBSD)

Electron Backscatter Diffraction (EBSD), also known as backscatter Kikuchi diffraction (BKD), is a microstructural characterization technique for the crystallographic characterization of materials. EBSD data is usually collected from cross-section samples under a SEM. This metrology is critical in diagnosing the root causes of electromigration (EM) induced open interconnect failures in advanced semiconductor packages [72–74].

As interconnects in advanced packages are getting smaller and denser, EM becomes a significant reliability concern. EM can lead to solder joint opens via void accumulation and propagation due to Sn self-diffusion. Additionally, the electron wind effect can accelerate the dissolution of the UBM on the cathode side, resulting in open interconnect failures due to UBM depletion. Studies have shown that this UBM dissolution is highly sensitive to the orientation of Sn grains [72], as the Sn crystal has strong anisotropic behavior. As an example, at a constant temperature and current density, the time required to dissolve a 2 μm Ni UBM can vary significantly, from 100 hours when the Sn *c*-axis is aligned with the direction of electron flow, to ~1,800 hours when the *c*-axis is perpendicular to the flow [73]. EBSD is employed to analyze Sn grain orientation distributions in solder joints to study the

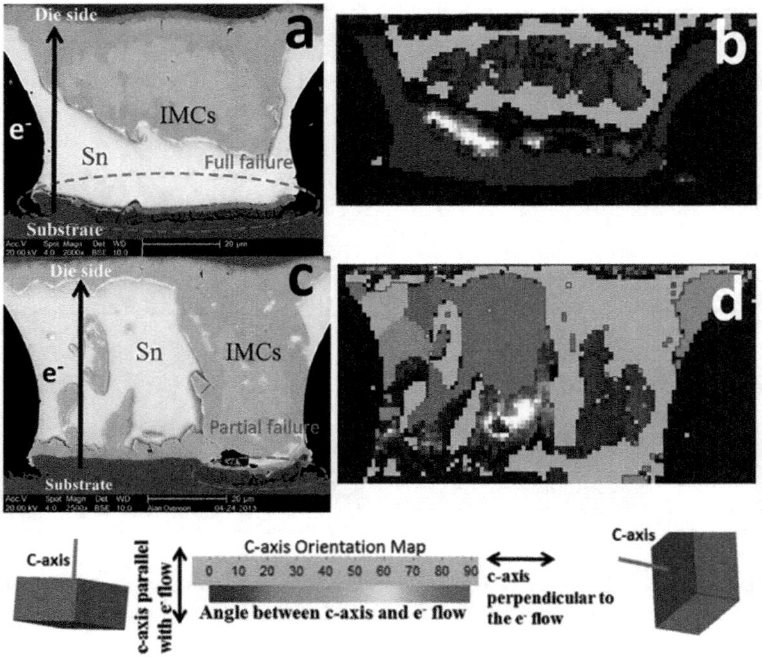

FIGURE 8.48 (a and b) The SEM image and the corresponding EBSD mapping of a failed solder interconnect after EM reliability tests with fast Ni dissolution when the c-axis of the single Sn grain of the bulk solder is in-line with the electron flow direction, respectively; (c and d) The SEM image and the corresponding EBSD mapping of a survived solder interconnect after EM reliability tests with slow Ni dissolution when the *c*-axis of some Sn grains in the bulk solder is perpendicular to the electron flow direction, respectively. (Adapted from Ref. [74].)

correlation between crystallographic alignment and EM-induced failures. As shown in Figure 8.48, solder joints with Sn crystals having their c-axis match with the direction of electron flow experience dramatically faster Ni dissolution, and are prone to open failures during EM stress testing [74].

EBSD is also very important in evaluating TSV extrusion, which is an important mechanical reliability concern in 3D packaging [75]. It is well-known that TSV extrusion is influenced by the mechanical properties of the Cu fill. EBSD is employed to compare the Cu grain size distributions in TSVs with different reliability performances. The studies reveal that TSVs with smaller, and more uniform grains exhibit a higher yield strength and are more resistant to extrusion, thereby enhancing structural integrity and reliability [75].

8.5 RELIABILITY TESTING OF MATERIALS IN ADVANCED SEMICONDUCTOR PACKAGING

In order to meet the AI and High Performance Computing (HPC) market needs, advanced packaging technologies need to incorporate increasingly complex designs, such as larger package footprints, finer and denser interconnects, compact 3D-stacked or embedded dies, and diverse multilayer materials. This complexity brings significant challenges to reliability, because the likelihood of failure increases as even minor defects can lead to significant performance degradation or system-level issues. To assure product long-term reliability and functionality, reliability testing guided by industry standards need to be conducted throughout the product development cycle. These tests can help with the reliability risk assessments of microelectronic devices to assure product performance under various environmental and operational stresses.

8.5.1 Interconnect Electromigration (EM) Test

EM and thermomigration are major reliability concerns for the interconnects in advanced semiconductor packages, including TSVs, RDLs, microbumps, C4 bumps, and SLI solder joints, because the interconnects in advanced packages are much smaller and denser with much higher TDP (>1,000 W), compared with conventional packages.

8.5.1.1 EM Testing Methodology

To evaluate EM reliability within a relatively short time, Accelerated Life Testing (ALT) is conducted at elevated temperatures and current densities. These reliability tests help build EM lifetime models and predict failure rates under actual use conditions. Reliability failures are commonly analyzed using various reliability statistics methods, including the Weibull Probability Density Function (PDF), showing in Equation (8.8) [11]:

$$F(t) = 1 - \exp\left[-\left(\frac{t}{\eta}\right)^{\beta}\right] \tag{8.8}$$

Where $F(t)$ is the cumulative failure rate over time t, η is the characteristic lifetime, and β is the shape factor indicating failure distribution. The experimental data of EM failure rate in the ALT can be plotted and fitted with Weibull PDF to calculate η and β at each EM ALT condition. The Mean Time to Failure (MTTF) at each EM ALT condition can be calculated following the equations [76]:

$$\text{MTTF} = \eta . \Gamma\left(\frac{1}{\beta} + 1\right) \tag{8.9}$$

Where

$$\Gamma\left(\frac{1}{\beta} + 1\right)$$

Is the gamma function evaluated at the value of :

$$\left(\frac{1}{\beta} + 1\right)$$

The gamma function is defined as:

$$\Gamma(n) = \int_0^{\infty} e^{-x} x^{n-1} \, dx \tag{8.10}$$

The semiconductor industry typically uses Black's equation to model the EM in interconnects of advanced packages [11]:

$$\text{MTTF} = A\left(j^{-n}\right)\exp\left(\frac{E_a}{kT}\right) \tag{8.11}$$

Where A is the pre-factor, j is the current density, n is the power factor, and E_a is the activation energy and is specific to the interconnects under EM testing.

Typical EM testing conditions include three different temperatures at a fixed current density, as well as three current density settings at a fixed temperature. Experimental MTTF data from EM ALT at different temperatures and current densities can be fitted to Black's equation, in order to extract the unknown parameters (A, n, and E_a) for a given interconnect system. With the experimentally determined parameters, Black's equation can be used to predict MTTF under use conditions, which involve lower temperatures and current density levels, compared to the accelerated life test environments.

8.5.1.2 EM Failures Caused by Sn Diffusion

EM typically leads to open failures in interconnects through a void nucleation and propagation mechanism. Voids typically form near the electron entry point, where atoms of primary diffusing species are displaced. This displacement is due to the momentum transfer between the "electron wind" and metal ions. The EM driving force can be described in the following equation [11]:

$$F_{em} = Z^{*}eE = \left(Z^{*}_{el} + Z^{*}_{wd}\right)eE \tag{8.12}$$

Where e is the charge of an electron and E is the electric field, F_{em} stands for the EM driving force, Z^* is the effective charge number representing transfer of momentum from electrons to atoms, Z^{*}_{el} is the nominal valence of the diffusing ion for the direct force, and Z^{*}_{wd} is the effective valence for the electron wind force.

In lead-free Sn-based solder joints, such as BGA SLI solder joints, C4 solder joints, and Cu pillars with solder caps, the dominant diffusing species under EM is Sn. Figure 8.49 demonstrates a common failure mode involving void formation at the solder to IMC interface on the cathode side. When current flows from the organic substrate toward the Cu pillar, divergence of Sn flux occurs at the substrate side Sn to IMC interface, triggering void formation and growth along that boundary.

Solder joints are metallurgically complex systems, because solute elements, such as Cu, Ni, Pd, and Au from substrate surface finishes and Cu pillars dissolve into the solder, creating a multi-component thermodynamic system. These solute atoms can interstitially diffuse through the solder matrix in a fast pace. Additionally, solders are multiphase alloys that can be involved in numerous interfacial intermetallic

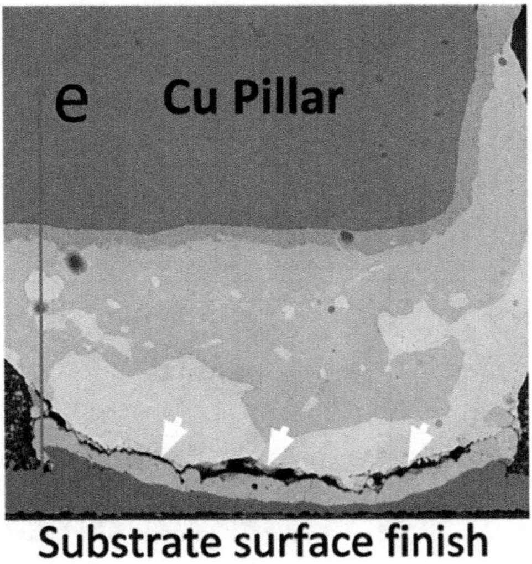

FIGURE 8.49 Typical solder joint EM failure induced by Sn diffusion flux divergence between IMC and Sn solder at the cathode side. (Adapted from Ref. [11].)

reactions, making the EM behavior of solder joints more complicated than that in Al or Cu interconnects. There are multiple factors affecting Sn diffusion-driven EM failures, such as solder alloy composition, IMC formation, current crowding effects, dissolution of metallization layers, and solder joint geometry. All these conditions are equally important for microbumps in advanced packaging technologies.

8.5.1.3 EM Failures Caused by Metallization Dissolution

Electromigration (EM) can significantly accelerate the dissolution of cathode side metallization (UBM) and promote the growth of thicker IMCs on the anode side. If the cathode side UBM layer is sufficiently thick, the solder joint can evolve into a full IMC joint or a joint with a very small remaining Sn height, having much higher EM resistance. However, if the UBM layer is too thin, EM can result in complete dissolution of the UBM layer, causing solder joint open failures [11,77,78]. This failure mechanism is different from IMC to solder interface separation induced by Sn diffusion discussed in the previous Section. A distinguished feature of this failure mode is its strong dependence on the Sn grain orientation [77–79]. It has been shown that the time required to dissolve a 2 μm Ni UBM layer may vary significantly, from roughly 100 hours (when the c-axis of the Sn crystal is parallel to the electron flow), to about 1,800 hours (when it is perpendicular to the electron flow), under identical temperature and current density conditions.

As an example, Figure 8.50a demonstrates an EM induced failure in a solder joint where the electroless Ni-P surface finish on the cathode side is dissolved. The corresponding EBSD analysis shown in Figure 8.50a reveals that the *c*-axis of the Sn grans aligns well with the current flow direction. In contrast, Figure 8.50b shows another solder joint subjected to the same EM stress conditions, exhibiting no signs of damage. The corresponding EBSD analysis of this solder joint shown in

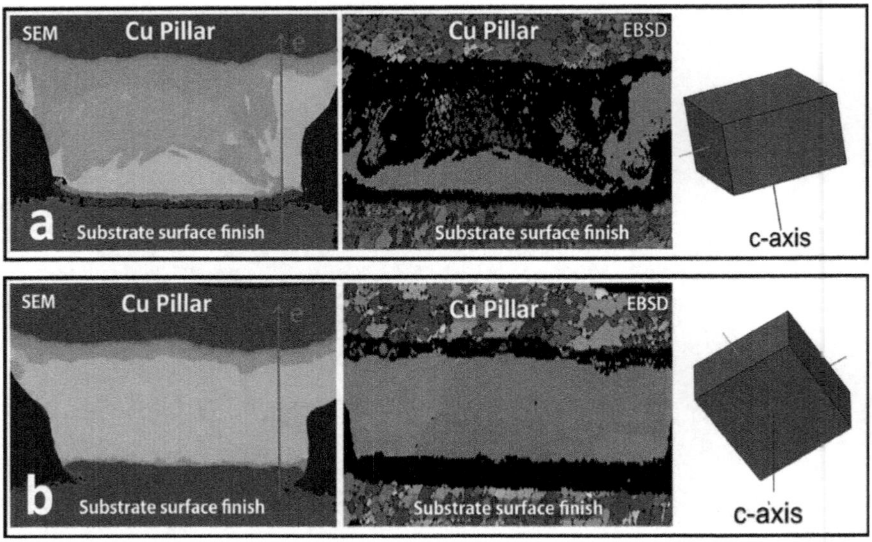

FIGURE 8.50 SEM and EBSD show very good correlation between *c*-axis of Sn and cathode-side EM failure due to Ni dissolution during EM. (a) Fast Ni dissolution when *c*-axis is in-line with the electron flow direction; (b) slow Ni dissolution when c-axis is perpendicular to the electron flow direction. (Adapted from Ref. [11].)

Figure 8.50b confirms that the Sn *c*-axis of this solder joint is perpendicular to the electron flow. These observations clearly indicate that the Sn grain orientation plays a crucial role in determining the susceptibility of UBM to dissolution under the EM stress. Additionally, it has been found that key assembly process parameters, including the cooling rate during solidification and the extent of Cu dissolution into the solder, can influence the Sn grain orientation distribution [11], and thus impact the EM reliability of solder joints.

8.5.1.4 EM in Microbumps, TSVs, RDLs, and Cu-Cu Interconnects

Microbumps, usually consisting of thick Cu or Ni pillars topped with thin solder caps, show unique EM behaviors compared to larger solder joints. Because of their unique bump geometry and solder volume ratio, they typically have better resistance to EM induced failures, for both the Sn flux divergence and UBM dissolution failure modes. First of all, microbumps are shorter, compared with the conventional C4 solder joints. Shorter solder joints in microbumps generate a larger back stress during EM. This back stress counteracts the electron wind force driven Sn atomic diffusion, thus reduces the risk of Sn flux divergence at the IMC to solder interface. Secondly, the volume ratio of solder to UBM layer in microbumps is generally high enough to form a full IMC joint before the UBM is entirely consumed, therefore avoiding the metallization dissolution failure mode that usually impacts larger C4 solder joints, as full IMC joints are inherently more robust against EM. In summary, although microbumps in 3D packaging experience larger current densities and higher Joule heating, their EM reliability risks are relatively low, assuming that there are no manufacturing defects and the UBM layer is sufficiently thick [80].

EM in TSVs, RDLs, and Cu-Cu interconnects is similar to EM in Cu damascene interconnects found in the backend of line layers of Si dies, which has been extensively studied [81–83]. EM degradation in Cu lines is usually interface driven, with failures often initiating at fast diffusion paths, such as via sidewalls or Cu to barrier interfaces. Key EM failure modes and mechanisms revealed in Cu damascene interconnects are also applicable to TSVs, RDLs, and Cu-Cu bonding, and are listed below.

Current crowding induced EM failures are very common in Cu interconnects, because of the non-uniform current density in the irregular interconnect geometries and complex routing. The EM failing locations are often at the inner corners of bent conductors with the highest current density [84]. Current crowding can also raise local temperatures, inducing thermal gradients and thermomechanical stress, both of which further drive Cu diffusion and EM failures.

Passivation layers can affect EM performance of Cu interconnects, because Cu EM is largely dominated by interfacial diffusion. Improving the Cu to passivation layer adhesion and reducing interfacial diffusion can enhance Cu EM reliability. For example, Cu traces passivated with SiN_x have been reported to exhibit 12 times longer EM lifetime, compared to those with SiC passivation [82], which aligns well with the much lower debonding energy measured at the Cu to SiC interface compared with that at the Cu to SiN interface [85].

The presence of metal line extensions, such as overhangs or reservoirs, near via connections, can delay EM induced failures by serving as void reservoirs [86]. Simulation studies [83] show that beyond a certain critical extension length, further increases in the extension size do not lead to any additional EM improvement due to current density saturation in the reservoir corners.

According to the Blech effect, thermomechanical stress gradients can serve as a driving force for atomic diffusion [87], because of the vacancy accumulation differences between the cathode and the anode under the electron wind force. Furthermore, process induced thermal mismatches, due to the design complexity, fabrication methods, and dielectric materials, can introduce significant thermomechanical stress gradients during high temperature EM testing. Factors, such as the passivation layer thickness [88] and interconnect processing temperatures [89], have been shown to largely impact stress distribution, and consequently, EM reliability.

8.5.2 Package Level Reliability Tests

Package level reliability, also referred to as component level reliability, is very important in the advanced semiconductor packaging technology development, as new materials and processes introduced for emerging products must undergo and pass standard component reliability tests to achieve qualification.

8.5.2.1 Package Level Reliability Test Methodologies

Industry-standard reliability tests for semiconductor packages encompass a range of stress tests designed to simulate real world operating and environmental conditions [65]. These include:

Preconditioning (PC) Tests: Incorporate temperature and humidity soak and multiple solder reflow cycles to replicate conditions experienced during shipping storage, and surface mounting of packages onto PCBs.

Temperature Cycling (TC) Tests: Expose devices to repeated transitions between extreme low and high temperatures to simulate power cycling (device "on" and "off" states).

Baking Tests: Apply prolonged high temperatures to simulate extended operational durations.

Highly Accelerated Stress Tests (HAST): Combine high temperature, high humidity, high pressure and time to assess package reliability either with or without electrical bias in a controlled environment, to mimic exposing devices to hash environments. Biased HAST (BHAST) stands for HAST test with electrical bias, while Unbiased HAST (UHAST) is the HAST test without electrical bias [11,90].

Specific conditions for each test type, depending on the use conditions and market segments, are described in the JEDEC standards. Failure rate vs testing time t in an ALT can be plotted, typically presenting as a bathtub curve, with a schematic illustration shown in Figure 8.51 [11]. ALT failures generally fall into three categories, each can be modeled using a Weibull Probability Density Function (PDF) (refer to Equations 8.8–8.10), with a distinct characteristic life (η) and shape parameter (β).

Infant mortality failures are due to manufacturing defects, which fail early in the test cycle. The shape factor $\beta < 1$, indicates a decreasing failure rate as defective units fail early, leaving behind more robust packages. The intrinsic failures have a constant failure rate during reliability tests, with the shape parameter $\beta = 1$. Wear-out failures emerge after prolonged testing or use, which are indicated by $\beta > 1$, reflecting an increasing failure rate as package age and degrade [76].

Different ALT types require different acceleration models. For instance, EM testing utilizes Black's equation, as mentioned in the previous section. The acceleration factor (AF) quantifies how the stress conditions accelerate failures compared to the use conditions, and can be calculated using the following equation:

$$\mathrm{AF} = \frac{L(\mathrm{ref})}{L(\mathrm{accel})} \tag{8.13}$$

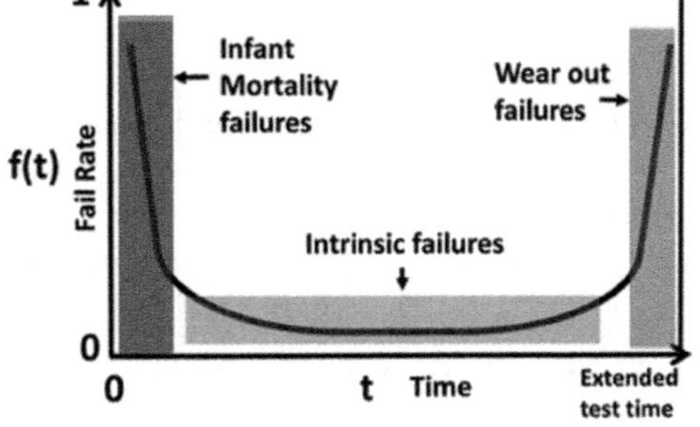

FIGURE 8.51 Schematic of the Bathtub curve showing failure rate versus reliability test time. (Adapted from Ref. [11].)

Where L(ref) is the lifetime of the product at reference, or use condition, and L(accel) is the lifetime of the device at an accelerated stress. Once the appropriate acceleration model is established, AF can be leveraged to project device lifetime under the use condition from the accelerated test data [90].

8.5.2.2 Arrhenius Model

Temperature is one of the most influential stress factors impacting the reliability of semiconductor packages. The Arrhenius model, which is derived from the Arrhenius equation, is commonly used to analyze the impact of temperature in reliability tests [91] It is expressed as the following Equation (8.14):

$$L(T) = Ae^{\frac{E_a}{KT}} \tag{8.14}$$

where $L(T)$ is the lifetime of the device as a function of temperature T, A is a constant that is design specific, K is the Boltzmann constant (8.617×10^{-5} electron volts/degree Kelvin), T is the absolute temperature in Kelvin, and E_a is the activation energy in electron volts. AF can then be calculated as:

$$\mathrm{AF} = e^{\frac{E_a}{K}\left(\frac{1}{T_{\mathrm{use}}} - \frac{1}{T_{\mathrm{test}}}\right)} \tag{8.15}$$

Where T_{test} is the test temperature in Kelvin and T_{use} is the use condition temperature in Kelvin. Larger E_a values and temperature differences lead to higher acceleration factors. The activation energy is specific to each failure mechanism and needs to be determined experimentally or inferred from literature [90].

8.5.2.3 Coffin-Manson Model

The Coffin-Manson model is usually employed to analyze thermal cycling test results, where failures are caused by thermomechanical stresses due to CTE mismatches between joined materials. The model is defined by:

$$N(\Delta T) = \frac{C}{(\Delta T)^m} \tag{8.16}$$

Where $N(\Delta T)$ is the number of cycles as a function of delta temperature (ΔT), C is a design specific constant, ΔT is the delta between the maximum and minimum temperatures, and m is the Coffin-Manson exponent. AF is then calculated to be:

$$\mathrm{AF} = \left(\frac{\Delta T_{\mathrm{test}}}{\Delta T_{\mathrm{use}}}\right)^m \tag{8.17}$$

Where ΔT_{test} is the delta temperature during test, and ΔT_{use} is the delta temperature during field use. A larger temperature swing and a higher exponent value result in higher AFs. The exponent m relates to material properties, and is typically in the range of 1–3 for ductile materials, such as solder [92, 93].

8.5.2.4 Basquin Model

The Basquin model is used to predict failures caused by mechanical fatigue, especially under cyclic stress strain conditions. It is particularly relevant for semiconductor packages in automotive applications, which are frequently subject to vibration under the use conditions. The model can be expressed as:

$$N(S) = CS^{\frac{1}{b}} \tag{8.18}$$

Where $N(S)$ is the number of fatigue cycles as a function of stress S, C is a constant that is design specific, S is the reversing stress magnitude, and b is the fatigue strength exponent (and it is always negative). AF is then calculated to be:

$$\text{AF} = \left(\frac{S_{\text{use}}}{S_{\text{test}}}\right)^{\frac{1}{b}} \tag{8.19}$$

Where S_{test} is the stress amplitude during test, and S_{use} is the stress amplitude during field use. A larger stress difference between the test and the use conditions, along with a larger absolute value of $1/b$, can lead to a higher acceleration factor. The exponent b varies by different designs and materials and needs to be defined based on testing data [94].

8.5.2.5 Multi-Stress Models

For ALT with accelerated temperature and humidity, such as HAST, temperature acceleration follows the Arrhenius model, and humidity acceleration follows the inverse power law. The acceleration model for lifetime calculation thus becomes:

$$L(T,\text{RH}) = Ce^{\frac{E_a}{\text{KT}}}\left(\frac{1}{\text{RH}}\right)^m \tag{8.20}$$

Where $L(T, RH)$ is the package lifetime as a function of temperature (T) and relative humidity (RH), C is a constant that is design specific, K is the Boltzmann constant, T is the absolute temperature in Kelvin, E_a is the activation energy in electron volts, and m is the humidity exponent. AF can be calculated as:

$$\text{AF} = e^{\frac{E_a}{K}\left(\frac{1}{T_{\text{use}}} - \frac{1}{T_{\text{test}}}\right)}\left(\frac{\text{RH}_{\text{test}}}{\text{RH}_{\text{use}}}\right)^m \tag{8.21}$$

Where T_{test} is the test temperature in Kelvin, T_{use} is the field temperature (use condition) in Kelvin, RH_{test} is the test relative humidity in percent, and RH_{use} is the field relative humidity in percent. This multi-stress model approach can be extended to more stressors, although it is rare to model more than two simultaneously [90].

8.5.3 Failure Modes in Package Level Reliability Tests

Understanding the failure rate distribution in reliability tests is critical in identifying whether failures are defect driven, intrinsic, or wear-out modes [95]. The well-known bathtub curve shown in Figure 8.51 demonstrates three distinct phases of failures over time during reliability tests. Infant mortality failures happen early during the tests and are usually caused by manufacturing defects, exhibiting an initially high failure rate that decreases over time. Intrinsic failures maintain a roughly constant failure rate throughout the entire test, and are due to the inherent limitation of package design, materials selection, or assembly processes. Wear-out failures emerge later in the testing and are characterized by an increasing failure rate, indicating the "end of life" of the product.

Each failure type has a different mechanism and root cause, therefore requiring distinct mitigation strategies. Infant mortality issues are typically resolved by optimizing the assembly process to reduce defects or by implementing effective testing methods to screen out defective units. Intrinsic failures are dependent on design, and may require significant changes in materials, layout, or component selection. Wear-out failures typically indicate the end of usable life of a semiconductor package, typically happening during extensive reliability testing [96].

Each reliability test is designed to trigger specific failure modes. For example, PC tests can cause failures like "popcorning" or delamination between interfaces, due to high internal vapor pressures from the vaporization or outgassing of the absorbed moisture or organic contaminants, built up during elevated temperature reflow cycles [97]. Figure 8.52a illustrates a Through Transmission Scanning Acoustic Microscopy (TSAM) image, and Figure 8.52b shows a the cross section along the white dash line in Figure 8.52b, revealing a significant substrate blistering due to excess moisture prior to PC reliability testing. TC tests primarily cause thermomechanical stresses within semiconductor packages, resulting in failures, such as solder joint fatigue cracks and substrate trace opens [98,99]. Figure 8.52c shows a representative solder joint fatigue crack after a prolonged TC test [11]. Baking tests can lead to materials degradation, including IMC aging and cracking in solder resist layers, which can lead to open interconnects. Figure 8.52e and f present cracks in the solder resist layer and the subsequent trace cracks within the substrate post extended bake testing [100]. UHAST can typically reveal corrosion related failures [101], while BHAST can also cause electrochemical migration, besides corrosion, because of the electrical bias applied during testing. Figure 8.52d demonstrates Cu dendrite growth between two adjacent pads of opposite polarity, resulting from electrochemical reactions after extended BHAST exposure [102].

8.6 FAULT ISOLATION (FI) AND FAILURE ANALYSIS (FA) OF ADVANCED SEMICONDUCTOR PACKAGES

Advanced semiconductor packages are highly complex systems comprising multiple dies, substrates, and interconnects ranging from submicron to a few mm in size. As a result, it is very common to have multiple failing structures within the same semiconductor package after the reliability tests. Performing accurate and comprehensive failure analysis for each failure in such a complex system is very challenging. High resolution, non-destructive techniques for fault isolation (FI), imaging, and materials characterization can be applied to address this challenge. In addition to the

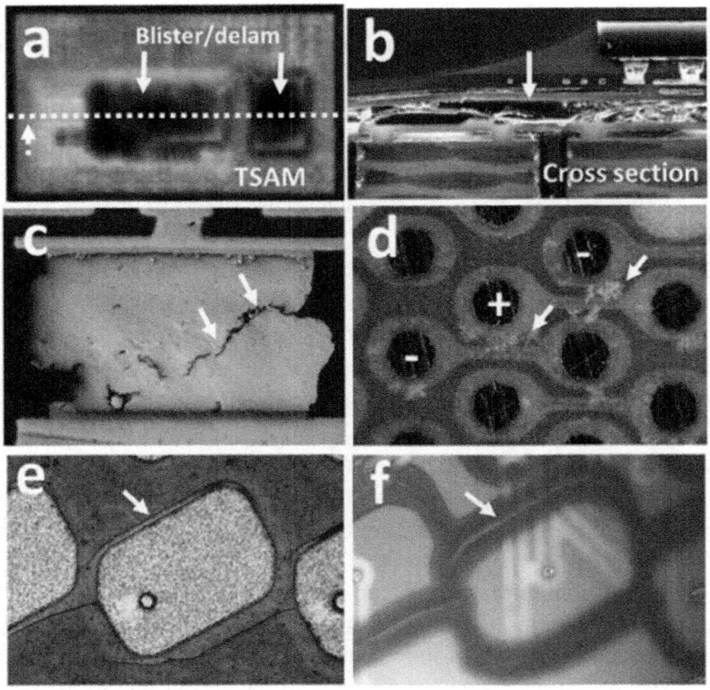

FIGURE 8.52 (a and b) The TSAM and cross-sectional optical image, respectively, of large substrate blisters in a package post PC test due to "popcorn". (c) A typical fatigue crack in solder joints post extended TC tests. (d) Cu dendrite between the adjacent Cu pads of FLI solder interconnects with different polarities post extended BHAST tests. (e and f) Substrate solder resist layer crack post extended Bake Tests, which lead to substrate trace cracks. (Adapted from Ref. [11].)

complexity of multiple failures, each electrical failure in an advanced semiconductor package may originate from different locations within the package, such as various dies, interconnect layers, or assembly levels. Precisely locating each failure within this complex multi-layered architecture is very demanding.

Fault isolation techniques, which can detect both open and short failures in three dimensions, are vital for effective failure analysis in advanced semiconductor packages. After a defect is located, physical failure analysis (PFA) is required to reveal the defect and investigate its root cause. However, the high aspect ratio of certain interconnects, such as TSV, further complicates the PFA. The aspect ratio of TSV can go up to 1:20, with small diameters (2–10 μm), and extended lengths (40–200 μm). In such cases, artifact-free cross sectioning sample preparation techniques with short TPT are critical for detecting and analyzing subtle defects across large cross sectional areas. Conventional cross sectioning methods, including mechanical polishing or traditional Ge ion FIB techniques, could not meet the demand in terms of either precision or TPT. As a result, advanced cross sectioning methods that offer higher milling rates while preserving accurate control over location and surface integrity are more suitable for analyzing advanced packaging failures.

Additionally, the subtle imperfections within semiconductor packages, including surface finish plating irregularities, interfacial contaminations, interconnect impurities, or abnormal grain structures in small interconnects, can develop into failures, such as delamination, non-wet joints, and interconnect cracking during the assembly process or reliability testing. Therefore, the innovative application of materials analysis techniques that are capable of providing elemental composition, chemical state, depth profiling, and crystallographic orientation near the defect sites, is vital for in-depth root cause analysis [11,90].

8.6.1 Failure Analysis Flow

Failure analysis (FA) flow is a structured diagnostic process created to locate defects responsible for electrical failures in complex semiconductor packages. As illustrated in Figure 8.53, the FA flow typically includes a series of steps: problem statement review, failure verification, non-destructive imaging, electrical failure analysis (EFA), physical failure analysis, and materials analysis [103].

Although sometimes being overlooked by rushed FA analysts, reviewing the problem statement is the most important step for a successful FA [103]. A well-defined problem statement should contain the following factors.

1. **Failure Conditions and History**: Understanding whether the failure occurred at EOL, during reliability testing, or at a customer site, can provide valuable insights about the possible failure mode and failure mechanism. EOL failures usually relate to process defects, and details about process steps and access to a baseline defect library can significantly help diagnosis.

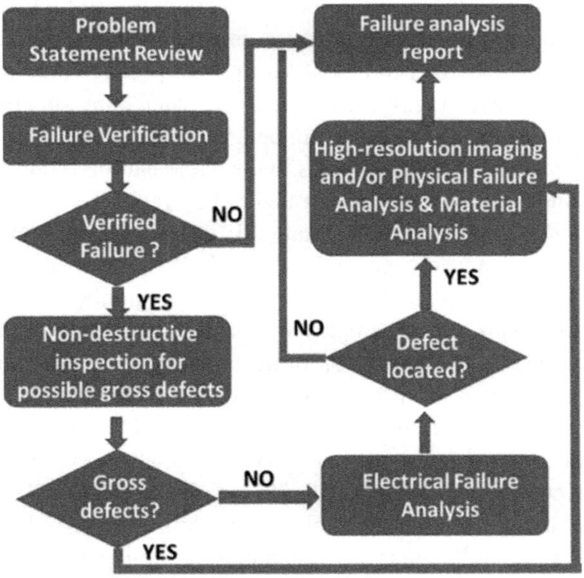

FIGURE 8.53 Schematic of failure analysis flow. (Adapted from Ref. [90].)

Failures which occurd during reliability tests are usually initiated or accelerated under stringent conditions, requiring test specific FA approaches [11]. For failing units in customer returns, communication regarding the sample's operational and thermal history is essential for a successful failure analysis.

2. **Failure Rate Analysis**: A high EOL failure rate indicates a systemic process issue with extensive defects, whereas a low EOL failure rate may suggest random and minor defects. In reliability testing, failure rates generally follow the "bathtub curve", as shown in Figure 8.51, encompassing the defect-driven infant mortality failures, intrinsic failures that are inherent to the package design, and wear-out failures that are the end-of-life indicators of semiconductor packages. For customer returned failing parts, failure rate analysis at the customer site can reveal information about possible process or test issues.
3. **Leveraging Historical Data**: FA data from previous failing parts with similar packaging structures can guide the current FA work. In some cases, known failure patterns or locations can help signature close cases without actual failure analysis work [11,90].

Before moving to extensive FA, failure verification is needed to rule out issues, including test setup errors, probe contacting problems, and intermittent failures [103]. Parametric failures, including opens and shorts, can be verified by collecting *I–V* curves, resistance, or current values from the failing structures, and comparing them to those of known good samples. Functional failures can be confirmed using debug tools and test programs, and the confirmed failures usually exhibit clear deviations when compared with test good units.

After verification, large field of view, non-destructive imaging techniques can be employed to examine the AOI for defect detection. For instance, optical microscopy can detect handling-induced damages on semiconductor packages. Scanning Acoustic Microscopy (SAM) can be used to inspect package and die interfaces for delamination. 2D X-ray imaging helps evaluate solder joint integrity in the entire bump area. Infrared (IR) microscopy can be used to detect internal die cracks. If a gross defect is located by these methods, the sample can be followed up with PFA directly, bypassing the EFA, as demonstrated in Figure 8.53.

If no gross failures are identified, detailed EFA becomes critical to locate subtle defects in complex advanced semiconductor packages. For open and short failures, Time Domain Reflectometry (TDR), Lock in Thermography (LIT), and Magnetic Field Imaging (MFI) are usually employed. For functional failures, techniques such as optical fault isolation tools, electron beam imaging, as well as micro probing or nano probing, can be utilized. In most cases, multiple fault isolation techniques need to be used to pinpoint the failure locations in semiconductor packages.

Once the defects are located, PFA and materials analysis are conducted to investigate the root cause. For large size defects, such as those in PCBs or large solder joints, mechanical cross sectioning or planar grinding is preferred for its lower cost and quick TPT. For subtle defects in organic substrates, interconnects, or die layers, a combination of multiple sample preparation techniques, including mechanical methods, laser ablation, ion milling, and FIB, is usually employed to conduct artifact-free cross sectioning with short TPT.

The final step is to write a comprehensive failure analysis report, including the initial problem statement, the FA techniques used, FA results, and final conclusions. This documentation is important because it facilitates collaboration with manufacturing and design partners, supports root cause investigations, and serves as a reference for future similar failures.

8.6.2 Non-Destructive Imaging Techniques

8.6.2.1 Optical and IR Imaging

As illustrated in Figure 8.53, optical and IR imaging are commonly used to scan the entire package for gross defects before proceeding to more sophisticated investigations [103]. If a large defect is captured at this inspection step, it can significantly reduce FA TPT. IR imaging is especially useful to inspect the die active side circuitry from the die backside, since Si is transparent to IR wavelengths longer than ~ 1.1 μm [103]. However, doped Si is less transparent to IR than undoped Si, due to the bandgap narrowing and free carrier absorption effects. To address this challenge, advanced IR techniques, such as near-IR, shortwave infrared (SWIR) imaging with 1,300–1,500 nm wavelengths, and IR confocal imaging with enhanced signal-to-noise ratio, can be employed to non-destructively inspect semiconductor packages that comprise doped Si dies [104–106].

These IR based techniques enable real time visualization of defects, including internal die cracks, active circuitry area cracks, and other defects located under the die surface and are invisible if using visible light optical microscopy. As an example, Figure 8.54 presents an IR image taken from the die back side of a failed unit,

FIGURE 8.54 The IR image taken from die backside showing an internal die crack highlighted by the red arrows. (Adapted from Ref. [90].)

without any die thinning or sample preparation. The dark region, highlighted by the arrows, suggests an internal die crack that grows into the die active area, damaging Si circuitry and causing the associated electrical failure [90].

8.6.2.2 Scanning Acoustic Microscopy (SAM)

Scanning Acoustic Microscopy (SAM) is an extensively used non-destructive technique to detect delamination, voids, and other internal defects within semiconductor packages [107–110]. As demonstrated in Figure 8.55, SAM employs a piezoelectric-based acoustic transducer, which can emit an ultrasonic pulse into the sample. Water is the coupling medium used to efficiently transmit the acoustic waves from the transducer to the sample surface. The refection (*R*) and transmission (*T*) coefficients of the acoustic wave at materials interfaces can be calculated using the following equations:

$$R(90^\circ) = \frac{Z_2 - Z_1}{Z_2 + Z_1} \tag{8.22}$$

$$T(90^\circ) = \frac{2Z_2}{Z_1 + Z_2} \tag{8.23}$$

Where Z_1 and Z_2 are the acoustic impedance of medium 1 and medium 2, respectively. The acoustic impedance Z is expressed as:

$$Z = \rho c \tag{8.24}$$

Where “c” is the acoustic wave velocity in the medium, and “ρ” is the volumetric mass density of the medium. For longitudinal acoustic waves traveling in an isotropic and homogeneous medium, c can be expressed as

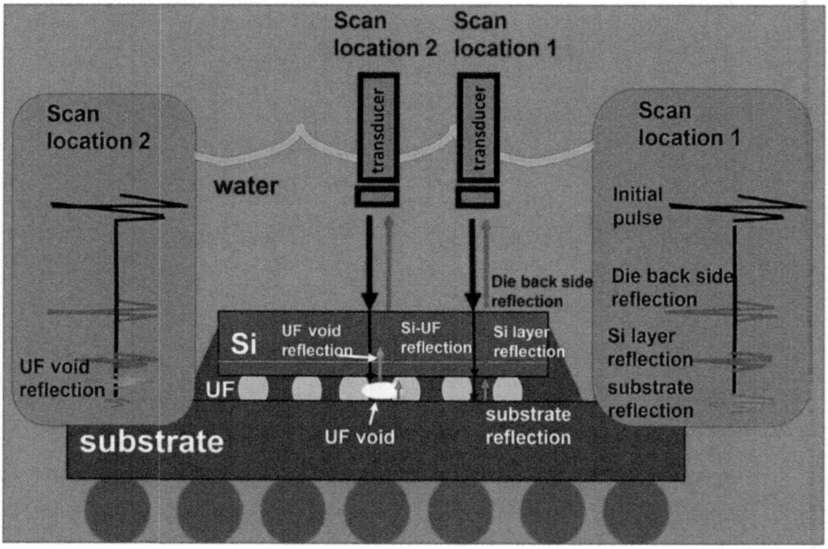

FIGURE 8.55 Schematic of the SAM technique. (Adapted from Ref. [11].)

$$c = \sqrt{\frac{E(1-\mu)}{\rho(1+\mu)(1-2\mu)}} \tag{8.25}$$

Where "μ" is the Poisson's ratio, "*E*" stands for the Young's modulus. Acoustic Impedance is a characteristic material property and is determined by Young's modulus, Poisson's ratio, and density [111,112]. According to Equations (8.22) and (8.23), assuming medium 2 is air, nearly 100% of the acoustic wave is reflected when transitioning from a solid material to air, since air has an extremely low acoustic impedance ($\sim 0.389 \times 10^{-3}$ kg/m^{2}s $\times 10^6$). Thus, when $Z_2 \approx 0$, the reflection coefficient $R \approx -1$ and $T \approx 0$, making SAM extremely sensitive to air gap type of defects, including delamination, cracks, voids, and blisters within semiconductor packages.

In the pulse echo mode shown in Figure 8.55, the same acoustic transducer sends and receives the acoustic pulse. Reflected echoes have both amplitude and time of flight (ToF) data, revealing the *x*–*y* location and depth (*z* dimension) of internal features [113–115]. As demonstrated in Figure 8.55, the echo from an underfill (UF) void located at scan location 2 has a distinct ToF compared to that from the die backside, enabling the defect *z* dimensional localization using B scan with a cross-sectional view, or C scan with a top-down view. Tomographic Acoustic Micro Imaging (TAMI) enhances C scan imaging by providing data from multiple depths using a series of time gates [116–118].

The lateral (*x*–*y* dimention) resolution of a CSAM scan is determined by the transducer beam spot size, the smaller the higher resolution. The axial (*z* dimention) resolution, which is crucial for detecting narro delamination, can be improved with higher frequency transducers. However, high frequency acoustic waves have increased attenuation, not good for units with thick Si. In order to achieve high resolution and non-destructive acoustic imaging for semiconductor pacakges, acoustic transducer frequency and focal length need to be optimized [118,119].

For deep embedded defects, the Through Transmission SAM (TSAM) technique can be employed. As illustrated in Figure 8.56, it has two aligned transducers, one to emit and the other to receive the acoustic puls [113]. When the acoustic pulse meets a defect, such as a crack or a delamination, the transmitted acoustic signal is

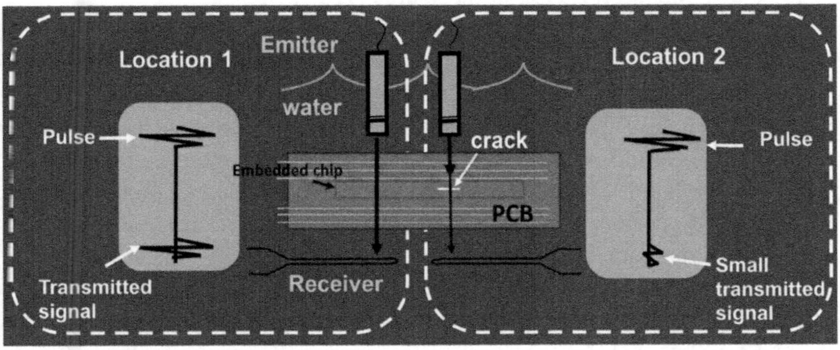

FIGURE 8.56 Schematic of the TSAM technique. (Adapted from Ref. [11].)

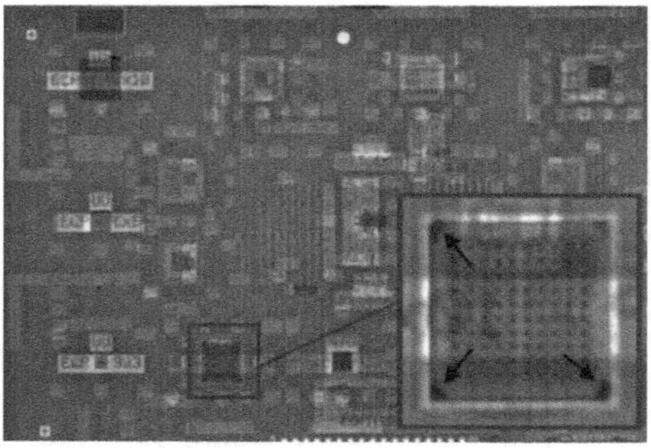

FIGURE 8.57 TSAM of a PCB with embedded chips showing defects in one embedded chip. (Adapted from Ref. [90].)

dramatically attenuated, generating a dark contrast in the TSAM image. In contrast, there is much less energy loss for the acoustic pulse transmitting to the defect-free areas. The TSAM configuration can be improved by using emission transducers operated at 20–50 MHz and receiving transducers with 10 MHz frequency. The improved TSAM configuration can elevate the spatial resolution up to 50 μm, and has been applied to successfully detect die cracks for the Si dies embedded in a PCB, as shown in Figure 8.57. The defects located at the embedded die corners are shown as dark spots highlighted by the red arrows in Figure 8.57, demonstrating the high resolution defect detection capability of TSAM in complex semiconductor multilayer systems [90].

8.6.2.3 2D X-Ray Radiography

2D X-ray Radiography is an extensively employed non-destructive inspection technique in the semiconductor industry, especially for detecting solder joint defects in semiconductor packages. As illustrated in Figure 8.58, the setup has the sample placed between a point source X-ray emitter and a detector [11,120]. The geometric magnification (M) is determined as the ratio of the distance from the source to the detector (D) over the distance from the source to the sample (*d*). To achieve high resolution imaging, the X-ray beam spot size needs to be minimized, and the sample needs to be positioned very close to the source [120,121].

As 2D X-ray radiography images a 2D projection of a 3D object, the depth and volume information of samples could be obtained by capturing sample images from multiple sample tilting angles. As shown in Figure 8.59, real time 2D X-ray can detect a range of solder joint defects in semiconductor packages, such as solder bridging, non-wet joints, and open contacts, by optimizing the sample tilt and rotation angles. The correlation of 2D X-ray image features and defects of solder joints can be established by performing physical cross section analysis of the same sample after the X-ray imaging, as illustrated in Figure 8.59. By comparing the contrast and geometry of 2D X-ray images taken from failing solder joints with those from a known good reference, solder joint defects, including partial or complete non-wet, opens, and bridging, can be clearly identified in real time using high resolution 2D X-ray imaging technique [122].

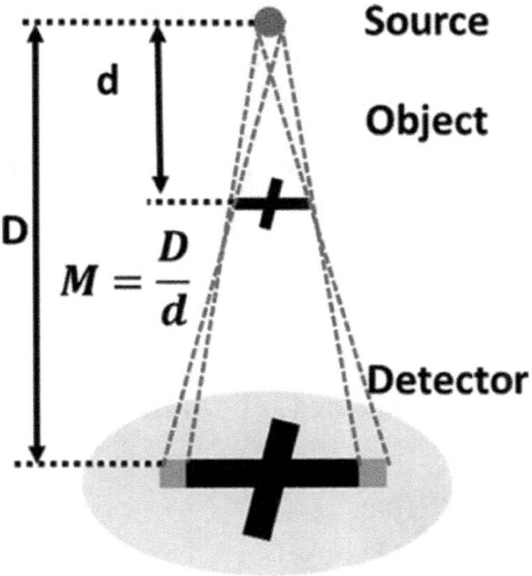

FIGURE 8.58 Schematic of a 2D X-ray radiography setup. (Adapted from Ref. [11].)

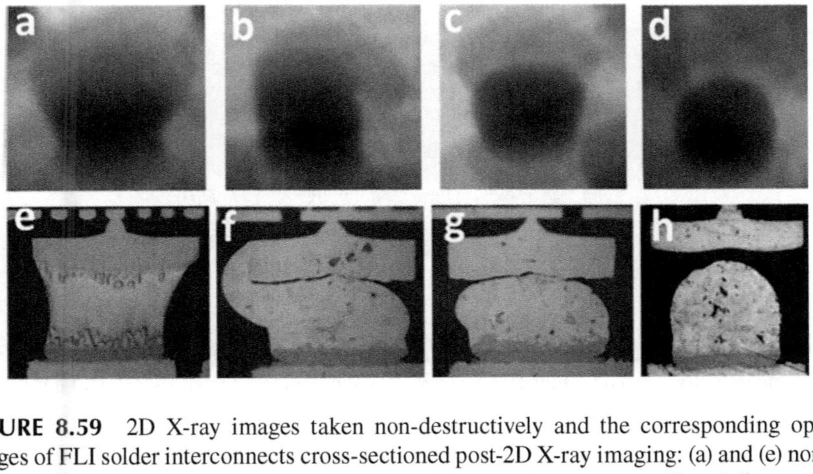

FIGURE 8.59 2D X-ray images taken non-destructively and the corresponding optical images of FLI solder interconnects cross-sectioned post-2D X-ray imaging: (a) and (e) normal or defect free, (b) and (f) partial non-wet, (c) and (g) complete non-wet, (d) and (h) non-contact open. (Adapted from Ref. [11].)

The non-destructive and real time capabilities of the 2D X-ray technique can support in situ failure analysis during reliability testing, such as thermal cycling or reflow process. The in situ FA can directly observe failure initiation and propagation, helping pinpoint failure root causes and guide corrective actions [123]. The technique has proven effective in monitoring solder void formation during multiple reflow cycles [121].

However, 2D X-ray imaging has limitations in detecting subtle defects, such as micro voids in Cu vias, cracks in RDL traces, or tiny solder extrusion, because of the overlapping layers and limited contrast in complex3D multi-layer systems [124].

8.6.2.4 3D X-Ray Computed Tomography (CT)

3D X-ray CT has become a powerful non-destructive technique for detecting micron or sub-micron scale defects in advanced packaging [125–127]. As displayed in Figure 8.60a, the system has an X-ray source illuminating the sample at multiple angles, a rotating stage that provides controlled angular displacement, and a detector that captures a series of 2D radiographic projections. As illustrated schematically in Figure 8.60b, the acquired 2D images are reconstructed using computational algorithms to create a volumetric 3D dataset. FA analysts can generate 3D views or extract virtual cross sections or planar views at any depth or orientation of the sample from this 3D dataset.

As illustrated in Figure 8.61, 3D X-ray CT has been successfully adopted for the detection of micron scale features, such as TSV voiding, microbump integrity, solder

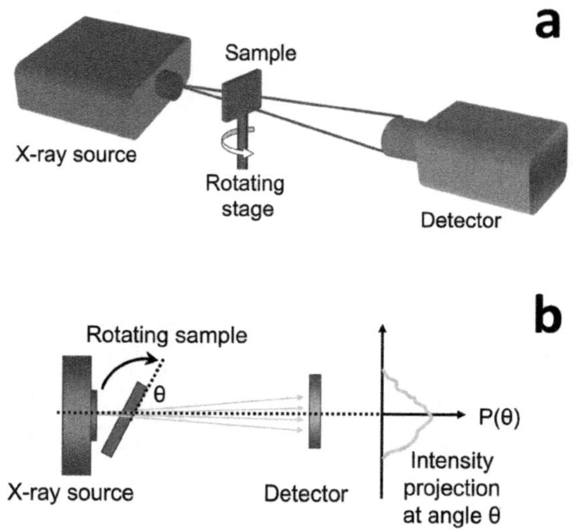

FIGURE 8.60 (a) Schematic of the 3D X-ray CT setup consisting of X-ray source, rotating stage, and detector. (b) The intensity projection at each angle is superimposed and mathematically processed to generate the three-dimensional image. (Adapted from Ref. [11].)

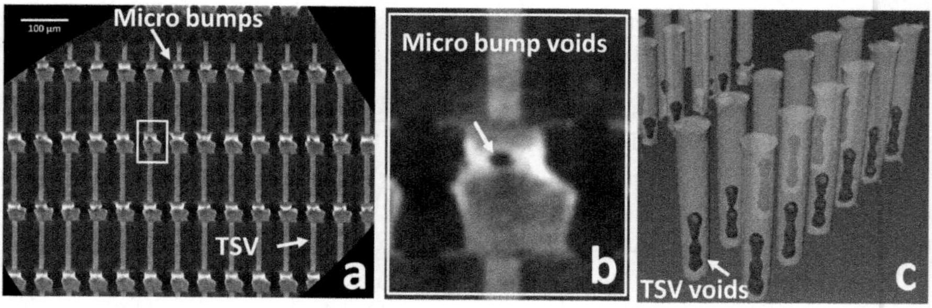

FIGURE 8.61 (a and b) 3D X-ray CT of microbumps and TSVs in a 3D package with stacked die configuration. (c) 3D X-ray CT of TSVs showing distinctly the TSV voids. (Adapted from Ref. [90].)

joint partial non-wet and voids, as well as Cu via cracks. Figure 8.61a shows a virtual cross sectional image by 3D X-ray CT, revealing TSVs and microbumps in a 3D package. A zoomed-in view of the microbump within the white frame shown in Figure 8.61a is illustrated in Figure 8.61b, which clearly reveals the small void in this microbump, while Figure 8.61c presents a 3D view of TSVs, having small voids distinctly detected, demonstrating the high spatial resolution of 3D CT [128,129].

8.6.3 Fault Isolation (FI) Tools

8.6.3.1 Time-Domain Reflectometry (TDR) and Electro-Optic Terahertz Pulse Reflectometry (EOTPR)

Non-destructive fault isolation (FI) techniques, such as TDR and EOTPR, are commonly used to help locate open and high-resistance failures in advanced semiconductor packages [130]. Compared to EOTPR, TDR is a more conventional and cost effective technique. TDR can inject a step electrical pulse with a rise time of 35–40 ps into the sample. The reflected signal is analyzed to define impedance variations along the interconnect path. Failure locations can be pinpointed by comparing the reflection profiles of the failing unit against the known good sample.

The spatial resolution of TDR is usually estimated to be 1/10 to 1/5 of the pulse rise time [131], and can be calculated by the following equation [132]:

$$\lambda_{\text{resolution}} = \frac{0.35}{\text{BW}_{\text{system}}} \cdot \frac{c}{2\sqrt{\varepsilon_{\text{eff}}}} \tag{8.26}$$

Where c is the speed of light in free space, ε_{eff} is the effective dielectric constant of the device, and BW_{system} is the total system bandwidth. The spatial resolution is about 500 µm for flip chip packages with system bandwidths of 18–20 GHz and pulse rise time of 35–40 ps [130].

As demonstrated in Figure 8.62, TDR can non-destructively locate an "open" failure in a semiconductor package. A couple of reference units, including a bare

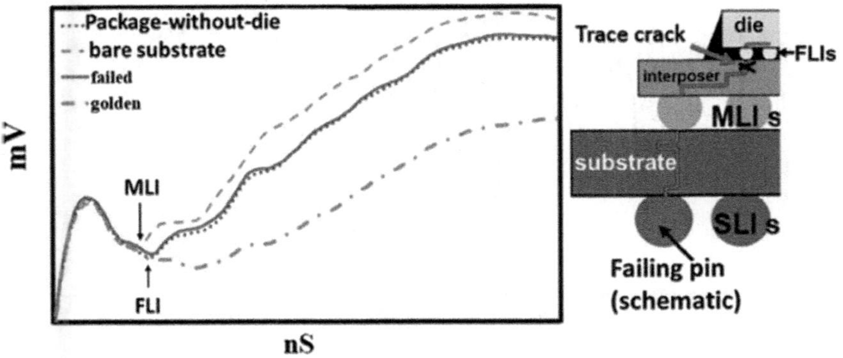

FIGURE 8.62 TDR spectra from a failed unit, a golden unit, a bare SiP substrate, and a SiP package without FPGA die, showing the open failure close to the FLI solder joint in the FPGA package of a SiP package. (Adapted from Ref. [11].)

organic substrate, a good package without the die, and a known good unit, are used to collect the reference TDR spectra, in order to define the locations of the middle level interconnect (MLI) (solder joints connecting the substrate and the interposer), and the first level interconnect (FLI) (solder joints connecting the interposer and the die). The TDR spectrum taken from the failed unit overlaps with that collecting from the good package without the die, indicating that the failure is located near the FLI.

EOTPR, as an advanced successor of TDR, provides significantly higher spatial resolutions. It employs a sharp electrical pulse, in the range of 40 GHz–4 THz, to probe the sample [130]. Because of its extremely fast rise time (~ 5.7 ps) and terahertz range bandwidth, EOTPR can achieve ~ 10 μm resolution [133]. In EOTPR spectra, open failures show as "peaks" and short failures appear as "valleys". As an example, EOTPR is used to analyze the same samples studied in Figure 8.63. The zoom-in plot of Figure 8.63 reveals that the failure is likely located in the interposer very close to the FLI, by comparing the EOTPR spectrum taken from the failing unit with that from a reference unit without the die (defining the FLI location). The time-domain EOTPR data can be converted to the distance domain using the phase velocity of electromagnetic waves, which is expressed as [134]:

$$v_p = \frac{c}{\sqrt{\varepsilon_{\text{eff}}}} \tag{8.27}$$

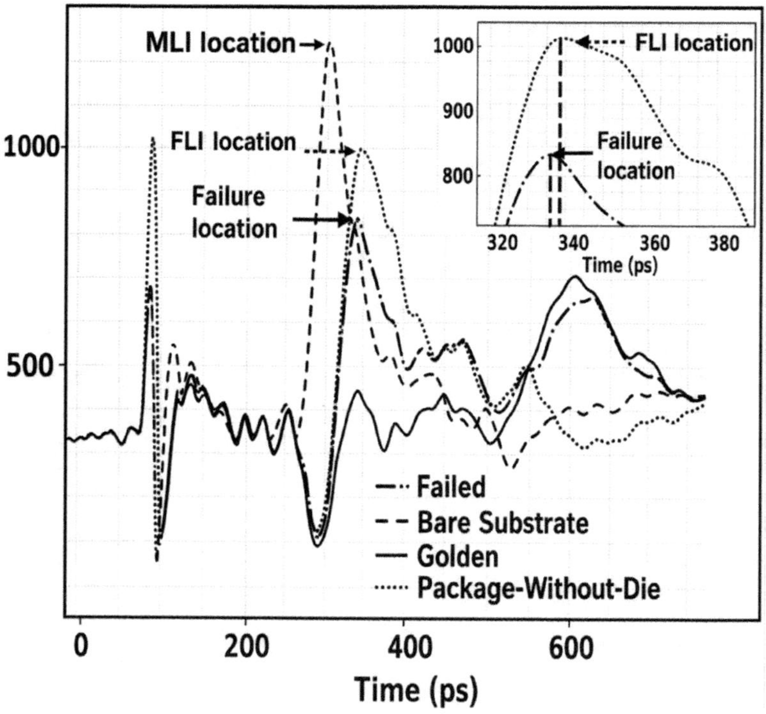

FIGURE 8.63 EOTPR spectra from a failed unit, a golden unit, a bare substrate, and a package without die. (adapted from Ref. [11].)

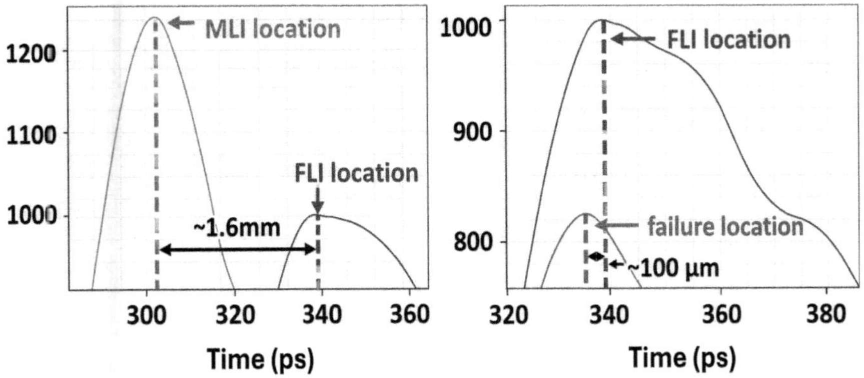

FIGURE 8.64 Failure location can be estimated using known package structure locations in EOTPR spectra. (Adapted from Ref. [11].)

Where c is the velocity of light in free space, ν_p is the phase velocity, and ε_{eff} is the effective dielectric constant of the package, which depends on the package design. Figure 8.64 demonstrates how ε_{eff} is estimated using the time domain EOTPR data between MLI and FLI, and the known physical distance (1.6 mm) from the design file. The distance from the FLI to the failure location is calculated to be roughly 100 µm [11].

8.6.3.2 Lock-In Thermography (LIT)

LIT is a powerful non-destructive fault isolation technique for identifying the 3D location of shorts and resistive defects in semiconductor devices. The real-time graphical lock-in methodology, capable of detecting small signals buried in random noises of higher magnitude, is utilized in LIT to locate the dissipated heat generated by resistive failures with input power [135]. As the thermal propagation is time dependent, LIT can also provide the z dimensional information of the defect, besides the x and y dimensional information. The phase shift between the thermal response and the excitation signal reveals the z axis information of the heat source. The phase shift is influenced by the lock-in frequency, thermal properties, and the layer thickness of the unit.

The z dimensional information of the hotspot from the defect can be defined by comparing the phase vs frequency characteristics of known defect locations. As an example, Figure 8.65 presents both experimental and simulated LIT results, illustrating how thermal emissions from different stacked dies can be differentiated. It confirms that LIT can effectively provide 3D defect locations (x, y, and z) [11,135,136]. However, LIT could not be used to analyze “open” failures or “shorts” with resistance below 1 ohm [11].

8.6.3.3 Magnetic Field Imaging (MFI) or Magnetic Current Imaging (MCI)

MFI (or MCI) detects the magnetic field generated by current flowing through a failing structure, using sensitive magnetic sensors [137–139]. The failing location can be determined by converting the magnetic field data into a current density map and

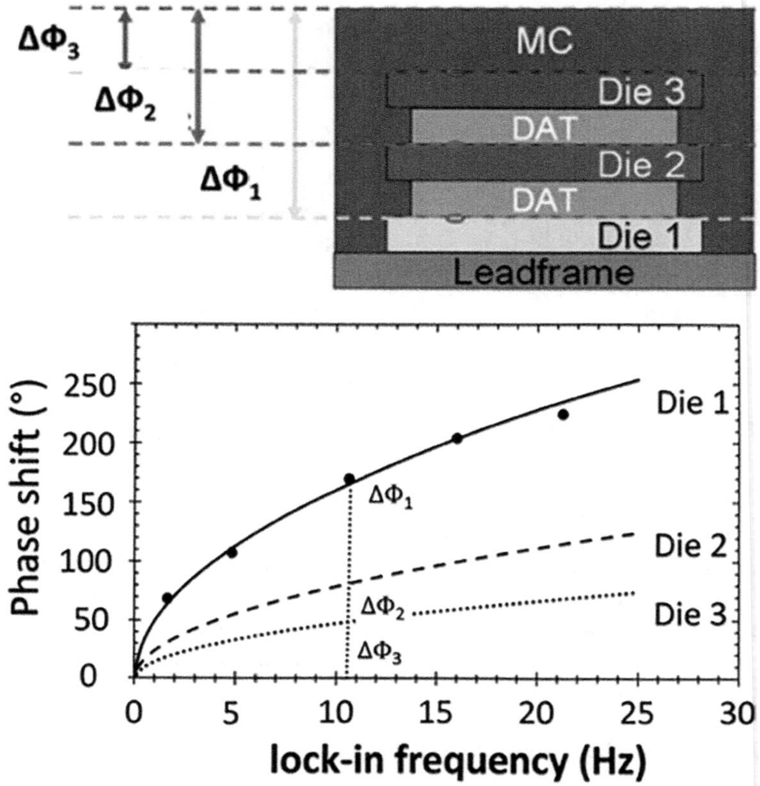

FIGURE 8.65 Measurement results of lock-in frequency vs phase shift curves from reference units with short failures at die 1, die 2, and die 3 of a package with stacked die configuration (shown in dots) compared to a calibrated simulation model (lines). (Adapted from ref. [135,136].)

comparing it with reference maps or circuit schematics. Multiple types of magnetic field detection techniques can be used for MFI, including Superconducting Quantum Interference Device (SQUID) [103], Giant Magnetoresistance (GMR) sensors [103], Tunnel Magnetoresistance (TMR) sensors [140], and Quantum Diamond Microscopy (QDM) using nitrogen vacancy centers [141,142].

Scanning SQUID Microscopy (SSM) has been used for short and high resistance failure localization for over two decades [11,137]. It can also isolate open failures using high frequency (RF) space domain reflectometry (SDR) [138]. The magnetic field generated by a current within the sample is captured, and a Fourier transform is applied to compute the current distribution. By comparing the current path of the failing unit with a reference design file, the x and y coordinates of the failure location can be determined. Additionally, because the distance between the minimum and maximum magnetic fields along a current path equals twice the distance from the SQUID sensor to the current, the z dimensional information of the failure can also be estimated.

As an example, Figure 8.66 demonstrated how SSM is used to isolate a short failure inside an advanced semiconductor package with stacked die configuration. Figure 8.66a shows the current path calculated from the magnetic field generated by the short failure when inputting a small power in the failing sample. The *x* and *y*

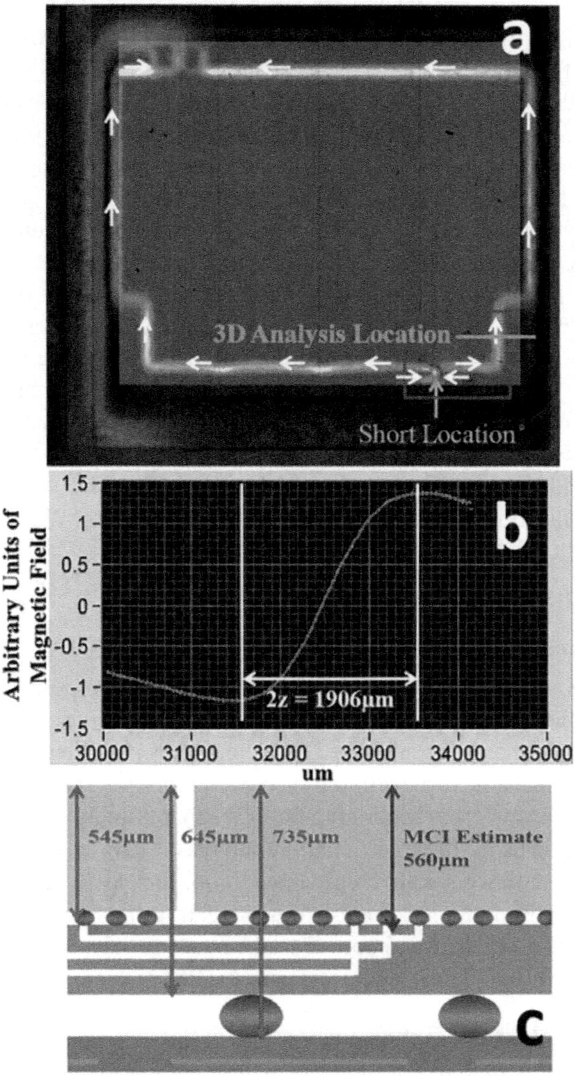

FIGURE 8.66 (a) SSM analysis of a short failure in a package with stacked die configuration. *x* and *y* locations of the short can be determined by the current analysis. (b) and (c) Z dimension analysis of the current path location shown in (a). The distance between the minimum and maximum magnetic fields along the current path in (a) is measured as 1,906 µm. The distance between the sample surface and the failure location is then estimated to be 560 µm, by subtracting the sensor to sample distance (393 µm) from the sensor to current path distance (953 µm, half of 1,906 µm. (Adapted from Ref. [139].)

dimensional information of the short failure can be determined by performing current analysis and comparing with design. Figure 8.66b and c demonstrate the 3D analysis along the green line shown in Figure 8.66a. The distance between the minimum and the maximum magnetic fields around the current path is measured to be around 1,906 μm as illustrated in Figure 8.66b. The sensor to current path distance, which is half of 1,906 μm, is calculated to be 953 μm. The fixed sensor to sample distance is 393 μm for this SSM set up. The z dimensional location of the defect, which is the distance from the current to the Si surface, is estimated to be around 560 μm, by subtracting the sensor to sample distance (393 μm) from the sensor to current path distance (953 μm). As illustrated in Figure 8.66c, the failure is estimated to be in the microbump area between the two stacked dies, based on the feature dimensions of the sample, as illustrated schematically in Figure 8.66c [139].

Unlike conventional MFI, which calculates only 2D current paths, recently developed algorithms can enable 3D current path reconstruction from the detected magnetic data. It has shown potential for full 3D fault isolation in advanced semiconductors with stacked die configuration [143]. MFI is a complementary technique to LIT, because it provides current mapping instead of detecting thermal emissions from defects. It can excel in scenarios where LIT is inefficient, such as shorts with extremely low resistance, short failures shadowed by routings with high resistance, and multiple shorts that can distort thermal signatures from each failure. In these cases, MFI could be a good alternative technique to effectively provide three dimensional location information of defects in advanced packages [90].

8.6.4 Sample Preparation Techniques

8.6.4.1 Mechanical Polishing

Mechanical polishing is a conventional approach in PFA to perform cross sectioning, planar grinding, de-layering, and Si die back side thinning. However, as the diameter of interconnects in advanced semiconductor packages decreases to a few microns, using mechanical polishing to achieve precise alignment and artifact-free surface finishes during cross sectioning becomes increasingly challenging. It is very time consuming if using the traditional Ge ion based FIB system to perform cross sectioning of interconnects with a high aspect ratio, such as 100 μm long TSVs [144]. To address these limitations, a couple of advanced sample preparation techniques, including plasma FIB, femtosecond (fs) and nanosecond (ns) laser ablation, and broad beam ion milling, have been developed for PFA.

8.6.4.2 Laser Ablation Techniques

The laser ablation method has the advantage of fast materials removal rates and short TPT. However, because of the limited spatial resolution and the potential of laser induced thermal damage, the laser ablation technique typically cannot serve as a standalone cross sectioning method [145]. FIB or ion milling is usually used as a post laser step to remove the heat affected zone in the sample, and enhance resolution. The integration of fs laser ablation with conventional FIB and SEM has proven to be highly effective in PFA. This hybrid approach provides improved location precision, minimal thermal damage, and a high material removal rate of roughly 10^{3} μm³/s. It is capable of making artifact-free TSV cross sections, with TPT comparable with plasma FIB systems [145].

8.6.4.3 Plasma-FIB

Traditional Ga ion FIB systems are limited to a maximum beam current of about 65 nA. In contrast, plasma FIB systems, which utilize an inductively coupled plasma (ICP) source, can generate focused xenon (Xe) beams with beam currents ranging from a few picoamperes (pA) to as high as 2 μA [146]. Therefore, sample preparation using plasma FIB can be 20–100 times faster than that using conventional FIB while maintaining a similar surface finishing quality. Plasma FIB combined with SEM enables fast and artifact-free visualization of subtle structural defects over large cross-sectional areas, including voids, cracks, and delamination in TSVs, microbumps, and stacked dies. Figure 8.67b illustrates a plasma FIB cross section showing voids in TSVs [146]. Additionally, plasma FIB has been used extensively in die backside sample preparation, including Si de-layering, cross sectioning, and dry etching for die level PFA [103,147].

8.6.4.4 Broad-Beam Ion Milling

Broad-beam ion milling is a cost effective alternative to plasma FIB and fs laser systems in PFA of semiconductor packages. Prior to ion milling, samples are typically pre-sectioned via mechanical polishing or laser ablation to expose regions close to the AOI. The ion milling technique can be used to "clean up" the sample and produce artifact-free cross sections over areas up to 1 mm^2 within tens of minutes to a few hours. The broad beam ion milling technique also minimizes typical FIB artifacts, such as curtaining, due to the use of varying incident angles to generate uniform ion beam over a large area. This method has proven effective in producing high

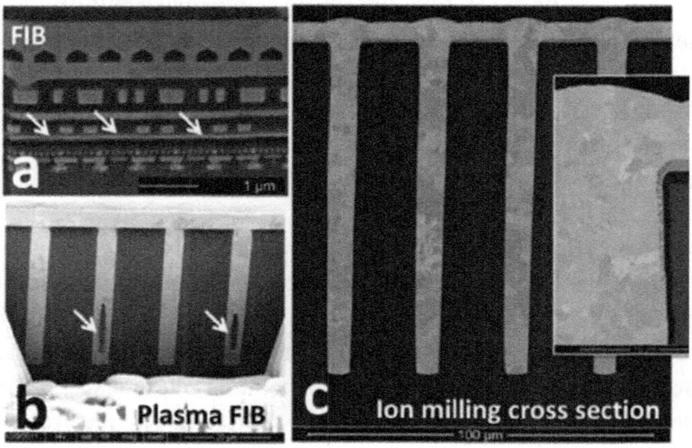

FIGURE 8.67 (a) SEM image of Si layer ILD delamination by FIB cross section. (b) and (c) SEM images of TSVs by plasma-FIB and ion milling cross section, respectively. (Adapted from Ref. [90].)

quality TSV cross sections with fast TPT [11,148]. As illustrated in Figure 8.67c, SEM images of TSVs cross sectioned using the ion milling technique reveal clear structural details, such as the TSV seed layer. In this example, ion milling provides a clean, large area cross section of the semiconductor package within 1 hour, offering a high resolution view without milling artifacts [148].

8.6.5 Defect Imaging Techniques

8.6.5.1 Scanning Electron Microscopy (SEM)

Once defects are exposed during PFA, imaging techniques are employed for further examination. Optical microscopy is usually used to analyze larger defects. For submicron level defects, SEM is routinely used. Different SEM imaging modes serve specific purposes: secondary electron imaging reveals fine morphological details; backscattered electron imaging highlights elemental contrast; and low energy SEM is critical for imaging low k interlayer dielectric (ILD) layers in Si, which are vulnerable to electron beam induced damage [103].

8.6.5.2 Transmission Electron Microscopy (TEM)

Although TEM has a very small field of view, typically at the submicron scale, it can deliver atomic level spatial resolution and is a key tool in die level PFA to detect nanoscale defects. TEM sample preparation used to be a labor intensive process, prior to the advent of the FIB technology. The integration of FIB-SEM has dramatically shortened the time needed for sample preparation. A thin lamella (< 100 nm) containing the defect can be extracted using FIB-SEM for subsequent TEM analysis.

8.6.6 In-Depth Failure Mechanism and Root Cause Investigation

The ultimate objective of failure analyses is to identify actionable solutions. Identifying a defect is only the first step in comprehensive failure mechanism and root cause analysis. Common strategies for root cause analysis include:

1. **Commonality Analysis**: Failure arising from process related defects can usually be traced through the commonality investigations, including components, materials, suppliers, Si wafers, assembly times, tools, environments, etc.
2. **Failure Rate Distribution and Reliability Testing**: as discussed in previous sections, analyzing failure distributions helps categorize failure types. Understanding reliability tests of the failing units can further determine if failures are related to environmental factors, such as moisture, thermal mechanical stress, corrosion, or electrochemical reactions.
3. **Detailed Physical Defect Analysis**: Artifact-free, progressive cross-sectional studies of the defects can reveal the three dimensional structure of defects. Materials characterization of the defects can help determine their chemical composition. Crack tip analysis of the defects can assist in the understanding of defect initiation and propagation.

4. **Direct Observation Using Non-Destructive and In Situ Techniques**: Approaches such as 2D X-ray imaging with a heated stage [123] or progressive reliability testing paired with non-destructive imaging [149] facilitate the real time observation of defect development. These approaches can help with the creation of kinetic models and provide deep insights into failure mechanisms and potential solutions.

8.7 SUMMARY

Materials, including interconnects, TIMs, dielectric layers, underfill, molding compounds, flux, and organic substrates play an important role in the development of advanced semiconductor packaging technologies. Testing methodologies for crucial bulk materials properties, including T_g, CTE, and modulus, are discussed in Section 8.2.1. CTE mismatches among materials in advanced packages can lead to warpage during assembly processes, which poses significant yield and reliability concerns. Metrologies for measuring dynamic warpage are covered in Section 8.2.2. Interfacial properties are crucial in defining adhesion strength between materials, and the corresponding test metrologies are reviewed in Section 8.2.3. The bonding strength of interconnects directly influences the reliability of millions of connections in advanced packages. Test metrologies for interconnect bonding are introduced in Section 8.2.4, which could be used for quick materials down selection. Surface characteristics, such as roughness, surface energy, and wettability, are essential for improving assembly yields in processes such as interconnect bonding, underfill dispensing, and lamination. These surface measurement techniques are detailed in Section 8.2.5. Thermal management remains one of the major challenges in advanced packaging. TTC and TTVs are usually used to evaluate and qualify TIM materials, as discussed in Section 10.2.6.

Accurate materials property data under various process and reliability conditions are essential for simulating thermal, mechanical, signal, power, and reliability behaviors during the design phase of advanced packages. As materials properties can vary non-linearly with temperature, extensive measurements are often required. When direct measurement at specific conditions is not feasible, simulation techniques, such as molecular dynamics, can be employed to predict properties, such as T_g, CTE, and modulus. These simulation methodologies are described in Section 8.3, along with a discussion on the application of AI in materials simulation and its comparison with molecular dynamic approaches.

Foreign materials, contamination, impurities, and variations in interconnect grain size and orientation, can cause yield loss or reliability failures. Section 8.4 introduces a range of advanced materials characterization techniques used for the materials analysis, including EDX, WDX, XRF, FTIR, O-PTIR, AFM-IR, XPS, TOF-SIMS, and EBSD. As advanced semiconductor packages become increasingly complex, with higher interconnect densities and larger form factors, ensuring quality and reliability of semiconductor packages presents enormous challenges. Section 8.5 reviews accelerated life testing methodologies, reliability modeling, acceleration factor calculations, risk assessment strategies, and common failure modes observed in interconnect EM tests and package level reliability assessments.

FI and FA provide important feedback for troubleshooting during advanced semiconductor packaging development. Due to the complexity of advanced packages, detecting subtle defects is getting very difficult, and determining root causes can be even more challenging. Section 8.6 offers a detailed overview of FI and FA process, covering non-destructive imaging, fault isolation tools, sample preparation, defect imaging techniques, and root cause investigation methodologies.

ACKNOWLEDGEMENTS

The author would like to thank Dr. Dongkai Shangguan for his detailed review of this chapter.

REFERENCES

1. M. Sandor, S. Agarwal, D. Peters, and M. S. Cooper, Investigation of Low Glass Transition Temperature on COTS PEMs Reliability for Space Applications, 2003. https://nepp.nasa.gov/docuploads/C73D74B0-DF69-4E1D-80A1938F30E4840E/Glass%20transition%20temperature%20presentation.pdf
2. Polymer Science Learning Center, Differential scanning calorimetry, 2005. https://pslc.ws/macrog/dsc.htm
3. W.J. Sichina, PerkinElmer Instruments, Measurement of Tg by DSC, 2000. https://thermalsupport.com/wp-content/uploads/2018/05/PETech-09.pdf
4. S. Delsante, R. Novakovic, and G. Borzone, Synthesis, characterization and thermal stability of SnAg and SnAgCu nanoparticles, *Journal of Alloys and Compounds*, vol. 747, p. 385, 2018.
5. M. M. Konarski, Effects of T_g and CTE on Semiconductor Encapsulants, 1999. https://smtnet.com/library/files/upload/semiconductors-encapsulants-tg-cte.pdf
6. Hitachi, Principle of Thermomechanical Analysis (TMA), 2001. https://www.hitachi-hightech.com/global/en/knowledge/analytical-systems/thermal-analysis/basics/tma.html
7. Hitachi, Principle of Dynamic Mechanical Analysis (DMA), 2001. https://www.hitachi-hightech.com/global/en/knowledge/analytical-systems/thermal-analysis/basics/dma.html
8. Anton Paar, Basics of Dynamic Mechanical Analysis (DMA), 2026. https://wiki.anton-paar.com/nz-en/basics-of-dynamic-mechanical-analysis-dma/
9. J. Gotro, Practical Tips for Curing Thermosets Part Eight: The Glass Transition Temperature Measured Using DMA, 2016. https://polymerinnovationblog.com/
10. D. Shangguan, Development of Innovative Advanced Packaging Materials for System in Package, 2022. https://www.indium.com/blog
11. Y. Li, and D. Goyal, *3D Microelectronic Packaging: From Architectures to Applications*, New York: Springer, 2021.
12. TA Instruments, Measurement of the Coefficient of Hygroscopic Expansion (CHE) Using Controlled Humidity Dynamic Mechanical Analysis, 2007. https://www.tainstruments.com/pdf/literature/TA370.pdf
13. E. H. Wong, S. W. Koh, R. Rajoo, and T. B. Lim, Underfill Swelling and Temperature-Humidity Performance of Flip Chip PBGA Package, *Electronics Packaging Technology Conference (ECTC)*, San Diego, CA, p. 258, 2000.
14. Y. He, Moisture absorption and hygroscopic swelling behavior of an underfill material, Thermochimica Acta, vol. 546, p. 143, 2012.

15. R. W. Kultermanl, W. K. Loh, H. Fu, and M. Tsuriya, Comparison of Advanced Package Warpage Measurement Metrologies, *International Electronic Manufacturing Technology (IEMT) Conference*, Austin, TX, 2016.
16. P. Zhong, S. Chen, and L. Nian, Method of Extracting High-resolution Digital Moire Fringe in Warpage Measurement, *International Symposium on the Physical and Failure Analysis of Integrated Circuits (IPFA)*, Penang, Malaysia, 2009.
17. X. Huang, C. Ou, S. Zhu, and Y. Huang, Warpage Measurement of Substrates and Printed Circuit Boards with Shadow Moiré, *International Conference on Electronic Packaging Technology (ICEPT)*, Xi'an, China, 2021.
18. N. Hubble, and L. Weaver, *Comparing Shadow Moiré and Digital Fringe Projection Warpage Metrology Technique*, Rosemont, IL: Surface Mount Technology International (SMTAI), 2017. https://akrometrix.com/wp-content/uploads/2016/02/DFPvsSM.pdf
19. M. Ishihara, and H. Sasaki, High-Speed 3D Shape Measurement Using a Nonscanning Multiple-Beam Confocal Imaging System, *Laser Interferometry IX: Techniques and Analysis. Vol. 3478. SPIE's International Society for Optics and Photonics*, San Diego, CA, 1998.
20. C. Cai, K. Pan, J. Yang, and S. Park, Comparative Analysis of Package Warpage Using Confocal Method and Digital Image Correlation, *19th IEEE The Intersociety Conference on Thermal and Thermomechanical Phenomena in Electronic Systems, (ITHERM) Conference*, Orlando, FL, p. 945, 2020.
21. S. Park, H. Zhang, X. Zhang, S. L. Ng, and H. C. Lee Temperature Dependency of Coefficient of Hygroscopic Swelling of Molding Compound, *Electronic Components and Technology Conference (ECTC)*, San Diego, CA, p. 172, 2009.
22. X. Dai, M. V. Brillhart, and P. S. Ho, Adhesion measurement for electronic packaging applications using double cantilever beam method, *IEEE Transactions on Components and Packaging Technologies*, vol. 23(1), p. 101, 2000, https://doi.org/10.1109/6144.833049
23. D. Liu, F. Ding, M. Su, R. Ma, C. Chen, and L. Cao, Characterization and Simulation of Interfacial Adhesion in Embedded Device Package, *23rd International Conference on Electronic Packaging Technology (ICEPT)*, Xi'an, China, 2022, https://doi.org/10.1109/ICEPT56209.2022.9872614
24. C. K. Hsu, P. Y. Lin, W. Y. Lin, M. C. Yew, S.-S. Yeh, K. C. Lee, J. H. Wang, P. C. Lai, C. C. Yang, and S. P. Jeng, Characterization Methodologies for Copper/Polymer Interfacial Strength, *IEEE International Microsystems, Packaging, Assembly and Circuits Technology conference (IMPACT)*, Taipei, Taiwan, p. 202, 2017.
25. D. J. Kim, S. Kang, S. W. Lee, I. Lee, S. Park, J. Lee, J. J. Kim, and T. S. Kim, Optimization of Cu Interconnects – SiCN Interfacial Adhesion by Surface Treatment, *IEEE 73rd Electronic Components and Technology Conference (ECTC)*, San Diego, CA, p. 367, 2023.
26. T. S. Kim, Adhesion measurement methods for thin films in microelectronics, *Journal of Welding and Joining*, vol. 30(3), pp. 15–20, 2012, https://doi.org/10.5781/KWJS.2012.30.3.213
27. M. Genanul, B. Arfaei, E. J. Cotts, F. Mutuku, E. Perfecto, S. Pollard, and A. Shorey, *Microstructure and Performance of Micro Cu Pillars Assemblies*, Rosemont, IL: Surface Mount Technology International (SMTAI), 2016.
28. Y. J. Chen, C. K. Chung, C. R. Yang, and C. R. Kao, Single-joint shear strength of micro Cu pillar solder bumps with different amounts of intermetallics, *Microelectronics Reliability*, vol. 53(1), p. 47, 2013.
29. S. G. Garreignot, N. Benzima, E. Benmussa, C. Moutin, P. O. Bouchard, V. Fiori, and C. Tavernier, Qualification of bumping processes: Experimental and numerical investigations on mechanical stress and failure modes induced by shear test, *Microelectronics Reliability*, vol. 55, p. 980, 2015.

30. F. Song, and S. W. R. Lee, Investigation of IMC Thickness Effect on the Lead-free Solder Ball Attachment Strength: Comparison between Ball Shear Test and Cold Bump Pull Test Results, *Electronic Components and Technology Conference (ECTC)*, San Diego, CA, p. 1196, 2006.
31. Xyztec, Pull testing bumped wafers, 2026. https://www.xyztec.com/pull-testing-bumped-wafers/
32. T. You, Y. Kim, J. Kim, J. Lee, B. Jung, J. Moon, and H. Choe, Predicting the drop performance of solder joints by evaluating the elastic strain energy from high-speed ball pull tests, *Journal of Electronic Materials*, vol. 38(3), p. 410, 2009.
33. A. A. Tseng, and J. S. Park, Effects of surface roughness and contact pressure on wafer bonding strength using transmission laser bonding technique, *Journal of Microlithography, Microfabrication, and Microsystems*, vol. 5(4), p. 043013, 2006.
34. S. P. S. Lim, S. C. Chong, and V. Chidambaram, Comprehensive Study on Chip to Wafer Hybrid Bonding Process for Fine Pitch High Density Heterogeneous Applications, *IEEE 71st Electronic Components and Technology Conference (ECTC)*, San Diego, CA, p. 438, 2021.
35. F. Niu, X. Wang, S. Yang, S. Xu, Y. Zhang, T. Suga, and C. Wang, Low-temperature Cu/SiO_2 hybrid bonding based on Ar/H_2 plasma and citric acid cooperative activation for multi-functional chip integration, *Applied Surface Science*, vol. 648(1), p. 159074, 2024.
36. S. C. Kang, C. Kim, J. Muncy, M. Schmidt, and D. Baldwin, Experimental Wetting Dynamics Study of Eutectic And Lead-Free Solders Varying Flux, Temperature and Surface Finish Metallization, *IEEE/CPMT/SEMI 29th International Electronics Manufacturing Technology (IEMT) Symposium*, Austin, TX, p. 56, 2004.
37. G. R. Ahmed, H. Wang, S. Ramalingam, N. Karunakaran, K. Pan, S. B. Park, A. Soundarajan, S. K. Kanaran, C. K. Chung, and Y. L. Huang, Lidless and lidded Flip Chip Packages for Advanced Applications, *IEEE 22nd Electronics Packaging Technology Conference (EPTC)*, Singapore, p. 104, 2020.
38. C. Wang, Y. Tang, Y. Hang, C. Yang, P. Kallassi, R. Agarwal, D. Shangguan, and B. Wu, Thermal Test Vehicle for Transient Thermal Characterization of Large Body Size Chips, *Proceedings of IEEE Electronics Packaging Technology Conference (EPTC)*, Singapore, p. 1214, 2024.
39. Y. Liu, R. Roy, D. J. Apigo, et al, High Power Thermal Test Vehicle with 2-Phase Cooling for AI Datacenters, *5G RAN, and EDGE Compute Nodes, 2024 IEEE 74th Electronic Components and Technology Conference (ECTC)*, San Diego, CA, p. 1030, 2024, https://doi.org/10.1109/ECTC51529.2024.00165
40. H. Oprins, V. Cherman, C. Torregiani, M. Stucchi, B. Vandevelde, and E. Beyne, Thermal Test Vehicle for the Validation of Thermal Modelling of Hot Spot Dissipation in 3D Stacked ICs, *3rd Electronics System Integration Technology Conference (ESTC)*, Berlin, Germany, p. 1, 2010, https://doi.org/10.1109/ESTC.2010.5642955
41. E.Ouyang, B.Ahn, R. Bornoff, W. He, N. Islam, G. Kim, K. Kim, and A. Vass-Varnai, "Transient thermal characterization of a FCBGA-H device", 29th IEEE Semiconductor Thermal Measurement and Management Symposium, San Jose, CA, USA, 17-21 March 2013, 2013. DOI: 10.1109/SEMI-THERM.2013.6526809
42. Y. Li, and W. Kim, The Applications of Simulation and Artificial Intelligence in Advanced Packaging, *2024 IEEE International Symposium on the Physical and Failure Analysis of Integrated Circuits (IPFA)*, Penang, Malaysia, p. 1, 2024.
43. Y. Li, S. Y. Kim, W. Kim, M. A. F. Afzal, H. S. Kwak, and D. A. Nicholson, Material property simulation for advanced packaging, *Journal of Electronic Materials*, vol. 55, pp. 1664–1673, 2025.
44. M. A. F. Afzal, A. R. Browning, A. Goldberg, M. D. Halls, J. L. Gavartin, T. Morisato, T. F. Hughes, D. J. Giesen, and J. E. Goose, High-throughput molecular dynamics simulations and validation of thermophysical properties of polymers for various applications, *ACS Applied Polymer Materials*, vol. 3, p. 620, 2021.

45. Schrödinger Release 2024-4: Materials Science Suite, New York: Schrödinger, LLC, 2024.
46. V. N. Novikov, and E. A. Rossler, Correlation between glass transition temperature and molecular mass in non–polymeric and polymer glass formers, *Polymer*, vol. 54, p. 6987, 2013.
47. C. Lu, C. Wu, D. Ghoreishi, W. Chen, L. Wang, and W. Damm, OPLS4: Improving force field accuracy on challenging regimes of chemical space, *Journal of Chemical Theory and Computation*, vol. 17(7), pp. 4291–4300, 2021, https://pubs.acs.org/doi/abs/10.1021/acs.jctc.1c00302
48. K. J. Bowers, D. E. Chow, H. Xu, R. O. Dror, M. P. Eastwood, and B. A. Gregersen, Scalable Algorithms for Molecular Dynamics Simulations on Commodity Clusters, *ACM/IEEE Conference on Supercomputing*, Tampa, FL, p. 43, 2006.
49. M. A. F. Afzal, C. Cheng, and J. Hachmann, Combining first-principles and data modeling for the accurate prediction of the refractive index of organic polymers, *Journal of Chemical Physics*, vol. 148(24), p. 241712, 2018.
50. W. G. Hoover, Canonical dynamics: Equilibrium phase-space distributions, *Physical Review A*, vol. 31, p. 1695, 1985.
51. G. J. Martyna, D. J. Tobias, and M. L. Klein, Constant pressure molecular dynamics algorithms, *Journal of Chemical Physics*, vol. 101, p. 4177, 1994.
52. P. N. Patrone, A. Dienstfrey, A. R. Browning, S. Tucker, and S. Christensen, Uncertainty quantification in molecular dynamics studies of the glass transition temperature, *Polymer*, vol. 87, p. 246, 2016.
53. J. Bicerano, *Prediction of Polymer Properties*, 3rd edition, Boca Raton, FL: CRC Press, 2002, https://doi.org/10.1201/9780203910115
54. P. G. M. Mileo, C. M. Krauter, J. M. Sanders, A. R. Browning, and M. D. Halls., Molecular-scale exploration of mechanical properties and interactions of poly (lactic acid) with cellulose and chitin, *ACS Omega*, vol. 8(45), p. 42417, 2023.
55. Dupont, Dupont Kapton Summary of Properties, 2025. https://www.dupont.com/content/dam/electronics/amer/us/en/electronics/public/documents/en/EI-10142_Kapton-Summary-of-Properties.pdf
56. Schrödinger Release 2025-1: DeepAutoQSAR, New York: Schrödinger, LLC, 2024.
57. Z. Kaplan, S. Ehrlich, and K. Leswing, Benchmark study of DeepAutoQSAR, ChemProp, and DeepPurpose on the ADMET subset of the Therapeutic Data Commons, 2024. https://www.schrodinger.com/life-science/learn/white-papers/benchmark-study-deep-autoqsar-chemprop-and-deeppurpose-admet-subset-therapeutic-data/
58. H. Xu, H. Zhang, and M. Xue, 58th Electronic Components and Technology Conference (ECTC), *Electronic Components and Technology Conference (ECTC)*, San Diego, CA, p. 403, 2013.
59. Bruker, What is WDS, 2024. https://www.bruker.com/en/landingpages/bna/technology/what-is-wds.html
60. M. Ohgaki, Methods of fluorescent x-ray analysis for measuring the thickness of multi-layered plating, *The Hitachi Scientific Instrument News*, vol. 13, p. 47, 2019.
61. A. Centrone, Infrared imaging and spectroscopy beyond the diffraction limit, *Annual Review of Analytical Chemistry*, vol. 8, p. 101, 2015.
62. J. Y. Kim, H. Kim, D. W. Hahn, and S. M. Hwang, ISTFA 2017: Conference Proceedings from the 43rd International Symposium for Testing and Failure Analysis, *The 43rd International Symposium for Testing and Failure Analysis (ISTFA)*, Pasadena, CA, p. 40, 2017.
63. A. Marchetti, V. Beltran, G. Nuyts, F. Borondics, S. D. Meyer, M. V. Bos, J. Jaroszewicz, E. Otten, M. Debulpaep, and K. D. Wael, Novel optical photothermal infrared (O-PTIR) spectroscopy for the noninvasive characterization of heritage glass-metal objects, *Science Advances*, vol. 8, p. 1, 2022, https://doi.org/10.1126/sciadv.abl6769

64. S. Zulkifli, B. Zee, and M. K. F. Lo, Novel Submicron Spatial Resolution Infrared Microspectroscopy for Failure Analysis of Semiconductor Components, *IEEE International Symposium on the Physical and Failure Analysis of Integrated Circuits (IPFA)*, Penang, Malaysia, 2022.
65. M. K. F. Lo, Rapid and Specific Failure Analysis by Fluorescence Guided Optical Photothermal Infrared Spectroscopy (FL-OPTIR), *IEEE International Symposium on the Physical and Failure Analysis of Integrated Circuits (IPFA)*, Penang, Malaysia, 2023.
66. A. Dazzi, and C. B. Prater, AFM-IR: Technology and applications in nanoscale infrared spectroscopy and chemical imaging, *Chemical Reviews*, vol. 117, p. 5146, 2017.
67. P. Van der Heide, *X-Ray Photoelectron Spectroscopy: An Introduction to Principles and Practices*. New York: John Wiley & Sons, Inc., 2011.
68. P. Van der Heide, *Secondary Ion Mass Spectrometry: An Introduction to Principles and Practices*. New York: John Wiley & Sons, Inc., 2014.
69. F. Yang, S. Yiqiang, L. H. Sheng, and F. Chao, Applications of XPS and TOF-SIMS in the Investigation of PCB Package Delamination, *Electronic Packaging Technology Conference (EPTC)*, Singapore, p. 37, 2013.
70. B. Pathangey, A. Proctor, Z. Wang, Z. Fu, and R. Tanikella, Application of TOFSIMS for contamination issues in the assembly world, *IEEE Transactions on Device and Materials Reliability*, vol. 7(1), p. 11, 2007.
71. Y. Liu, J. Wang, L. Yin, P. Kondos, C. Parks, P. Borgesen, D. W. Henderson, E. J. Cotts, and N. Dimitrov, Effect of copper grain size on the interfacial microstructure of a Sn/Cu joint, *Journal of Applied Electrochemistry*, vol. 38, p. 1695, 2008.
72. M. H. Lu, D. Y. Shih, P. Lauro, C. Goldsmith, and D. W. Henderson, Effect of the anisotropic characteristics of B-Sn on current-induced solder evolution, *Applied Physics Letters*, vol. 92, p. 211909, 2008.
73. Y. Wang, and P. S. Ho, Mode II electromigration failure mechanism in sn-based pb-free solder joints with Ni under-bump metallization, *Applied Physics Letters*, vol. 103, p. 121909, 2013.
74. P. Liu, A. Overson, and D. Goyal, 2015 IEEE 65th Electronic Components and Technology Conference, *Electronic Components and Technology Conference (ECTC)*, San Diego, CA, p. 99, 2015.
75. C. Wu, T. Jiang, J. IM, K. M. Liechti, R. Huang, and P. S. Ho, IPFA is a Cornerstone Conference for Semiconductor Failure Analysis and Reliability Assessment, *International Symposium on the Physical and Failure Analysis of Integrated Circuits (IPFA)*, Penang, Malaysia, p. 312, 2014.
76. Reliawiki, The Weibull distribution, 2023. https://www.reliawiki.com/index.php/The_Weibull_Distribution
77. M. H. Lu, D. Y. Shih, P. Lauro, C. Goldsmith, and D. W. Henderson, Effect of the anisotropic characteristics of B-Sn on current-induced solder evolution, *Applied Physics Letters*, vol. 92, p. 211909, 2008.
78. Y. Wang, K. H. Lu, V. Gupta, L. Stiborek, D. Shirley, S. H. Chae, J. Im, and P. S. Ho, Effects of Sn grain structure on the electromigration of Sn-Ag solder joints, *Journal of Materials Research*, vol. 27, p. 1131, 2012.
79. Y. Wang, and P. S. Ho, Mode II electromigration failure mechanism in sn-based pb-free solder joints with Ni under-bump metallization, *Applied Physics Letters*, vol. 103, p. 121909, 2013.
80. P. L. Liu, O. Alan, and D. Goyal, X-ray Microscopy and Automatic Detection of Defects in Through Silicon Vias in Three-Dimensional Integrated Circuits, *Electronic Components and Technology Conference (ECTC)*, San Diego, CA, p. 95, 2015.
81. C. M. Tan, and A. Roy, Electromigration in ULSI interconnects, *Materials Science and Engineering: R*, vol. 58, p. 1, 2007.

82. C. K. Hu, L. Gignac, E. Liniger, B. Herbst, D. L. Rath, S. T. Chen, S. Kaldor, A. Simon, and W. T. Tseng, Comparison of Cu electromigration lifetime in Cu interconnects coated with various caps, *Applied Physics Letters*, vol. 83, p. 869, 2003.
83. Z. H. Gan, W. Shao, S. G. Mhaisalkar, Z. Chen, H. Li, K. N. Tu, and A. Gusak, Reservoir effect and the role of low current density regions on electromigration lifetimes in copper interconnects, *Journal of Materials Research*, vol. 21, p. 2241, 2006.
84. C. M. Tan, A. Roy, A. V. Vairagar, A. Krishnamoorthy, and S. G. Mhaisalkar, Current crowding effect on copper dual damascene via bottom failure for ULSI applications, *IEEE Transactions on Device and Materials Reliability*, vol. 5, p. 198, 2005.
85. M. W. Lane, E. G. Liniger, and J. R. Lloyd, Relationship between interfacial adhesion and electromigration in Cu metallization, *Journal of Applied Physics*, vol. 93, p. 1417, 2003.
86. K. N. Tu, C. C. Yeh, C. Y. Liu, and C. Chen, Comment on effect of current crowding on vacancy diffusion and void formation in electromigration, *Applied Physics Letters*, vol. 76, p. 988, 2000.
87. J. R. Lloyd, Electromigration and mechanical stress, *Microelectronic Engineering*, vol. 49, p. 51, 1999.
88. J. R. Lloyd, and P. M. Smith, The effect of passivation thickness on the electromigration lifetime of Al/Cu thin film conductors, *Journal of Vacuum Science and Technology A*, vol. 1, p. 455, 1983.
89. L. Kisselgof, S. P. Baranowski, M. C. Broomfield, T. Spooner, L. Elliott, L. Brooke, and J. R. Lloyd, Thermally induced stress and electromigration failure, *SPIE Submicrometer Metallization*, vol. 1805, p. 154, 1992.
90. Y. Li, and H. Shi, *Advanced Driver Assistance Systems and Autonomous Vehicles From Fundamentals to Applications*, New York: Springer Nature, 2022, https://doi.org/10.1007/978-981-19-5053-7
91. D. Crowe, and A. Feinberg, *Design for Reliability*, Boca Raton, FL: CRC press LLC, 2001.
92. L. Coffin, A study of the effects of cyclic thermal stresses on a ductile metal, *Transactions of the ASME*, vol. 76, p. 931, 1954.
93. JEDEC Solid State Technology Association, *Failure Mechanisms and Models for Semiconductor Devices*, JEDEC JEP122F. Arlington, VA: JEDEC Solid State Technology Association, 2010.
94. https://www.asminternational.org/documents/10192/1849770/05224G_Chapter14.pdf.
95. R. R. Tummala, *Fundamentals of Microsystems Packaging P878*, New York: McGraw-Hill, 2001.
96. M. H. Lu, D. Y. Shih, P. Lauro, C. Goldsmith, and D. W. Henderson, Effect of Sn grain orientation on electromigration degradation mechanism in high Sn-based Pb-free solders, *Applied Physics Letters*, vol. 92, p. 211909, 2008.
97. S. Atkins, L. Teems, W. Rowe, P. Selby, and R. Vaughters, Use of C-SAM acoustical microscopy in package evaluations and failure analysis, *Microelectronics Reliability*, vol. 38, p. 773, 1998.
98. T. Akutsu, Q. Yu, and Y. Nishimura, *Electronic Packaging Technology Conference (EPTC)*, Singapore, p. 550, 2010.
99. M. M. Basit, M. Motalab, J. C. Suhling, Z. Hai, J. Evans, M. J. Bozack, and P. Lall, High Temperature Mechanical Behavior of SAC and SAC+X Lead Free Solders, *Electronic Components and Technology Conference (ECTC)*, San Diego, CA, p. 106, 2015.
100. D. Yang, and Z. Cui, International Conference on Electronic Packaging Technology and High Density Packaging, *International Conference on Electronic Packaging Technology and High Density Packaging (ICEPT)*, Xi'an, China, p. 1030, 2011.
101. S. H. Kim, J. W. Park, S. J. Hong, and J. T. Moon, The Interface Behavior of the Cu-Al Bond System in High Humidity Conditions, *Electronic Packaging Technology Conference (EPTC)*, Singapore, p. 545, 2010.

102. K. Lee, S. Barbeau, F. Racicot, D. Powell, C. Arvin, T. Wassick, and J. Ross, Electronic Components and Technology Conference (ECTC), *2012 IEEE 62nd Electronic Components and Technology Conference (ECTC 2012)*, San Diego, CA, p. 2138, 2013.
103. T. Gandhi, *Microelectronics Failure Analysis Desk Reference*, 7th edition. New York: ASM International, 2019.
104. S. F. Lin, C. H. Chen, and C. Y. Lo, Near-infrared imaging system for nondestructive inspection of micro-crack in wafer through dicing tape, *Applied Optics*, vol. 54(28), pp. E123–E128, 2015.
105. M. H. Ettenberg, and D. Malchow, InGaAs SWIR Imagers Optimize Semiconductor Inspection Photonic Tech Briefs Magazine, July 1, 2007.
106. M. Matysiak, J. P. Parry, F. Albri, J. G. Crowder, N. Jones, K. Jonas, N. Weston, D. P. Hand, and J. D. Shephard, Infrared confocal imaging for inspection of flaws in yttria stabilized tetragonal zirconia polycrystal (Y-TZP), *Measurement Science and Technology*, vol. 22, p. 125502, 2011.
107. W. Lawton, and J. Barrett, Characterization of chip-on-board and flip chip packaging technologies by acoustic microscopy, *Microelectronics Reliability*, vol. 36, p. 1803, 1996.
108. J. E. Semmens, Flip chips and acoustic micro imaging: An overview of past applications, present status, and roadmap for the future, *Microelectronics Reliability*, vol. 40, p. 1539, 2000.
109. D. A. Hutt, D. P. Webb, K. C. Hung, C. W. Tang, P. P. Conway, D. C. Whalley, and Y. C. Chan, Scanning Acoustic Microscopy Investigation of Engineered Flip-Chip Delamination, *IEEE/CPMT International Electronics Manufacturing Technology (IEMT) Symposium*, Austin, TX, 2000.
110. J. E. Semmens, and L. W. Kessler, Application of acoustic frequency domain imaging for the evaluation of advanced micro electronic packages, *Microelectronics Reliability*, vol. 42, p. 1735, 2002.
111. B. T. Khuri-Yakub, Scanning acoustic microscopy, *Ultrasonics*, vol. 31(5), p. 361, 1993.
112. G. A. D. Briggs, and O. V. Kolosov, *Acoustic Microscopy*, 2nd edition, Oxford University, Oxford, 2009.
113. T. M. Moore, and C. D. Hartfield, Trends in Nondestructive Imaging of IC Packages, *Conference Proceedings from American Institute of Physics (AIP)*, Heraklion, Greece, 1998.
114. A. J. Komrowski, L. A. Curiel, Q. Nguyen, D. J. D. Sullivan, and L. Logan-Willams, Backside Application of Acoustic Micro Imaging (AMI) on Plastic Ball Grid Array (PBGA) and Plastic Quad Flat Pack (PQFP) Packages, *International Symposium for Testing and Failure Analysis (ISTFA)*, Pasadena, CA, 2004.
115. L. Angrisani, L. Bechou, D. Dallet, P. Daponte, and Y. Ousten, Detection and location of defects in electronic devices by means of scanning ultrasonic microscopy and the wavelet transform, *Measurement*, vol. 31, p. 77, 2002.
116. J. Sigmund, and M. Kearney, TAMI analysis of flip chip packages, Advanced Packaging, no. July/August, 1998.
117. S. Haque, G. Lu, J. Goings, and J. Sigmund, Characterization of interfacial thermal resistance by acoustic micrography imaging, *Microelectronics Reliability*, vol. 40, p. 465, 2000.
118. T. S. Leng, J. C. P. McKeon, and H. S. Jang, Correlation Study of Delamination Gap Detection Capability of SAM and Cross Section/SEM Analysis, *International Symposium for Testing and Failure Analysis (ISTFA)*, Pasadena, CA, 2006.
119. Y. Li, L. Hu, G. Li, R. Dias, and D. Goyal, High Resolution C-Mode Scanning Acoustic Microscope Techniques for the Failure Analysis of Microelectronic Packages, *International Symposium for Testing and Failure Analysis (ISTFA)*, Pasadena, CA, 2013.
120. M. Feser, J. Gelb, H. Chang, H. Cui, F. Duewer, S. H. Lau, A. Tkachuk, and W. Yun, Submicron resolution CT for failure analysis and process development, *Measurement Science and Technology*, vol. 19, p. 094001, 2008.

121. Y. Li, J. S. Moore, B. Pathangey, R. C. Dias, and D. Goyal, Lead-free solder joint void evolution during multiple subsequent high-temperature reflows, *IEEE Transactions on Device and Materials Reliability*, vol. 12(2), p. 494, 2012.
122. Y. Li, P. K. M. Srinath, and D. Goyal, A review of failure analysis methods for advanced 3D microelectronic packages, *Journal of Electronic Materials*, vol. 45, p. 116, 2016.
123. Y. Li, R. Panat, B. Li, R. Mulligan, P. K. M. Srinath, and A. Raman, The application of two-dimensional X-ray hot stage in flip chip package failure analysis, *IEEE Transactions on Device and Materials Reliability*, vol. 11(1), p. 141, 2011.
124. Y. Li, Y. Cai, M. Pacheco, R. C. Dias, and D. Goyal, Non Destructive Failure Analysis of 3D Electronic Packages Using Both Electro Optical Terahertz Pulse Reflectometry and 3D X-ray Computed Tomography, *International Symposium for Testing and Failure Analysis (ISTFA)*, Pasadena, CA, 2012.
125. M. Pacheco, and D. Goyal, Detection and Characterization of Defects in Microelectronic Packages and Boards by Means of High-Resolution X-Ray Computed Tomography (CT), *Electronic Components and Technology Conference (ECTC)*, San Diego, CA, 2011.
126. C. Schmidt, C. Hartfield, S. T. Kelly, L. England, and S. Kannan, Nanoscale 3D X-Ray Microscopy for High Density Multi-Chip Packaging FA, *International Symposium for Testing and Failure Analysis (ISTFA)*, Pasadena, CA, 2018.
127. Y. Li, M. Pacheco, D. Goyal, J. W. Elmer, H. D. Barth, and D. Parkinson, High Resolution and Fast Throughput-time X-Ray Computed Tomography for Semiconductor Packaging Applications, *Electronic Components and Technology Conference (ECTC)*, San Diego, CA, 2014.
128. E. Zschech, M. Löffler, J. Gluch and M. J. Wolf, *Micro and Nano X-ray Tomography of 3D IC Stacks*, New York: Materials Research Society (MRS)/Spring, 2016.
129. W. Yun, M. Feser, J. Gelb, and L. Hunter, Multiscale 3-D X-Ray Imaging for 3-D IC, *International Conference on Frontiers of Characterization and Metrology for Nanoelectronics (FCMN)*, Gaithersburg, 2013.
130. Y. Li, Y. Cai, M. Pacheco, R. C. Dias, and D. Goyal, Non-Destructive Failure Analysis of 3D Electronic Packages Using Both Electro Optical Terahertz Pulse Reflectometry and 3D X-ray Computed Tomography, *International Symposium for Testing and Failure Analysis (ISTFA)*, Pasadena, CA, 2012.
131. D. Smolyansky, TDR Testing Primer, Printed Circuit Design Magazine, p. 20, 2002.
132. D. Abessolo-Bidzo, P. Poirier, P. Descamps, and B. Domenges, Failure Localization in IC Packages Using Time Domain Reflectometry: Technique Limitations and Possible Improvements, *International Symposium on the Physical and Failure Analysis of Integrated Circuits (IPFA)*, Penang, Malaysia, 2005.
133. Y. Cai, Z. Wang, R. C. Dias, and D. Goyal, Electro Optical Terahertz Pulse Reflectometry – An Innovative Fault Isolation Tool, *Electronic Components and Technology Conference (ECTC)*, San Diego, CA, 2010.
134. S. H. Hall, G. W. Hal, and J. A. McCall, *High-Speed Digital System Design: A Handbook of Interconnect Theory and Design Practices*, New York: John Wiley & Sons, Inc., 2000.
135. R. Schlangen, S. Motegi, T. Nagatomo, C. Schmidt, F. Altmann, H. Murakami, S. Hollingshead, and J. West, Use of Lock-In Thermography for Non-Destructive 3D Defect Localization on System in Package and Stacked-Die Technology, *International Symposium for Testing and Failure Analysis (ISTFA)*, Pasadena, CA, 2011.
136. F. Naumann, F. Altmann, C. Grosse, and R. Herold, Efficient Non-Destructive 3D Defect Localization by Lock-in Thermography utilizing Multi Harmonics Analysis, *International Symposium for Testing and Failure Analysis (ISTFA)*, Pasadena, CA, 2014.
137. R. Dias, L. Skoglund, Z. Wang, and D. Smith, Integration of SQUID Microscopy into FA Flow, *International Symposium for Testing and Failure Analysis (ISTFA)*, Pasadena, CA, 2001.

138. M. Xie, Z. Qian, M. Pacheco, Z. Wang, R. Dias, and V. Talanov, Fault Isolation of Open Defects Using Space Domain Reflectometry, *International Symposium for Testing and Failure Analysis (ISTFA)*, Pasadena, CA, 2012.
139. J. Gaudestad, D. Nuez, and P. Tan, Short Localization in 2.5D Microchip with Interposer Using Magnetic Current Imaging, *International Symposium for Testing and Failure Analysis (ISTFA)*, Pasadena, CA, 2014.
140. M. Kögel, F. Altmann, S. Tismer, and S. Brand, Magnetic field and current density imaging using off-line lock-in analysis, *Microelectronics Reliability*, vol. 64, p. 346, 2016.
141. M. J. Turner, N. Langellier, R. Bainbridge, D. Walters, S. Meesala, T. M. Babinec, P. Kehayias, A. Yacoby, E. Hu, M. Loncar, R. L. Walsworth, and E. V. Levine, Magnetic field fingerprinting of integrated-circuit activity with a quantum diamond microscope, *Physical Review Applied*, vol. 14(1), p. 014097, 2020.
142. E. V. Levine, M. J. Turner, N. Langellier, T. M. Babinec, M. Lonˇcar, and R. L. Walsworth, Backside Integrated Circuit Magnetic Field Imaging with a Quantum Diamond Microscope, *International Symposium for Testing and Failure Analysis (ISTFA)*, Pasadena, CA, 2020.
143. A. Orozco, N. E. Gagliolo, C. Rowlett, E. Wong, A. Moghe, J. Gaudestad, V. Ta-lanov, A. Jeffers, K. Torkashvan, F. C. Wellstood, S. Dobritz, M. Boettcher, A. B. Cawthorne, and F. Infante, 3D IC/Stacked Device Fault Isolation Using 3D Magnetic Field Imaging, *International Symposium for Testing and Failure Analysis (ISTFA)*, Pasadena, CA, 2014.
144. F. Altmann, J. Beyersdorfer, J. Schischka, M. Krause, G. Franz, and L. Kwakman, Cross Section Analysis of Cu Filled TSVs based on High Throughput Plasma-FIB Milling, *International Symposium for Testing and Failure Analysis (ISTFA)*, Pasadena, CA, 2012.
145. L. Kwakman, M. Straw, G. Coustillier, M. Sentis, J. Beyersdorfer, J. Schischka, F. Naumann, and F. Altmann, Sample Preparation Strategies for Fast and Effective Failure Analysis of 3D Devices, *International Symposium for Testing and Failure Analysis (ISTFA)*, Pasadena, CA, 2013.
146. C. Boit, U. Kerst, R. Schlangen, A. Kabakow, E. L. Roya, T. Lundquista, and S. Pauthner, Impact of Back Side Circuit Edit on Active Device Performance in Bulk Silicon ICs, *International Test Conference*, Washington, DC, 2005.
147. E. Principe, N. Asadizanjani, D. Forte, M. Tehranipoor, R. Chivas, M. DiBattista, and S. Silverman, Plasma FIB deprocessing of integrated circuits from the backside, *Electronic Device Failure Analysis*, vol. 19(4), p. 36, 2017.
148. A. Meyer, G. Grimm, M. Hecker, M. Weisheit, and E. Langer, Challenges for Physical Failure Analysis of 3D-Integrated Devices—Sample Preparation and Analysis to Support Process Development of TSVs, *International Symposium for Testing and Failure Analysis (ISTFA)*, Pasadena, CA, 2013.
149. Y. Li, L. Xu, P. Liu, B. Pathangey, M. Pacheco, M. Hossain, L. Hu, R. Dias, and D. Goyal, Electronic Components and Technology Conference (ECTC), *2021 IEEE 71st Electronic Components and Technology Conference (ECTC)*, San Diego, CA, p. 432, 2015.

Index

For Product Safety Concerns and Information please contact our EU representative GPSR@taylorandfrancis.com Taylor & Francis Verlag GmbH, Kaufingerstraße 24, 80331 München, Germany

Batch number: 10799746

Printed by Printforce, the Netherlands